Arbuscular Mycorrhizal Fungi and Higher Plants

Golam Jalal Ahammed • Roghieh Hajiboland
Editors

Arbuscular Mycorrhizal Fungi and Higher Plants

Fundamentals and Applications

Springer

Editors
Golam Jalal Ahammed
Department of Horticulture
Henan University of Science and Technology
Luoyang, Henan, China

Roghieh Hajiboland
University of Tabriz
Tabriz, Iran

ISBN 978-981-99-8219-6 ISBN 978-981-99-8220-2 (eBook)
https://doi.org/10.1007/978-981-99-8220-2

This Springer imprint is published by the registered company Springer Nature Singapore Pte Ltd.
The registered company address is: 152 Beach Road, #21-01/04 Gateway East, Singapore 189721, Singapore

Paper in this product is recyclable.

Preface

Arbuscular mycorrhizal fungi (AMF) are ubiquitous, soilborne, endophytic, obligate biotrophs that colonize the roots of most terrestrial plants to establish mutually beneficial relationships. The colonization of host roots by AMF improves the access of plant roots to nutrients, mainly phosphorus and nitrogen. In return, host plants provide the colonizing fungi with carbon-rich photosynthates such as sugar and lipids. This nutrient exchange strongly impacts plant and microbial ecosystems by influencing plant fitness, core soil processes, and the carbon cycle. Crucially, both mutualism and ecosystem services largely rely on the bidirectional flow of resources. The AMF symbiosis not only provides plants with nutrients but also supplies water and enhances their resistance to environmental and biotic challenges.

In recent years, AMF have been the focus of interdisciplinary research, and the knowledge regarding physiological and molecular functions and the underlying mechanism is accumulating. Accumulating evidence suggests that AMF act as the key convergence point of intra- and inter-organismal communications in response to environmental signals. The physiological and molecular processes by which AMF aid plants in nutrient foraging and stress resilience have been extensively studied. Recent work has established the framework for nutrient exchange and regulatory interactions between host plants and AMF. Despite a long history of AMF research relating to enhanced tolerance to environmental stress, the accurate biological mechanisms and different biochemical, physiological, and molecular pathways that lead to improved stress tolerance still remain unclear.

The utilization of AMF in modern agriculture is of great significance in the context of global climate change. Bioaugmentation of AMF can alter crop yield under favorable and/or unfavorable environmental conditions. AMF have potential applications in crop production, forestry management, bioremediation, and ecological restoration as biofertilizers and bioprotectors. Therefore, it is important to further explore how utilization of AMF can potentially secure good harvest under challenging environmental conditions. In this book, we compiled the multifaceted roles of AMF in plant growth, development, and stress management to reach a broad audience of plant and environmental sciences including researchers and advanced university students. The present book is composed of 14 chapters on various aspects of AMF in the context of higher plants from fundamental to applications. The book provides new insights into our understanding of AMF-mediated plant growth regulation and stress tolerance covering the most recent biochemical, physiological, molecular, environmental, and ecological studies.

Despite a few minor emendations, the authors' entire premise was preserved in every chapter. However, it is indeed possible that there are still some flaws in the book, so any input from readers would be greatly appreciated for future editions. We would like to thank all the authors who contributed to this book. Thank you to Springer, especially the in-house editors and production staff involved in this book project, for your gracious assistance in completing this project!

Luoyang, Henan, China Golam Jalal Ahammed
Tabriz, Iran Roghieh Hajiboland

The original version of the book has been revised. A correction to this book can be found at https://doi.org/10.1007/978-981-99-8220-2_15

Acknowledgments

This work was supported by the Ministry of Science and Technology of the People's Republic of China (Grant numbers DL2022026004L, QNJ2021026001, QNJ20200226001), National Key Research and Development Program of China (Grant number 2018YFD1000800), National Natural Science Foundation of China (Grant numbers 31950410555, 31872092, 31872157), Henan International Joint Laboratory of Stress Resistance Regulation and Safe Production of Protected Vegetables, Henan Engineering Technology Research Center for Horticultural Crop safety and Disease Control, Natural Science Foundation of Henan (Grant numbers 202300410152, 222102110078), Henan University of Science and Technology Research Start-up Funds for New Faculty (Grant numbers 13480058, 13480070), Innovative Research Team (Science and Technology) in University of Henan Province (Grant number 23IRTSTHN024), and the University of Tabriz, Iran, through a grant to R.H. (Grant No. 3/408447; 07/06/2023).

Contents

Editors and Contributors

About the Editors

Golam Jalal Ahammed is an Associate Professor at the Department of Horticulture, College of Horticulture and Plant Protection, Henan University of Science and Technology, Luoyang, China. Dr. Ahammed received a Ph.D. in Olericulture with a major focus on Plant Stress Physiology and Hormonal Regulation in 2012 from Zhejiang University, China. Afterward, he completed two consecutive postdoctoral programs at the Institute of Pesticide and Environmental Toxicology and the Institute of Crop Science of Zhejiang University. His major research interests include plant stress physiology, phytohormones, climate change effect on plants, biological control of plant diseases, and environmental pollution. Dr. Ahammed authored over 160 peer-reviewed papers in internationally reputed journals. He is a Senior Editorial Board Member of *Scientific Reports* and an Associate Editor of *BMC Plant Biology*, *AoB Plants*, and the *Journal of Plant Growth Regulation*. Dr. Ahammed has been recognised on the annual Highly Cited Researchers 2023 list from Clarivate.

Roghieh Hajiboland is a Professor at the Department of Plant, Cell and Molecular Biology, University of Tabriz, Tabriz, Iran. Dr. Hajiboland received a Ph.D. in Plant Physiology from the University of Hohenheim, Germany, in 2000. Afterward, she started working as academic staff at the University of Tabriz and completed two international research programs through working as visiting scientist in the Faculty of Biosciences, Autonomous University of Barcelona, Spain, in 2009 and in the Faculty of Agriculture and forestry, University of Helsinki, in 2010. Her major research interests include plant nutritional physiology, plant stress physiology, and plant-microbe interactions. Dr. Hajiboland authored over 150 peer-reviewed papers in internationally reputed journals.

Contributors

Mayar Abdellatef Department of Botany and Microbiology, Faculty of Science, Suez Canal University, Ismailia, Egypt

Niraj Agarwala Department of Botany, Gauhati University, Guwahati, Assam, India

Golam Jalal Ahammed Department of Horticulture, Henan University of Science and Technology, Luoyang, Henan, China

Mohd Ali Department of Zoology, Guru Nanak Dev University, Amritsar, Punjab, India

Nasser Aliasgharzad Department of Soil Science, Faculty of Agriculture, University of Tabriz, Tabriz, Iran

Aya G. A. Asran Department of Botany and Microbiology, Faculty of Science, Suez Canal University, Ismailia, Egypt

Mohamed A. M. Atia Genome Mapping Department, Agricultural Genetic Engineering Research Institute (AGERI), Agricultural Research Center (ARC), Giza, Egypt

Hamideh Bakhshayeshan-Agdam Department of Plant, Cell and Molecular Biology, Faculty of Natural Sciences, University of Tabriz, Tabriz, East Azerbaijan, Iran

Harish Changotra Department of Molecular Biology and Biochemistry, Guru Nanak Dev University, Amritsar, Punjab, India

Heikham Evelin Department of Botany, Rajiv Gandhi University, Rono Hills, Doimukh, Arunachal Pradesh, India

Linda Yuya Gorim Department of Agricultural, Food and Nutritional Sciences, 410 Agriculture/Forestry Centre, University of Alberta, Edmonton, AB, Canada

Monika A. Gorzelak Agriculture and Agri-Food Canada (AAFC), Lethbridge Current Research and Development Centre, Lethbridge, AB, Canada

Roghieh Hajiboland University of Tabriz, Tabriz, Iran

B. M. Herath Department of Biological Sciences, Faculty of Applied Sciences, Rajarata University of Sri Lanka, Mihintale, Sri Lanka

Mahdieh Houshani Department of Plant, Cell and Molecular Biology, Faculty of Natural Sciences, University of Tabriz, Tabriz, East Azerbaijan, Iran

Michelle Hubbard Agriculture and Agri-Food Canada (AAFC), Swift Current Research and Development Centre, Swift Current, SK, Canada

I. K. Jayamanna Department of Bioprocess Technology, Faculty of Technology, Rajarata University of Sri Lanka, Mihintale, Sri Lanka

Li Ji School of Forestry, Central South University of Forestry and Technology, Changsha, People's Republic of China

K. W. Y. R. Kalamulla Department of Biological Sciences, Faculty of Applied Sciences, Rajarata University of Sri Lanka, Mihintale, Sri Lanka

Priyakshi Kashyap Chaiduar College, Gohpur, Gohpur, Assam, India

Sampurna Kashyap Department of Botany, Gauhati University, Guwahati, Assam, India

Sandeep Kour Department of Zoology, Guru Nanak Dev University, Amritsar, Punjab, India

Deepak Kumar Department of Zoology, Guru Nanak Dev University, Amritsar, Punjab, India

Elham Malekzadeh Department of Soil Science, Gorgan University of Agricultural Sciences and Natural Resources, Gorgan, Iran

Rajesh Kumari Manhas Department of Microbiology, Guru Nanak Dev University, Amritsar, Punjab, India

T. A. N. Mayadunna Department of Biological Sciences, Faculty of Applied Sciences, Rajarata University of Sri Lanka, Mihintale, Sri Lanka

Kana Miyata Department of Applied Biosciences, Toyo University, Itakura, Gunma, Japan

Puja Ohri Department of Zoology, Guru Nanak Dev University, Amritsar, Punjab, India

Parkirti Department of Zoology, Guru Nanak Dev University, Amritsar, Punjab, India

M. A. E. G. Perera Department of Bioprocess Technology, Faculty of Technology, Rajarata University of Sri Lanka, Mihintale, Sri Lanka

Somayeh Rahmat Miandoab, Iran

Melissa Robdrup Agriculture and Agri-Food Canada (AAFC), Lethbridge Current Research and Development Centre, Lethbridge, AB, Canada

Seyed Yahya Salehi-Lisar Department of Plant, Cell and Molecular Biology, Faculty of Natural Sciences, University of Tabriz, Tabriz, East Azerbaijan, Iran

Vinay Shankar Department of Botany, Gaya College, Magadh University, Bodh Gaya, Bihar, India

Indrani Sharma Department of Botany, Gauhati University, Guwahati, Assam, India

Roohi Sharma Department of Zoology, Guru Nanak Dev University, Amritsar, Punjab, India

Maryluce Albuquerque da Silva Campos University of Pernambuco (UPE) Campus Petrolina, Petrolina, Pernambuco, Brazil

Zhaleh Soheilikhah Department of Biology, Faculty of Science, Razi University, Kermanshah, Iran

Amanso Tayang Department of Botany, Rajiv Gandhi University, Rono Hills, Doimukh, Arunachal Pradesh, India

Mikihisa Umehara Department of Applied Biosciences, Toyo University, Itakura, Gunma, Japan

Vikram Department of Zoology, Guru Nanak Dev University, Amritsar, Punjab, India

Sara Fareed Mohamed Wahdan Department of Botany and Microbiology, Faculty of Science, Suez Canal University, Ismailia, Egypt

P. N. Yapa Department of Biological Sciences, Faculty of Applied Sciences, Rajarata University of Sri Lanka, Mihintale, Sri Lanka

Chapter 1
Introduction to Arbuscular Mycorrhizal Fungi and Higher Plant Symbiosis: Characteristic Features, Functions, and Applications

Golam Jalal Ahammed and Roghieh Hajiboland

Abstract Arbuscular mycorrhizal fungi (AMF) are ubiquitous, soil-borne, endophytic, obligate biotrophs that colonize the roots of most terrestrial plants to establish mutually beneficial relationships. The colonization of host roots by AMF improves the access of plant roots to nutrients, mainly phosphorus (P) and nitrogen (N). In return, host plants provide the colonizing fungus with carbon (C)-rich photosynthates such as sugar and lipids. This nutrient exchange strongly impacts plant and microbial ecosystems by influencing plant fitness, core soil processes, and the C cycle. Crucially, both mutualism and ecosystem services largely rely on the bidirectional flow of resources. The AMF symbiosis not only provides plants with nutrients but also supplies water and enhances their resistance to environmental and biotic challenges. The physiological and molecular processes by which AMF aids plants in nutrient foraging and stress resilience have been extensively studied. Recent work has established the framework for nutrient exchange and regulatory interactions between host plants and AMF, but the mechanisms underlying AMF-mediated host plant tolerance to stress still remain unclear. AMF are of potential applications in crop production, forestry management, bioremediation, and ecological restoration as biofertilizers and bioprotectors. In the first chapter, we introduce AMF, plant-AMF symbioses and their characteristic features, functions, and potential applications.

Keywords Arbuscule · Mycorrhizal fungi · Mycorrhizal symbiosis · Phosphate acquisition · Stress tolerance

G. J. Ahammed (✉)
Department of Horticulture, Henan University of Science and Technology, Luoyang, Henan, China
e-mail: ahammed@haust.edu.cn

R. Hajiboland (✉)
University of Tabriz, Tabriz, Iran
e-mail: ehsan@tabrizu.ac.ir

G. J. Ahammed, R. Hajiboland (eds.), *Arbuscular Mycorrhizal Fungi and Higher Plants*, https://doi.org/10.1007/978-981-99-8220-2_1

1.1 Introduction

Arbuscular mycorrhizal fungi (AMF) are ubiquitous, soil-borne, endophytic, obligate biotrophs that colonize the roots of 70–90% of terrestrial plants in various soil types and environmental conditions to establish mutually beneficial relationships (Branco et al. 2022; Shi et al. 2023). The evolutionary trajectory of AMF has been intrinsically tied to the history of plant evolution with the appearance of the first terrestrial plant around 400 MYA ago (Salvioli di Fossalunga and Novero 2019). Taxonomically, AMF belong to the phylum *Glomeromycota* (Giovannini et al. 2020). The name AMF signifies the peculiar relationship between plant roots and mycorrhizal fungi and its key features, wherein "Arbuscule" stands for a special treelike fungal (hyphal) structure in the inner cortex of the root that is used for resource-exchange between "plant root" and "AMF," whereas "mycorrhiza" derived from two words "myco," meaning fungus, and "rhiza," meaning "root" (Shi et al. 2023).

In plant-AMF symbiosis, AMF benefit from the plant-supplied carbon (C) substances such as sugars and lipids derived from photosynthesis (Jiang et al. 2017), while the AMF provide the plant host with mineral nutrients, particularly phosphorus (P) and nitrogen (N) (Wipf et al. 2019). This nutrient exchange strongly impacts plant and microbial ecosystems by influencing plant fitness, core soil processes, and the C cycle (Diagne et al. 2020; Giovannini et al. 2020). AMF assist the roots of the host plant with their extensive hyphal network in absorbing water and nutrients from faraway soils, thus decreasing the requirement for chemical fertilizers and irrigation (Kakouridis et al. 2022). AMF are found in different kinds of environments including severely disturbed ecosystems containing soils polluted with excessive salt, xenobiotics, and heavy metals (Boorboori and Zhang 2022; Branco et al. 2022; Ahammed et al. 2023).

The symbiosis between plant roots and AMF hyphae results in the recruitment of specialized microorganisms into the mycorrhizosphere (Giovannini et al. 2020; Phour et al. 2020). AMF stimulate the mycorrhizosphere microbiome and help plants develop, take up nutrients, and resist abiotic and biotic hazards (Philippot et al. 2013; Zhang et al. 2022). By facilitating nutrient uptake, AMF aid plant development and stress tolerance including tolerance to drought stress (Begum et al. 2019), salinity (Cui et al. 2022; Li et al. 2022), nutrient deficiency (Shi et al. 2021), or heavy metal stress (Dhalaria et al. 2020; Alam et al. 2019). The physiological, biochemical, microbial, and molecular processes by which AMF aid plants in nutrient foraging and stress resilience have been extensively studied (Salvioli di Fossalunga and Novero 2019). Recent work has established the framework for nutrient exchange and regulatory interactions between host plants and AMF, but mechanisms underlying AMF-mediated host plant tolerance to stress still remain unclear (Shi et al. 2023). AMF have potential applications in crop production, forestry management, bioremediation, and ecological restoration as biofertilizers and bioprotectors (Phour et al. 2020; Boorboori and Zhang 2022; Zhu et al. 2022). In this chapter, we provide a brief introduction to AMF, their characteristic features, functions, and potential applications in crop production and environmental management.

1.2 Arbuscular Mycorrhizal Symbiosis in Higher Plants: Characteristic Features

An estimated 250,000 plant species develop mycorrhizal symbiosis with around 50,000 fungal species of four major mycorrhizal types, such as arbuscular mycorrhizal, ectomycorrhizal, ericoid mycorrhizal, and orchid mycorrhizal fungi (van der Heijden et al. 2015). About 70–90% of land plant species develop arbuscular mycorrhizal symbiosis (AMS) with the obligate symbiotic fungi of the phylum *Glomeromycota* (Shi et al. 2023). This includes the vast majority of herbs, grasses, and trees, which act as hosts of AMF in AMS. Importantly, AMF establish symbiotic relationships with around 90% of agricultural plants, encompassing a wide variety of agricultural crops such as cereals, vegetables, and horticultural plants (Diagne et al. 2020). Non-mycorrhizal plants with regard to food crops are limited to genera and species that fall under the families Brassicaceae and Chenopodiaceae (Giovannini et al. 2020).

Although species of host plants are found across the phylogeny of terrestrial plants (van der Heijden et al. 2015), only approximately 345 species of AMF from the phylum *Glomeromycota* were documented on the basis of DNA barcoding (http://www.amf-phylogeny.com/, accessed on 30 June 2023), indicating that AMF are distinguished by limited diversity in species and largely lacking host specificity as opposed to a huge diversity of host plants (Shi et al. 2023). However, a substantial degree of variation exists within the AMF species itself (intraspecific diversity). The significant intraspecific variation among AMF may lead to considerable functional differences such as extraradical hyphal expansion, spore formation, root colonization, nutrient and water fluxes, as well as varying mycorrhizal growth responses. It was proposed that the concurrent evolution of plant and fungal communities might lead to a high level of genetic diversity among AMF isolates (Wipf et al. 2019).

Despite the considerable diversity observed in AMF, limited research has been conducted on their functional diversity, aiming to identify and select the most suitable isolates for agricultural applications (Edlinger et al. 2022; Powell and Rillig 2018). The majority of commercially available inocula are formulated using *Rhizoglomus irregulare* (previously named *Rhizophagus irregularis*, previously classified as *Glomus intraradices*) and *Funneliformis mosseae* (previously named *Glomus mosseae*), which are common symbiotic organisms with a broad range of host plants and are extensively distributed in various soil types and climatic regions worldwide (Giovannini et al. 2020).

Arbuscular mycorrhizal fungi live in dual habits, one is the root, from where they obtain organic C and transfer nutrients, and the other is the soil from where they acquire nutrients, mainly N and P (van der Heijden et al. 2015; Yang et al. 2022; Shi et al. 2021). AMF primarily forage for nutrients derived from plants and saprotrophic microorganisms (Shi et al. 2023). While soil fungal communities are very abundant, AMF make up just a tiny fraction of their abundance (<0.5%) (Fierer 2017). The initiation of AMS involves a communication process between host plants and AMF during the pre-contact phase (López-Ráez and Pozo 2013; Zhou et al. 2019). This interaction progresses to the infection phase and ultimately leads to the development of fully formed arbuscule within the inner root cortical cells of the host plant, facilitating nutrient exchange. Before the development of appressoria, the roots of host plants release strigolactones (SLs, a group of plant hormones commonly known for controlling bud outgrowth) into the rhizosphere in order to induce the germination of fungal spores and hyphal branching of AMF, thereby facilitating the establishment of physical contact between AMF and roots (Zhou et al. 2019). Notably, SL biosynthesis elevates under P-deprived conditions. The regulation of P uptake in host plants is governed by transcription factors involved in the phosphate starvation response (Ho-Plágaro and García-Garrido 2022). In particular, a group of the MYB-CC family of transcription factors, phosphate starvation responses (PHRs), serve as key regulators of AMS (Shi et al. 2021; Das et al. 2022). This implies that the primary driving force behind the establishment of AMS is the host plants' need for phosphate (Wang et al. 2022). AMF release specific substances known as mycorrhizal factors, which include lipo-chitooligosaccharides (Myc-LCOs) as well as short-chain chitin oligomers CO4 and CO5 (Maillet et al. 2011; Phour et al. 2020). These substances serve as signals that are detected by host plants (Genre et al. 2013). Following the mutual exchange of respective signals between AMF and host plants, the hyphae of AMF exhibit growth toward the roots of the host plant, utilizing the SL gradients as a navigational cue within the rhizosphere (Kretzschmar et al. 2012). Upon contact with the epidermis of root cells, the hyphae proceed to extensively branch out, resulting in the formation of a structure known as an appressorium, which is also commonly referred to as a hyphopodium (Fig. 1.1). Subsequently, the hyphae of AMF originating from the appressoria proceed to infiltrate the epidermis of the host plant root and the cells of the cortex. The hyphal infiltration is facilitated by the pre-penetration apparatus (PPA), a complex consisting of cytoskeletal and endoplasmic reticulum components, which guides the migration of the hyphae in a manner directed by the nucleus (Genre et al. 2013; Bonfante and Genre 2010). After traversing the epidermis and outer root cortical cells, the AMF hyphae proceed to expand intercellularly along the longitudinal direction of the root and subsequently infiltrate the inner cortex (Shi et al. 2023). The hyphae of AMF undergo multiple instances of dichotomous branching, leading to the development of arbuscules, which are treelike structures found within the root cells of the cortex (Bonfante and Genre 2010). Arbuscules are enveloped by a layer of plant membranes known as the periarbuscular membrane (PAM). Arbuscules serve as major sites of resource exchange between the AMF and host plants and are the functional hub of such mutually beneficial relationships (Salvioli di Fossalunga and Novero 2019). The arbuscules exhibit a remarkable degree of dynamism, characterized by a relatively short life span lasting only a few days prior to their eventual collapse (Giovannini

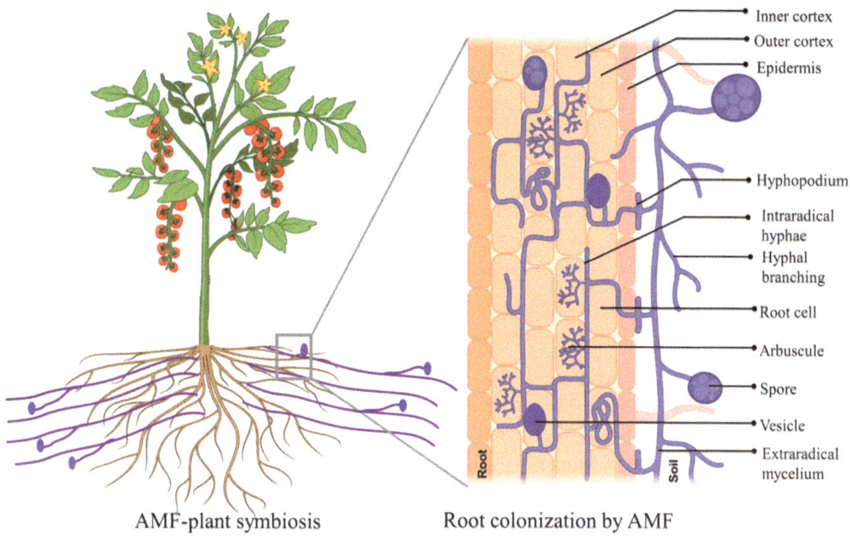

Fig. 1.1 Plant root colonization by arbuscular mycorrhizal fungi (AMF). Created with BioRender.com.

et al. 2020). The regulation of arbuscule degeneration is affected by both the AMF and host plants, indicating the existence of molecular communication between the cortex and AMF hyphae prior to the degeneration of arbuscules over time (Shi et al. 2023).

The life cycle of AMF is dependent on the existence of host plants due to their obligate biotrophic nature (Ahammed et al. 2023). It commences with an initial phase characterized by the absence of symbiotic interactions. During this phase, spores undergo germination in response to various physical stimuli, including the availability of moisture, optimal temperature, and pH. This germination process leads to the formation of hyphae, which have a short life span (Giovannini et al. 2020). Subsequently, a sequence of events occurs, commencing with the physical interaction between the AMF hyphae and the root. This interaction as mentioned above leads to the penetration of the root cells by the AMF hyphae, ultimately resulting in the formation of arbuscules. This marks the initiation of the symbiotic phase. There are two distinct forms of root colonization: (1) Arum type and (2) Paris type (Tominaga et al. 2022). In the Arum type, the AMF pass on intercellularly through the cells of the cortex, resulting in the development of terminal arbuscules on intracellular hyphal branches (Smith and Smith 1997). The Paris type of root colonization is characterized by the direct growth of the fungus from one root cell to another root cell within the cortex. This growth pattern gives rise to intracellular hyphal coils and intercalary arbuscules (i.e., the arbusculate coil), which are distributed along the coils (Tominaga et al. 2019). Nevertheless, the majority of the available data concerning AMF come from studies conducted on the Arum-type root colonization, which is extensively found in both agricultural and natural ecosystems (Smith and Smith 1997). In addition to the formation of arbuscules, numerous species of AMF exhibit the ability to generate intraradical vesicles. The vesicles are specialized storage structures resembling fungal spores and containing lipid reserves (Giovannini et al. 2020). Upon receiving C from the host plant root, the fungal symbiont can grow outside of the root system, establishing itself in the surrounding soil. This colonization enables the absorption of mineral nutrients, which can then be delivered to the host plant. Additionally, AMF interact with microbes present in the rhizosphere and soil and are capable of colonizing the roots of neighboring plants, even those belonging to different species, genera, and families than their original host plants (Zhang et al. 2022). Furthermore, AMF are capable of transferring nutrients from one host plant to another (van der Heijden et al. 2015). The life cycle of AMF is completed through the production of asexual spores by the extraradical mycelium, which plays a key role in sustaining a high level of mycorrhizal potential in the soil (Giovannini et al. 2020). This, in turn, contributes to the overall biological fertility of the soil.

The existing body of literature suggests that plants have the ability to regulate the extent of root colonization by AMF based on their needs for specific nutrients particularly P and N (Shi et al. 2021). It is widely postulated that the nutrients transported to the cortical cells of the root are responsible for initiating a signaling mechanism that regulates the release of C-rich photoassimilates such as sugars to the partner AMF (López-Ráez and Pozo 2013). The underlying reasons for employing this particular approach rely on the premise that the symbiotic fungal partner (AMF) that is incapable of providing substantial quantities of essential nutrients (e.g., Pi and N) would only be able to obtain limited amounts of C from the apoplast of roots (Wipf et al. 2019).

Plants colonized with AMF (AM plants) employ two distinct pathways for nutrient acquisition, namely, the direct pathway and the mycorrhizal pathway (Salvioli di Fossalunga and Novero 2019). In the first pathway, nutrient uptake takes place directly through the epidermis of root cells and root hairs, while in the latter pathway, nutrient acquisition is facilitated by the AMF hyphae, which enable the transport of nutrients that are acquired from the surrounding soil to the root cells (Ma et al. 2021; van der Heijden et al. 2015). The choice of AM plants between the direct pathway and mycorrhizal pathway is contingent upon the level of phosphorus availability in the rhizosphere (Zhang et al. 2021b). The hyphae of AMF exhibit the ability to extend their growth beyond the zone of phosphorus (Pi, inorganic P) depletion, thereby enabling them to access Pi resources that are otherwise unavailable to plant roots (Fig. 1.2).

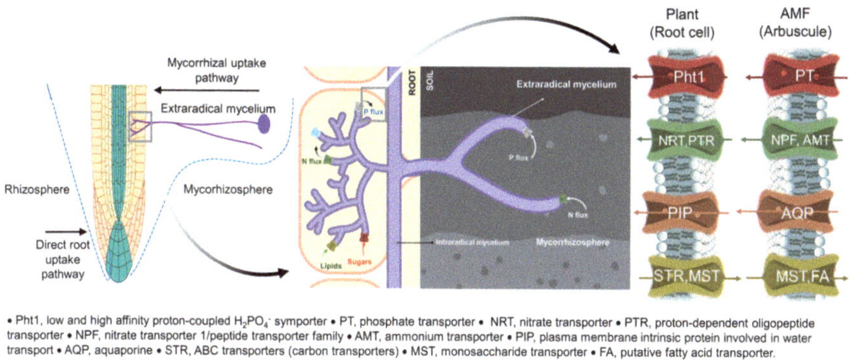

Fig. 1.2 Mycorrhizal pathways of nutrient acquisition and resource exchange in arbuscules. Created with BioRender.com.

1.3 Functions of AMS in Plant Growth, Development, and Stress Tolerance

1.3.1 Plant Growth, Development, and Nutrient Acquisition

The growth and reproductive development of plants were found to be significantly influenced by the nutrient availability in the soil and the presence of AMF (Wang et al. 2022). The plant root colonization by AMF enhances the photosynthetic rate and elevates the levels of chlorophyll concentrations and biomass accumulation of the partner plants (Chen et al. 2017b). A recent study found that AMS had a direct positive impact on the growth and nutrient content of cherry tomato plants (Wang et al. 2022). The study confirmed that AMS played a role in promoting flower and seed formation in cherry tomato plants, and the positive outcomes of AMS are thought to be primarily due to the improved nutrient status of the host plants. For instance, phosphorus plays a vital role in the growth, development, and survival of plants; however, its availability to plants is limited in common soil types (Zhang et al. 2021b). The presence of AMS has been shown to enhance the efficiency of phosphate uptake by the host plant (Liao et al. 2022; Ma et al. 2021). Plants allocate approximately 10–20% of their photosynthetic carbon to AMF. In reciprocation, AMF are known to provide ~90% of the phosphorus required by plants (van der Heijden et al. 2015). AMF possess the ability to enter nutrient-rich areas of soil located beyond the rhizosphere through their broad extraradical hyphal network, thereby enabling them to acquire P that would otherwise be inaccessible to plants. It is worth mentioning that extraradical hyphae are capable of releasing low-molecular-weight organic acids, which have the ability to dissolve phosphorus from sources that are typically not accessible to biological processes (Andrino et al. 2021). AMF acquire phosphate (Pi) from the immediate surroundings in soil by means of phosphate transporters (PHT1), which are specifically located on the extraradical hyphae (Wipf et al. 2019).

AMF can potentially contribute around 33% of the root protein nitrogen (Govindarajulu et al. 2005). The uptake of nitrogen is facilitated by a range of transport systems, which include the transport of inorganic nitrogen in the forms of nitrate (NO_3^-) and ammonium (NH_4^+), as well as the transport of organic nitrogen in the forms of peptides and amino acids (Wipf et al. 2019). AMF primarily acquire inorganic N from the soil by utilizing ammonium transporters (AMTs), and they exhibit a preference for NH_4^+ over NO_3^- as a nitrogen source (Hui et al. 2022). This preference arises from the fact that NO_3^- is to be converted to NH_4^+ before it can be assimilated into organic compounds, a process that requires energy expenditure (Shi et al. 2023). Nitrogen transport through the mycorrhizal pathway is significantly impacted by AMTs (Breuillin-Sessoms et al. 2015). Nevertheless, the mutualistic transport of NH_4^+ may not be a necessary prerequisite for the formation of AMS or the development of arbuscules. Furthermore, it has been observed that apart from NH_4^+, there is a symbiotic pathway for NO_3^- uptake. Nitrate is acquired through an uptake mechanism that relies on energy expenditure and is facilitated by specialized transporters belonging to the large family of nitrate and peptide transporters, namely, NPF (NRT1/PTR), NRT2, and NRT3 families (Wipf et al. 2019).

1.3.2 Abiotic and Biotic Stress Tolerance

Plants in natural environments face a multitude of challenges, encompassing both biotic factors, such as pathogenic fungi, bacteria, nematodes, and insect pests, and abiotic factors, including salinity, drought, and toxic metal contamination (Sanchez-Bermudez et al. 2022; Li and Ahammed 2023). Crucially, AMF have the potential to significantly augment the resistance of the plant partner to various forms of stress (Fig. 1.3). AMS has been shown to increase the acquisition of mineral nutrients in plants in both non-stressful and stressful environments (Chen et al. 2017b). However, the impact of AMS on nutrient uptake is particularly significant under conditions of stress (Diagne et al. 2020). Moreover, plant roots colonized by AMF have the ability to absorb more water from the rhizosphere, and AMF have the potential to ameliorate drought stress (Begum et al. 2019; Moradtalab et al. 2019; Puschel et al. 2020). It was found that the water transported by AMF *Rhizophagus intraradices* through the air gap constituted approximately 34.6% of the total water transpired by the host plant *Avena barbata*, commonly known as wild oat (Kakouridis et al. 2022). This study presents empirical evidence demonstrating that AMF can function as extensions of the root system within the water movement continuum encompassing the soil, plant, and air. The enhanced uptake of mineral nutrients and increased water content in AM plants can potentially be associated with the indirect effect of AMF in mitigating stress (Hajiboland et al. 2019; Puschel et al. 2020; Diagne et al. 2020).

The colonization of plant roots by AMF can alleviate the detrimental effects associated with drought and salinity through enhancements in nutrient absorption, reduction of oxidative stress, augmentation of osmotic adjustment, and alterations in stress responses (Shi et al. 2023). AMF colonization assists host plants in

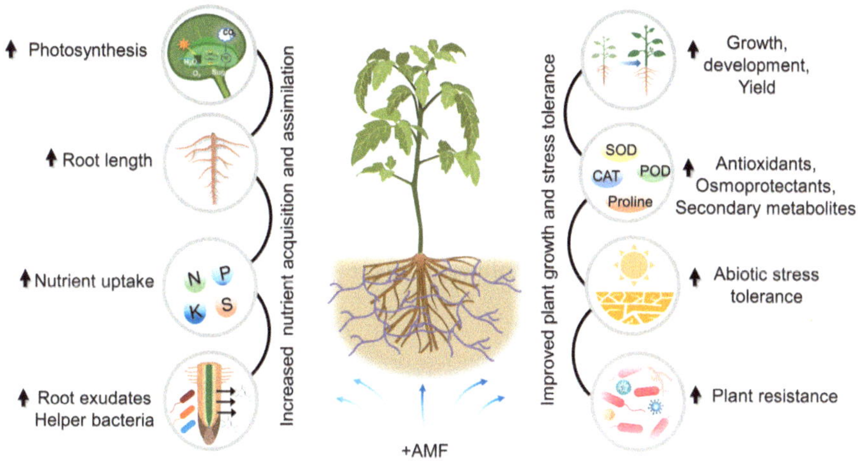

Fig. 1.3 Effects of plant root colonization by arbuscular mycorrhizal fungi (AMF) on plant growth, development, and stress tolerance. Created with BioRender.com.

maintaining appropriate ion homeostasis when subjected to salinity stress (Chen et al. 2017a; Porcel et al. 2016). Studies also indicate that the presence of AMS has a beneficial impact on host plants subjected to flooding stress (Diagne et al. 2020). AMF have a positive impact on plant performance in conditions of heavy metal stress (Boorboori and Zhang 2022).

In plants, abiotic stress often coincides with oxidative stress, as a result of the excessive production of reactive oxygen species (ROS) causing elevated levels of ion leakage and lipid peroxidation (Sanchez-Bermudez et al. 2022). The colonization of AMF has been observed to mitigate oxidative stress in AM plants (Diagne et al. 2020). This is achieved through the upregulation in the activity and transcripts of antioxidant enzymes, such as superoxide dismutase (SOD), peroxidase (POD), catalase (CAT), and glutathione reductase (GR) in AM plants (Zhu et al. 2022). The modulation of enzyme activity leads to a reduction in the accumulation of ROS and a decrease in oxidative damage to lipids and other biomolecules. In addition, the colonization of AMF has been found to have a significant impact on the accumulation of osmoregulatory substances such as soluble sugars and free proline in AM plants when subject to stress (Yooyongwech et al. 2016; Pasbani et al. 2020).

Symbiotic associations with AMF provide beneficial effects to host plants in mitigating biotic stress as well, particularly in the context of pathogen infections (Zhang et al. 2021a). The beneficial impacts of AMS in mitigating pathogen infections can be ascribed to several factors, including enhancements in plant nutrient availability, alterations in root exudate compositions, influences on the rhizosphere microbiome, improvement of plant resistance (both local and systemic resistance), and priming of defense responses (Wu et al. 2021). Enhanced resistance of AM plants is achieved through the direct activation of genes (pathogenesis-related protein *PR1*) that are involved in the host's defense responses against pathogen infections (Fiorilli et al. 2018). After being infected by pathogens, the roots that are colonized by AMF exhibit the deposition of callose in the spaces between cells (Sanmartin et al. 2020). Additionally, they form structures resembling papillae at the locations where fungal pathogens penetrate, effectively limiting the growth of pathogens both within and between cells (Shi et al. 2023). The pre-inoculation of tomato plants with AMF leads to a significantly enhanced defense response compared to control plants that are not colonized by AMF when infected with *Alternaria solani* (Song et al. 2015). AMF can assist host plants in mounting a systemic resistance against pathogens (Fiorilli et al. 2018).

1.3.3 Soil Fertility, Nutrient Cycling, and Ecosystem Services

The AMF-plant symbiosis is of great significance in the carbon cycle of soil ecosystems (Fierer 2017; van der Heijden et al. 2015). Soil organic matter (SOM) represents the most substantial C reservoir on land, and its composition is influenced by many factors such as plant biomass, primary productivity, and the transfer of photosynthetic C to underground biomass (Phour et al. 2020). The primary role played by AMF in soil is to serve as a carbon sink for plant-derived carbon, which is subsequently allocated to AMF hyphae and other microorganisms present in the soil (Giovannini et al. 2020; Ossowicki et al. 2021; Hermans et al. 2023). AMF and the diverse soil microbial communities play a decisive role in the release or intake of carbon from the soil through processes such as respiration, decomposition, mineralization, and immobilization (Parihar et al. 2020; Philippot et al. 2013). Consequently, the absorption of nutrients by AMF and subsequent transfer to the host plant can enhance the assimilation of C by plants, resulting in a greater supply of C to mycorrhizal fungi. This establishes a positive feedback loop involving host plants, AMF, and carbon. A substantial portion of this carbon is subsequently deposited in the soil in the form of glomalin-related soil protein (GRSP) and fungal chitin (Genre et al. 2013). Glomalin is a glycoprotein synthesized by AMF that exhibits a significant degree of homology with heat shock protein 60 (Gao et al. 2019). There exists a strong positive correlation between GRSP and soil aggregate stability, as GRSP represents an important constituent of soil organic matter. In fact, GRSP alone constitutes approximately 30–40% of the total organic carbon present in undisturbed soils (Parihar et al. 2020). Thus, AMF-released glomalin exhibits potential functionality within soil ecosystems, encompassing the facilitation of soil organic carbon storage, enhancement of soil aggregate structure, augmentation of plant resistance, and mitigation of plant metal toxicity (Riaz et al. 2021; Ahammed et al. 2023; Gao et al. 2019). Moreover, it should be noted that the extraradical hyphae of AMF make up a significant amount of the microbial biomass in the soil. These hyphae account for approximately 54–900 kg ha^{-1} of SOM (Parihar et al. 2020). Additionally, the presence of extraradical mycelia and GRSP is crucial in the development of soil aggregates, which are crucial in safeguarding organic carbon against decomposition (Parihar et al. 2020).

In addition to mycorrhizae, various fungi and bacteria residing in the rhizosphere of host plants are drawn to and consume the rhizo-deposits, thereby enhancing plant growth, nutrient absorption, and stress resilience (Zhang et al. 2022; Philippot et al. 2013; Hermans et al. 2023). Bacteria that contribute positively to the ecological health, establishment, and functions of AMS are referred to as mycorrhizae helper bacteria (MHBs) (Sangwan and Prasanna 2022). MHBs exert their promotional effects through the production of volatile compounds that stimulate growth and hydrolytic enzymes. Additionally, MHBs enhance plant nutrient uptake by AMF (Sangwan and Prasanna 2022). Previous studies have reported that MHBs exhibit an affinity for the external surface of mycorrhizal fungi, thereby augmenting mycorrhizal symbiotic interactions. This is exemplified by their ability to enhance nutrient uptake (Sangwan and Prasanna 2022). Plant growth-promoting rhizobacteria (PGBs) are currently recognized as MHBs, exhibiting promising prospects for practical implementation in the fields of agriculture and forestry (Giovannini et al. 2020; Shi et al. 2023).

1.4 Applications of AMF in Crop Production, Crop Protection, and Bioremediation

Despite the incomplete understanding of the mechanistic underpinnings of chemical signaling and regulation in the responses of plants during the AMS, it is of great importance to exploit this signaling process for the purposes of enhancing crop protection and increasing yields (López-Ráez and Pozo 2013; Ho-Plágaro and García-Garrido 2022). The mutualistic association between AMF and plants has the potential to significantly enhance crop productivity and promote the long-term stability of ecosystems (Diagne et al. 2020). AMF not only exhibit nutritional benefits, such as enhanced water and nutrient absorption, but also possess non-nutritional advantages, including soil aggregate stabilization, erosion prevention, and mitigation of biotic and abiotic stressors (Parihar et al. 2020). Given the ongoing elucidation of the nature and chemistry of the major molecules implicated in the establishment of AMS in the subterranean environment, there exists significant potential for the manipulation of their production to enhance AMS and its associated advantages, thereby offering promising prospects for the advancement of sustainable agriculture (Phour et al. 2020).

Notably, plant tolerance to heavy metals such as chromium (Cr) is improved by the direct influence of AMF on the transformation and stabilization of Cr, as well as the symbiotic relationship between AMF and plants, which indirectly enhances nutrient acquisition and physiological regulation in plants (Ahammed et al. 2023; Riaz et al. 2021). Studies provide clear evidence that AMF have the potential to not only enhance crop production in heavy metal-contaminated soils but also improve the effectiveness of phytoremediation (Boorboori and Zhang 2022). For AMF-assisted phytoremediation, it is imperative to explore AMF isolates that exhibit a notable capacity for tolerating heavy metals, as well as plant species that demonstrate both an increased growth rate and the ability to accumulate substantial biomass (Dhalaria et al. 2020; Ahammed et al. 2023).

AMF have been employed as biofertilizers to augment plant growth and yields in agricultural practices, albeit their utilization remains relatively limited in comparison to conventional approaches (Gelvez-Pardo et al. 2023). The positive and beneficial effects of the symbiotic relationship on the plant host are contingent upon the specific combination of plant and AMF (Chen et al. 2017b). Hence, the crucial aspect lies in the careful selection of suitable AMF isolates that are well adapted to the specific local conditions and possess optimal functional characteristics for the particular plant species under consideration (Gianinazzi et al. 2010; Giovannini et al. 2020). Gaining a comprehensive understanding of the mechanisms that govern colonization, symbiosis efficiency, and competence is of utmost importance in order to maximize the benefits derived from mycorrhizal associations (Wipf et al. 2019). The comprehensive understanding of the molecules implicated in the interaction between host plants and AMF will undoubtedly enhance the effective application of "mycorrhizal technology" in agricultural systems (Gelvez-Pardo et al. 2023). There exists a debate regarding the suitability of AMF in agricultural systems, primarily stemming from the potential counteractive impact of excessive chemical fertilizer application, a prevalent practice in numerous countries, on the beneficial outcomes of AMF on crop growth (Giovannini et al. 2020). Nevertheless, AMF do possess pivotal roles in augmenting the efficacy and durability of agroecosystems (Gianinazzi et al. 2010; Piliarová et al. 2019; Powell and Rillig 2018).

1.5 Conclusions

In natural ecosystems, terrestrial plants establish symbiotic associations with mycorrhizal fungi. Subterranean symbionts AMF are of paramount importance in terrestrial ecosystems due to their crucial role in the regulation of nutrients and the carbon cycle, as well as their influence on soil structure and the multifaceted functions of ecosystems. AMF are integral components of earth ecosystems and exert significant influence on global C and nutrient cycles. AMF play a vital role in the provision of nitrogen and phosphorus to plants. Numerous vascular plant species rely on these symbiotic associations with AMF for their growth and overall survival.

The fundamental requirement for any AMF isolate intended for agricultural application is a strong capacity for root colonization. This ability is crucial as it enables the isolate to effectively compete with native AMF, which are known to be highly competitive. AMF offer mineral nutrient provision and protective benefits to their host plants in the face of biotic and abiotic stressors, suggesting their potential application as biofertilizers within the context of sustainable agricultural practices.

The role of bacterial communities in the mycorrhizosphere as a potential mediator of the root colonization capacity of AMF has been well established in scientific literature. The symbiotic relationship between host plants, AMF symbionts, and related bacteria exhibits advantageous emergent characteristics that can be effectively harnessed in the context of sustainable agriculture. The colonization capacity, efficiency, and stress tolerance of AMF exhibit significant variation among different isolates. The extensive inter- and intraspecific diversity of AMF can be effectively utilized by carefully selecting AMF inocula based on their colonization capacity and efficiency. These characteristics are influenced by the genotypes of both fungi and host plants, as well as various environmental factors.

Biotechnological tools encompass processes of biological importance that leverage the chemistry of living organisms to advance the development of novel and enhanced methods for the production of conventional products all while upholding the integrity of the natural environment. AMS as a biotechnological tool plays a crucial role in shaping plant growth and facilitating stress adaptation. The effective management and assessment of ecosystem services offered by AMF is imperative for enhancing plant production both in terms of quality and quantity within the framework of agriculture that relies on the limited use of synthetic chemical inputs. To effectively manage AMF in ecological engineering and crop production and in selecting plant species that enhance the benefits of AMF, it is crucial to comprehend the intricate mechanisms that govern the development and functions of AMS.

References

Ahammed GJ, Shamsy R, Liu A, Chen S (2023) Arbuscular mycorrhizal fungi-induced tolerance to chromium stress in plants. Environ Pollut. https://doi.org/10.1016/j.envpol.2023.121597

Alam MZ, Hoque MA, Ahammed GJ, Carpenter-Boggs L (2019) Arbuscular mycorrhizal fungi reduce arsenic uptake and improve plant growth in Lens culinaris. PLoS ONE 14(5):e0211441. https://doi.org/10.1371/journal.pone.0211441

Andrino A, Guggenberger G, Kernchen S, Mikutta R, Sauheitl L, Boy J (2021) Production of organic acids by arbuscular mycorrhizal fungi and their contribution in the mobilization of phosphorus bound to iron oxides. Front Plant Sci 12:661842. https://doi.org/10.3389/fpls.2021.661842

Begum N, Ahanger MA, Su Y, Lei Y, Mustafa NSA, Ahmad P, Zhang L (2019) Improved drought tolerance by amf inoculation in maize (Zea mays) involves physiological and biochemical implications. Plants 8(12):579. https://doi.org/10.3390/plants8120579

Bonfante P, Genre A (2010) Mechanisms underlying beneficial plant-fungus interactions in mycorrhizal symbiosis. Nat Commun 1:48. https://doi.org/10.1038/ncomms1046

Boorboori MR, Zhang HY (2022) Arbuscular mycorrhizal fungi are an influential factor in improving the phytoremediation of arsenic, cadmium, lead, and chromium. J Fungi 8(2):176. https://doi.org/10.3390/jof8020176

Branco S, Schauster A, Liao HL, Ruytinx J (2022) Mechanisms of stress tolerance and their effects on the ecology and evolution of mycorrhizal fungi. New Phytol 235(6):2158–2175. https://doi.org/10.1111/nph.18308

Breuillin-Sessoms F, Floss DS, Gomez SK, Pumplin N, Ding Y, Levesque-Tremblay V, Noar RD, Daniels DA, Bravo A, Eaglesham JB, Benedito VA, Udvardi MK, Harrison MJ (2015) Suppression of arbuscule degeneration in Medicago truncatula phosphate transporter4 mutants is dependent on the ammonium transporter 2 family protein AMT2;3. Plant Cell 27(4): 1352–1366. https://doi.org/10.1105/tpc.114.131144

Chen J, Zhang H, Zhang X, Tang M (2017a) Arbuscular mycorrhizal symbiosis alleviates salt stress in black locust through improved photosynthesis, water status, and K(+)/Na(+) homeostasis. Front Plant Sci 8:1739. https://doi.org/10.3389/fpls.2017.01739

Chen S, Zhao H, Zou C, Li Y, Chen Y, Wang Z, Jiang Y, Liu A, Zhao P, Wang M, Ahammed GJ (2017b) Combined inoculation with multiple arbuscular mycorrhizal fungi improves growth, nutrient uptake and photosynthesis in cucumber seedlings. Front Microbiol 8:2516. https://doi.org/10.3389/fmicb.2017.02516

Cui X, Jia B, Diao F, Li X, Xu J, Zhang Z, Li FY, Guo W (2022) Transcriptomic analysis reveals the molecular mechanisms of arbuscular mycorrhizal fungi and nitrilotriacetic acid on Suaeda salsa tolerance to combined stress of cadmium and salt. Process Saf Environ Prot 160:210–220. https://doi.org/10.1016/j.psep.2022.02.019

Das D, Paries M, Hobecker K, Gigl M, Dawid C, Lam HM, Zhang J, Chen M, Gutjahr C (2022) Phosphate starvation response transcription factors enable arbuscular mycorrhiza symbiosis. Nat Commun 13(1):477. https://doi.org/10.1038/s41467-022-27976-8

Dhalaria R, Kumar D, Kumar H, Nepovimova E, Kuča K, Torequl Islam M, Verma R (2020) Arbuscular mycorrhizal fungi as potential agents in ameliorating heavy metal stress in plants. Agronomy 10(6):815. https://doi.org/10.3390/agronomy10060815

Diagne N, Ngom M, Djighaly PI, Fall D, Hocher V, Svistoonoff S (2020) Roles of arbuscular mycorrhizal fungi on plant growth and performance: importance in biotic and abiotic stressed regulation. Diversity 12(10):370. https://doi.org/10.3390/d12100370

Edlinger A, Garland G, Hartman K, Banerjee S, Degrune F, García-Palacios P, Hallin S, Valzano-Held A, Herzog C, Jansa J, Kost E, Maestre FT, Pescador DS, Philippot L, Rillig MC, Romdhane S, Saghaï A, Spor A, Frossard E, van der Heijden MGA (2022) Agricultural management and pesticide use reduce the functioning of beneficial plant symbionts. Nat Ecol Evol 6(8):1145–1154. https://doi.org/10.1038/s41559-022-01799-8

Fierer N (2017) Embracing the unknown: disentangling the complexities of the soil microbiome. Nat Rev Microbiol 15(10):579–590. https://doi.org/10.1038/nrmicro.2017.87

Fiorilli V, Vannini C, Ortolani F, Garcia-Seco D, Chiapello M, Novero M, Domingo G, Terzi V, Morcia C, Bagnaresi P, Moulin L, Bracale M, Bonfante P (2018) Omics approaches revealed how arbuscular mycorrhizal symbiosis enhances yield and resistance to leaf pathogen in wheat. Sci Rep 8(1):9625. https://doi.org/10.1038/s41598-018-27622-8

Gao WQ, Wang P, Wu QS (2019) Functions and application of glomalin-related soil proteins: a review. Sains Malaysiana 48(1):111–119. https://doi.org/10.17576/jsm-2019-4801-13

Gelvez-Pardo I, Lobo-Berbesi L, Santos-Díaz A (2023) Biological efficacy of plant growth–promoting bacteria and arbuscular mycorrhizae fungi: assessments in laboratory and greenhouse conditions. Curr Protocols 3(4):e732. https://doi.org/10.1002/cpz1.732

Genre A, Chabaud M, Balzergue C, Puech-Pagès V, Novero M, Rey T, Fournier J, Rochange S, Bécard G, Bonfante P, Barker DG (2013) Short-chain chitin oligomers from arbuscular mycorrhizal fungi trigger nuclear Ca2+ spiking in Medicago truncatula roots and their production is enhanced by strigolactone. New Phytol 198(1):190–202. https://doi.org/10.1111/nph.12146

Gianinazzi S, Gollotte A, Binet MN, van Tuinen D, Redecker D, Wipf D (2010) Agroecology: the key role of arbuscular mycorrhizas in ecosystem services. Mycorrhiza 20(8):519–530. https://doi.org/10.1007/s00572-010-0333-3

Giovannini L, Palla M, Agnolucci M, Avio L, Sbrana C, Turrini A, Giovannetti M (2020) Arbuscular mycorrhizal fungi and associated microbiota as plant biostimulants: research strategies for the selection of the best performing inocula. Agronomy 10(1):106. https://doi.org/10.3390/agronomy10010106

Govindarajulu M, Pfeffer PE, Jin H, Abubaker J, Douds DD, Allen JW, Bücking H, Lammers PJ, Shachar-Hill Y (2005) Nitrogen transfer in the arbuscular mycorrhizal symbiosis. Nature 435(7043):819–823. https://doi.org/10.1038/nature03610

Hajiboland R, Joudmand A, Aliasgharzad N, Tolrá R, Poschenrieder C (2019) Arbuscular mycorrhizal fungi alleviate low-temperature stress and increase freezing resistance as a substitute for acclimation treatment in barley. Crop Pasture Sci 70(3):218–233

Hermans SM, Lear G, Case BS, Buckley HL (2023) The soil microbiome: an essential, but neglected, component of regenerative agroecosystems. Science 26(2):106028. https://doi.org/10.1016/j.isci.2023.106028

Ho-Plágaro T, García-Garrido JM (2022) Molecular regulation of arbuscular mycorrhizal symbiosis. Int J Mol Sci 23(11):5960. https://doi.org/10.3390/ijms23115960

Hui J, An X, Li Z, Neuhäuser B, Ludewig U, Wu X, Schulze WX, Chen F, Feng G, Lambers H, Zhang F, Yuan L (2022) The mycorrhiza-specific ammonium transporter ZmAMT3;1 mediates mycorrhiza-dependent nitrogen uptake in maize roots. Plant Cell 34(10):4066–4087. https://doi.org/10.1093/plcell/koac225

Jiang Y, Wang W, Xie Q, Liu N, Liu L, Wang D, Zhang X, Yang C, Chen X, Tang D, Wang E (2017) Plants transfer lipids to sustain colonization by mutualistic mycorrhizal and parasitic fungi. Science 356(6343):1172–1175. https://doi.org/10.1126/science.aam9970

Kakouridis A, Hagen JA, Kan MP, Mambelli S, Feldman LJ, Herman DJ, Weber PK, Pett-Ridge J, Firestone MK (2022) Routes to roots: direct evidence of water transport by arbuscular mycorrhizal fungi to host plants. New Phytol. https://doi.org/10.1111/nph.18281

Kretzschmar T, Kohlen W, Sasse J, Borghi L, Schlegel M, Bachelier JB, Reinhardt D, Bours R, Bouwmeester HJ, Martinoia E (2012) A petunia ABC protein controls strigolactone-dependent symbiotic signalling and branching. Nature 483(7389):341–344. https://doi.org/10.1038/nature10873

Li Z, Ahammed GJ (2023) Salicylic acid and jasmonic acid in elevated CO2-induced plant defense response to pathogens. J Plant Physiol. https://doi.org/10.1016/j.jplph.2023.154019

Li X, Zhang Z, Luo J, Cui X, Xu J, Hou Y, Hao B, Guo W (2022) Arbuscular mycorrhizal fungi and nitrilotriacetic acid regulated Suaeda salsa growth in Cd-contaminated saline soil by driving rhizosphere bacterial assemblages. Environ Exp Bot 193:104669. https://doi.org/10.1016/j.envexpbot.2021.104669

Liao D, Sun C, Liang H, Wang Y, Bian X, Dong C, Niu X, Yang M, Xu G, Chen A, Wu S (2022) SlSPX1-SlPHR complexes mediate the suppression of arbuscular mycorrhizal symbiosis by phosphate repletion in tomato. Plant Cell 34(10):4045–4065. https://doi.org/10.1093/plcell/koac212

López-Ráez JA, Pozo MJ (2013) Chemical signalling in the arbuscular mycorrhizal symbiosis: biotechnological applications. In: Symbiotic endophytes. Soil biology. Springer, Cham, pp 215–232. https://doi.org/10.1007/978-3-642-39317-4_11

Ma X, Li X, Ludewig U (2021) Arbuscular mycorrhizal colonization outcompetes root hairs in maize under low phosphorus availability. Ann Bot 127(1):155–166. https://doi.org/10.1093/aob/mcaa159

Maillet F, Poinsot V, André O, Puech-Pagès V, Haouy A, Gueunier M, Cromer L, Giraudet D, Formey D, Niebel A, Martinez EA, Driguez H, Bécard G, Dénarié J (2011) Fungal lipochitooligosaccharide symbiotic signals in arbuscular mycorrhiza. Nature 469(7328):58–63. https://doi.org/10.1038/nature09622

Moradtalab N, Hajiboland R, Aliasgharzad N, Hartmann TE, Neumann G (2019) Silicon and the association with an arbuscular-mycorrhizal fungus (Rhizophagus clarus) mitigate the adverse effects of drought stress on strawberry. Agronomy 9(1):41

Ossowicki A, Raaijmakers JM, Garbeva P (2021) Disentangling soil microbiome functions by perturbation. Environ Microbiol Rep 13(5):582–590. https://doi.org/10.1111/1758-2229.12989

Parihar M, Rakshit A, Meena VS, Gupta VK, Rana K, Choudhary M, Tiwari G, Mishra PK, Pattanayak A, Bisht JK, Jatav SS, Khati P, Jatav HS (2020) The potential of arbuscular mycorrhizal fungi in C cycling: a review. Arch Microbiol 202(7):1581–1596. https://doi.org/10.1007/s00203-020-01915-x

Pasbani B, Salimi A, Aliasgharzad N, Hajiboland R (2020) Colonization with arbuscular mycorrhizal fungi mitigates cold stress through improvement of antioxidant defense and accumulation of protecting molecules in eggplants. Sci Hortic 272:109575

Philippot L, Raaijmakers JM, Lemanceau P, van der Putten WH (2013) Going back to the roots: the microbial ecology of the rhizosphere. Nat Rev Microbiol 11(11):789–799. https://doi.org/10.1038/nrmicro3109

Phour M, Sehrawat A, Sindhu SS, Glick BR (2020) Interkingdom signaling in plant-rhizomicrobiome interactions for sustainable agriculture. Microbiol Res 241:126589. https://doi.org/10.1016/j.micres.2020.126589

Piliarová M, Ondreičková K, Hudcovicová M, Mihálik D, Kraic J (2019) Arbuscular mycorrhizal fungi – their life and function in ecosystem. Agriculture 65(1):3–15. https://doi.org/10.2478/agri-2019-0001

Porcel R, Aroca R, Azcon R, Ruiz-Lozano JM (2016) Regulation of cation transporter genes by the arbuscular mycorrhizal symbiosis in rice plants subjected to salinity suggests improved salt tolerance due to reduced Na(+) root-to-shoot distribution. Mycorrhiza 26(7):673–684. https://doi.org/10.1007/s00572-016-0704-5

Powell JR, Rillig MC (2018) Biodiversity of arbuscular mycorrhizal fungi and ecosystem function. New Phytol 220(4):1059–1075. https://doi.org/10.1111/nph.15119

Puschel D, Bitterlich M, Rydlova J, Jansa J (2020) Facilitation of plant water uptake by an arbuscular mycorrhizal fungus: a Gordian knot of roots and hyphae. Mycorrhiza 30(2-3): 299–313. https://doi.org/10.1007/s00572-020-00949-9

Riaz M, Kamran M, Fang Y, Wang Q, Cao H, Yang G, Deng L, Wang Y, Zhou Y, Anastopoulos I, Wang X (2021) Arbuscular mycorrhizal fungi-induced mitigation of heavy metal phytotoxicity in metal contaminated soils: a critical review. J Hazard Mater 402:123919. https://doi.org/10.1016/j.jhazmat.2020.123919

Salvioli di Fossalunga A, Novero M (2019) To trade in the field: the molecular determinants of arbuscular mycorrhiza nutrient exchange. Chem Biol Technol Agric 6(1):12. https://doi.org/10.1186/s40538-019-0150-7

Sanchez-Bermudez M, Del Pozo JC, Pernas M (2022) Effects of combined abiotic stresses related to climate change on root growth in crops. Front Plant Sci 13:918537. https://doi.org/10.3389/fpls.2022.918537

Sangwan S, Prasanna R (2022) Mycorrhizae helper bacteria: unlocking their potential as bioenhancers of plant-arbuscular mycorrhizal fungal associations. Microb Ecol 84(1):1–10. https://doi.org/10.1007/s00248-021-01831-7

Sanmartin N, Pastor V, Pastor-Fernandez J, Flors V, Pozo MJ, Sanchez-Bel P (2020) Role and mechanisms of callose priming in mycorrhiza-induced resistance. J Exp Bot 71(9):2769–2781. https://doi.org/10.1093/jxb/eraa030

Shi J, Zhao B, Zheng S, Zhang X, Wang X, Dong W, Xie Q, Wang G, Xiao Y, Chen F, Yu N, Wang E (2021) A phosphate starvation response-centered network regulates mycorrhizal symbiosis. Cell 184(22):5527–5540. https://doi.org/10.1016/j.cell.2021.09.030

Shi J, Wang X, Wang E (2023) Mycorrhizal symbiosis in plant growth and stress adaptation: from genes to ecosystems. Annu Rev Plant Biol 74(1):569–607. https://doi.org/10.1146/annurev-arplant-061722-090342

Smith FA, Smith SE (1997) Structural diversity in (vesicular)–arbuscular mycorrhizal symbioses. New Phytol 137(3):373–388. https://doi.org/10.1046/j.1469-8137.1997.00848.x

Song Y, Chen D, Lu K, Sun Z, Zeng R (2015) Enhanced tomato disease resistance primed by arbuscular mycorrhizal fungus. Front Plant Sci 6:786. https://doi.org/10.3389/fpls.2015.00786

Tominaga T, Miura C, Takeda N, Kanno Y, Takemura Y, Seo M, Yamato M, Kaminaka H (2019) Gibberellin promotes fungal entry and colonization during Paris-type arbuscular mycorrhizal symbiosis in eustoma grandiflorum. Plant Cell Physiol 61(3):565–575. https://doi.org/10.1093/pcp/pcz222

Tominaga T, Yao L, Saito H, Kaminaka H (2022) Conserved and diverse transcriptional reprogramming triggered by the establishment of symbioses in tomato roots forming arum-type and Paris-type arbuscular mycorrhizae. Plants 11(6):747. https://doi.org/10.3390/plants11060747

van der Heijden MGA, Martin FM, Selosse MA, Sanders IR (2015) Mycorrhizal ecology and evolution: the past, the present, and the future. New Phytol 205(4):1406–1423. https://doi.org/10.1111/nph.13288
Wang L, Chen X, Du Y, Zhang D, Tang Z (2022) Nutrients regulate the effects of arbuscular mycorrhizal fungi on the growth and reproduction of cherry tomato. Front Microbiol 13: 843010. https://doi.org/10.3389/fmicb.2022.843010
Wipf D, Krajinski F, van Tuinen D, Recorbet G, Courty PE (2019) Trading on the arbuscular mycorrhiza market: from arbuscules to common mycorrhizal networks. New Phytol 223(3): 1127–1142. https://doi.org/10.1111/nph.15775
Wu M, Yan Y, Wang Y, Mao Q, Fu Y, Peng X, Yang Z, Ren J, Liu A, Chen S, Ahammed GJ (2021) Arbuscular mycorrhizal fungi for vegetable (VT) enhance resistance to Rhizoctonia solani in watermelon by alleviating oxidative stress. Biol Control 152:104433. https://doi.org/10.1016/j.biocontrol.2020.104433
Yang H, Fang C, Li Y, Wu Y, Fransson P, Rillig MC, Zhai S, Xie J, Tong Z, Zhang Q, Sheteiwy MS, Li F, Weih M (2022) Temporal complementarity between roots and mycorrhizal fungi drives wheat nitrogen use efficiency. New Phytol. https://doi.org/10.1111/nph.18419
Yooyongwech S, Samphumphuang T, Tisarum R, Theerawitaya C, Cha-um S (2016) Arbuscular mycorrhizal fungi (AMF) improved water deficit tolerance in two different sweet potato genotypes involves osmotic adjustments via soluble sugar and free proline. Sci Hortic 198: 107–117. https://doi.org/10.1016/j.scienta.2015.11.002
Zhang C, He J, Dai H, Wang G, Zhang X, Wang C, Shi J, Chen X, Wang D, Wang E (2021a) Discriminating symbiosis and immunity signals by receptor competition in rice. Proc Natl Acad Sci 118(16):e2023738118. https://doi.org/10.1073/pnas.2023738118
Zhang L, Chu Q, Zhou J, Rengel Z, Feng G (2021b) Soil phosphorus availability determines the preference for direct or mycorrhizal phosphorus uptake pathway in maize. Geoderma 403: 115261. https://doi.org/10.1016/j.geoderma.2021.115261
Zhang L, Zhou J, George TS, Limpens E, Feng G (2022) Arbuscular mycorrhizal fungi conducting the hyphosphere bacterial orchestra. Trends Plant Sci 27(4):402–411. https://doi.org/10.1016/j.tplants.2021.10.008
Zhou Y, Ge S, Jin L, Yao K, Wang Y, Wu X, Zhou J, Xia X, Shi K, Foyer CH, Yu J (2019) A novel CO2-responsive systemic signaling pathway controlling plant mycorrhizal symbiosis. New Phytol 224(1):106–116. https://doi.org/10.1111/nph.15917
Zhu B, Gao T, Zhang D, Ding K, Li C, Ma F (2022) Functions of arbuscular mycorrhizal fungi in horticultural crops. Sci Hortic 303:111219. https://doi.org/10.1016/j.scienta.2022.111219

Chapter 2
Non-host Plant Species: Definition, Description, and Mechanisms of Interaction with Arbuscular Mycorrhizal Fungi

Somayeh Rahmat

Abstract About 18–29% of vascular plant species, e.g., important crops, are not involved in a symbiotic interaction with arbuscular mycorrhizal fungi (AMF). These species are known as "non-host" plants. However, these plants can be colonized under certain conditions and develop rudimentary AM (RAM) phenotypes, which often results in a decrease in plant growth and an increase in resistance to pathogens and insects. Several "symbiosis toolbox" genes that are present in the genome of host plants are conserved in non-host plants. Moreover, specific non-symbiotic genes can be brought into play for symbiosis in non-host plants. Altogether, the results show that molecular connections with the symbiosis machinery have still remained in non-host plants, which can enhance the prospect of engineering this functional trait for agricultural plants.

Keywords AMF · Non-host plant · Rudimentary arbuscular mycorrhiza · Ectomycorrhiza · Glomeromycotina

2.1 Introduction

Roots of the majority of vascular plants host mycorrhizal fungi, so mycorrhizal roots, as organs absorbing nutrients in nature, are found much more in abundance, compared to non-mycorrhizal roots (Tester et al. 1987). Up to now, considering the interaction between fungi and host roots, especially the structure of mycorrhizal hyphae (Brundrett 2004), six types of mycorrhizal associations have been identified as arbuscular mycorrhiza (AM), ericoid mycorrhiza, ectomycorrhiza (ECM), arbutoid mycorrhiza, orchid mycorrhiza, and monotropoid mycorrhiza (Smith and Read 2008). Among these associations, AM and ECM are regarded as the most common and most economically important symbiosis in natural and agricultural ecosystems, which are colonizing about 80% and 2% of all investigated species in different plant families, respectively (He et al. 2003; Wang and Qiu 2006; Smith and

S. Rahmat (✉)
Miandoab, Iran

G. J. Ahammed, R. Hajiboland (eds.), *Arbuscular Mycorrhizal Fungi and Higher Plants*, https://doi.org/10.1007/978-981-99-8220-2_2

Read 2008; Brundrett 2009). Ectomycorrhizae consists of a hyphal sheath (mantle) that wraps around the plant root and a network of hyphae (Hartig net) that surrounds the cells of the root cortex. In AM type, fungal hyphae penetrate the cell lumen, but do not form a mantle (Genre et al. 2020).

AMF have greatly attracted researchers' attention due to their powerful function and wide range of hosts (Lanfranco et al. 2018). AM multifunctional symbiosis is established between plant roots throughout the radiation of plant diversity and soil fungi belonging to the subphylum *Glomeromycotina* (Van Der Heijden et al. 2015; Brundrett and Tedersoo 2018). The extensive extraradical mycelium of AMF leads to the increased exploratory capacity of the host root for water and nutrients in exchange for photosynthates (Ferrol et al. 2019; Smith and Read 2008; Leake et al. 2004). Besides the nutritional effects, AM symbiosis brings about other benefits such as protection against disease (Smith and Read 2008; Wehner et al. 2010) and resistance to abiotic stresses (Hajiboland 2013; Mnasri et al. 2017; Latef et al. 2016; Begum et al. 2019) for the host plant and influences soil chemical and biological properties and its microbial community structure (Rillig and Mummey 2006; Iffis et al. 2016; Dagher et al. 2020).

Under natural conditions, each fungus is likely to interact with several plants, and in turn, each plant is linked to more than one fungus species. The grafting of individual plants from different or similar species via mycorrhizal fungi leads to the formation of common mycorrhizal networks (CMNs) (Van Der Heijden et al. 2015), which serve as communication channels for transmitting signals, transporting nutrients, and organizing the plant community, thus bringing about a competitive advantage for plants belonging to the network compared to other plants (Hoeksema 2015; Gilbert and Johnson 2017). However, in most terrestrial ecosystems, there are a large number of plant species described as "non-host" that do not allow AM symbioses (Brundrett 2009; Cosme et al. 2018). This is a remarkable fact, as mycorrhizal infection usually leads to increased plant growth in ecosystems under nutrient deficiency, giving rise to the expectation that non-host plants will grow in ecosystems without such stresses or develop alternative mechanisms to increase nutrient uptake (Tester et al. 1987).

In spite of the absence of AM symbiosis, due to the pervasiveness of AMF and their host plants in almost all habitats, non-host plant species typically grow in the presence of AM mycelium network and are likely to be influenced by them (Cosme et al. 2018; Zhang et al. 2019). Accordingly, collecting thorough systematic information about the nature of non-host plants, the mechanisms of the two-way AMF-non-host interaction and the host-AMF-non-host tripartite interaction can play an important role in understanding the establishment and stability of the plant community as well as development of intercropping systems (Wang et al. 2022). In this chapter, we first present a clear definition of non-host plants and discuss some difficulties in finding a clear delimitation for the non-host feature and then make an attempt to provide an updated summary of the mechanisms of mutual AMF-non-host-plant interactions.

2.2 Non-host Plants

Contrary to plant species having mycorrhizal symbiosis with fungi, the roots of some so-called "non-mycorrhizal" or "non-host" plant species are very resistant to colonization and do not have symbiosis with mycorrhizal fungi (Wang et al. 2022). It is apparent that if the plant roots are totally clear of any infection, the plant's mycorrhization status can be easily determined with sufficient evidence of absence (Cosme et al. 2018). However, the reality is that it is very rare for a non-host root to be consistently clear of mycorrhizal fungi, which can lead to some diagnostic problems in the plant's mycorrhizal status (Brundrett 2017). For example, a non-mycorrhizal root may sometimes be recognized as an ECM root due to the surface growth of fungal hyphae, despite the absence of any sign of Hartig net (Brundrett 2017). Therefore, a clear demarcation between host and non-host plant species is very difficult, and failure to exercise sufficient caution can lead to misclassification (Cosme et al. 2018). For instance, consider *Arabidopsis thaliana* (hereafter *Arabidopsis*). Although there is strong evidence that it can be colonized by the AM fungus *Piriformospora indica*, according to mycorrhizal structure analysis and genome comparison, it can be said that *Arabidopsis* is not a true AM host (Mandyam et al. 2013; Keim et al. 2014). Or, for example, *Buddleja* spp. was first taxonomized into the host plant class but later confirmed as the true AM host (Dickie et al. 2007).

With the assumption that a target species is likely to be a non-host species if it belongs to a family with mostly non-host members, some researchers used taxonomic extrapolation in order to enhance the taxonomy of non-hosts (Brundrett 2009). However, non-host specificity can be considered as a feature of some plant species as well, such as *Arabidopsis*, where plant roots can be colonized, but there is no matching between the resulting mycorrhizal structure and any of the typical functional mycorrhizal patterns (Tester et al. 1987; Brundrett 2009; Brundrett and Tedersoo 2018). The criteria involved in the determination of functional mycorrhizal colonization are associated with an evolutionary process (Tester et al. 1987). For instance, in the past, a plant was considered as a vesicular arbuscular plant only when vesicles or arbuscules or both (Reeves et al. 1979) can be found in the root. Introducing the arbuscule as the main site of nutrient exchange, the presence of the arbuscule is considered as the main criterion for accepting a plant species as a mycorrhizal species. Accordingly, fungal infection lacking arbuscules (despite shortage of experimental evidence for lack of a physiological function) is regarded as characteristic of a non-host status (Glenn et al. 1985; Demars and Boerner 1996; Brundrett 2009; Brundrett and Tedersoo 2018). Although the arbuscule is not the only functional structure in AM symbiosis (Dickson 2004), it is always entirely absent in non-host plants (Tommerup 1984; Allen et al. 1989; Regvar et al. 2003; Vogel-Mikuš et al. 2006).

Furthermore, even mycorrhizal plants are not capable of hosting all mycorrhizal fungi. Plants that are not able to have a symbiotic interaction with a specific mycorrhizal fungus are also taxonomized into the non-host class. For example, a plant species that cannot be colonized by one type of AMF is likely to be infected by ECM fungi or even another type of AMF, which makes this plant species non-host to a specific AM fungus (Veiga et al. 2013; Cosme et al. 2018; Fernández et al. 2019).

Sometimes, plant species are not infected due to other reasons except the plant not being susceptible. It is likely that the plant looks like a non-host because it was studied in a wrong growth stage or at a wrong time of the year (Tester et al. 1987; Orłowska et al. 2002; Vogel-Mikuš et al. 2006). Another possibility is that the plant is sensitive but it grows in an unsuitable habitat, e.g., under flooded conditions, that prevents it from being infected or destroys the infection. It is also possible that the plant is susceptible to infection but does not come into contact with mycorrhizal fungi (Tester et al. 1987). Any of these cases can result in very low levels of infection or even no infection at all. Potted cultivation of plants under favorable conditions and exposure to fungal inoculation is the only way to ensure an understanding of the inherent immunity or non-immunity of a plant species against mycorrhization (Tester et al. 1987). In contrast, given the fact that AMF is capable of colonizing soil organic matter, sometimes, the reported infection of non-host plants may be a separated colonized root or even an old root (Tommerup 1984; Glenn et al. 1985) uncapable of removing mycorrhizal fungi. It is of high importance to determine whether the interactions found in non-host species exposed to mycorrhizal inoculation are just a non-specific infection of old roots or a "true mycorrhiza" (Tester et al. 1987).

2.3 AMF Non-host Plants

According to the estimations, 71–82% of all vascular plants host AMF (hereafter hosts), and the remaining 18–29% (including important crops), i.e., non-hosts of AMF (hereafter non-hosts) (Brundrett 2009; Cosme et al. 2018), are either non-mycorrhizal (NM) or plants producing non-AM mycorrhizae. The majority of non-AM mycorrhizal species never provide a functional AM symbiosis. Pyrolaceae, Monotropaceae, Ericaceae, and Orchidaceae are exclusive hosts of monotropoid, arbutoid, ericoid, and orchid, respectively. Among ECM host plants, only a few genera are capable of hosting AMF (Giovannetti and Sbrana 1998).

Sometimes, plants have roots that may be mycorrhizal or not, depending on the soil or habitat conditions. These are considered NM-AM plants. They typically grow in habitats where AMF is inhibited, which makes them unable to be continuously AM (Brundrett 2017). NM and NM-AM plants both occupy similar habitats (mainly cold montane and polar regions), which makes it difficult to distinguish them (Brundrett 2009; Newsham et al. 2009). As mentioned earlier, this is because NM plants are not consistently clear of mycorrhizal fungi.

Currently, researchers have introduced two classes of non-host plant families that grow at both ends of the soil fertility spectrum (Lambers and Teste 2013). The first class includes plants that have evolved highly specialized feeding strategies. As examples of these plants, we can mention parasites, carnivores, and species with specialized roots that often grow in soils with very low phosphorus levels (Brundrett 2009; Lambers and Teste 2013). These specialized root structures include cluster roots (Shane and Lambers 2005) or dauciform roots (Shane et al. 2006), which release organic acids to extract soil unavailable phosphorus, and a lesser-known type, namely, sand-binding and capillaroid roots (Lambers et al. 2006, 2013). These non-host species are known as Proteaaceae type and appear abundantly in the Cyperaceae, Restionaceae, Proteaaceae, and Hamodoraceae families (Lambers and Teste 2013). Members of the second class are usually found in very cold, salty, dry, wet, and disturbed habitats with high availability of soil phosphorus and low competition with other plants (Brundrett 2009; Lambers and Teste 2013). These species, also known as Brassicaceae type, are often regarded as agricultural weeds (Jordan et al. 2000) and are especially abundant in families such as Brassicaceae, Urticaceae, Polygonaceae, Cenopodiaceae, Caryophyllaceae, and Amaranthaceae (Lambers and Teste 2013). The arbuscular mycorrhizal habit is likely to provide disadvantages for the both classes and develop a selective force against mycorrhization (Lambers and Teste 2013). Broad groups of non-host species represent a significant ecological distribution (Lambers and Teste 2013). Nevertheless, as with most general rules in biology, there are some exceptions, e.g., non-host species such as *Daviesia* and *Kennedia* (Fabaceae; Brundrett and Abbott 1991) or even genera such as *Lupinus* (Fabaceae; Lambers et al. 2013b) and *Cicer arytinum* (Fabaceae; Jones 1924) appear in typical host families. Similarly, host species are found in non-host families as well (Boulet and Lambers 2005; Lagrange et al. 2011).

2.4 Can AMF Colonize Non-host Plants?

In the absence of a functional symbiosis, AMF has little saprophytic capacity to produce hyphae, depending on limited energy resources, which are mainly supplied by spores and vesicles. Thus, in terms of the AMF, due to its dependence on the carbon source (Smith and Read 2008), it is crucial to select a suitable plant symbiont to complete the obligate biotrophic life cycle and ensure their survival in nature (Kiers et al. 2011). The AMF life cycle involves the following key stages: spore germination, presymbiotic mycelium development, differential branching, appressorium differentiation, root cortical penetration, intraradical hyphal proliferation, and arbuscule development. This cycle is not complete when dealing with non-host plants (Giovannetti et al. 1994).

In spite of the wide specificity in the interaction between AMF and their host, mycorrhizal interaction, like other symbiotic interactions, involves host-microorganism mutual recognition and developmental steps, which ultimately results in a close morphological and physiological interaction between the two partners (Tester et al. 1987). The interaction between AMF and the host root begins with the exchange of diffusible signals before the establishment of physical contact between the two symbiotic partners (Bonfante and Genre 2015). In response to the presence of the fungus, the host root excretes phenolic compounds and strigolactones, which leads to the increased possibility of hyphae contact with the host root (Besserer et al. 2006) by stimulating the hyphae growth and inducing the branching of the presymbiotic fungus (López-Ráez et al. 2011; Steinkellner et al. 2007). The host roots, in turn, can sense the signaling molecules released by the microsymbiont. According to some reports, the reason for the low sensitivity of non-host plants to AMF is attributed to the presence of fungal compounds in root exudates or root cortex, including aromatic glucosinolates (Vierheilig et al. 2000; Pongrac et al. 2008). However, there is no clear evidence that AMF can differentiate between the host and non-host roots (Tester et al. 1987; Giovannetti and Sbrana 1998).

Appressorium differentiation at the root surface by AMF is regarded as the most important feature for potential host recognition by the fungal partner, which is established regardless of the interaction outcome, but cannot develop on the non-host root surface (Giovannetti and Sbrana 1998). Thus, at this stage, the distinction between non-host and host species becomes more apparent with the initial efforts of the fungal partner to penetrate the root (Giovannetti and Sbrana 1998). It has been suggested that an important control point for AMF entry into the root cortex can be the interaction between the two symbionts at the level of the middle lamella and/or the root cell wall (Tester et al. 1987). When exposed to non-host roots, AMF displays active responses including lack of development of penetration pegs (Tommerup 1984), development of runner hyphae that move away from lignified areas of the root in search of other areas (Allen et al. 1989), retraction of fungal cytoplasm and senescence, appearing as septate hyphae in and on roots (Glenn et al. 1985), and regrowth of a vegetative tip, instead of a penetrating peg, on hyphopodia previously attached to roots (Tommerup 1984). The mechanism by which AMF detects the non-host status of roots is largely unknown (Cosme et al. 2018).

Following appressoria formation, AMF colonizes host roots by forming intracellular hyphae and arbuscules (Giovannetti and Sbrana 1998). Depending on the host genome, two morphological classes, i.e., Arum and Paris, are identified in arbuscular mycorrhizae. The former is associated with the formation of normal arbuscules and vesicles and the latter with the differentiation of intracellular hyphae, coils or arbuscular coils, and vesicles (Fig. 2.1) (Dickson 2004). Despite the apparent inconsistency, depending on species (Regvar et al. 2003), developmental stage (Orłowska et al. 2002), time of year (Regvar et al. 2003; Vogel-Mikuš et al. 2006), experimental conditions (presence or absence of host plants) (Veiga et al. 2013), plant age (Regvar et al. 2003), etc., numerous evidence show that roots of non-host plants are mostly colonized with hyphae and vesicles (Matsumura et al. 2007; Mnasri et al. 2017) and even arbuscules or arbusculate coils (Tommerup 1984; Lekberg et al. 2015) (Fig. 2.1), especially when AMF are, at the same time, part of a mycorrhizal network supported by neighboring host plants (Veiga et al. 2013). This raises a fundamental uncertainty about considering such plants as non-hosts. Nevertheless, even if formation of arbuscule, coil, or arbusculate coil is considered as a measure of a symbiotic phenotype (Tester et al. 1987; Brundrett 2009), the evidence for the benefits of "true mycorrhizae" in the view of nutrient transfer from the fungus

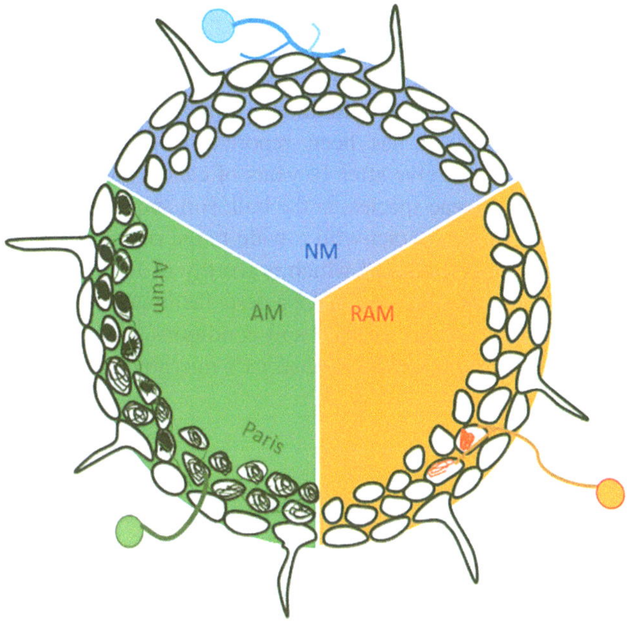

Fig. 2.1 Types of AM symbiotic morphologies in the Brassicaceae family. NM are never colonized by AMF; RAM form rudimentary AM phenotype under certain conditions; AM accommodates two main morphological types that create functional symbiosis of AM, Arum type and Paris type. Redrawn from Cosme et al. (2018)

is almost nil (Cosme et al. 2018; Fernández et al. 2019). However, the absence of nutrients exchange does not exclude other physiological effects of arbuscular mycorrhiza (Hajiboland 2013), as one study showed that colonization with hyphae promotes growth through an unknown mechanism (Van der Heijden et al. 1998). For such plants that have lost the potential to form a prominent AM symbiosis but are able to differentiate multiple symbiotic structures in their roots under certain conditions, an alternative classification has been suggested, namely, the RAM (rudimentary AM) phenotype (Cosme et al. 2018).

It has been proposed that a neighboring host plant can increase the possibility of infecting its non-host neighbor through some mechanisms such as exudation of infection-stimulating nutrients, which are not present in non-host exudates, production of signaling compounds stimulating AMF mycelial growth and root colonization by acting as a direct source of nutrients for AMF, and possibly the non-host neighbor and through increasing density of propagules at the interface between the two species (Tester et al. 1987). Genre et al. reported that non-host plants are not able to recognize bioactive molecules excreted by AMF (Genre et al. 2013). But the bioactive molecules excreted by the host reach the root surface of the non-host plants, which results in their colonization (Lekberg et al. 2015). Hence, they suggested that in this type of colonization, the non-host plant is a passive partner. There are only a few reports that non-host plant colonization does not require presence of a neighboring host plant (Zhang et al. 2019). For example, a significant level of colonization was observed in *Carax capillacea*, especially under alpine conditions in the absence of a host. The results indicate that the main factor in *C. capillacea* colonization is the environmental conditions, not the presence of the host (Zhang et al. 2019). It has been reported that the AMF belonging to *Glomeraceae* family may survive after 10 years of canola monoculture (absence of host), a non-host Brassicaceae species, in the bulk soil and rhizosphere of the plant, and interestingly, they may interact with a wide range of soil bacteria. These data raise fascinating hypotheses, each of which is worthy of further examination: first, bacteria can act as hosts for AMF; second, bacteria facilitate the interaction between AMF and non-host plants under adverse soil conditions. It can be concluded that AMF is an important component of *Arabidopsis* microbiome despite its non-host nature (Floc'h et al. 2022).

2.5 Effects of AMF on Non-host Plant Growth

Despite the absence of AM symbiosis, non-host plants can also be affected by AMF. One of the many benefits provided by AMF to their host partners is growth promotion; however, host-supported AMF impose a different effect on the growth of non-host plants. Host-supported AM mycelium network often imposes an adverse impact on the growth of non-host species, especially on Brassicaceae type (Francis and Read 1994; Veiga et al. 2011; Lambers and Teste 2013). Whether and how AMF benefits from interaction with non-host plants and which mechanisms involve in these effects are largely unknown (Cosme et al. 2018). As reported by Allen and Allen (1990), Lekberg et al. (2015), and Raven et al. (2018), one possibility is that intraradical hyphae have the capability of absorbing photosynthetic products directly from the host plant. Another possibility is that they benefit indirectly by providing competitive advantages to host plants as members of a community (Francis and Read 1994; Lekberg et al. 2015; Zhang et al. 2019).

What mechanism is responsible for the growth reduction observed in non-host plants in the presence of AM mycelium? The evidence indicates that AMF and the host plant both are able to suppress the non-host plant development by producing allelochemicals (Francis and Read 1994; Veiga et al. 2012). However, it seems that a mechanism other than allelopathy is responsible for this process (Veiga et al. 2012, 2013). According to the majority of previous research, non-host plants have weaker competitive potential in the presence of host species (Francis and Read 1995; Veiga et al. 2012). Studies have reported the depletion of nutrients, especially phosphorus, and its transfer to the neighboring host plant at the sacrifice of the non-host one (Lekberg et al. 2015). Moreover, it is hypothesized that root contact or AMF infection reduces total phosphorus uptake and plant growth by disabling the direct pathway of root phosphorus uptake (Neumann and George 2005; Smith et al. 2009; Facelli et al. 2010). However, competition for nutrients between host and non-host plants does not seem to sufficiently explain the suppression of non-host plant growth because the biomass of non-host plants was always negatively affected (Wang et al. 2021). On the other hand, studies have suggested that a shift from the beneficial status of AMF to pathogenic, observed during the colonization of non-host species, is likely to activate defense responses that divert resources and lead to suppression of growth (Lambers and Teste 2013). Similar to a hypersensitive response, Allen et al. (1989) observed that inoculation of the non-host *Salsula coli* initially leads to the arbuscules formation but then caused browning and cell death in the invaded areas and the disappearance of the fungus. Symptoms such as swelling and distortion of the meristems of several other non-host roots in the presence of AMF mycelium also resemble the plant's hypersensitive response (Francis and Read 1995). Furthermore, Hajiboland et al. reported a distinct pattern of defense responses in the non-host plant sugar beet (Chenopodiaceae) co-cultured with the host plant barley. In sugar beet, the decrease in the level of endogenous salicylic acid, the tissue concentration, and the excretion of phenolic acids indicated that the mycelium of the fungus AM *Rhizophagos intraradis* did not cause a hypersensitive response; instead, it stimulated antioxidant and structural defense (lignin deposition) (Hajiboland 2013). It seems that due to such balanced defense cost, sugar beet growth did not decrease in the presence of AMF mycelia (Hajiboland 2013), an evidence which indirectly supports the role of costly defense responses in reducing the growth of non-host plants.

The role of direct absorption of carbon from non-host plants via rudimentary arbuscules (which may no longer exist in the later stages of plant development) has also been suggested as a mechanism for growth suppression of non-host plants (Allen and Allen 1990). They also suggest rapid degeneration of arbuscules as a potential factor in the rare detection of arbuscules in non-host plants (Allen and Allen 1990).

It is noteworthy that such defense responses are activated in AMF-host symbiosis as well. However, in host plants, nutritional effects generally outweigh defense costs (Ryan et al. 2012). Allocation of fewer resources to growth also occurs following host plant colonization, particularly at early colonization stages, and likely is a result of the same defense responses that are responsible for persistent growth limitation in non-host plants (Lambers and Teste 2013). Although imposing such defense responses is likely to reduce mycorrhizal infection and the growth of non-host species, it stimulates the induced systemic resistance. Such increased disease resistance has been observed in colonized non-host plants against pathogens and insects, which highlights the multifunctional role of AMF in non-host-host interactions (Fernández et al. 2019).

In a similar way, non-host plants, e.g., Brassicaceae type, have an adverse impact on the growth characteristics and acquisition of nutrients of neighboring host species, without negatively influencing the non-host neighbors (Francis and Read 1994; García-Garrido and Ocampo 2002). There appears to be intense competition between host and non-host species in the overlapping area of relatively nutrient-poor and relatively fertile habitats. There is evidence indicating that host plants employ biological means (AMF), while non-host plants use chemical factors (allelopathy) to compete with each other (Lambers and Teste 2013). However, the nature of the interactions between the supported mycelium network with host and non-host species of the Protaceae type is mainly beneficial (Bruno et al. 2003). Considering the capability of non-host plants to mobilize nutrients (Muler et al. 2013), evidence for negative effects is scant (Lambers and Teste 2013).

2.6 Molecular Cross-Talks Between AMF-Non-host Plants

Contrary to much information on the molecular mechanisms underlying the interactions between host plants and AMF, the molecular interaction between non-host plants and AMF is still poorly understood. Recently, the interaction between *Arabidopsis*, a model and non-host species, and the host-supported AM fungus *Rhizophagus irregularis* (hereafter *Rhizophagus*) has enhanced our understanding of the molecular cross-talk between non-host plants and AMF (Fernández et al. 2019).

A set of conserved genes called "Symbiotic Toolkit" determine the ability of host plants to establish a functional symbiosis (Delaux et al. 2013). Although there is little experimental evidence about the possible cause of AM loss, such as carbon cost, changes in resistance to pathogens and insects, and changes in lifestyle and root morphology, significant genetic evidences have been found for evolutionary loss of AM colonization (Delaux et al. 2014; Radhakrishnan et al. 2020; Hornstein et al. 2023). In all non-mycorrhizal plants belonging to diverse taxonomic groups, a common subset of specific genes, e.g., the "common symbiosis pathway" (CPS) genes involved in signal perception and transduction, are absent (Table 2.1) (Genre et al. 2020; Radhakrishnan et al. 2020). It is noteworthy that some genes responsible

Table 2.1 Some genes associated with AM symbiosis lost in non-host plants (Cosme et al. 2018)

Symbiotic gene	Affected colonization step	Gene absence phenotype
PT4	Arbuscule formation	Reduced colonization, increased arbuscule degeneration and reduced phosphate uptake
STR2	Intraradical hyphal colonization and arbuscule formation	Reduced colonization and stunted arbuscules
STR	Intraradical hyphal colonization and arbuscule formation	Reduced colonization and stunted arbuscules
VAPYRIN	Intraradical hyphal colonization and arbuscule formation	Reduced colonization and no arbuscules
RAM2	Fungal penetration into the root	No hyphopodia
RAM1	Fungal penetration into the root	No hyphopodia
IPD3	Presymbiotic	Reduced colonization and no arbuscules
CASTOR	Presymbiotic	Reduced colonization
DMI3	Presymbiotic	No colonization
DMI2	Presymbiotic	Reduced colonization
NFP	Presymbiotic	Reduced number of arbuscules

for the initial interaction between AMF and the host root have been conserved in non-host *Arabidopsis* (Delaux et al. 2013, 2014). For example, similar to the compatible species, *Arabidopsis* roots responded to the presence of *Rhizophagus* within 24 h by overexpressing the strigolactone biosynthetic genes, *CCD7* and *CCD8*, which indicates that the primary interactions of host colonization are not completely disrupted in non-host plants (Delaux et al. 2014; Fernández et al. 2019). This is a confirmation of previous findings that during presymbiotic stages, AMF cannot differentiate between non-host and host species (Tester et al. 1987; Giovannetti and Sbrana 1998).

For successful colonization and achieving a functional symbiosis, the fungus must reduce host defense responses by modulating defense signals (Fernández et al. 2019). Genes acting in defense, e.g., *PR1*, *ERF4*, *MYB51*, etc., which are enabled in the initial stages of interaction between host roots and AMF, as well as between *Arabidopsis* and endophyte or pathogenic fungi, were not expressed in *Arabidopsis* roots in the presence of *Rhizophagus* fungi (Fernández et al. 2019). It indicates that when exposed to *Rhizophagus* hyphae, *Arabidopsis* roots activate the primary detection stages of host-AMF, rather than host-endophytic fungi or host-pathogenic fungi interactions. In other words, AMF is not recognized as an antagonist for the non-host (Wang et al. 2021). In addition, contrary to the functional colonization of the roots of host plants, in *Arabidopsis* roots colonized with *Rhizophagus*, there was no evidence of overexpression of symbiosis-related genes, like *GintMST4*, *GintMST2*, *GintAMT2*, and *GintPT*. Also, the gene related to nutrient transport and carbohydrate metabolism had not been upregulated (Zouari et al. 2014; Fernández et al. 2019; Wang et al. 2021), Furthermore, the defense-related gene such as flavin monooxygenase S-oxygenase 3 (*FMO GS-OX3*) and the genes related to salicylic acid biosynthesis and systemic acquired resistance are upregulated (Fernández et al. 2019).

In general, the early stages of the interaction between AMF supported by host and non-host *Arabidopsis* are similar to some of the presymbiotic stages of the AMF-host interaction (Fig. 2.2). However, the fungus is not then recognized as a symbiotic partner but as an unwanted invader. Thus, the exchange deviates from the usual symbiotic pathway of AM toward the activation of defense responses and growth-defense exchange (Fernández et al. 2019; Wang et al. 2021). Possibly, the transfer of signaling molecules from the host colonized roots (AMF and/or host plant) to the non-hosts leads to the activation of defense responses (Stringlis et al. 2018), which can explain part of the negative impact of AMF colonization on the growth characteristics of non-host plants.

It is intriguing that there are other molecular components, in addition to the conventional symbiotic toolkit present in the host genome, that play a role in supporting the AM colonization of non-host roots (Cosme et al. 2018). For example, in a recent study, the upregulation of *MYB72* and *BGLU42* in *Arabidopsis* increased the AM infection of roots in the presence of a host plant, through the pathway of scopoletin production and excretion (Cosme et al. 2018). It could be concluded that scopoletin, as a plant signal, can positively influence the plant-fungus interaction in the non-host *Arabidopsis* before penetration stage (Cosme et al. 2018). Scopoletin,

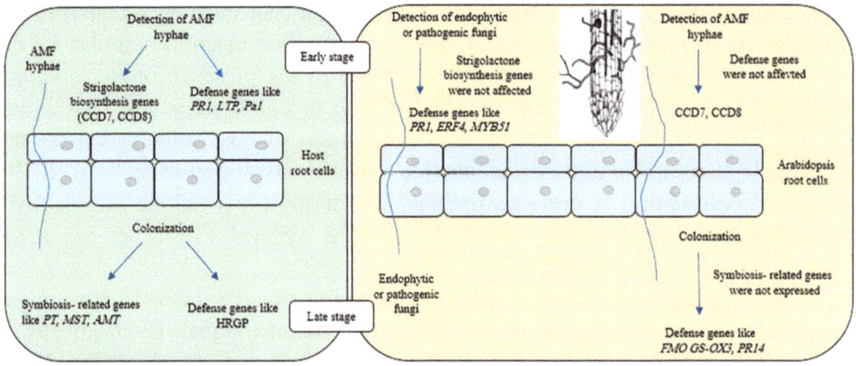

Fig. 2.2 Molecular dialogue between *Arabidopsis thaliana*-AMF and its comparison with Arabidopsis-endophytic or pathogenic fungi interaction. *CCD7 8* carotenoid cleavage dioxygenase 7, 8, *PR1 14* pathogenesis-related protein1, 14, *Pal* phenylalanine ammonia-lyase, *LTP* lipid transfer protein, *PT* phosphate transporter, *AMT* ammonium transporter, *MST* monosaccharide transporter, *HRGP* hydroxyproline-rich glycoproteine, *MYB51* Myb domain protein 51, *CYP71A12* cytochrome P450, family 71, polypeptide 12, subfamily A, *ERF4* ethylene responsive element binding factor 4. Redrawn from Wang et al. (2022)

independent from the root and in a dose-dependent manner, stimulates the elongation of the hyphae of germinated *rhizophagus* spores and also induces metabolic processes in the *rhizophagus* extraradical mycelium (Cosme et al. 2018). Furthermore, it has been observed that non-orthologs genes can be potential alternatives for symbiotic genes. For example, AMF can colonize *Arabidopsis* disrupted in the AM pre-symbiosis signal receptor (*NFP*) gene, which suggests that other receptors are likely to be present (Bonfante and Requena 2011). Also, the ortholog of the *GPAT* gene (a key enzyme in the lipid biosynthesis pathway) (*RAM2*) is not present in *Arabidopsis*, and implementation of *Arabidopsis GPAT* genes in the *ram2* mutant of *Medicago truncatula* (an AM host species) restores colonization of *ram2* plants, which indicates that the *RAM2* function can be restored by the non-orthologous *GPAT* gene (Wang et al. 2012; Cosme et al. 2018).

A very recent study addressed the intriguing question of whether restoring gene expression to non-host plants can restore gene function or not (Hornstein et al. 2023). One of the important CSP genes lost along with the AM trait in non-host plants is *IPD3*, a key transcription factor that mediates CSP signaling and regulation of AM response genes (Hornstein et al. 2023). *IPD3* mutants have been almost completely inhibited in AM development (Watts-Williams and Cavagnaro 2015). In the study, researchers found that restoring the expression of *IPD3* in *Arabidopsis*, in the form of its DNA binding domain, leads to the regulation of the same symbiotic gene networks even in the absence of the fungal signal (Hornstein et al. 2023). This indicates that despite the long history of being a non-host and the deletion of majority of the related genes, the molecular links to the symbiosis machinery in *Arabidopsis* are still preserved. This is promising that AM symbiosis, as a beneficial agricultural trait, could be reintroduced to the non-host plant species with significant economic benefits (Cosme et al. 2018).

2.7 Conclusion and Future Perspectives

AM symbiosis is known as a complicated molecular interaction between fungus and host plants, with mutual benefits in terms of nutrition, protection from stresses, and growth. The potential to form a functional symbiosis has been lost in different plant species named non-host plants. However, under certain conditions, host-supported AM mycelium has the potential to colonize non-host roots and develop RAM phenotype that is different from endophytic or pathogenic fungi. The early stages of pre-symbiotic interaction between AMF and non-host species are similar with the processes observed during pre-symbiotic cross-talk in AMF-host interactions. In the later stages of the interaction, AMF colonize the root cortex without establishing a functional symbiosis (in view of nutrients exchange) and instead activate plant defense responses, which are mainly associated with decreased plant growth and increased resistance to pathogens and insects. Future studies are suggested to investigate the function of RAM colonization not only in terms of nutrients uptake but also considering all the established benefits of AM symbiosis. Furthermore, the underlying mechanisms of AMF impact on the growth of host and non-host plant species is also an important open problem to study.

The mechanisms by which the non-host nature of plant species is determined at the molecular level are not well understood. Some AM symbiosis-specific genes are conserved in non-host plants. Furthermore, specific non-symbiotic genes could have roles in the symbiosis of non-host plants. These additional molecular components and RAM phenotypes are likely to be the missing links that can help us better understand the molecular constraints and mechanisms leading to the evolution of the non-host nature of species. Making an attempt to understand such control mechanisms can increase our understanding of the biological factors controlling AM symbiosis in soils while shedding more light on the development of future crops with higher productivity and less need for agrochemical inputs.

References

Allen MF, Allen EB (1990) Carbon source of VA mycorrhizal fungi associated with Chenopodiaceae from a semiarid shrub-steppe. Ecology 71(5):2019–2021

Allen MF, Allen EB, Friese CF (1989) Responses of the non7hyphen; mycotrophic plant Salsola kali to invasion by vesicular–arbuscular mycorrhizal fungi. New Phytol 111(1):45–49

Begum N, Qin C, Ahanger MA, Raza S, Khan MI, Ashraf M, Zhang L (2019) Role of arbuscular mycorrhizal fungi in plant growth regulation: implications in abiotic stress tolerance. Front Plant Sci 10:1068

Besserer A, Puech-Pagès V, Kiefer P, Gomez-Roldan V, Jauneau A, Roy S, Séjalon-Delmas N (2006) Strigolactones stimulate arbuscular mycorrhizal fungi by activating mitochondria. PLoS Biol 4(7):e226
Bonfante P, Genre A (2015) Arbuscular mycorrhizal dialogues: do you speak 'plantish'or 'fungish'? Trends Plant Sci 20(3):150–154
Bonfante P, Requena N (2011) Dating in the dark: how roots respond to fungal signals to establish arbuscular mycorrhizal symbiosis. Curr Opin Plant Biol 14(4):451–457
Boulet FM, Lambers H (2005) Characterisation of arbuscular mycorrhizal fungi colonisation in cluster roots of shape Hakea verrucosa F. Muell (Proteaceae), and its effect on growth and nutrient acquisition in ultramafic soil. Plant Soil 269:357–367
Brundrett M (2004) Diversity and classification of mycorrhizal associations. Biol Rev 79(3): 473–495
Brundrett MC (2009) Mycorrhizal associations and other means of nutrition of vascular plants: understanding the global diversity of host plants by resolving conflicting information and developing reliable means of diagnosis. Plant Soil 320:37–77
Brundrett MC (2017) Global diversity and importance of mycorrhizal and nonmycorrhizal plants. In: Biogeography of mycorrhizal symbiosis. Springer, Cham, pp 533–556
Brundrett MC, Abbott LK (1991) Roots of jarrah forest plants. I. Mycorrhizal associations of shrubs and herbaceous plants. Aust J Bot 39(5):445–457
Brundrett MC, Tedersoo L (2018) Evolutionary history of mycorrhizal symbioses and global host plant diversity. New Phytol 220(4):1108–1115
Bruno JF, Stachowicz JJ, Bertness MD (2003) Inclusion of facilitation into ecological theory. Trends Ecol Evol 18(3):119–125
Cosme M, Fernández I, Van der Heijden MG, Pieterse CM (2018) Non-mycorrhizal plants: the exceptions that prove the rule. Trends Plant Sci 23(7):577–587
Dagher DJ, de la Providencia IE, Pitre FE, St-Arnaud M, Hijri M (2020) Arbuscular mycorrhizal fungal assemblages significantly shifted upon bacterial inoculation in non-contaminated and petroleum-contaminated environments. Microorganisms 8(4):602
Delaux PM, Séjalon-Delmas N, Bécard G, Ané JM (2013) Evolution of the plant–microbe symbiotic 'toolkit'. Trends Plant Sci 18(6):298–304
Delaux PM, Varala K, Edger PP, Coruzzi GM, Pires JC, Ané JM (2014) Comparative phylogenomics uncovers the impact of symbiotic associations on host genome evolution. PLoS Genet 10(7):e1004487
Demars BG, Boerner RE (1996) Vesicular arbuscular mycorrhizal development in the Brassicaceae in relation to plant life span. Flora 191(2):179–189
Dickie IA, Thomas MM, Bellingham PJ (2007) On the perils of mycorrhizal status lists: the case of Buddleja davidii. Mycorrhiza 17:687–688
Dickson S (2004) The Arum-Paris continuum of mycorrhizal symbioses. New Phytol 163:187–200
Facelli E, Smith SE, Facelli JM, Christophersen HM, Andrew Smith F (2010) Underground friends or enemies: model plants help to unravel direct and indirect effects of arbuscular mycorrhizal fungi on plant competition. New Phytol 185(4):1050–1061
Fernández I, Cosme M, Stringlis IA, Yu K, de Jonge R, van Wees SM, van der Heijden MG (2019) Molecular dialogue between arbuscular mycorrhizal fungi and the nonhost plant Arabidopsis thaliana switches from initial detection to antagonism. New Phytol 223(2):867–881
Ferrol N, Azcón-Aguilar C, Pérez-Tienda J (2019) Arbuscular mycorrhizas as key players in sustainable plant phosphorus acquisition: an overview on the mechanisms involved. Plant Sci 280:441–447
Floc'h JB, Hamel C, Laterrière M, Tidemann B, St-Arnaud M, Hijri M (2022) Long-term persistence of arbuscular mycorrhizal fungi in the rhizosphere and bulk soils of non-host Brassica napus and their networks of co-occurring microbes. Front Plant Sci 13:828145
Francis R, Read DJ (1994) The contributions of mycorrhizal fungi to the determination of plant community structure. Plant Soil 159:11–25

Francis RM, Read DJ (1995) Mutualism and antagonism in the mycorrhizal symbiosis, with special reference to impacts on plant community structure. Can J Bot 73(S1):1301–1309
García-Garrido JM, Ocampo JA (2002) Regulation of the plant defence response in arbuscular mycorrhizal symbiosis. J Exp Bot 53(373):1377–1386
Genre A, Chabaud M, Balzergue C, Puech-Pagès V, Novero M, Rey T, Barker DG (2013) Short-chain chitin oligomers from arbuscular mycorrhizal fungi trigger nuclear Ca^{2+} spiking in M edicago truncatula roots and their production is enhanced by strigolactone. New Phytol 198(1): 190–202
Genre A, Lanfranco L, Perotto S, Bonfante P (2020) Unique and common traits in mycorrhizal symbioses. Nat Rev Microbiol 18(11):649–660
Gilbert L, Johnson D (2017) Plant–plant communication through common mycorrhizal networks. In: Advances in botanical research, vol 82. Academic, New York, pp 83–97
Giovannetti M, Sbrana C (1998) Meeting a non-host: the behaviour of AM fungi. Mycorrhiza 8(3): 123–130
Giovannetti M, Sbrana C, Logi C (1994) Early processes involved in host recognition by arbuscular mycorrhizal fungi. New Phytol 127(4):703–709
Glenn MG, Chew FS, Williams PH (1985) Hyphal penetration of Brassica (Cruciferae) roots by a vesicular–arbuscular mycorrhizal fungus. New Phytol 99(3):463–472
Hajiboland R (2013) Role of arbuscular mycorrhiza in amelioration of salinity. In: Salt stress in plants: signalling, omics and adaptations. Springer, Cham, pp 301–354
He XH, Critchley C, Bledsoe C (2003) Nitrogen transfer within and between plants through common mycorrhizal networks (CMNs). Crit Rev Plant Sci 22(6):531–567
Hoeksema JD (2015) Experimentally testing effects of mycorrhizal networks on plant-plant interactions and distinguishing among mechanisms. In: Mycorrhizal networks. Springer, Cham, pp 255–277
Hornstein ED, Charles M, Franklin M, Edwards B, Vintila S, Kleiner M, Sederoff H (2023) Re-engineering a lost trait: IPD3, a master regulator of arbuscular mycorrhizal symbiosis, affects genes for immunity and metabolism of non-host Arabidopsis when restored long after its evolutionary loss. bioRxiv, 2023-03
Iffis B, St-Arnaud M, Hijri M (2016) Petroleum hydrocarbon contamination, plant identity and arbuscular mycorrhizal fungal (AMF) community determine assemblages of the AMF spore-associated microbes. Environ Microbiol 18(8):2689–2704
Jones FR (1924) A mycorrhizal fungus in the roots of legumes and some other plants. J Agric Res 29:459–470
Jordan NR, Zhang J, Huerd S (2000) Arbuscular-mycorrhizal fungi: potential roles in weed management. Weed Res 40(5):397–410
Keim J, Mishra B, Sharma R, Ploch S, Thines M (2014) Root-associated fungi of Arabidopsis thaliana and Microthlaspi perfoliatum. Fungal Divers 66:99–111
Kiers ET, Duhamel M, Beesetty Y, Mensah JA, Franken O, Verbruggen E, Bücking H (2011) Reciprocal rewards stabilize cooperation in the mycorrhizal symbiosis. Science 333(6044): 880–882
Lagrange A, Ducousso M, Jourand P, Majorel C, Amir H (2011) New insights into the mycorrhizal status of Cyperaceae from ultramafic soils in New Caledonia. Can J Microbiol 57(1):21–28
Lambers H, Teste FP (2013) Interactions between arbuscular mycorrhizal and non-mycorrhizal plants: do non-mycorrhizal species at both extremes of nutrient availability play the same game. Plant Cell Environ 36(11):1911–1915
Lambers H, Shane MW, Cramer MD, Pearse SJ, Veneklaas EJ (2006) Root structure and functioning for efficient acquisition of phosphorus: matching morphological and physiological traits. Ann Bot 98(4):693–713
Lambers H, Ahmedi I, Berkowitz O, Dunne C, Finnegan PM, Hardy GESJ, Teste FP (2013a) Phosphorus nutrition of phosphorus-sensitive Australian native plants: threats to plant communities in a global biodiversity hotspot. Conserv Physiol 1(1):cot010

Lambers H, Clements JC, Nelson MN (2013b) How a phosphorus-acquisition strategy based on carboxylate exudation powers the success and agronomic potential of lupines (Lupinus, Fabaceae). Am J Bot 100(2):263–288

Lanfranco L, Fiorilli V, Gutjahr C (2018) Partner communication and role of nutrients in the arbuscular mycorrhizal symbiosis. New Phytol 220(4):1031–1046

Latef AAHA, Hashem A, Rasool S, Abd Allah EF, Alqarawi AA, Egamberdieva D, Ahmad P (2016) Arbuscular mycorrhizal symbiosis and abiotic stress in plants: a review. J Plant Biol 59: 407–426

Leake J, Johnson D, Donnelly D, Muckle G, Boddy L, Read D (2004) Networks of power and influence: the role of mycorrhizal mycelium in controlling plant communities and agroecosystem functioning. Can J Bot 82(8):1016–1045

Lekberg Y, Rosendahl S, Olsson PA (2015) The fungal perspective of arbuscular mycorrhizal colonization in 'nonmycorrhizal' plants. New Phytol 205(4):1399–1403

López-Ráez JA, Pozo MJ, García-Garrido JM (2011) Strigolactones: a cry for help in the rhizosphere. Botany 89(8):513–522

Mandyam KG, Roe J, Jumpponen A (2013) Arabidopsis thaliana model system reveals a continuum of responses to root endophyte colonization. Fungal Biol 117(4):250–260

Matsumura A, Horii S, Ishii T (2007) Effects of arbuscular mycorrhizal fungi and intercropping with bahiagrass on growth and antioxidative enzyme activity of radish. J Jpn Soc Hortic Sci 76(3):224–229

Mnasri M, Janoušková M, Rydlová J, Abdelly C, Ghnaya T (2017) Comparison of arbuscular mycorrhizal fungal effects on the heavy metal uptake of a host and a non-host plant species in contact with extraradical mycelial network. Chemosphere 171:476–484

Muler AL, Oliveira RS, Lambers H, Veneklaas E (2013) Does cluster-root activity of Banksia attenuata (Proteaceae) benefit phosphorus or micronutrient uptake and growth of neighbouring shrubs. Oecologia 174:23–31

Neumann E, George E (2005) Does the presence of arbuscular mycorrhizal fungi influence growth and nutrient uptake of a wild-type tomato cultivar and a mycorrhiza-defective mutant, cultivated with roots sharing the same soil volume? New Phytol 166(2):601–609

Newsham KK, Upson R, Read DJ (2009) Mycorrhizas and dark septate root endophytes in polar regions. Fungal Ecol 2(1):10–20

Orłowska E, Zubek S, Jurkiewicz A, Szarek-Łukaszewska G, Turnau K (2002) Influence of restoration on arbuscular mycorrhiza of Biscutella laevigata L. (Brassicaceae) and Plantago lanceolata L. (Plantaginaceae) from calamine spoil mounds. Mycorrhiza 12:153–159

Pongrac P, Vogel-Mikuš K, Regvar M, Tolrà R, Poschenrieder C, Barceló J (2008) Glucosinolate profiles change during the life cycle and mycorrhizal colonization in a Cd/Zn hyperaccumulator Thlaspi praecox (Brassicaceae). J Chem Ecol 34:1038–1044

Radhakrishnan GV, Keller J, Rich MK, Vernié T, Mbadinga Mbadinga DL, Vigneron N, Delaux PM (2020) An ancestral signalling pathway is conserved in intracellular symbioses-forming plant lineages. Nat Plants 6(3):280–289

Raven JA, Lambers H, Smith SE, Westoby M (2018) Costs of acquiring phosphorus by vascular land plants: patterns and implications for plant coexistence. New Phytol 217(4):1420–1427

Reeves FB, Wagner D, Moorman T, Kiel J (1979) The role of endomycorrhizae in revegetation practices in the semi-arid west. I. A comparison of incidence of mycorrhizae in severely disturbed vs. natural environments. Am J Bot 66(1):6–13

Regvar M, Vogel K, Irgel N, Wraber T, Hildebrandt U, Wilde P, Bothe H (2003) Colonization of pennycresses (Thlaspi spp.) of the Brassicaceae by arbuscular mycorrhizal fungi. J Plant Physiol 160(6):615–626

Rillig MC, Mummey DL (2006) Mycorrhizas and soil structure. New Phytol 171(1):41–53

Ryan MH, Tibbett M, Edmonds-Tibbett T, Suriyagoda LDB, Lambers H, Cawthray GR, Pang J (2012) Carbon trading for phosphorus gain: the balance between rhizosphere carboxylates and arbuscular mycorrhizal symbiosis in plant phosphorus acquisition. Plant Cell Environ 35(12): 2170–2180

Shane MW, Lambers H (2005) Cluster roots: a curiosity in context. Plant Soil 274:101–125
Shane MW, Cawthray GR, Cramer MD, Kuo J, Lambers H (2006) Specialized 'dauciform' roots of Cyperaceae are structurally distinct, but functionally analogous with 'cluster' roots. Plant Cell Environ 29(10):1989–1999
Smith SE, Read DJ (2008) Mycorrhizal symbiosis, 3rd edn. Academic, New York
Smith FA, Grace EJ, Smith SE (2009) More than a carbon economy: nutrient trade and ecological sustainability in facultative arbuscular mycorrhizal symbioses. New Phytol 182(2):347–358
Steinkellner S, Lendzemo V, Langer I, Schweiger P, Khaosaad T, Toussaint JP, Vierheilig H (2007) Flavonoids and strigolactones in root exudates as signals in symbiotic and pathogenic plant-fungus interactions. Molecules 12(7):1290–1306
Stringlis IA, Proietti S, Hickman R, Van Verk MC, Zamioudis C, Pieterse CM (2018) Root transcriptional dynamics induced by beneficial rhizobacteria and microbial immune elicitors reveal signatures of adaptation to mutualists. Plant J 93(1):166–180
Tester M, Smith SE, Smith FA (1987) The phenomenon of "nonmycorrhizal" plants. Can J Bot 65(3):419–431
Tommerup IC (1984) Development of infection by a vesicular–arbuscular mycorrhizal fungus in Brassica napus L. and Trifolium subterraneum L. New Phytol 98(3):487–495
Van der Heijden MG, Boller T, Wiemken A, Sanders IR (1998) Different arbuscular mycorrhizal fungal species are potential determinants of plant community structure. Ecology 79(6): 2082–2091
Van Der Heijden MG, Martin FM, Selosse MA, Sanders IR (2015) Mycorrhizal ecology and evolution: the past, the present, and the future. New Phytol 205(4):1406–1423
Veiga RS, Jansa J, Frossard E, van der Heijden MG (2011) Can arbuscular mycorrhizal fungi reduce the growth of agricultural weeds? PLoS One 6(12):e27825
Veiga RS, Howard K, van der Heijden MG (2012) No evidence for allelopathic effects of arbuscular mycorrhizal fungi on the non-host plant Stellaria media. Plant Soil 360:319–331
Veiga RS, Faccio A, Genre A, Pieterse CM, Bonfante P, van der Heijden MG (2013) Arbuscular mycorrhizal fungi reduce growth and infect roots of the non-host plant A rabidopsis thaliana. Plant Cell Environ 36(11):1926–1937
Vierheilig H, Bennett R, Kiddle G, Kaldorf M, Ludwig-Müller J (2000) Differences in glucosinolate patterns and arbuscular mycorrhizal status of glucosinolate-containing plant species. New Phytol 146(2):343–352
Vogel-Mikuš K, Pongrac P, Kump P, Nečemer M, Regvar M (2006) Colonisation of a Zn, Cd and Pb hyperaccumulator Thlaspi praecox Wulfen with indigenous arbuscular mycorrhizal fungal mixture induces changes in heavy metal and nutrient uptake. Environ Pollut 139(2):362–371
Wang B, Qiu YL (2006) Phylogenetic distribution and evolution of mycorrhizas in land plants. Mycorrhiza 16:299–363
Wang E, Schornack S, Marsh JF, Gobbato E, Schwessinger B, Eastmond P, Oldroyd GE (2012) A common signaling process that promotes mycorrhizal and oomycete colonization of plants. Curr Biol 22(23):2242–2246
Wang Y, Wang R, Lu B, Guerin-Laguette A, He X, Yu F (2021) Mycorrhization of Quercus mongolica seedlings by Tuber melanosporum alters root carbon exudation and rhizosphere bacterial communities. Plant Soil 467(1-2):391–403
Wang Y, He X, Yu F (2022) Non-host plants: are they mycorrhizal networks players? Plant Div 44(2):127–134
Watts-Williams SJ, Cavagnaro TR (2015) Using mycorrhiza-defective mutant genotypes of non-legume plant species to study the formation and functioning of arbuscular mycorrhiza: a review. Mycorrhiza 25(8):587–597
Wehner J, Antunes PM, Powell JR, Mazukatow J, Rillig MC (2010) Plant pathogen protection by arbuscular mycorrhizas: a role for fungal diversity? Pedobiologia 53(3):197–201

Zhang H, Qin Z, Chu Y, Li X, Christie P, Zhang J, Gai J (2019) Interactions between arbuscular mycorrhizal fungi and non-host Carex capillacea. Mycorrhiza 29:149–157

Zouari, I., Salvioli, A., Chialva, M., Novero, M., Miozzi, L., Tenore, G. C., ... & Bonfante, P. (2014). From root to fruit: RNA-Seq analysis shows that arbuscular mycorrhizal symbiosis may affect tomato fruit metabolism. Bmc Genomics, 15, 1–19.

Chapter 3
Interactions Between Arbuscular Mycorrhizal Fungi and Other Microorganisms in the Rhizosphere and Hyphosphere

Deepak Kumar, Sandeep Kour, Mohd Ali, Roohi Sharma, Parkirti, Vikram, Harish Changotra, Rajesh Kumari Manhas, and Puja Ohri

Abstract The ecosystem of the earth is fascinating and intricate. The interactions that occur in the soil affect its characteristics as a substrate for development and activity of soil microorganisms and plants. The majority of terrestrial plants have a symbiotic, mutually beneficial interaction with the soil fungi known as arbuscular mycorrhizal fungi (AMF). These microorganisms expand the root's absorption region, which improves the plant's ability to absorb nutrients. The symbiont gets plant carbohydrates in return for accomplishing its life cycle. Additionally, AMF aids in the adaptation of plants to biotic and abiotic challenges including salt, drought, extremely high or low temperatures, heavy metals, diseases and infections. AMF are constantly interacting with ample variety of microbes, including endobacteria, rhizobacteria that encourage plant growth and mycorrhiza helper bacteria, plant parasitic nematodes, fungi and other microbes inhabiting the rhizosphere and hyphosphere. Their interactions may be of utmost significance and might affect agriculture. The present chapter summarizes the main microbial community groups and their interaction with AMF in the rhizosphere and hyphosphere along with the advantageous effects of AMF on plants.

Keywords AMF · Mycorrhiza helper bacteria · Rhizobacteria · Rhizosphere · Inter-organismal cross-talk

---

D. Kumar · S. Kour · M. Ali · R. Sharma · Parkirti · Vikram · P. Ohri (✉)
Department of Zoology, Guru Nanak Dev University, Amritsar, Punjab, India
e-mail: puja.zoology@gndu.ac.in

H. Changotra
Department of Molecular Biology and Biochemistry, Guru Nanak Dev University, Amritsar, Punjab, India

R. K. Manhas
Department of Microbiology, Guru Nanak Dev University, Amritsar, Punjab, India

G. J. Ahammed, R. Hajiboland (eds.), *Arbuscular Mycorrhizal Fungi and Higher Plants*, https://doi.org/10.1007/978-981-99-8220-2_3

3.1 Introduction

Understanding the dynamic processes that occur during the plant-soil interactions requires an intense study of various functional groups of the soil microflora. Among these, the interaction of arbuscular mycorrhizal fungi (AMF) with other microbes is especially interesting because they serve as a living connection between the soil and the roots. AMF are soil microorganisms that interact with 80–90% of vascular plant species and 90% of agricultural plants such as vegetables, cereals and horticultural plants (Chen et al. 2018; Diagne et al. 2020). This symbiosis predates the earliest appearance of terrestrial plants by 400–450 million years, according to fossil records and molecular evidence (Diagne et al. 2020). AMF are widely distributed in various ecosystems, which are essentially determined by the distribution of recognized plant hosts worldwide (Yang et al. 2012). AMF belongs to the phylum *Glomeromycota* and subkingdom *Mucoromyceta*, which includes three classes (*Archaeosporomycetes*, *Glomeromycetes* and *Paraglomeromycetes*) (Tedersoo et al. 2018; Sangwan and Prasanna 2022), consisting of 25 genera, 11 families and over 250 species (Husna et al. 2022). *Glomeromycota* are obligatory symbionts that derive around 20% of carbon from their host plants and, in return, favour host plant's delivery of water and nutrients like phosphorous and nitrogen through root apoplast, arbuscules and intraradical and extraradical hyphae. Arbuscular mycorrhizal symbiosis is undoubtedly one of the most familiar and advantageous relationships between plants and microbes. According to various investigations, they are crucial for plant nutrition and development under stress, and they also improve a variety of crucial ecological functions (Parihar et al. 2019; Chandrasekaran et al. 2021; Dowarah et al. 2021; Yang et al. 2023).

Arbuscular mycorrhizal symbiotic association was once thought to have a bipartite relationship. It is now understood that under normal circumstances, several microorganisms including bacteria, fungi, nematodes and arthropods interact with this fungal-plant symbiotic association and have an impact on the metabolic process of both AMF and host plant (Lu et al. 2023; Szili-Kovács and Takács 2023) (Fig. 3.1). For instance, the interaction of AMF and several bacterial genera including plant growth- promoting rhizobacteria (PGPR) and mycorrhiza helper bacteria (MHB) have been shown to support the health of AMF symbiosis, development of hyphae, germination of spore, colonization of root and the metabolous efficiency of AMF (Chen et al. 2023c). Some of the noteworthy interactions between AMF and bacteria include the binding of bacteria to the fungal spore, production of volatiles by bacteria, transfer of chemicals inside the fungal spore by bacteria and disintegration of the fungal cellular wall (Miransari 2011). It is also believed that some microbial communities in the hyphosphere help AMF acquire organic resources by breaking down complex organic molecules that AMF cannot break down on their own. In exchange, the AMF offers a hyphal exudate- and energy-rich microhabitat that promotes microbial proliferation and motility in the hyphosphere (Wang et al. 2022, 2023a; Faghihinia et al. 2023). This AMF-bacterial interaction can be exploited to increase crop output in sustainable and environmentally friendly way

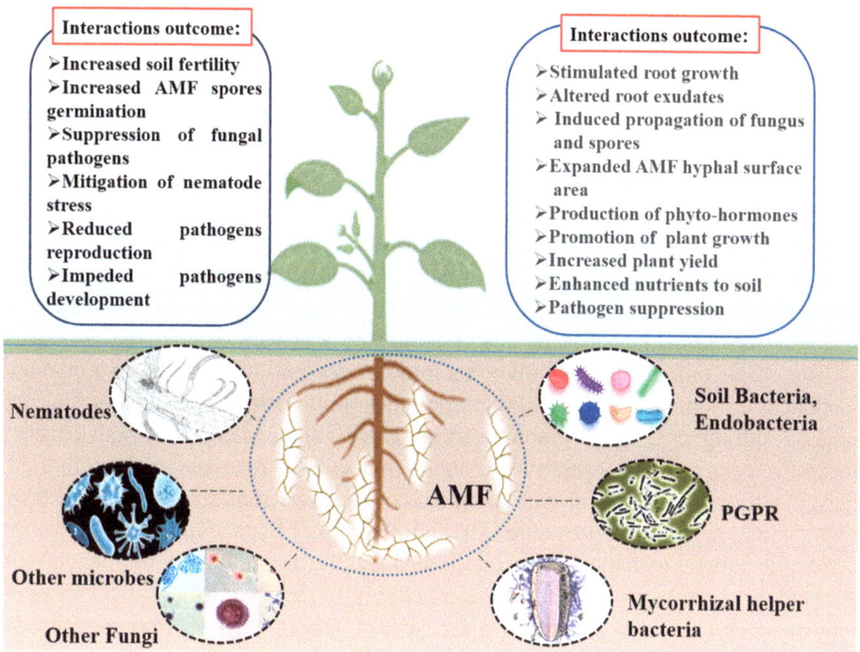

Fig. 3.1 Interaction of arbuscular mycorrhizal fungi (AMF) with other microorganisms in the rhizosphere and hyphosphere

by producing inoculum comprising the fungus and bacteria, also known as AMF bio-enhancers (Sangwan and Prasanna 2022).

Further, AMF also interacts with plant parasitic nematodes as they share the plant roots as source of food and habitat in the rhizosphere. AMF may promote resistance and plant tolerance by inhibiting the development of nematodes. The possibility of greater nematode tolerance or resistance of AMF-infected plants has sparked research in AMF-nematode interactions (Patel et al. 2022; Vishwakarma et al. 2022). Furthermore, the interaction of AMF with other fungi and various other microorganisms in rhizosphere and hyphosphere has also been studied and is being utilized for the sustainable management of plant pathogens (Bao et al. 2022). Consequently, it is crucial to carefully assess such interactions and take into account their application in agronomy. This may lead to some novel perceptions for future research, hastening the development of the discipline and yielding more effective agricultural strategies. Therefore, considering the outcomes of AMF interactions, the present chapter focuses on the interaction of AMF with other microorganisms inhabiting the rhizosphere and hyphosphere.

3.2 AMF and Plants

Plant roots are inhabited by numerous microbes that also include AMF, which are beneficial for plant health in many ways. They modify morphology and physiology of plant roots, affect their growth, influence their nutrient status and resist various plant stresses (Sharma et al. 2023; Thangavel et al. 2022). AMF interact with their wide range of hosts by penetrating into their root epidermis using their own fungal hyphae. In roots, they colonize in cortical cells and form arbuscules. Arbuscules consist of fungal hyphae enveloped in peri-arbuscular membrane, which is a modification of cortical cell plasma membrane. This interaction provides better use of soil nutrients and better resistance to variable stresses (Wipf et al. 2019).

The AMF colonization involves reciprocal recognition between two partners mediated by symbiotic molecules before they both come in contact with each other, resulting in the pre-symbiotic stage. During their interaction, both partners recognize each other via cross-talk of signal molecules synthesized by both individuals. As soon as the fungal hyphae make contact with root epidermis, they form hyphopodia, an adhesion structure. The introduction of hyphae in roots initiates the symbiotic phase, and the latter is followed by the formation of pre-penetration apparatus, which leads hyphae to the inner root tissue. From there, hyphae grow into the deep cortical cells and form arbuscules. Thus, root colonization proceeds, and extensive hyphal network explores more soil by forming extraradical hyphae. This extraradical mycelium extends to form common mycorrhizal networks to make contacts among root systems of two or more than two host plants (Lanfranco et al. 2016; Sarkar and Sadhukhan 2023).

At the time of colonization, the interaction between both partners is facilitated mainly by regulated signalling and carotenoid pathways. After establishing colonies, the AMF provides the host plant with ion and nutrient uptake in return for receiving photosynthates (such as carbohydrates). This carbon supply in AMF thus transports sugars from aerial parts of the plants to their roots. Along with nutrient exchange, AMF also strengthen defence response in plants by modifying biosynthesis of secondary metabolites. Thus, AMF modify signalling pathways and nutritional state of host plants, which enhance the production of protective compounds in plants for defence (Kaur and Suseela 2020; Sarkar and Sadhukhan 2023). However, during the interaction of AMF with non-host plants, the pattern-recognition receptors presenting on plasma membrane of plant act as a part of the innate immune system. Their function is to recognize conserved molecular patterns linked with pathogens or microbes as well as molecular patterns associated with damage and ultimately to activate pattern-triggered immunity in plants. Thus, when AMF come in contact with a non-host plant, the danger-associated peptide signalling pathway is activated; consequently, the trading of carbon and phosphorus between AMF and host plant does not occur, which negatively affects nutrient availability, causing photosynthetic suppression, oxidative stress and cell wall lignification, which eventually subdue the growth of the non-host plant (Wang et al. 2023b). In contrast to it, the symbiosis between AMF and host plant acquires numerous benefits, as the plant provides photosynthates to fungi in a controlled way, and the fungi also help in better survival of the host plant under stressful conditions. AMF colony provides a feedback mechanism for regulated transport of mineral nutrients in root cortical cells, thus, improving availability of all essential nutrients (Thangavel et al. 2022). AMF also modulates immunity of plants against plant pathogens by providing more nutrient to the plant to produce more metabolites by modulating plant metabolites (Sarkar and Sadhukhan 2023). AMF may modify plant physiological functions under various stresses to increase plant tolerance by altering various morphological features (Zhang et al. 2022; Jiménez-Pérez et al. 2022). AMF inoculation may also enhance expression of certain defensive genes in plants to elevate its immunity (Stratton et al. 2022). The spores, vesicles as well as the extra- and intra-radical mycelia of AMF also act as a sink in case of heavy metal accumulation in plants; thus, it also helps in detoxification (Adeyemi et al. 2021). Conclusively, more insight into the role of AMF in enhancing plant resistance toward various environmental stress along with its widespread use in agricultural field can minimize the use of chemical pesticides and fertilizers, thus promoting soil and plant health. Table 3.1 summarizes interaction between AMF and their host plants during various stress conditions.

3.3 Interaction of AMF with Bacteria

Populations of bacterial species in the rhizosphere have a remarkable effect on plant growth and productivity (Liu et al. 2020). PGPR, endobacteria, MHB, detrimental bacteria and other kinds of soil-based bacteria have been found to have interactions with AMF (Olsson et al. 2017). Enhancing the tripartite interaction of AMF, bacteria and plants that host it necessitates an understanding of how they interact (Noceto et al. 2021; Chang et al. 2017). Both types of bacterial species present in the rhizosphere and cell matric of some fungal strains are included in this group of bacteria (Miransari 2011). Positive, neutral or adverse associations can occur with the mycorrhizal association or other rhizospheric microbes, but AMF association with bacteria is predominantly based on their constructive outcomes (Hassani et al. 2018).

The root metabolites influence the biochemicals produced by microbes (bacteria); however, these soil microbes are also capable of releasing chemicals that increase the concentration of root exudates activating AM hyphae and expanding the degree of root inhabitation (Olanrewaju et al. 2019). In addition, plant hormones produced by soil microorganisms can affect spore and hyphae development as well as AM establishment (Miransari et al. 2014). Various elements exert an impact on the interaction between soil bacteria and AMF (Rillig et al. 2018). In fact, bacterial species have different capacities for adhering to AM hyphae, which is also influenced by the physiological stage of the hyphae. The research on bacterial attachment is scarce; however, information regarding other linked features including AM race and strains of bacteria, plant genera, rhizosphere and climatic parameters is available in the literature (Artursson et al. 2006; Sanon et al. 2009). Forthcoming

Table 3.1 Effect of arbuscular mycorrhizal fungi (AMF) on plants

AMF	Host plant	Stress condition	Reported effects	References
Rhizophagus irregularis	Pistachio (*Pistacia vera*)	Salinity stress	Enhanced activities of catalase, superoxide dismutase and glutathione reductase and increased content of tocopherol and carotenoid	Afshar and Abbaspour (2023)
R. irregularis	Soybean (*Glycine max*)	Drought stress	Increased root activity, plant biomass and photosynthesis, decreased ROS activity and increased membrane index stability	Begum et al. (2023)
F. mosseae	Shisham (*Dalbergia sissoo*)	Entisol soil	Enhanced stomatal conductance and photosynthesis rate and plant growth and biomass	Bhardwaj et al. (2023)
R. intraradices	Manchurian catalpa (*Catalpa bungei*)	Low, medium and high nitrogen levels	Enhanced growth of seedlings and N and P levels	Chen et al. (2023a)
F. mosseae and *C. etunicatum*	Camel grass (*Alhagi sparsifolia*)	Salt or drought stress	Enhanced root growth, phosphorus nutrient uptake, activated antioxidant enzymes and balanced hormonal levels	Chen et al. (2023b)
F. mosseae	Milkvetch (*Astragalus adsurgens*)	Cadmium-consisting saline soil	Improved phosphorus uptake and improved sodium/potassium ratio in shoot, also elevated shoot biomass observed	Jia et al. (2023)
R. irregularis	Flooded gum (*Eucalyptus grandis*)	Cadmium stress	AMF helped retaining Cd in fungal structure	Kuang et al. (2023)
F. mosseae	Milkvetch (*A. adsurgens*)	Cadmium- and lead-contaminated soil	Increased root mass and shoot mass, elevated rhizospheral bacterial community	Li et al. (2023a)

(continued)

Table 3.1 (continued)

AMF	Host plant	Stress condition	Reported effects	References
			thus enhancing phytoremediation efficiency	
G. mosseae	Barrelclover (*Medicago truncatula*)	Cadmium stress	Enhanced photosynthetic efficiency, increased plant biomass, decreased heavy metal contamination and increased phytohormone synthesis	Li et al. (2023b)
C. etunicatum	Green tea (*Camellia sinensis*)	Drought stress	Increased plant height, plant biomass and plant mineral content in roots	Liu et al. (2023a)
Paraglomus occultum	Tomato (*Solanum lycopersicum*)	Salt stress	Enhanced plant growth performance and gas exchange rate	Liu et al. (2023b)
Paraglomus species	Indian shot (*Canna indica*)	Flooding and lead stress	Alleviated Pb toxicity, alleviated flooding stress and enhanced growth of host plant	Lv et al. (2023)
Funneliformis sp., *Glomus* sp., *Rhizophagus* sp., *Claroideoglomus* sp. and *Acaulospora* sp.	Rice (*Oryza sativa*)	Phosphorus deficiency	Increased plant phosphorus uptake, increased plant growth, enhanced soil enzyme activities	Mitra et al. (2023)
R. intraradices and *F. mosseae*	Sunflower (*Helianthus annuus*)	Water stress	Elevated chlorophyll content and declined MDA and shoot proline content	Noroozi et al. (2023)
Funneliformis mosseae, *R. intraradices*, *F. geosporus*, *Claroideoglomus etunicatum*, *Glomus aggregatum* and *R. irregularis*	Wheat (*Triticum durum*)	Water deficiency	Increased height of aerial part, length of internodes, length of ear, plant dry weight and chlorophyll content	Ould et al. (2023)
R. irregularis		Salt stress	Salt stress was alleviated by	Wang et al. (2023c)

(continued)

Table 3.1 (continued)

AMF	Host plant	Stress condition	Reported effects	References
	Swamp oak (*Casuarina glauca*)		increasing plant biomass and accumulation of chlorine in roots; also salt exclusion process observed	
G. versiforme *F. mosseae*	*Snapdragon (Antirrhinum majus)*	Low temperature and weak light	Increased stomatal conductance, photosynthetic rate, transpiration rate and water use efficiency	Wei et al. (2023a)
R. intraradices	Sweet flag (*Acorus calamus*)	Chromium toxicity	Altered rhizospheric bacterial community, increased tolerance toward heavy metal and enhanced nutrient uptake and also the soil carbon input	Wei et al. (2023b)
P. brasilianum, *G. deserticola*, *G. monosporum*, *F. mosseae*, *R. irregularis*, *G. clarum*, *G. etunicatum*, *Gigaspora margarita* and *G. aggregatum*	Tulsi (*Ocimum sanctum*)	Saline stress	Enhanced plant growth and antioxidant enzymes	Yilmaz et al. (2023)
F. mosseae, *Acaulospora foveata* and *C. etunicatum*	Rice (*O. sativa*)	Water deficit	Increased phosphorus content and osmotic potential in plants	Yooyongwech et al. (2023)
F. mosseae (Fm) along with *Agrobacterium rhizogenes* and *Piriformis mosseae*	Rice (*O. sativa*)	Salt stress	Increased biomass, improved osmoregulation ability and maintained high ratio of K^+/Na^+	Zhang et al. (2023a)
R. intraradices and *C. etunicatum*	Maize (*Zea mays*)	Molybdenum stress	Localized heavy metals in root, enhanced nutrient uptake, plant biomass and photosynthetic pigments	Zhang et al. (2023b)

(continued)

Table 3.1 (continued)

AMF	Host plant	Stress condition	Reported effects	References
Glomus sp. consisting AMF complex	Tomato (*S. lycopersicum*)	Drought stress	Stimulated growth of plant in terms of increased length, greater number of leaves and more dry and fresh weight	Slimani et al. (2022)

investigation therefore needs to note this impact and explore whether such connections can affect crop yields commercially.

There are numerous instances of better communication between various bacterial strains such as *Bacillus*, *Paenibacillus*, *Pseudomonas* and *Rhizobia* and various AMF species like *G. clarum*, *G. intraradices*, *G. mosseae* and *G. versiforme* (Artursson et al. 2006). These encouraging responses encompass pathogenic suppression, phosphate solubilization, root colonization by AMF and development and propagation of fungus and spores (Singh et al. 2013; Emmanuel and Babalola 2020).

3.3.1 Endobacteria

The closest interaction occurs when bacteria exist as endobacteria inside the cells of fungus. Regardless of their genomic and functional differences, the communities of bacteria that are associated with fungi make up a brand-new form of microbiome called the fungal microbiome (Desirò et al. 2015). The relationship that exists between AMF and endosymbiotic bacteria can affect how well they function and how well they interact with microbes that are pathogenic as well as how effectively plants grow and develop (Nanjundappa et al. 2019). An endobacterial symbiosis improves a mycorrhizal fungus' viability and boosts its bioenergetic potential (Salvioli et al. 2016). There are several ways or mechanisms that depict that the bacteria anchored to a fungal spore can impair the activity of fungal organism, through release of chemicals that influence the expression of fungal genes or by introduction of chemicals in fungal spore or by adherence to the fungal exterior through the release of lectins and probably by the breakdown of the fungal cell wall and generation of volatile compounds that alter expression of fungal genes (Bonfante and Venice 2020).

Endobacteria have been found in a variety of fungal groups in numerous recent investigations (Olsson et al. 2017). The bacterial endosymbiont that has been studied most extensively is the *Glomeromycota* endobacteria (Bonfante and Desirò 2017). There are two different kinds of endosymbionts in *Glomeromycota*: (1) a gram-negative, rod-shaped beta-proteobacterium and such notable species of the *Gigasporaceae* family including Candidatus *Glomeribactergigasporarum* and (2) a homogeneous gram-positive, coccoid- shaped bacteria distributed among

Glomeromycota family that includes *Mollicutes*-related endobacteria (MRE) (Alabid et al. 2019; Naito et al. 2017). For instance, MRE was found in two species of liverwort, *Conocephalum conicum* and *Lunularia cruciata*, which was hosted by AMF in the order *Glomerales* (Desirò et al. 2013).

Some endobacteria found inside the cytoplasm of AMF are also thought to be PGPR, creating a close relationship with AMF such as actinobacteria (Qin et al. 2017). The results from Lasudee et al. (2018) demonstrated that rice crops cultivated in soil supplemented with PGPR, *Streptomyces thermocarboxydus* strain S3 stimulated growth in low-nutrient soil facing distress from a drought. Similarly, growth was enhanced, and growth parameters were eased with association of endobacteria *Bacillus pumilus* as PGPR and AMF *Glomus deserticola* (Medina et al. 2003). Endobacteria played a significant function in the pre-symbiotic phase of fungus by boosting its bioenergetic capability, according to an examination of RNA-seq data from the AMF *Gigaspora margarita* having endobacterium or in its absence.

Besides some reports that suggest interrelationship between endosymbiotic bacteria and *Glomus* species, majority of the studies focus on the relationship between endosymbiotic bacterium and mycorrhizal fungi *Gigaspora margarita* (Bennett and Groten 2022; Bonfante and Venice 2020). *Gammaproteobacteria* were found to be present in *Ascomycota* and *Basidiomycota* accordingly (Kramer et al. 2016; Liu et al. 2023c). Further reports showed that the presence of endobacterium *Mycoavidus cysteinexigens* may substantially impact *M. elongata*'s metabolic processes where the non-mutant strain having endobacteria displayed a slower rate of development in comparison to the strain that was lacking endobacteria, indicating that the fungus encounters a metabolic cost for dealing with *Mycoavidus* (Uehling et al. 2017).

3.3.2 Plant Growth-Promoting Rhizobacteria

Plant growth- promoting rhizobacteria improve plant growth while interacting with root systems. PGPRs are advantageous microbes that stimulate plant growth using various processes, such as (1) the creation of plant hormones like jasmonic acid, (2) improved solubility of mineral phosphate in soil, (3) pathogenic management, (4) mitigation of plant stress and (5) interacting with beneficial microbes like AMF, bacteria, etc. (Aalipour et al. 2020; Vocciante et al. 2022). AMF and PGPR also interact and modify plant growth, development and stress management. By influencing root colonization and boosting the absorption of nitrogen and phosphorus, PGPR have the power to promote AM fungal growth (Sagar et al. 2021). Although some PGPR have been shown to flawlessly inoculate plant roots, information regarding the treatment intensity of AMF hyphae by PGPR is scarce (Hartmann et al. 2008).

There are various microbial strains of PGPR and AMF variety combinations that could encourage rice establishment and growth in paddy fields. They have combinatorial stimulating effects that boosted grain yield, phosphate solubilization, nitrogen fixation, plant growth and pathogen suppression (Artursson et al. 2006). *Bacillus*, *Pseudomonas*, *Azospirillum* and *Herbaspirillum* are among the bacterial strains, whereas *Rhizophagus irregularis* (formerly known as *G. intraradices*) and *Funneliformis mosseae* are AMF species (Hoseinzade et al. 2016; Premkumari and Prabina 2017; Lasudee et al. 2018; Norouzinia et al. 2020; Beura et al. 2020). For instance, Beura et al. (2020) observed that the interaction of AMF (*F. mosseae*) with PGPR *A. brasilense* and *B. cepacia* together with rock phosphate-enriched compost was found to boost root length and dry root biomass due to hyphal extensions that expanded the surface size and growth of rice. Similarly, the pairing of urea and triple super phosphate nutrients with AMF (*F. mosseae*) and bacteria that fix nitrogen (*Herbaspirillum seropedicae*) enhanced the development of rice (Hoseinzade et al. 2016). According to Norouzinia et al. (2020), the treatment of *Pseudomonas putida* strain S34 and *Pseudomonas fluorescens* strain R167 together with AMF *R. irregularis* considerably boosted rice grain production while reducing the adverse effects of salinity stress.

In comparison to applying chemical fertilizers alone, dual treatment of *Schizolobium parahyba* with AMF and PGPR enhanced wood output by around 20% (Diagne et al. 2020). AMF and *B. subtilis* in dual treatment of *Acacia gerrardii* Benth (Talh trees) under salinity stress led to a substantially higher dry mass, number of nodules and leghemoglobin level compared to those treated with AMF or *B. subtilis* alone (Hashem et al. 2018). These researchers discovered a beneficial synergistic relationship between AMF and *B. subtilis* in terms of the activities of enzymes such as nitrogenase, nitrate and nitrite reductase enzymes as well as the amounts of total lipids, phenols, fibre and osmoprotectants including betaine and amino acids (glycine and proline).

AMF and phosphate- solubilizing bacteria (PSB) may interact in a beneficial way, as PSB with the aid of AM hyphae converts scarcely accessible phosphorous molecules into orthophosphate (inorganic as well as organic forms of phosphorus) that AMF might take in and transfer to the host plant (Nacoon et al. 2020). Plant treatment with AMF and such bacteria not only enhances phosphorus acquisition from nutrient-poor soils but additionally improves the activity of phosphatase, PSB number and mycorrhization (Wahid et al. 2022). AMF (*Funneliformis mosseae*), *Pseudomonas fluorescens* (PSB) and triple superphosphate addition are efficient and affordable techniques to enhance crop production in maize under salt-stress conditions (Ghorchiani et al. 2018). Inoculation with two local phosphate-solubilizing bacterial strain, *Pseudomonas fluorescens* BAM4 and *Burkholderia cepacia* BAM12, along with AMF (*Glomus etunicatum*) and plants supply with tricalcium phosphate in a phosphorus-poor soil remarkably boosted various growth parameters such as dry shoot and root yield and leaf surface area in wheat (Saxena et al. 2015). Under severe drought conditions, dual treatment with AMF (*Rhizophagus irregularis* or *Funneliformis mosseae* or a mixture of both) and PGPR (*Pseudomonas fluorescens*) enabled *Cupressus arizonica* to mitigate water scarcity damages and improved drought tolerance by means of an improved activity of ascorbate peroxidase and glutathione peroxidase as compared to those treated with a single microbe (Aalipour et al. 2020). Similar effect was observed under drought stress in *Melissa officinalis* L. (Lemon balm) when it was inoculated with PGPR and AMF, which enhanced its photosynthetic pigments; biomass; proline level; nitrogen, phosphorus and potassium content; and relative water content (Eshaghi Gorgi et al. 2022).

3.4 Interaction of AMF with Mycorrhiza Helper Bacteria

The existence of other microbes, notably bacteria in the rhizosphere, often influences mycorrhizal symbiosis. These microbes create mechanisms for discerning interplay with the rhizospheric microbiome, and they influence the growth of mycorrhizal connections in a neutral, unfavourable or advantageous way. The specialization of the bacterial and mycorrhizal (1) strain and species, (2) root of plants, (3) soil structure, (4) abiotic and biotic stressors, (5) nutrient competition and (6) soil microbial diversity all affect the relationship among the AMF and the MHB (Arruda et al. 2021). A group of bacteria known as mycorrhiza helper bacteria affects the symbiotic relationship of AMF with plant roots. These bacteria can assist the growth of hyphae, spores, root colonization and metabolic fitness of AMF, all of which are necessary for the symbiosis to occur (Richardson et al. 2009). They encourage the development of particular AMF symbiotically associated with their non-specific host plant (Bharadwaj et al. 2008). By influencing the spore wall (Boer et al. 2005), generating stimulants like carbon dioxide (Carpenter-Boggs et al. 1995), or influencing the AMF phosphorus uptake (Ruiz-Lozano and Bonfante 2000), they can affect spore germination. The majority of AM spores are colonized by a variety of microorganisms, including fungi, bacteria, actinomycetes and microbes, which alter spore germination and chitin mineralization (Ames et al. 1989).

AMF can also change the composition of microorganisms in the rhizosphere by competing for nutrients in soil. Saprophytes and symbionts are two forms of bacteria that communicate with AM species, a few of which are unfavourable, while others are neutral or even beneficial (Johansson et al. 2004). Although the amount of specificity was not recognized, a number of strains of *Rhizobium* and *Pseudomonas* had different degrees of adherence to germinating spores of AMF and hyphae under aseptic conditions (Bianciotto et al. 1996). The capacity of MHB to generate molecular signals that control and govern the gene networks of fungus having AMF symbiosis was also demonstrated in various investigations (Lies et al. 2018). Table 3.2 enlists some interactive effects of MHB on AMF.

3.5 Interaction of AMF with Plant Parasitic Nematodes

Various biotic stressors including plant parasitic nematodes, in addition to abiotic stresses, have impact on plant growth and development (Kaur et al. 2022; Kumar and Ohri 2023). PPNs are significant agricultural pests commercially, and nematicides that are most effective must be substituted with nematode control methods, which have less adverse effect on non-target organisms since they do not meet the

Table 3.2 Interaction of arbuscular mycorrhizal fungi (AMF) with mycorrhiza helper bacteria (MHB)

AMF	Associated MHB	Host plant	Effect on AMF	References
Rhizophagus irregularis	*Rhizobium tropici*	Bean (*Phaseolus vulgaris* L.)	Increased amount of infectious propagules such as vesicles and arbuscules and improved hyphal growth	Banuelos et al. (2023)
R. irregularis	*Paenibacillus validus*	Carrot (*Daucus carota*)	Improved the mycelium of the fungus, leading to the production of viable spores	Gupta and Chakraborty (2020)
Funneliformis mosseae	*Pseudomonas fluorescens*	Maize (*Zea mays*)	Significantly increased vegetative and reproductive traits, root colonization, yield of maize grain and nutrient content in plant tissue	Ghorchiani et al. (2018)
R. irregularis	*L. fusiformis*	Maize (*Z. mays*)	Enhanced root growth and uptake of nutrients by promoting AMF mycelium growth	Battini et al. (2017)
Gigaspora margarita	*Curtobacterium*, *Ensifer* and *Bacillus* sp.	Alfalfa (*Medicago sativa*)	Improved hyphal growth, solubilized degraded chitin and spore germination	Long et al. (2017)
Rhizophagus irregularis	*Burkholderia anthina*	Chicory (*Cichorium intybus*)	Enhanced the solubilization of phosphate	Taktek et al. (2015)
Gigaspora margarita	*Bacillus megaterium*	Neem (*Azadirachta indica*)	Increased mycorrhization effectiveness	Budi et al. (2012)
Glomus intraradices	*Paenibacillus favisporus*	Tomato (*Solanum lycopersicum*)	Increased IAA production, biomass and root	Bidondo et al. (2011)
Gigaspora rosea	*Pseudomonas putida*	Cucumber (*Cucumis sativus*)	Improved plant tolerance to stressful conditions	Gamalero et al. (2010)
G. geosporum	*Flexibacter*, *Lysobacter*, *Pseudomonas*, *Chondromyces* and *Cellvibrio*	*Plantago lanceolata*	Improved spore germination	Roesti et al. (2005)
F. mosseae	*Pseudomonas fluorescens* 92rk	Tomato (*S. lycopersicum*)	Improved root colonization and hyphal development	Gamalero et al. (2004)
G. clarum	*Bacillus pabuli*	Pea (*Pisum sativum*)	Increased intake of nutrients, AMF	

(continued)

Table 3.2 (continued)

AMF	Associated MHB	Host plant	Effect on AMF	References
			colonization, development of hyphae, spore emergence and root colonization	Xavier and Germida (2003)

requirements of the present environmental conditions (Sharma et al. 2023; Kumar et al. 2023). A possible alternative would be to use AMF to increase host resistance and/or tolerance (Pires et al. 2022). AMF, a subphylum of *Glomeromycotina*-related fungi that symbiotically coexist with over 85% of terrestrial plants, including agricultural or horticultural crops, have the potential to defend plants through biodefense (Malar et al. 2022). Numerous researchers have reported that AMF may act as alternative to pesticides for the growth and development of agricultural plants as well as nematode management (Table 3.3) (Wani et al. 2017; Ryan and Graham 2018; Bagyaraj et al. 2022). AMF found in aerial or underground plant parts have been reported to decrease nematode infection (Schouteden et al. 2015). Similarly, plants with mycorrhizae are more nematode-resistant. Mycorrhizae treatment changes the root metabolites, which has been found to minimize the penetration of nematodes in mycorrhizal plants and transiently paralyze nematodes (Dey and Ghosh 2022). AMF are particularly effective against *Meloidogyne* spp. in several hosts; however, the protective impact depends on the host species, the AMF or other factors (da Silva Campos 2020; Poveda et al. 2020). Additionally, the growth and development of host plants is facilitated by AMF colonization because it increases nutrient uptake in plants from the soil and increases the host plant's resilience or tolerance to root-knot nematode (RKN) infection (Begum et al. 2019; Wang et al. 2023d). In particular, by fostering plant growth and absorption of nutrients and competing with other microbes for photosynthate and infective areas, AMF might, to some extent, boost the host plant's disease resistance (Dowarah et al. 2021). However, host plant, AMF species and RKN species are just a few of the variables that affect AMF and plant RKN interaction. AMF has the potential to boost host plant resistance, particularly to RKN. In addition, interaction between *Pratylenchus* and AMF occurs in host roots, and the control and strength of impact that AMF have on the population densities of *Pratylenchus* spp. can vary depending on the order/genus of AMF and the species of hosts plants. *Pratylenchus* population densities may change due to AMF inhabitation via improved plant nutritional status, enhanced root biomass, resource competition or induced systemic plant responses (Gough et al. 2022; Rodrigues et al. 2021). Further, AMF decreases the amount and size of large cells in roots that have been infected by *Meloidogyne*. As a result, the nematodes lay down fewer eggs since they are smaller and require longer to mature into adults. Additionally, AMF has a positive impact on species that are resistant to *Meloidogyne*, and using AMF along with other microbes or nematicide substances lowers the chance of *Meloidogyne* infection and promotes plant growth (da Silva Campos 2020; da Silva Campos et al. 2013).

Table 3.3 Interaction of arbuscular mycorrhizal fungi (AMF) with plant parasitic nematodes

AMF	Nematode species	Host plant	Effect reported	References
Funneliformis mosseae, *Rhizophagus fasciculatus* and *R. intraradices*	*Meloidogyne graminicola*	Rice (*Oryza sativa*)	Reduced root gall development	Malviya et al. (2023)
Septoglomus deserticola and *F. mosseae*	*M. incognita*	Pepper (*Piper nigrum*)	Reduced egg masses	Udo et al. (2023)
F. mosseae, *R. intraradices* and *Glomus versiforme*	*M. incognita*	Tomato (*S. lycopersicum*)	Reduced density of nematodes in soil	Wang et al. (2023d)
G. mosseae	*M. incognita*	Tomato (*S. lycopersicum*)	Decreased gall index of root	Abo-Korahv and Yassin (2022)
F. mosseae	*M. incognita*	Tomato (*S. lycopersicum*)	Increased juvenile mortality	Alamri et al. (2022)
R. irregularis	*M. incognita*	Tomato (*S. lycopersicum*)	Increased number of galls and egg masses	Detrey et al. (2022)
F. mosseae	*Pratylenchus thornei*	Mung bean (*Vigna radiate*)	Reduced reproduction	Gough et al. (2022)
F. mosseae	*M. javanica*	Tomato (*S. lycopersicum*)	Reduced nematode population	Nafady et al. (2022)
R. irregularis	*M. incognita*	Tomato (*S. lycopersicum*)	Decreased galling index	Sedhupathi et al. (2022)
R. intraradices	*M. graminicola*	Tomato (*S. lycopersicum*)	Reduced egg masses by reducing number of giant cells	Shanthi et al. (2022)
G. mosseae	*M. incognita*	Sweet pepper (*Capsicum annuum*)	Decreased galling and egg mass production	Udo et al. (2022)
F. geosporum	*M. incognita*	Sweet pepper (*Capsicum annuum*)	Reduced galling index and nematode reproduction	Herrera-Parra et al. (2021)
G. mosseae	*M. javanica*	Eggplants (*S. melongela*)	Reduced egg masses and number of galls	Sharma et al. (2021)
R. intraradices	*Meloidogyne* spp.	Coffee (*Coffea arabica*)	Reduced nematode development	Vallejos-Torres et al. (2021)
Dentiscutata heterogama, *F. mosseae* and *Rhizophagus* sp.	*Heterodera glycines*	Soybean (*Glycine max*)	Reduced egg hatching by up to 62%	Pawlowski and Hartman (2020)

(continued)

Table 3.3 (continued)

AMF	Nematode species	Host plant	Effect reported	References
F. mosseae	*M. incognita*	Tomato (*S. lycopersicum*)	Decreased the density of nematodes in soil	Pham et al. (2020)
G. mosseae	*M. incognita*	Tomato (*S. lycopersicum*)	Reduced gall index and population of nematode	Shanthi (2019)
R. intraradices and *F. mosseae*	*Nacobbus aberrans*	Tomato (*S. lycopersicum*)	Reduced nematode penetration in root	Marro et al. (2018)
G. fasciculatum	*M. incognita*	Tomato (*S. lycopersicum*)	Reduced reproduction of nematode	Samal et al. (2018)
Gigaspora albida, Claroideoglomus etunicatum and *Acaulospora longula*	*M. arenaria*	Red ginger (*Alpinia purpurata*)	Decreased number of galls and egg mass production	da Silva Campos et al. (2017)
R. irregularis	*M. incognita*	Tomato (*S. lycopersicum*)	Reduced egg masses	Sharma and Sharma (2017)
G. mosseae	*M. javanica*	Tomato (*S. lycopersicum*)	Decreased egg number in egg masses	Sohrabi et al. (2017)
Trichoderma harzianum and *G. aggregatum*	*M. incognita*	Basil (*Ocimumbasilicum*)	Increased mortality	Tiwari et al. (2017)
F. geosporum	*M. incognita*	Cucumber (*Cucumis sativus*)	Reduced egg number per egg mass	Wang et al. (2017)
Acaulospora colombiana	*Meloidogyne* spp.	Cassava (*Manihot esculenta*)	Reduced nematode eggs and population density	Séry et al. (2016)

3.6 Interaction of AMF with Other Fungi

Another important group of microorganisms in the mycorrhizosphere is fungi. They help in promotion of plant growth by playing vital role in the ecosystem like the solubilization of phosphorus (P), fixation of atmospheric nitrogen and synthesis of indole-3 acetic acid (IAA) (Mendes et al. 2013). Interaction of AMF with other fungi has been studied in various plants like tomato, oil palm, soyabean, *Brassica juncea*, etc. (Bao et al. 2022). AMF are closely associated with the roots of their host plants, so they are bound to interact with plant pathogens that dwell in the mycorrhizosphere through competition, antibiosis and parasitism (Filion et al. 1999). Several pathogenic fungi affect plant growth and yield. However, it has been seen that mycorrhizal plants are less severely damaged from fungal infection as compared to non-mycorrhizal plants (Dehne 1982; Filion et al. 1999). The infection caused by fungal pathogens like *Ophiobolus*, *Phytophthora*, *Thielaviopsis*, *Fusarium*, *Macrophomina*, *Pyrenochaeta*, *Rhizoctonia*, *Phoma*, *Pythium*, *Sclerotium* and *Cylindrocarpon* in plants like onion, banana, red clover, peach, soyabean, barley, kidney bean, peanut, cotton, citrus, strawberries, tobacco, poplar and ginseng can be ameliorated by the inoculation of AMF like *Glomus margarita*, *G. fasciculatum*, *G. heterogama*, *G. etunicatum*, *G. calospora* and *G. macrocarpum* (Becker 1976; Steinkellner et al. 2012; Bubici et al. 2019; Eke et al. 2020; Aljawasim et al. 2020; Guzman et al. 2021). AMF defends plants from fungal pathogens by increasing the build-up of lignins and non-soluble polysaccharides in plant root cell walls (Jamiołkowska et al. 2017). Also, the mycorrhizal fungal hyphae present in the plant roots create physical barrier for the contagion of pathogenic fungi (Amer and Abou-El-Seoud 2008). Sudhasha et al. (2020) reported the interaction between AMF *G. intraradices* and a pathogenic fungus, *F. oxysporum*, and found that after inoculation with *G. intraradices*, the development of pathogenic fungus was inhibited. It was proposed that the growth and reproduction of fungal pathogen might have been subdued by the chemical balance of the mycorrhizae (Sudhasha et al. 2020).

Another group of fungi that can co-exist with AMF and improve plant growth is ectomycorrhizal fungi (EMF). There are certain plant species called 'dual mycorrhizal plants' that are capable of associating with both AMF and EMF, either in the same root system or at a distinct life stage or environment (Teste et al. 2020). In these plants, AMF and EMF compete with each other for root space, which results in vertical segregation with AMF abundance at deep soil and EMF at shallow horizons of soil (Albornoz et al. 2022). The proposed reason behind this vertical segregation is that AMF can only amass inorganic phosphorus that is plentiful at deeper soil horizons, whereas EMF can also obtain organic phosphorus that is abundant at shallow soil (Neville et al. 2002). Association of plants with both AMF and EMF depends on various environmental factors like climatic conditions, availability of nutrients, age of plants, etc. (Teste et al. 2020). Usually, AMF colonization occurs first and succeeded by EMF; however, in certain instances, ectomycorrhizas establish first and decrease AMF colonization by developing a mantle that serves as a barrier to AMF infection (Diagne et al. 2020). Contrastingly, EMF infection is not affected negatively by AMF establishment (Duponnois et al. 2003). Several studies have been conducted, which showed that combined application of AMF and EMF enhanced plant's growth. For instance, in a study, it was observed that simultaneous application of both AMF and EMF considerably enhanced phosphorus content and biomass of *Casuarina equisetifolia* plants as compared to plants applied solely with either EMF or AMF (Elumalai and Raaman 2009). Combined inoculation of AMF and EMF increased plant growth and improved rhizobial nodulation process in *Acacia crassicarpa* (Lesueur and Duponnois 2005). AMF, EMF and *Rhizobium* application at the same time showed similar results in *Robinia pseudoacacia* plants (Tian et al. 2003). In another study, inoculation of *Eucalyptus grandis* with mixtures of both AMF and EMF species enhanced below-ground dry weight of the plant but had negative effect on its above-ground dry weight (Holste et al. 2017). In dual mycorrhizal plants, different advantages of both AMF and EMF like drought resistance, pathogen defence and aggregation of soil have been hypothesized, but they still need full investigation (Teste et al. 2020).

3.7 Interaction of AMF with Other Rhizospheric and Hyphospheric Microorganisms

The zone influenced by root as well as mycorrhizal fungus is referred to as the mycorrhizosphere (Bao et al. 2022). In other words, it comprises the soil region governed by plant roots, i.e. rhizosphere, and the region surrounding, influenced by the individual fungal hyphae, i.e. hyphosphere (Priyadharsini et al. 2016). The organic substances exuded by the roots and mycorrhizae stimulate and enhance the microbial activity in mycorrhizosphere, which differs from the bulk soil (Linderman 1988; Frey-Klett et al. 2005). Apart from fungi and bacteria, AMF also intermingle with protozoa, arthropods and other soil fauna (Zhang et al. 2020; Geisen and Quist 2021) (Fig. 3.2). Soil protozoans and collembola (soil arthropods) are known to coexist and interact with AMF species in the mycorrhizosphere (Innocenti and Sabatini 2018).

Another group of rhizospheric microorganisms that interact with AMF are *Actinomycetes*. The spores of AMF often harbour *Actinomycetes*. The volatile compounds released by these *Actinomycetes* help in the germination of AMF spores

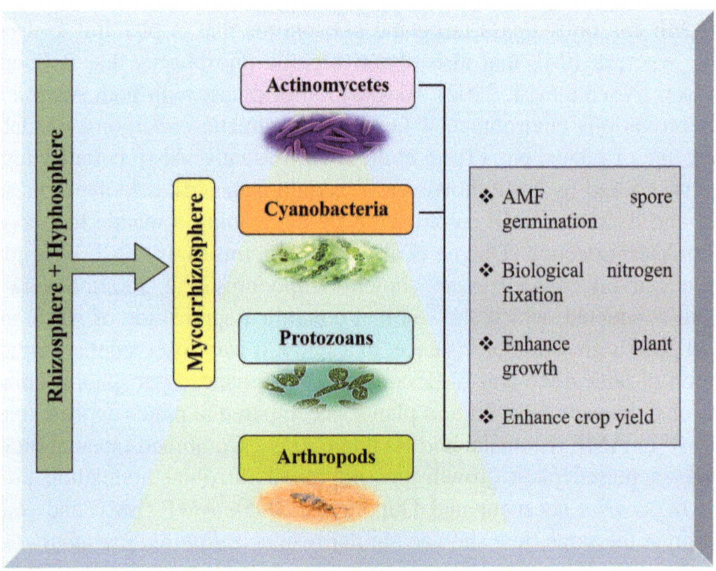

Fig. 3.2 Some other microorganisms that interact with AMF in the mycorrhizosphere

(Gryndler 2000). *Frankia*, a diazotrophic genus of *Actinomycetes*, lives in symbiotic association with roots of plants like *Alnus*, *Casuarina*, *Comptonia*, *Elaeagnus* and *Myrica* and form actinorhiza (specialized nodules) where nitrogen fixation takes place. These plants might also interact with fungi and form arbuscular mycorrhiza (Berliner and Torrey 1989). This tripartite interaction has variable results on plant growth. In an experiment, treatment of *A. cordata* with *Frankia* increased plants growth, but inoculation with *G. mosseae* and *G. fasciculatum* alone had no effect on the growth. However, when both AMF and *Frankia* were applied simultaneously, the plant growth was remarkably enhanced than when *Frankia* was applied alone (Lumini et al. 1994).

AMF also interact with blue-green algae (cyanobacteria) and play a crucial role in enhancing the fertility of agricultural fields. These autotrophic, occasionally motile and mostly filamentous organisms are capable of fixing carbon and feeding it to the soil ecosystem (Gryndler 2000). Cyanobacteria are key constituents of microbial consortia in rice fields (Ashmrita and Radha 2017). In the agricultural ecosystems, AMF-*Cyanobacterium* interaction shows promising effects on nitrogen fixation, growth and productivity of crops (Ojha et al. 2018). For instance, the combined application of cyanobacteria and AMF in paddy fields leads to biological nitrogen fixation by blue-green algae and the production of phytohormones by AMF, which brought favourable results and improved nutrient availability, grain and straw yield and soil structure (Panneerselvam et al. 2017; Bao et al. 2022).

3.8 Conclusion

Interactions of AMF and microorganisms in the rhizosphere and hyposphere are among the most significant and crucial factors affecting soil structure and characteristics, along with plant's growth. In the rhizosphere and hyphosphere, there are numerous kinds of microbial life that interact with AMF, which include bacteria, fungi and other microorganisms. For the tripartite symbiosis of plants, AMF and other microorganisms, nutritional exchange is crucial. The root exudates as well as microbial secretions can support the symbiotic interactions between different plants, AMF and soil microbes. Furthermore, the development of associative symbiosis and colonization are linked to various microbial-mediated proteins, events and pathways. Although some microbial consortia incorporating mycorrhizal inoculum are used in agriculture, our understanding regarding the complicated interaction between AMF and microorganisms in the rhizosphere and hyphosphere is still in its infancy. Therefore, utilizing various molecular approaches, future investigation may focus on the more in-depth exploration of relationship between the host plant, AMF and other microorganisms. This could result in a more effective use of soil microorganisms for crop production, which is also eco-friendly and agriculturally sustainable. New methods for molecular ecology and metagenomic investigations may provide more information about the relationships between various microbes and AMF.

References

Aalipour H, Nikbakht A, Etemadi N, Rejali F, Soleimani M (2020) Biochemical response and interactions between arbuscular mycorrhizal fungi and plant growth promoting rhizobacteria during establishment and stimulating growth of *Arizona cypress* (*Cupressus arizonica* G.) under drought stress. Sci Hortic 261:108923

Abo-Korahv M, Yassin A (2022) Effect of different bioagents on the population density of *Meloidogyne incognita* infected tomato plants. Egypt J Agron 21(1):14–22

Adeyemi NO, Atayese MO, Sakariyawo OS, Azeez JO, Sobowale SPA, Olubode A, Mudathir R, Adebayo R, Adeoye S (2021) Alleviation of heavy metal stress by arbuscular mycorrhizal symbiosis in *Glycine max* (L.) grown in copper, lead and zinc contaminated soils. Rhizosphere 18:100325

Afshar AS, Abbaspour H (2023) Mycorrhizal symbiosis alleviate salinity stress in pistachio plants by altering gene expression and antioxidant pathways. Physiol Mol Biol Plants 29(2):263–276

Alabid I, Glaeser SP, Kogel KH (2019) Endofungal bacteria increase fitness of their host fungi and impact their association with crop plants. Curr Issues Mol Biol 30(1):59–74

Alamri S, Nafady NA, El-Sagheer AM, El-Aal MA, Mostafa YS, Hashem M, Hassan EA (2022) Current utility of arbuscular mycorrhizal fungi and hydroxyapatite nanoparticles in suppression of Tomato Root-Knot Nematode. Agronomy 12(3):671

Albornoz FE, Prober SM, Ryan MH, Standish RJ (2022) Ecological interactions among microbial functional guilds in the plant-soil system and implications for ecosystem function. Plant Soil 476(1-2):301–313

Aljawasim BD, Khaeim HM, Manshood MA (2020) Assessment of arbuscular mycorrhizal fungi (*Glomus* spp.) as potential biocontrol agents against damping-off disease *Rhizoctoniasolani* on cucumber. J Crop Prot 9:141–147

Amer MA, Abou-El-Seoud II (2008) Mycorrhizal fungi and *Trichoderma harzianum* as biocontrol agents for suppression of *Rhizoctonia solani* damping-off disease of tomato. Commun Agric Appl Biol Sci 73(2):217–232

Ames RN, Mihara KL, Bayne HG (1989) Chitin-decomposing actinomycetes associated with a vesicular–arbuscular mycorrhizal fungus from a calcareous soil. New Phytol 111(1):67–71

Arruda B, George PB, Robin A, de Mescolotti D, Herrera WF, Jones DL, Andreote FD (2021) Manipulation of the soil microbiome regulates the colonization of plants by arbuscular mycorrhizal fungi. Mycorrhiza 31(5):545–558

Artursson V, Finlay RD, Jansson JK (2006) Interactions between arbuscular mycorrhizal fungi and bacteria and their potential for stimulating plant growth. Environ Microbiol 8(1):1–10

Ashmrita M, Radha S (2017) Blue-green algal biofertilizer and growth response of rice plants. Int J Plant Sci 12:68–71

Bagyaraj DJ, Sridhar KR, Revanna A (2022) Arbuscular mycorrhizal fungi influence crop productivity, plant diversity, and ecosystem services. In: Rajpal VR, Singh I, Navi SS (eds) Fungal diversity, ecology and control management. Fungal biology. Springer, Singapore, pp 345–362

Banuelos J, Martínez-Romero E, Montaño NM, Camargo-Ricalde SL (2023) *Rhizobium tropici* and riboflavin amendment condition arbuscular mycorrhiza colonization in *Phaseolus vulgaris* L. Agronomy 13(3):876

Bao X, Zou J, Zhang B, Wu L, Yang T, Huang Q (2022) Arbuscular mycorrhizal fungi and microbes' interaction in rice mycorrhizosphere. Agronomy 12:1277

Battini F, Grønlund M, Agnolucci M, Giovannetti M, Jakobsen I (2017) Facilitation of phosphorus uptake in maize plants by mycorrhizosphere bacteria. Sci Rep 7(1):4686

Becker WN (1976) Quantification of onion vesicular-arbuscular mycorrhizae and their resistance to *Pyrenochaeta terrestris*. University of Illinois at Urbana-Champaign, Champaign, p 7624041

Begum N, Qin C, Ahanger MA, Raza S, Khan MI, Ashraf M, Ahmed N, Zhang L (2019) Role of arbuscular mycorrhizal fungi in plant growth regulation: implications in abiotic stress tolerance. Front Plant Sci 10:1068

Begum N, Xiao Y, Wang L, Li D, Irshad A, Zhao T (2023) Arbuscular mycorrhizal fungus *Rhizophagus irregularis* alleviates drought stress in soybean with overexpressing the GmSPL9d gene by promoting photosynthetic apparatus and regulating the antioxidant system. Microbiol Res 273:127398

Bennett AE, Groten K (2022) The costs and benefits of plant–arbuscular mycorrhizal fungal interactions. Annu Rev Plant Biol 73:649–672

Berliner R, Torrey JG (1989) On tripartite *Frankia*-mycorrhizal associations in the Myricaceae. Can J Bot 67:1708–1712

Beura K, Pradhan AK, Ghosh GK, Kohli A, Singh M (2020) Root architecture, yield and phosphorus uptake by rice: response to rock phosphate enriched compost and microbial inoculants. Int Res J Pure Appl Chem 21(19):33–39

Bharadwaj DP, Lundquist PO, Alström S (2008) Arbuscular mycorrhizal fungal spore-associated bacteria affect mycorrhizal colonization, plant growth and potato pathogens. Soil Biol Biochem 40(10):2494–2501

Bhardwaj AK, Chandra KK, Kumar R (2023) Water stress changes on AMF colonization, stomatal conductance and photosynthesis of *Dalbergia sissoo* seedlings grown in entisol soil under nursery condition. For Sci Technol 19:47–58

Bianciotto V, Bandi CD, Minerdi M, Sironi H, Tichy V, Bonfante P (1996) An obligately endosymbiotic mycorrhizal fungus itself harbors obligately intracellular bacteria. Appl Environ Microbiol 62(1):3005–3010

Bidondo LF, Silvani V, Colombo R, Pérgola M, Bompadre J, Godeas A (2011) Pre-symbiotic and symbiotic interactions between *Glomus intraradices* and two *Paenibacillus* species isolated from AM propagules. In vitro and in vivo assays with soybean (AG043RG) as plant host. Soil Biol Biochem 43(9):1866–1872

Boer WD, Folman LB, Summerbell RC, Boddy L (2005) Living in a fungal world: impact of fungi on soil bacterial niche development. FEMS Microbiol Rev 29(4):795–811

Bonfante P, Desirò A (2017) Who lives in a fungus? The diversity, origins and functions of fungal endobacteria living in Mucoromycota. ISME J 8:1727–1735

Bonfante P, Venice F (2020) Mucoromycota: going to the roots of plant-interacting fungi. Fungal Biol Rev 34(2):100–113

Bubici G, Kaushal M, Prigigallo MI, Gómez-Lama Cabanás C, Mercado-Blanco J (2019) Biological control agents against fusarium wilt of banana. Front Microbiol 10:616

Budi SW, Bakhtiar Y, May NL (2012) Bacteria associated with arbuscular mycorrhizal spores *Gigaspora margarita* and their potential for stimulating root mycorrhizal colonization and neem (*Melia azedarach* Linn) seedling growth. Microbiol Indones 6(4):6–6

Carpenter-Boggs L, Loynachan TE, Stahl PD (1995) Spore germination of Gigaspora margarita stimulated by volatiles of soil-isolated actinomycetes. Soil Biol Biochem 27(11):1445–1451

Chandrasekaran M, Boopathi T, Manivannan P (2021) Comprehensive assessment of ameliorative effects of AMF in alleviating abiotic stress in tomato plants. J Fungi 7(4):303

Chang C, Nasir F, Ma L, Tian C (2017) Molecular communication and nutrient transfer of arbuscular mycorrhizal fungi, symbiotic nitrogen-fixing bacteria, and host plant in tripartite symbiosis. In: Sulieman S, Tran LSP (eds) Legume nitrogen fixation in soils with low phosphorus availability: adaptation and regulatory implication. Springer, Cham, pp 169–183

Chen M, Arato M, Borghi L, Nouri E, Reinhardt D (2018) Beneficial services of arbuscular mycorrhizal fungi–from ecology to application. Front Plant Sci 9:1270

Chen W, Mou X, Meng P, Chen J, Tang X, Meng G, Xin K, Zhang Y, Wang C (2023a) Effects of arbuscular mycorrhizal fungus inoculation on the growth and nitrogen metabolism of *Catalpa bungei* CA Mey. under different nitrogen levels. Front Plant Sci 14:1138184

Chen X, Aili Y, Ma X, Wang H, Dawuti M (2023b) Mycorrhizal fungal colonization promotes apparent growth and physiology of *Alhagi sparsifolia* seedlings under salt or drought stress at vulnerable developmental stage. Plant Growth Regul 2023:1–12

Chen D, Saeed M, Ali MNHA, Raheel M, Ashraf W, Hassan Z, Farooq U, Hakim MF, Rao MJ, Negm S (2023c) Plant growth promoting rhizobacteria (PGPR) and arbuscular mycorrhizal fungi combined application reveals enhanced soil fertility and rice production. Agronomy 13(2): 550

da Silva Campos MA (2020) Bioprotection by arbuscular mycorrhizal fungi in plants infected with *Meloidogyne* nematodes: a sustainable alternative. Crop Prot 135:105203

da Silva Campos MA, da Silva FSB, Yano-Melo AM, de Melo NF, Pedrosa EMR, Maia L (2013) Responses of guava plants to inoculation with arbuscular mycorrhizal fungi in soil infested with *Meloidogyne enterolobii*. Plant Pathol J 29(3):242

da Silva Campos MA, da Silva FSB, Yano-Melo AM, de Melo NF, Maia LC (2017) Application of arbuscular mycorrhizal fungi during the acclimatization of *Alpinia purpurata* to induce tolerance to *Meloidogyne arenaria*. Plant Pathol J 33(3):32

Dehne HW (1982) Interaction between vesicular mycorrhizal fungi and plant pathogens. Phytopathology 72:1115–1119

Desirò A, Naumann M, Epis S, Novero M, Bandi C, Genre A, Bonfante P (2013) Mollicutes-related endobacteria thrive inside liverwort-associated arbuscular mycorrhizal fungi. Environ Microbiol 15(3):822–836

Desirò A, Faccio A, Kaech A, Bidartondo MI, Bonfante P (2015) Endogone, one of the oldest plant-associated fungi, host unique mollicutes-related endobacteria. New Phytol 205(4):1464–1472

Detrey J, Cognard V, Djian-Caporalino C, Marteu N, Doidy J, Pourtau N, Vriet C, Maurousset L, Bouchon D, Clause J (2022) Growth and root-knot nematode infection of tomato are influenced by mycorrhizal fungi and earthworms in an intercropping cultivation system with leeks. Appl Soil Ecol 169:104181

Dey M, Ghosh S (2022) Arbuscular mycorrhizae in plant immunity and crop pathogen control. Rhizosphere 13:100524

Diagne N, Ngom M, Djighaly PI, Fall D, Hocher V, Svistoonoff S (2020) Roles of arbuscular mycorrhizal fungi on plant growth and performance: importance in biotic and abiotic stressed regulation. Diversity 12(10):370

Dowarah B, Gill SS, Agarwala N (2021) Arbuscular mycorrhizal fungi in conferring tolerance to biotic stresses in plants. J Plant Growth Regul 44:1429–1444

Duponnois R, Diédhiou S, Chotte J, Sy MO (2003) Relative importance of the endomycorrhizal and (or) ectomycorrhizal associations in *Allocasuarina* and *Casuarina* genera. Can J Microbiol 49: 281–287

Eke P, Adamou S, Fokom R, Nya VD, Fokou PVT, Wakam LN, Nwaga D, Boyom FF (2020) Arbuscular mycorrhizal fungi alter antifungal potential of lemongrass essential oil against *Fusarium solani*, causing root rot in common bean (*Phaseolus vulgaris* L.). Heliyon 6:e05737

Elumalai S, Raaman N (2009) In vitro synthesis of Frankia and mycorrhiza with *Casuarina equisetifolia* and ultrastructure of root system. Indian J Exp Biol 47:289–297

Emmanuel OC, Babalola OO (2020) Productivity and quality of horticultural crops through co-inoculation of arbuscular mycorrhizal fungi and plant growth promoting bacteria. Microbiol Res 239:126569

Eshaghi Gorgi O, Fallah H, Niknejad Y, Barari Tari D (2022) Effect of plant growth promoting rhizobacteria (PGPR) and mycorrhizal fungi inoculations on essential oil in *Melissa officinalis* l. under drought stress. Biologia 77:11–20

Faghihinia M, Jansa J, Halverson LJ, Staddon PL (2023) Hyphosphere microbiome of arbuscular mycorrhizal fungi: a realm of unknowns. Biol Fertil Soils 59(1):17–34

Filion MM, Arnaud S, Fortin JA (1999) Direct interaction between the arbuscular mycorrhizal fungus *Glomus intraradices* and different rhizosphere microorganisms. New Phytol 141:525–533

Frey-Klett P, Chavatte M, Clausse ML, Courrier S, Le Roux C, Raaijmakers J, Martinotti MG, Pierrat JC, Garbaye J (2005) Ectomycorrhizal symbiosis affects functional diversity of rhizosphere fluorescent pseudomonads. New Phytol 165:317–328

Gamalero E, Trotta A, Massa N, Copetta A, Martinotti MG, Berta G (2004) Impact of two fluorescent pseudomonads and an arbuscular mycorrhizal fungus on tomato plant growth, root architecture and P acquisition. Mycorrhiza 14:185–192

Gamalero E, Berta G, Massa N, Glick BR, Lingua G (2010) Interactions between *Pseudomonas putida* UW4 and *Gigaspora rosea* BEG9 and their consequences for the growth of cucumber under salt-stress conditions. J Appl Microbiol 108(1):236–245

Geisen S, Quist CW (2021) Microbial-faunal interactions in the rhizosphere. In: Gupta VVSR, Sharma AK (eds) Rhizosphere biology: interactions between microbes and plants. Springer, Singapore, pp 237–253

Ghorchiani M, Etesami H, Alikhani HA (2018) Improvement of growth and yield of maize under water stress by co-inoculating an arbuscular mycorrhizal fungus and a plant growth promoting rhizobacterium together with phosphate fertilizers. Agric Ecosyst Environ 258:59–70

Gough EC, Owen KJ, Zwart RS, Thompson JP (2022) The role of nutrients underlying interactions among root-nodule bacteria (*Bradyrhizobium* sp.), arbuscular mycorrhizal fungi (*Funneliformis mosseae*) and root-lesion nematodes (*Pratylenchus thornei*) in nitrogen fixation and growth of mung bean (*Vigna radiata*). Plant Soil 472(1-2):421–449

Gryndler M (2000) Interactions of arbuscular mycorrhizal fungi with other soil organisms. In: Kapulnik Y, Douds DD (eds) Arbuscular mycorrhizas: physiology and function. Springer, Dordrecht, pp 239–262

Gupta SK, Chakraborty AP (2020) Mycorrhiza helper bacteria: future prospects. Int J Res Rev 7(3): 387–391

Guzman A, Montes M, Hutchins L, DeLaCerda G, Yang P, Kakouridis A, Dahlquist-Willard RM, Firestone MK, Bowles T, Kremen C (2021) Crop diversity enriches arbuscular mycorrhizal fungal communities in an intensive agricultural landscape. New Phytol 231:447–459

Hartmann A, Schmid M, Tuinen DV, Berg G (2008) Plant-driven selection of microbes. Plant Soil 321:235–257

Hashem A, Alqarawi AA, Al-Huqail AA, AbdAllah EF (2018) Biodiversity of arbuscular mycorrhizal fungi associated with *Acacia gerrardii* Benth in different habitats of Saudi Arabia. Pak J Bot 50(3):1211–1217

Hassani M, Durán P, Hacquard S (2018) Microbial interactions within the plant holobiont. Microbiome 6(1):1–7

Herrera-Parra E, Ramos-Zapata J, Basto-Pool C, Cristóbal-Alejo J (2021) Sweet pepper (*Capsicum annuum*) response to the inoculation of native arbuscular mycorrhizal fungi and the parasitism of root-knot *Meloidogyne incognita*. Rev Bio Cien 8:E0071020

Holste EK, Kobe RK, Gehring CA (2017) Plant species differ in early seedling growth and tissue nutrient responses to arbuscular and ectomycorrhizal fungi. Mycorrhiza 27:211–223

Hoseinzade H, Ardakani MR, Shahdi A, Rahmani HA, Noormohammadi G, Miransari M (2016) Rice (Oryza sativa L.) nutrient management using mycorrhizal fungi and endophytic *Herbaspirillum seropedicae*. J Integr Agric 15(6):1385–1394

Husna H, Tuheteru FD, Albasri A, Arif A, Basrudin B, Nurdin WR, Safitri I (2022) Diversity of arbuscular mycorrhizal fungi of kalappiacelebica: an endemic and endangered plant species in Sulawesi, Indonesia. Biodivers J Biol Divers 23(10):5290–5297

Innocenti G, Sabatini MA (2018) Collembola and plant pathogenic, antagonistic and arbuscular mycorrhizal fungi: a review. Bull Insectol 71:71–76

Jamiołkowska A, Księżniak A, Hetman B, Kopacki M, Skwaryło-Bednarz B, Gałązka A, Thanoon AH (2017) Interactions of arbuscular mycorrhizal fungi with plants and soil microflora. Acta Sci Pol Hortorum Cultus 16(5):89–95

Jia B, Diao F, Ding S, Shi Z, Xu J, Hao L, Li FY, Guo W (2023) Differential effects of arbuscular mycorrhizal fungi on three salt-tolerant grasses under cadmium and salt stress. Land Degrad Dev 34(2):506–520

Jiménez-Pérez M, Morales-Manzo II, Ana FITA, Rodríguez-Burruezo A (2022) Mitigation of drought stress in solanaceae vegetables through symbiosis with plant growth-promoting bacteria and arbuscular mycorrhizal fungi. A review. Sci J 11:86

Johansson JF, Paul LR, Finlay RD (2004) Microbial interactions in the mycorrhizosphere and their significance for sustainable agriculture. FEMS Microbiol Ecol 48(1):1–13

Kaur S, Suseela V (2020) Unraveling arbuscular mycorrhiza-induced changes in plant primary and secondary metabolome. Metabolites 10(8):35

Kaur R, Sharma N, Tikoria R, Ali M, Kour S, Kumar D, Ohri P (2022) Insights into biosynthesis and signaling of cytokinins during plant growth, development and stress tolerance. In: Aftab T (ed) Auxins, cytokinin and gibberellins signaling in plants. Springer, Cham, pp 153–187

Kramer S, Dibbern D, Moll J, Huenninghaus M, Koller R, Krueger D, Marhan S, Urich T, Wubet T, Bonkowski M, Buscot F (2016) Resource partitioning between bacteria, fungi, and protists in the detritusphere of an agricultural soil. Front Microbiol 7:1524

Kuang Y, Li X, Wang Z, Wang X, Wei H, Chen H, Hu W, Tang M (2023) Effects of arbuscular mycorrhizal fungi on the growth and root cell ultrastructure of eucalyptus grandis under cadmium stress. J Fungi 9(2):140

Kumar D, Ohri P (2023) Say "NO" to plant stresses: unravelling the role of nitric oxide under abiotic and biotic stress. Nitric Oxide 130:36–57

Kumar D, Manhas RK, Ohri P (2023) Deciphering the growth promoting and stress curtailing role of nitric oxide in *Meloidogyne incognita* infested *Solanum lycopersicum* seedlings. Sci Hortic 319:112147

Lanfranco L, Bonfante P, Genre A (2016) The mutualistic interaction between plants and arbuscular mycorrhizal fungi. Microbiol Spectr 4(6):4–6

Lasudee K, Tokuyama S, Lumyong S, Pathom-Aree W (2018) Actinobacteria associated with arbuscular mycorrhizal *Funneliformis mosseae* spores, taxonomic characterization and their beneficial traits to plants: evidence obtained from mung bean (*Vigna radiata*) and Thai jasmine rice (Oryza sativa). Front Microbiol 9:1247

Lesueur D, Duponnois R (2005) Relations between rhizobial nodulation and root colonization of Acacia crassicarpa provenances by an arbuscular mycorrhizal fungus, *Glomus intraradices* Schenk and Smith or an ectomycorrhizal fungus, *Pisolithus tinctorius* Coker & Couch. Ann Sci 62:467–474

Li X, Wang Y, Guo P, Zhang Z, Cui X, Hao B, Guo W (2023a) Arbuscular mycorrhizal fungi facilitate Astragalus adsurgens growth and stress tolerance in cadmium and lead contaminated saline soil by regulating rhizosphere bacterial community. Appl Soil Ecol 187:104842

Li W, Chen K, Li Q, Tang Y, Jiang Y, Su Y (2023b) Effects of arbuscular mycorrhizal fungi on alleviating cadmium stress in *Medicago truncatula* Gaertn. Plan Theory 12(3):547

Lies A, Delteil A, Prin Y, Duponnois R (2018) Using mycorrhiza helper microorganisms (MHM) to improve the mycorrhizal efficiency on plant growth. In: Meena V (ed) Role of rhizospheric microbes in soil. Springer, Singapore, pp 277–298

Linderman RG (1988) Mycorrhizal interaction with the rhizosphere microflora: the mycorrhizosphere effect. Phytopathology 78:366–371

Liu L, Huang X, Zhang J, Cai Z, Jiang K, Chang Y (2020) Deciphering the relative importance of soil and plant traits on the development of rhizosphere microbial communities. Soil Biol Biochem 148:107909

Liu CY, Hao Y, Wu XL, Dai FJ, Abd-Allah EF, Wu QS, Liu SR (2023a) Arbuscular mycorrhizal fungi improve drought tolerance of tea plants via modulating root architecture and hormones. Plant Growth Regul 2023:1–10

Liu MY, Li QS, Ding WY, Dong LW, Deng M, Chen JH, Tian X, Haashem A, Al-Arjani AF, Alenazi MM, Abd-Allah EF, Wu QS (2023b) Arbuscular mycorrhizal fungi inoculation impacts expression of aquaporins and salt overly sensitive genes and enhances tolerance of salt stress in tomato. Chem Biol Technol Agric 10(1):5

Liu H, Zhang J, Zhang L, Zhang X, Yang R (2023c) *Funneliformis mosseae* influences leaf decomposition by altering microbial communities under saline-alkali conditions. Sci Total Environ 23:165079

Long L, Lin Q, Yao Q, Zhu H (2017) Population and function analysis of cultivable bacteria associated with spores of arbuscular mycorrhizal fungus *Gigaspora margarita*. 3 Biotech 7(1):8

Lu C, Zhang Z, Guo P, Wang R, Liu T, Luo J, Hao B, Wang Y, Guo W (2023) Synergistic mechanisms of bioorganic fertilizer and AMF driving rhizosphere bacterial community to improve phytoremediation efficiency of multiple HMs-contaminated saline soil. Sci Total Environ 883:163708

Lumini E, Bosco M, Puppi G, Isopi R (1994) Field performance of *Alnus cordata* Loisel (Italian alder) inoculated with *Frankia* and VA-mycorrhizal strains in mine-spoil afforestation plots. Soil Biol Biochem 26:659–661

Lv Y, Liu J, Fan Z, Fang M, Xu Z, Ban Y (2023) The function and community structure of arbuscular mycorrhizal fungi in ecological floating beds used for remediation of Pb contaminated wastewater. Sci Total Environ 872:162233

Malar CM, Wang Y, Stajich JE, Kokkoris V, Villeneuve-Laroche M, Yildirir G, Corradi N (2022) Early branching arbuscular mycorrhizal fungus *Paraglomus occultum* carries a small and repeat-poor genome compared to relatives in the Glomeromycotina. Microb Genom 8(4): 000810

Malviya D, Singh P, Singh UB, Paul S, Bisen PK, Rai JP, Verma RL, Fiyaz RA, Kumar A, Kumari P, Dei S, Ahmed MR, Bagyaraj DJ, Singh HV (2023) Arbuscular mycorrhizal fungi-mediated activation of plant defense responses in direct seeded rice (*Oryza sativa* L.) against root-knot nematode *Meloidogyne graminicola*. Front Microbiol 14:1104490

Marro N, Caccia M, Doucet ME, Cabello M, Becerra A, Lax P (2018) Mycorrhizas reduce tomato root penetration by false root-knot nematode *Nacobbus aberrans*. Appl Soil Ecol 124:262–265

Medina A, Probanza A, Mañero FG, Azcón R (2003) Interactions of arbuscular-mycorrhizal fungi and *Bacillus* strains and their effects on plant growth, microbial rhizosphere activity (thymidine and leucine incorporation) and fungal biomass (ergosterol and chitin). Appl Soil Ecol 22(1): 15–28

Mendes R, Garbeva P, Raaijmakers JM (2013) The rhizosphere microbiome: Significance of plant beneficial, plant pathogenic, and human pathogenic microorganisms. FEMS Microbiol Rev 37: 634–663

Miransari M (2011) Interactions between arbuscular mycorrhizal fungi and soil bacteria. Appl Microbiol Biotechnol 89:917–930

Miransari M, Abrishamchi A, Khoshbakht K, Niknam V (2014) Plant hormones as signals in arbuscular mycorrhizal symbiosis. Crit Rev Biotechnol 34(2):123–133

Mitra D, Panneerselvam P, Senapati A, Chidambaranathan P, Nayak AK, Mohapatra PKD (2023) Arbuscular mycorrhizal fungi response on soil phosphorus utilization and enzymes activities in aerobic rice under phosphorus-deficient conditions. Lifestyles 13(5):1118

Nacoon S, Jogloy S, Riddech N, Mongkolthanaruk W, Kuyper TW, Boonlue S (2020) Interaction between phosphate solubilizing bacteria and arbuscular mycorrhizal fungi on growth promotion and tuber inulin content of *Helianthus tuberosus* L. Sci Rep 10:4916

Nafady NA, Sultan R, El-Zawahry AM, Mostafa YS, Alamri S, Mostafa RG, Hashem M, Hassan EA (2022) Effective and promising strategy in management of tomato root-knot nematodes by *Trichoderma harzianum* and arbuscular mycorrhizae. Agronomy 12(2):315

Naito M, Desirò A, González JB, Tao G, Morton JB, Bonfante P, Pawlowska TE (2017) 'Candidatus *Moeniiplasma glomeromycotorum*', an endobacterium of arbuscular mycorrhizal fungi. Int J Syst Evol Microbiol 5:1177–1184

Nanjundappa A, Bagyaraj DJ, Saxena AK, Kumar M, Chakdar H (2019) Interaction between arbuscular mycorrhizal fungi and *Bacillus* spp. in soil enhancing growth of crop plants. Fungal Biol Biotechnol 6:1–10

Neville J, Tessier J, Morrison I, Scarratt J (2002) Soil depth distribution of ecto-and arbuscular mycorrhizal fungi associated with *Populus tremuloides* within a 3-year-old boreal forest clear-cut. Appl Soil 19:209–216

Noceto PA, Bettenfeld P, Boussageon R, Hériché M, Sportes A, van Tuinen D, Courty PE, Wipf D (2021) Arbuscular mycorrhizal fungi, a key symbiosis in the development of quality traits in crop production, alone or combined with plant growth-promoting bacteria. Mycorrhiza 31(6): 655–669

Noroozi N, Mohammadi G, Ghobadi M (2023) Some physio-biochemical traits of sunflower (*Helianthus annuus* L.) as affected by arbuscular mycorrhizal fungi inoculation under different irrigation treatments. Ital J Agron 18:1

Norouzinia F, Ansari MH, Aminpanah H, Firozi S (2020) Alleviation of soil salinity on physiological and agronomic traits of rice cultivars using Arbuscular mycorrhizal fungi and *Pseudomonas* strains under field conditions. Rev Agric Neotrop 7(1):25–42

Ojha SK, Benjamin JC, Singh AK (2018) Role on biofertilizer (blue green algae) in paddy crop. J Pharmacogn Phytochem 7:830–832

Olanrewaju OS, Ayangbenro AS, Glick BR, Babalola OO (2019) Plant health: feedback effect of root exudates-rhizobiome interactions. Appl Microbiol Biotechnol 103:1155–1166

Olsson S, Bonfante P, Pawlowska TE (2017) Ecology and evolution of fungal-bacterial interactions. In: Dighton T, White JF (eds) The fungal community its organization and role in the ecosystem. CRC Press, Boca Raton, pp 563–584

Ould AS, Aliat T, Kucher DE, Bensaci OA, Rebouh NY (2023) Investigating the potential of arbuscular mycorrhizal fungi in mitigating water deficit effects on durum wheat (*Triticum durum* Desf.). Agriculture 13(3):552

Panneerselvam P, Kumar U, Sugitha TCK, Parameswaran C, Sahoo S, Binodh AK, Jahan A, Anandan A (2017) Arbuscular mycorrhizal fungi (AMF) for sustainable rice production. In: Adhya T, Mishra B, Annapurna K, Verma D, Kumar U (eds) Advances in soil microbiology: recent trends and future prospects. Springer, Singapore, pp 99–126

Parihar M, Meena VS, Mishra PK, Rakshit A, Choudhary M, Yadav RP, Rana K, Bisht JK (2019) Arbuscular mycorrhiza: a viable strategy for soil nutrient loss reduction. Arch Microbiol 201: 723–735

Patel HK, Jhala YK, Raghunandan BL, Solanki JP (2022) Role of mycorrhizae in plant-parasitic nematodes management. In: Soni R, Suyal DC, Yadav AN, Goel R (eds) Trends of applied microbiology for sustainable economy. Academic Press, New York, pp 225–251

Pawlowski ML, Hartman GL (2020) Impact of arbuscular mycorrhizal species on *Heterodera glycines*. Plant Dis 104(9):2406–2410

Pham TT, Giang BL, Nguyen NH, Dong Yen PN, Minh Hoang VD, Lien Ha BT, Le NTT (2020) Combination of mycorrhizal symbiosis and root grafting effectively controls nematode in replanted coffee soil. Plan Theory 9(5):555

Pires D, Vicente CS, Menéndez E, Faria JM, Rusinque L, Camacho MJ, Inácio ML (2022) The fight against plant-parasitic nematodes: Current status of bacterial and fungal biocontrol agents. Pathogens 11(10):1178

Poveda J, Abril-Urias P, Escobar C (2020) Biological control of plant-parasitic nematodes by filamentous fungi inducers of resistance: *Trichoderma*, mycorrhizal and endophytic fungi. Front Microbiol 11:992

PremKumari SM, Prabina BJ (2017) Impact of the mixed consortium of indigenous arbuscular mycorrhizal fungi (AMF) on the growth and yield of rice (*Oryza sativa* L.) under the system of rice intensification (SRI). Int J Environ Agric Biotechnol 2(2):238743

Priyadharsini P, Rojamala K, Ravi RK, Muthuraja R, Nagaraj K, Muthukumar T (2016) Mycorrhizosphere: the extended rhizosphere and its significance. In: Choudhary D, Varma A, Tuteja N (eds) Plant-microbe interaction: an approach to sustainable agriculture. Springer, Singapore, pp 97–124

Qin S, Feng WW, Wang TT, Ding P, Xing K, Jiang JH (2017) Plant growth-promoting effect and genomic analysis of the beneficial endophyte *Streptomyces* sp. KLBMP 5084 isolated from halophyte *Limonium sinense*. Plant Soil 416:117–132

Richardson AE, Barea JM, McNeill AM, Prigent-Combaret C (2009) Acquisition of phosphorus and nitrogen in the rhizosphere and plant growth promotion by microorganisms. Plant Soil 321(8):305–339

Rillig MC, Lehmann A, Lehmann J, Camenzind T, Rauh C (2018) Soil biodiversity effects from field to fork. Trends Plant Sci 23(1):17–24

Rodrigues MT, Calandrelli A, Miamoto A, Rinaldi LK, Pereira Moreno B, da Silva C, Dias-Arieira CR (2021) Pre-inoculation with arbuscular mycorrhizal fungi affects essential oil quality and the reproduction of root lesion nematode in *Cymbopogon citratus*. Mycorrhiza 31:613–623

Roesti D, Ineichen K, Braissant O, Redecker D, Wiemken A, Aragno M (2005) Bacteria associated with spores of the arbuscular mycorrhizal fungi *Glomus geosporum* and *Glomus constrictum*. Appl Environ Microbiol 71(11):6673–6679

Ruiz-Lozano JM, Bonfante P (2000) A Burkholderia strain living inside the arbuscular mycorrhizal fungus *Gigaspora margarita* possesses the vacB gene, which is involved in host cell colonization by bacteria. Microb Ecol 39(2):137–144

Ryan MH, Graham JH (2018) Little evidence that farmers should consider abundance or diversity of arbuscular mycorrhizal fungi when managing crops. New Phytol 220(4):1092–1107

Sagar A, Rathore P, Ramteke PW, Ramakrishna W, Reddy MS, Pecoraro L (2021) Plant growth promoting rhizobacteria, arbuscular mycorrhizal fungi and their synergistic interactions to counteract the negative effects of saline soil on agriculture: Key macromolecules and mechanisms. Microorganisms 9(7):1491

Salvioli A, Ghignone S, Novero M, Navazio L, Venice F, Bagnaresi P, Bonfante P (2016) Symbiosis with an endobacterium increases the fitness of a mycorrhizal fungus, raising its bioenergetic potential. ISME J 10(1):130–144

Samal BK, Mahalik JK, Dash BK, Patasani J (2018) Effect of interaction between Arbuscular mycorrhizal fungi (*Glomus fasciculatum*) and root knot nematode (*Meloidogyne incognita*) on tomato. Ann Plant Prot Sci 26(2):355–359

Sangwan S, Prasanna R (2022) Mycorrhizae helper bacteria: unlocking their potential as bioenhancers of plant–arbuscular mycorrhizal fungal associations. Microb Ecol 84(1):1–10

Sanon A, Andrianjaka ZN, Prin Y, Bally R, Thioulouse J, Comte G, Duponnois R (2009) Rhizosphere microbiota interfers with plant-plant interactions. Plant Soil 321:259–278

Sarkar AK, Sadhukhan S (2023) Unearthing the alteration in plant volatiles induced by mycorrhizal fungi: a shield against plant pathogens. Physiol Plant 175(1):e13845

Saxena J, Saini A, Ravi I, Chandra S, Garg V (2015) Consortium of phosphate-solubilizing bacteria and fungi for promotion of growth and yield of chickpea (*Cicer arietinum*). J Crop Improv 29(3):353–369

Schouteden N, De Waele D, Panis B, Vos CM (2015) Arbuscular mycorrhizal fungi for the biocontrol of plant-parasitic nematodes: a review of the mechanisms involved. Front Microbiol 6:1280

Sedhupathi K, Kennedy ZJ, Shanthi A (2022) Interaction of arbuscular fungus (*Rhizophagus irregularis*), *Bacillus subtilis* and *Purpureocillium lilacinum* against root-knot nematode (*Meloidogyne incognita*) in tomato. Indian J Nematol 52(1):92–103

Séry DJM, Kouadjo ZC, Voko BR, Zeze A (2016) Selecting native arbuscular mycorrhizal fungi to promote cassava growth and increase yield under field conditions. Front Microbiol 7:2063

Shanthi A (2019) Arbuscular mycorrhizal (AM) fungi for management of root knot Nematode, *Meloidogyne incognita* on Tomato. Ann Plant Prot Sci 27(2):257–260

Shanthi A, Arun A, Shandeep SG (2022) Studies on histopathological changes in the tomato roots colonized by arbuscular mycorhizal fungus, *Rhizophagus intraradices* and infested by root knot nematodes, *Meloidogyne incognita*. Ann Plant Prot Sci 30(2):91–96

Sharma IP, Sharma AK (2017) Physiological and biochemical changes in tomato cultivar PT-3 with dual inoculation of mycorrhiza and PGPR against root-knot nematode. Symbiosis 71(3): 175–183

Sharma M, Saini I, Kaushik P, Aldawsari MM, Al Balawi T, Alam P (2021) Mycorrhizal fungi and *Pseudomonas fluorescens* application reduces root-knot nematode (*Meloidogyne javanica*) infestation in eggplant. Saudi J Biol Sci 28(7):3685–3691

Sharma N, Khanna K, Jasrotia S, Kumar D, Bhardwaj R, Ohri P (2023) Metabolites and chemical agents in the plant roots: an overview of their use in plant-parasitic nematode management. Nematology 25(3):243–257

Singh R, Soni SK, Kalra A (2013) Synergy between *Glomus fasciculatum* and a beneficial Pseudomonas in reducing root diseases and improving yield and forskolin content in Coleus forskohlii Briq. under organic field conditions. Mycorrhiza 23:35–44

Slimani A, Harkousse O, Mazri MA, Zouahri A, Ouahmane L, Koussa T, Al Feddy MN (2022) Impact of a selected mycorrhizal complex and a rhizobacterial species on tomato plants' growth under water stress conditions. Indian J Agric Res 56(6):696–704

Sohrabi F, Sheikholeslami M, Heydari R, Rezaee S, Sharifi R (2017) Study on combined application of arbuscular mycorrhizal fungi isolates and plant growth promoting rhizobacteria in controlling root-knot nematode *Meloidogyne javanica* in tomato under greenhouse conditions. Iranian J Plant Pathol 53(4):449–462

Steinkellner S, Hage-Ahmed K, García-Garrido JM, Illana A, Ocampo JA, Vierheilig H (2012) A comparison of wild-type, old and modern tomato cultivars in the interaction with the arbuscular mycorrhizal fungus *Glomus mosseae* and the tomato pathogen *Fusarium oxysporum* f. sp. *lycopersici*. Mycorrhiza 22:189–194

Stratton CA, Ray S, Bradley BA, Kaye JP, Ali JG, Murrell EG (2022) Nutrition vs association: plant defenses are altered by arbuscular mycorrhizal fungi association not by nutritional provisioning alone. BMC Plant Biol 22(1):1–10

Sudhasha S, Balabaskar P, Sivakumar T, Sanjeev KK, Kannan R, Venkatesh KL (2020) Constructiveness of the biocontrol agents on Fusarial wilt of tomato incited by the destructive pathogen *Fusarium oxysporum* f. sp. *lycopersici*. In: Current research and innovations in plant pathology. AkiNik Publications, Delhi, pp 4–8

Szili-Kovács T, Takács T (2023) Advanced research of rhizosphere microbial activity. Agriculture 13(4):911

Taktek S, Trépanier M, Servin PM, St-Arnaud M, Piché Y, Fortin JA, Antoun H (2015) Trapping of phosphate solubilizing bacteria on hyphae of the arbuscular mycorrhizal fungus *Rhizophagus irregularis* DAOM 197198. Soil Biol Biochem 90:1–9

Tedersoo L, Sánchez-Ramírez S, Koljalg U, Bahram M, Döring M, Schigel D, May T, Ryberg M, Abarenkov K (2018) High-level classification of the fungi and a tool for evolutionary ecological analyses. Fungal Divers 90:135–159

Teste FP, Jones MD, Dickie IA (2020) Dual-mycorrhizal plants: their ecology and relevance. New Phytol 225(5):1835–1851

Thangavel P, Anjum NA, Muthukumar T, Sridevi G, Vasudhevan P, Maruthupandian A (2022) Arbuscular mycorrhizae: natural modulators of plant–nutrient relation and growth in stressful environments. Arch Microbiol 204(5):264

Tian C, He X, Zhong Y, Chen J (2003) Effect of inoculation with ecto- and arbuscular mycorrhizae and *Rhizobium* on the growth and nitrogen fixation by black locust, *Robinia pseudoacacia*. New For 25:125–131

Tiwari S, Pandey S, Chauhan PS, Pandey R (2017) Biocontrol agents in co-inoculation manages root knot nematode [*Meloidogyne incognita* (Kofoid & White) Chitwood] and enhances essential oil content in *Ocimumbasilicum* L. Ind Crop Prod 97:292–301

Udo IA, Uko AE, Obok EE, Ubi JO, Umoetok SB (2022) Management of Meloidogyne incognita and salinity on sweet pepper (*Capsicum annuum* L.) with different arbuscular mycorrhizal fungus species. J Appl Biol Biotechnol 10(4):66–72

Udo IA, Akpan JF, Bello OS, Otie VO (2023) Arbuscular mycorrhizal fungus species differed in bio-control potential on pepper genotypes infected with *Meloidogyne incognita*. Int J Veg Sci 2023:1–14

Uehling J, Gryganskyi A, Hameed K, Tschaplinski T, Misztal PK, Wu S, Desirò A, Vande Pol N, Du Z, Zienkiewicz A, Zienkiewicz K (2017) Comparative genomics of *Mortierella elongata* and its bacterial endosymbiont *Mycoavidus cysteinexigens*. Environ Microbiol 19(8): 2964–2983

Vallejos-Torres G, Espinoza E, Marín-Díaz J, Solis R, Arévalo LA (2021) The role of arbuscular mycorrhizal fungi against root-knot nematode infections in coffee plants. J Soil Sci Plant Nutr 21(1):364–373

Vishwakarma SK, Ilyas T, Malviya D, Shafi Z, Shahid M, Yadav B, Singh UB, Rai JP, Singh HB, Singh HV (2022) Arbuscular mycorrhizal fungi (AMF) as potential biocontrol agents. In: Singh UB, Sahu PK, Singh HV, Sharma PK, Sharma SK (eds) Rhizosphere microbes: biotic stress management. Springer, Singapore, pp 197–222

Vocciante M, Grifoni M, Fusini D, Petruzzelli G, Franchi E (2022) The role of plant growth-promoting rhizobacteria (PGPR) in mitigating plant's environmental stresses. Appl Sci 12(3): 1231

Wahid F, Sharif M, Shah F, Ali A, Adnan M, Shah S, Danish S, Ali MA, Ahmed N, Arslan H, Arslan D (2022) Mycorrhiza and phosphate solubilizing bacteria: potential bioagents for sustainable phosphorus management in agriculture. Phyton 91(2):257

Wang W, Xu L, Liu R (2017) Effects of combined inoculation with various arbuscular mycorrhizal fungi on plant resistance to root-knot nematode disease in cucumber. Mycosystema 36(7): 1010–1017

Wang F, Zhang L, Zhou J, Rengel Z, George TS, Feng G (2022) Exploring the secrets of hyphosphere of arbuscular mycorrhizal fungi: processes and ecological functions. Plant Soil 481(1-2):1–22

Wang G, Jin Z, George TS, Feng G, Zhang L (2023a) Arbuscular mycorrhizal fungi enhance plant phosphorus uptake through stimulating hyphosphere soil microbiome functional profiles for phosphorus turnover. New Phytol. https://doi.org/10.1111/nph.18772

Wang Y, Chen H, Shao M, Zhu T, Li S, Olsson PA, Hammer EC (2023b) Arbuscular mycorrhizal fungi trigger danger-associated peptide signaling and inhibit carbon–phosphorus exchange with nonhost plants. Plant Cell Environ. https://doi.org/10.1111/pce.14600

Wang Y, Dong F, Chen H, Xu T, Tang M (2023c) Effects of arbuscular mycorrhizal fungus on sodium and chloride ion channels of *Casuarina glauca* under salt stress. Int J Mol Sci 24(4): 3680

Wang L, Chen X, Tang Z (2023d) Arbuscular mycorrhizal symbioses improved biomass allocation and reproductive investment of cherry tomato after root-knot nematodes infection. Plant Soil 482(1-2):513–527

Wani KA, Manzoor J, Shuab R, Lone R (2017) Arbuscular mycorrhizal fungi as biocontrol agents for parasitic nematodes in plants. In: Varma A, Prasad R, Tuteja N (eds) Mycorrhiza-nutrient uptake, biocontrol, ecorestoration. Springer, Cham, pp 195–210

Wei LI, Zhai Y-L, Xing H-S, Xing L-J, Guo S-X (2023a) Arbuscular mycorrhizal fungi promote photosynthesis in *Antirrhinum majus* L. under low-temperature and weak-light conditions. Not Bot Horti Agrobot Cluj-Napoca 51(1):13012–13012

Wei Z, Sixi Z, Xiuqing Y, Guodong X, Baichun W, Baojing G (2023b) Arbuscular mycorrhizal fungi alter rhizosphere bacterial community characteristics to improve Cr tolerance of Acorus calamus. Ecotoxicol Environ Saf 253:114652

Wipf D, Krajinski F, van Tuinen D, Recorbet G, Courty PE (2019) Trading on the arbuscular mycorrhiza market: from arbuscules to common mycorrhizal networks. New Phytol 223(3): 1127–1142

Xavier LJ, Germida JJ (2003) Bacteria associated with Glomus clarum spores influence mycorrhizal activity. Soil Biol Biochem 35(3):471–478

Yang H, Zang Y, Yuan Y, Tang J, Chen X (2012) Selectivity by host plants affects the distribution of arbuscular mycorrhizal fungi: evidence from ITS rDNA sequence metadata. BMC Evol Biol 12:1–13

Yang S, Imran, Ortas I (2023) Impact of mycorrhiza on plant nutrition and food security. J Plant Nutr 2023:1–26

Yilmaz A, Yildirim E, Yilmaz H, Soydemir HE, Güler E, Ciftci V, Yaman M (2023) Use of arbuscular mycorrhizal fungi for boosting antioxidant enzyme metabolism and mitigating saline stress in sweet basil (*Ocimumbasilicum* L.). Sustainability 15(7):5982

Yooyongwech S, Tisarum R, Samphumphuang T, Phisalaphong M, Cha-Um S (2023) Integrated strength of osmotic potential and phosphorus to achieve grain yield of rice under water deficit by arbuscular mycorrhiza fungi. Sci Rep 13(1):5999

Zhang Y, Li S, Li H, Wang R, Zhang K, Xu J (2020) Fungi-nematode interactions: diversity, ecology, and biocontrol prospects in agriculture. J Fungi 6:206

Zhang H, Li L, Ren W, Zhang W, Tang M, Chen H (2022) Arbuscular mycorrhizal fungal colonization improves growth, photosynthesis, and ROS regulation of split-root poplar under drought stress. Acta Physiol Plant 44(6):62

Zhang B, Shi F, Zheng X, Pan H, Wen Y, Song F (2023a) Effects of AMF compound inoculants on growth, ion homeostasis, and salt tolerance-related gene expression in *Oryza sativa* L. under salt treatments. Rice 16(1):1–18

Zhang M, Shi Z, Lu S, Wang F (2023b) AMF inoculation alleviates molybdenum toxicity to maize by protecting leaf performance. J Fungi 9(4):479

Chapter 4
Signaling Events During the Establishment of Symbiosis Between Arbuscular Mycorrhizal Fungi and Plant Roots

Roghieh Hajiboland and Golam Jalal Ahammed

Abstract The most prevalent microorganism association in terrestrial plants is the symbiosis between arbuscular mycorrhizal fungi (AMF) and plant roots. This implies that the genetic background for establishing this symbiosis was developed in the early phases of land plant evolution. A symbiosis faces several challenges, such as penetrating plant cells and overcoming their defense mechanisms. At the same time, it must activate some developmental pathways for symbiotic structures along with membrane transporters required to exchange nutrients and metabolites between two partners. This chapter discusses the response of plants to fungal signals, the function of receptor molecules, and other actors that play a crucial role in the signaling pathways. Ultimately, these pathways result in the expression of symbiosis-specific genes and the formation of symbiosis-specific structures.

Keywords Arbuscule · Mycorrhizal fungi · Mycorrhizal symbiosis · Common symbiotic pathway · Ca^{2+} signaling pathway · Strigolactones · Phosphate acquisition · Plant immune response

4.1 Introduction

The most prevalent microorganism association in terrestrial plants is the symbiosis between arbuscular mycorrhizal fungi (AMF) and plant roots. This implies that the genetic background for establishing this symbiosis was developed in the early periods of land plant evolution. A symbiosis faces several challenges, such as penetrating plant cells and overcoming their defense mechanisms. At the same time, it must activate some developmental pathways for symbiotic structures along

R. Hajiboland (✉)
University of Tabriz, Tabriz, Iran
e-mail: ehsan@tabrizu.ac.ir

G. J. Ahammed
Department of Horticulture, Henan University of Science and Technology, Luoyang, Henan, China

G. J. Ahammed, R. Hajiboland (eds.), *Arbuscular Mycorrhizal Fungi and Higher Plants*, https://doi.org/10.1007/978-981-99-8220-2_4

with membrane transporters required to exchange nutrients and metabolites between two partners. This chapter will discuss the response of plants to fungal signals, the function of receptor molecules, and other actors that play a crucial role in the signaling pathways. Ultimately, these pathways result in the expression of symbiosis-specific genes and the formation of symbiosis-specific structures.

4.2 The Common Symbiotic Pathway

Evidence from two significant beneficial plant–microbe interactions, namely, arbuscular mycorrhiza symbiosis (AMS) and root nodule symbiosis (RNS), revealed the molecular components that transduce the microorganism-derived signals to the plants, resulting in the establishment of a compatible plant-microbe symbiosis. Genetic and mutagenic approaches using legume species that are able to form both AMS and RNS interactions indicated that there are single plant mutations that could block the penetration of both bacteria and fungi to the root. These studies led to the hypothesis that AMS and RNS depend on a shared genetic toolkit as the apparatuses of a common symbiosis pathway (CSP). This pathway is believed to operate downstream of the perception of fungal and rhizobial signals but upstream of the activation of the plant's response to the symbiotic bacteria or fungi.

4.2.1 Receptors and Associated Proteins in the CSP

The initial step in establishing a compatible interaction between legumes and rhizobia is associated with a molecular dialog between two partners. This dialogue requires receptors on the plant side to sense the microorganism's signals.

SYMRK SYMBIOSIS RECEPTOR-LIKE KINASE (SYMRK), a leucine-rich repeat (LRR) receptor-like kinase (Does Not make Infections2, DMI2), constitutes the entry point of CSP and is essential for both RNS and AMS. SYMRK functions as a downstream module of Nod Factor Receptors (NFRs) and Myc Factor Receptors (MFRs) and is a principal component of symbiotic signaling that is functionally conserved. SYMRK is widespread in plant kingdoms but has been subjected to some diversification during evolution (Markmann et al. 2008). *Arabidopsis* lacking RNS or AMS contains SYMRK homologs, ShRK1 and ShRK2 (Shiu and Bleecker 2003), which promote the reproduction ability of the obligate biotroph, oomycete *Hyaloperonospora arabidopsis* (Hpa), within plant cells (Ried et al. 2019). This indicates a widespread function of SYMRK in the interactions between plants and microbes and an overlap between the signaling events related to intracellular symbionts and pathogens.

The knockout mutations of SYMRK abolish the formation of infection threads and inhibit nodulation, while overexpression of the full-length SYMRK or its kinase domain results in the appearance of spontaneous nodules in the absence of rhizobia (Saha et al. 2014; Ried et al. 2014). This suggests that the kinase domain of SYMRK plays a determining role in nodulation. Many plant RLKs, such as SYMRK, are characterized by the presence of tyrosine (Tyr) in the "gatekeeper" position close to the hinge region of the kinase domain. This gatekeeper Tyr in SYMRK (Y^{670}) is essential for orchestrating epidermal/cortical responses in RNS (Saha et al. 2016) and is predominantly auto-phosphorylated in vitro and in planta (Samaddar et al. 2013).

The extra-cytoplasmic region of SYMRK consists of three leucine-rich repeats (LRRs) and a malectin-like ectodomain (MLD) (Chiu and Paszkowski 2020). Both LRRs and MLD elements are linked via the GDPC motif that is conserved in the majority of MLD-LRR RLKs. This area is cleaved to release the MLD without symbiotic stimulation. Although the truncated version of SYMRK (SYMRKΔMLD that lacks the MLD) is subject to high turnover, it is able to strongly and specifically associate with NFR5 and transduce the signal to downstream targets (Antolín-Llovera et al. 2014). However, the role of MLD release in the function of SYMRK in the AM symbiosis is obscure.

HMGR1 SYMRK interacts, through its cytoplasmic domain, with the enzyme HMGR1 (3-Hydroxy-3-Methylglutaryl CoA Reductase 1), a mevalonate (MVA) biosynthetic enzyme. It has been postulated that symbiotic signal perception by SYMRK leads to the induction of HMGR1, resulting in the localized formation of MVA. This MVA signal is then transmitted to the nucleus, which activates cation channels, thereby initiating nuclear-associated Ca^{2+} spiking (Venkateshwaran et al. 2015). Our knowledge of the players acting between the plasma membrane and the nucleus is extremely limited, and it must be determined whether the signal transition from the cytosol to the nucleus is solely mediated through MVA and its derivatives. Ca^{2+} spiking or Ca^{2+} oscillation is a key component of the interaction between plants and microbes. This is one of the most common events discriminating mutations of the CSP and is used in screening various mutants defective in establishing symbiosis.

4.2.2 Components of Ca^{2+} Signaling Pathway

Nuclear Pore Complexes The downstream components of the SYM pathway are localized in the nucleus, suggesting the trafficking of signaling molecules through the nuclear envelope. The involvement of several nucleoporins (NUPs), e.g., NUP85, NUP133, and NENA, in symbiotic signaling has been evidenced in genetic studies (Kanamori et al. 2006; Saito et al. 2007; Groth et al. 2010). Mutations in these nucleoporins lead to defective Ca^{2+} spiking and aborted symbiosis (Parniske 2008). It has been proposed that NUPs likely contribute to the protein translocation between the nuclear envelope's inner membrane and outer membrane (Tamura and Hara-Nishimura 2013).

Ion Channels The ion channels and the calcium pump that are localized in the nuclear envelope and necessary for generating calcium (Ca^{2+}) spiking have been identified in model legume plants. These transporters include (1) CASTOR and POLLUX (DMI1 in *Medicago truncatula*) that are potassium-permeable channels (Charpentier et al. 2008), (2) a PII-type Ca^{2+}-ATPase (MCA8) (Capoen et al. 2011), (3) and the Ca^{2+} channel, cyclic nucleotide gated channel 15 (CNGC15) (Charpentier et al. 2016). These three components (CASTOR/POLLUX, MCA8, and CNGC15) interact together and generate Ca^{2+} oscillation in the nucleus. It has been assumed that a cyclic nucleotide (CN) binds to and triggers the activation of CNGC15, leading to the release of Ca^{2+}. The mobility of potassium (K^+) ions that balances the transmembrane charge is mediated by CASTOR/POLLUX, whereas the return of Ca^{2+} to the store is mediated by MCA8. Since CASTOR/POLLUX, MCA8, and CNGC15 are localized to both the outer and inner nuclear membranes, Ca^{2+} spiking is produced on one or simultaneously on both sides of the nuclear envelope (Charpentier 2018).

Decoding Ca^{2+} Signature In the nucleoplasm, CCaMK (DMI3), a Ca calmodulin-dependent protein kinase, is potentially involved in decoding Ca^{2+} oscillations (Miller et al. 2013). CCaMK binds to and phosphorylates another nuclear protein, CYCLOPS (Interacting Protein of DMI3, IPD3) (Yano et al. 2008). Together with other transcription factors, CYCLOPS (IPD3) regulates the expression of symbiotic genes (Oldroyd 2013) (Fig. 4.1).

4.3 Signaling in the AMS

The AM fungi and the host plants must communicate at the molecular levels to establish symbiosis. First, the host root produces signaling molecules and releases them into the rhizosphere. Then, these signaling molecules induce the germination of fungal spores and branching of the fungal hyphae and subsequently induce the fungus to produce and release fungal factors responsible for the modifications in the host roots' gene expression.

4.3.1 Establishment of Symbiosis

The AMF spores may germinate without receiving any signal from plants. After germination, the germ tube extends simultaneously with the consumption of triacylglyceride and glycogen reserves to support growth. However, if the fungus fails to find a host root, the hyphal growth will cease to prevent the depletion of spore reserves. This allows the fungus to re-germinate and find a root of the host plant. In contrast, germ tube growth increases significantly near a root, and the hyphae undergo profuse branching, indicating the presence of specific signaling molecules

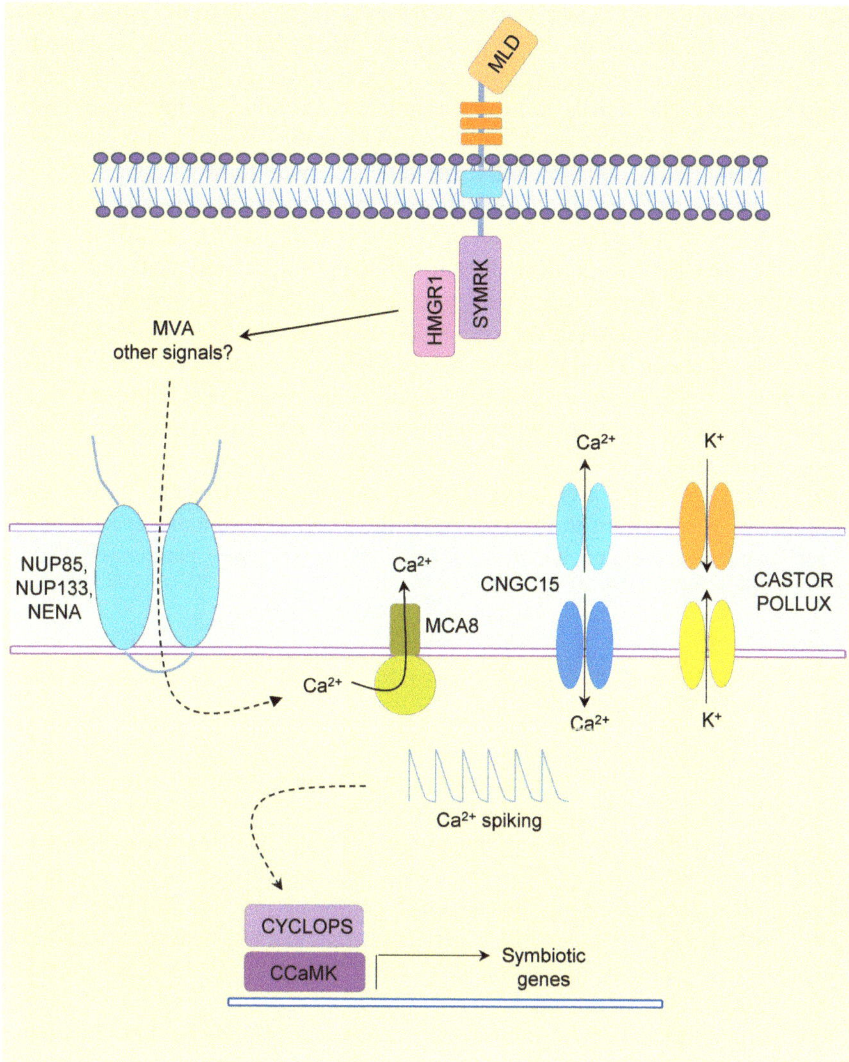

Fig. 4.1 The common signaling pathway. SYMRK is a common receptor complex member involved in the RNS and AMS signaling. The truncated form of SYMRK (the SYMRK version that remains after MLD release, SYMRK-ΔMLD) percepts NFR or MFR signal and produces MVA, as a second messenger, after its interaction with HMGR1. MVA, its derivative, or other unknown signals transduce the message to the nucleus through a nuclear pore complex (NUP85, NUP133, and NENA). Ca^{2+} spiking is initiated through a coordinated function of three transporter proteins, CASTOR/POLLUX (K^+ channels), MCA8 (Ca^{2+}-ATPase), and CNGC15D (Ca^{2+} channel). Ca^{2+} signal activates CCaMK, a Ca calmodulin-dependent protein kinase, leading to the phosphorylation and activation of its downstream target, CYCLOPS. As a transcription factor, CYCLOPS interacts with other transcriptional regulators (mostly RNS and AMS-specific), binds to the upstream elements of the symbiotic gene, and activates their transcription

(e.g., strigolactones (SLs); see below) in the exudate of host roots. This extensive branching, which maximizes the possibility of contact with the root, is accompanied by a significant increase in respiratory activity, which persists until spore reserves are depleted. Indeed, the signals from host plants result in initiating the "presymbiotic growth phase" in that the fungus is committed to starting an association with plant roots (Harrison 2005). Subsequently, the fungal hyphae form an appressoria-like structure named "hyphopodium" that attaches to the root epidermis and acts as the point of entry for the fungus into the root epidermis (Murray et al. 2013). At this phase, the AM hyphae produce fungal factors (Myc factors) that increase the expression of several symbiotic plant genes and Ca^{2+} spiking (Genre et al. 2013).

Following the formation of the hyphopodium, a specific structure called the "pre-penetration apparatus" (PPA) forms in the epidermal and outer root cortical cell. This structure is a broad cytoplasmic bridge that guides the hypha toward the cortical cells (Genre et al. 2008). Fungal hyphae enter the cell and begin to form arbuscules in the inner roots of cortical cells. A plant-derived membrane subsequently surrounds the intracellular hyphae and arbuscules. The "periarbuscular membrane," which separates the arbuscules from the symplasm of the plant cell, contains specific transporters required for the exchange of metabolites between two partners (Balestrini and Bonfante 2005) (Fig. 4.2).

4.3.2 Strigolactones

It has been demonstrated that SLs serve as the initial point of communication between the fungus and the root of the host plants before direct physical contact. SLs are a group of apocarotenoids, the products of oxidative cleavage of carotenoid precursors (Giuliano et al. 2003). They were first identified in the rhizosphere of parasitic plant hosts, allowing their seeds to germinate in close proximity to their hosts (Cook et al. 1966). Later, SLs were identified as the root signals that allow AM fungi to form a symbiotic association with a host (Akiyama et al. 2005; Akiyama and Hayashi 2006).

The natural SLs have a tricyclic lactone structure containing an ABC-ring and a D-ring butenolide group coupled with an enol-ether bridge. Based on the stereochemistry of the B–C-ring junction, the natural SLs are classified into two groups, strigol and orobanchol (canonical SLs). Both of these groups possess a conserved R-configuration at the C-2 position that connects the D-ring to the core and is liable for different bioactivities of different SLs. Conversely, non-canonical SLs generally lack the typical ABC-rings but comprise an enol-ether bridge and D-ring moieties, such as methyl carlactonoate (MeCLA), avenaol, and zealactone (Mashiguchi et al. 2021) (Fig. 4.3).

The pathways of SLs biosynthesis have been investigated, and the enzymes and genes involved have been primarily characterized (Mashiguchi et al. 2021). It has been observed that the carotenoid precursor is subjected to isomerization and

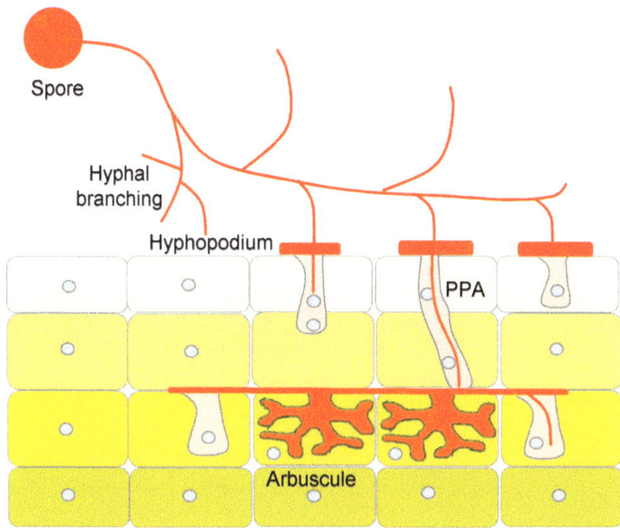

Fig. 4.2 An illustration of the root colonization process by AM fungi. The germination of a resting spore leads to the formation of a short mycelium. The perception of plant exudates induces hyphal branching. Fungi produce and release factors in response to changes in their metabolism. These fungal exudates (Myc factors) are perceived by the root, where they trigger calcium spiking through the activation of the common SYM pathway. In the meantime, a specific fungal structure, "hyphopodium," is formed and adheres to the root surface. This triggers the formation of a particular structure in the epidermal cell outer cortical cell named pre-penetration apparatus (PPA). The intercellular hyphae develop along the root axis, and highly branched arbuscules are formed and occupy the inner cortex cells

cleavage and generates carlactone. The latter compound is the precursor of various SLs (Mori et al. 2020; Wakabayashi et al. 2019, 2020) (Fig. 4.4).

4.3.2.1 Response of AM Fungus to SL

Sub-nanogram levels of SLs in root exudates enhance spore germination and promote AM hyphal branching, most likely by activating lipid metabolic pathways (Lanfranco et al. 2018). SLs detection by AM fungus results in the induction of its oxidative metabolism, ATP production, and generation of the necessary energy for stimulation of growth and hyphal branching and preparation of AM fungus to establish symbiosis (Lanfranco et al. 2018). SLs augment the fungal-derived production of biochemical signals, such as chitin oligomers (COs; see below) (Genre et al. 2013), which induce the Ca^{2+} spiking in plants as the first indication of the interaction between two partners (Bonfante and Genre 2015). Applying COs to plants enhances the expression of an SLs biosynthesis gene (CCD7) and other symbiotic marker genes (Giovannetti et al. 2015), suggesting a synergistic effect of SLs and COs in the establishment of communication between two partners.

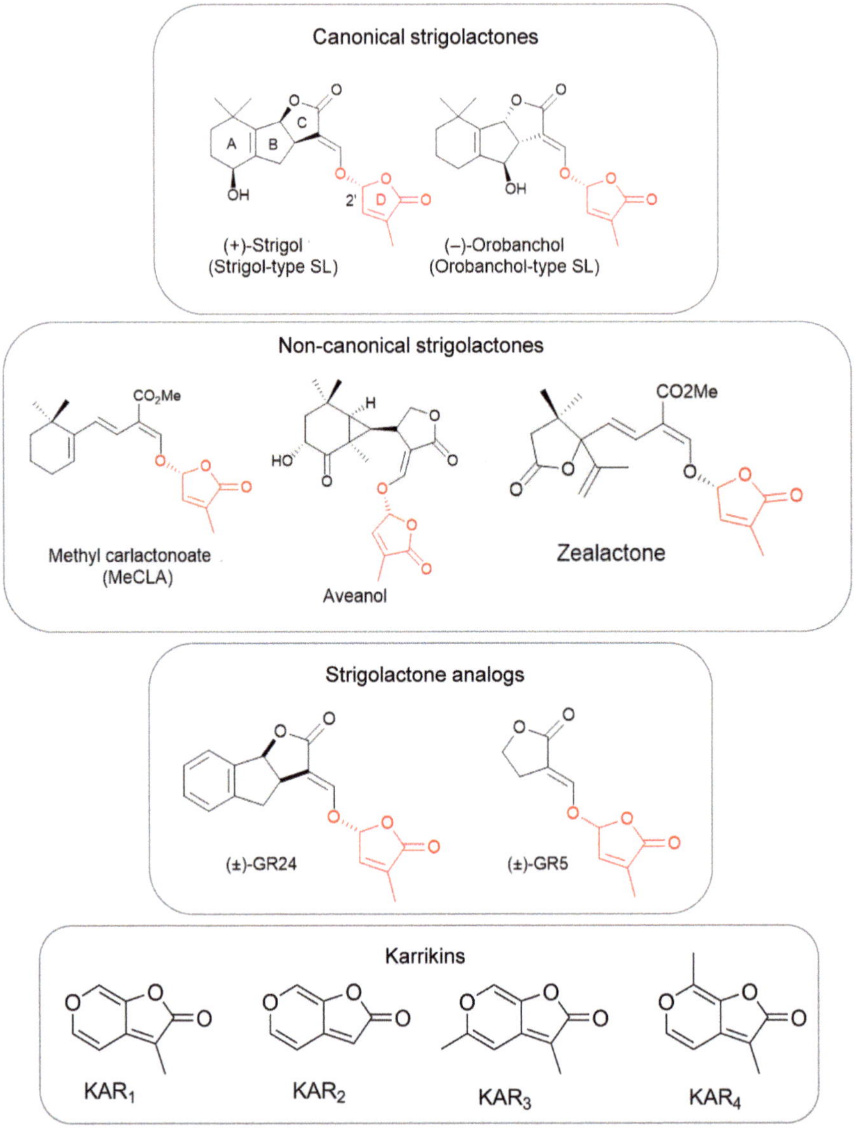

Fig. 4.3 Structures of SLs and SL-related compounds. Naturally occurring SLs are classified into canonical and non-canonical SLs. Canonical SLs possess ABC rings that link to the D-ring via an enol–ether bond. Canonical SLs are further classified into strigol and orobanchol types by the stereochemistry of the B/C-ring junction. In non-canonical SLs, such as methyl carlactonate 3 (MeCLA), avenaol 4, and zealactone 5, the D-ring and an enol–ether bond are conserved, whereas the typical ABC-ring structure is absent. GR24 is a widely used synthetic SL analog. Karrikins (KAR) are smoke-derived chemicals that can induce seed germination of weeds

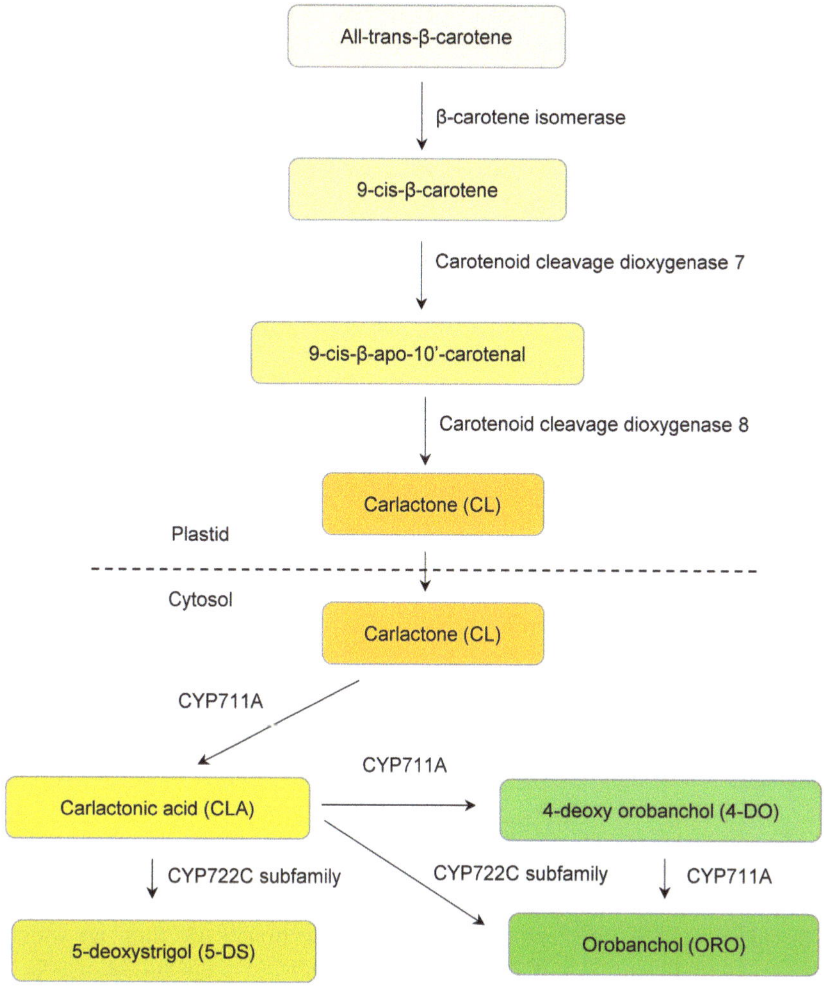

	Arabidopsis	Rice	Pea	Petunia
β-carotene isomerase	AtD27	D27		
Carotenoid cleavage dioxygenase 7 (CCD7)	MAX3	HTD1/D17	RMS5	DAD3
Carotenoid cleavage dioxygenase 8 (CCD8)	MAX4	D10	RMS1	DAD1
CYP711A (A Cytochrome p450 family member)	MAX1	OsMAX1		PhMAX1
CYP722C subfamily	Was found in *Vigna unguiculata* (cowpea); *Solanum lycopersicum* (tomato); *Gossypium arboreum* (cotton); *Lotus japonicas* (Wakabayashi et al., 2019, 2020; Mori et al., 2020)			

Fig. 4.4 The pathways involved in the biosynthesis of SL. The carotenoid precursor undergoes isomerization by plastid-localized β-carotene isomerase (D27) and oxidation by carotenoid cleavage dioxygenases 7 (CCD7) and 8 (CCD8) to produce carlactone (CL) as the precursor of various SLs. CL is further oxidized in the cytosol by the CYP711A family to yield carlactonic acid (CLA). Some members of the CYP711A and CYP722C families can produce orobanchol (ORO), a canonical SL, from CLA, while GaCYP722C and LjCYP722C are responsible for the generation of 5-deoxystrigol, a strigol-type canonical SL (5-DS). The following table details the enzymes of *Arabidopsis*, rice, pea, and petunia

Finally, SLs regulate the expression of several proteins secreted by the fungal partner (Tsuzuki et al. 2016) that positively regulate host plant colonization during presymbiotic and symbiotic stages (see below). Mutations that lead to defects in the biosynthesis and export of SLs reduce AMF hyphal branches and causes significantly lower colonization rates than wild-type plants. After establishing the AM fungus in the root, SL biosynthesis is reduced, presumably as a mechanism to prevent the plant from becoming over-colonized (Koltai et al. 2010; Lanfranco et al. 2018).

The effect of SLs on the AM fungus depends on its concentration and structural features. The structure-activity relationship of SLs has been comprehensively studied for a variety of physiological functions, including as an endogenous hormone in the suppression of shoot branching and as a regulator of plant–plant interactions in the stimulation of parasitic plant seed germination (Boyer et al. 2012; Zwanenburg and Pospíšil 2013; Zwanenburg et al. 2013; Sanchez et al. 2018) or as a hyphal branching agent for AM fungi (Akiyama et al. 2010). Compared to their function in plants, the structural requirements for an optimal effect of SLs on hyphal branching differ. This may be due to the action of distinct receptor molecules in these three primary functional effects of SLs. The function of SLs as endogenous plant hormones requires receptors with α,β-hydrolase activity, which belongs to the D14 clade in higher plants (Waters et al. 2017).

In contrast, its function for seed germination stimulation in parasitic plants depends on its perception by phylogenetically distinct proteins, D14-like receptors (Lumba et al. 2017). The nature of receptor molecules in the AMF is obscure. No homologs of the D14 proteins have been identified within the only available genome of AM fungi belonging to *Rhizophagus irregularis* (Tisserant et al. 2013).

4.3.2.2 SL Signal Perception by AMF

The mechanisms of SLs perception and signal transduction in AM fungi are widely unknown. The synthetic SL (GR24) evokes a rapid increase in the intracellular Ca^{2+} concentration in the fungus (Moscatiello et al. 2014), which is a characteristic response to stress factor (Zhivotovsky and Orrenius 2011), suggesting that AM fungi primarily sense SLs as foreign molecules.

Previous studies suggest that SLs are perceived by both the AM fungi and their associated bacterial communities (Lanfranco et al. 2018). It has been observed that *Candidatus Glomeribacter gigasporarum* (*Ca*Gg), the endobacterium of *G. margarita*, improves the efficiency of the fungus in responding to SLs. In addition, the bacterial scavenging system specifically metabolizes excess ROS generated due to the SLs-mediated increase in fungal respiration. Thus, the fungal microbiota plays a crucial role in the presymbiotic phase of this AM fungus (Salvioli et al. 2016).

4.3.2.3 SLs Signaling Pathway and AMF

Intriguingly, SLs do not appear to act as endogenous signals in plants during AM development, as the development of arbuscules is unaffected in SL-deficient or export mutants (Liao et al. 2018). Furthermore, SL-insensitive D14 rice mutants devoid of the SL receptor do not exhibit a decrease in AM colonization (Yoshida et al. 2012). These results imply that the effect of SL released into the rhizosphere is limited to the early association stage, i.e., the presymbiotic stage.

4.3.3 Karrikins and Karrikin-Like Signals

In contrast to the unaffected AMF colonization in the SL receptor rice mutant *d14*, in *max2/rms4/d3*, another mutant of the SL signaling pathway, AM development is severely reduced (Yoshida et al. 2012). MAX2/RMS4/D3 is an F-box protein that forms a complex with D14 (SL receptor) and a repressor protein, SMXL6/7/8, resulting in its degradation as the initial step in the activation of SL-responsive genes (Lumba et al. 2017; Rehman et al. 2021; Fig. 4.5). This suggests that an element downstream of the SL signaling pathway, and not its receptor, is involved in the AM signaling in plants. Furthermore, D14L (DWARF 14 LIKE) was identified as an essential factor for establishing rice AM association because, in the *d14l* rice mutant, the transcriptional response to AMF is completely absent, implying the role of D14L in AMF recognition (Gutjahr et al. 2015a, b).

Relationship with Karrikins Pathway D14L is homologous to *Arabidopsis* KAI2 (KARRIKIN INSENSITIVE), a receptor for Karrikins, butanolide molecules derived from wildfire smoke and responsible for post-fire germination of seeds (Waters et al. 2012). The Karrikins signaling pathway has been identified in *Arabidopsis* and shares some common elements with SL signaling (De Cuyper et al. 2017; Fig. 4.5). In this pathway, activation of KAI2 leads to the recruitment of MAX2 and the removal of a negative regulator, SMAX1 (Suppressor of MAX2-1) (De Cuyper et al. 2017; Hull et al. 2021). SMAX1 in rice has been identified as an AM association suppressor that negatively regulates root colonization and symbiotic gene transcription (Choi et al. 2020).

This evidence demonstrated a significant overlap between the AM signaling and karrikins pathways; however, the endogenous ligand of D14L/KAI2 that initiates this response is unknown. KAI2 also responds to the N-substituted phthalimides (cotylimides) (Tsuchiya et al. 2010) and non-naturally occurring SLs (Flematti et al. 2016). Since these compounds are not found naturally in plants, it has been hypothesized that D14L/KAI2 recognizes a "yet-to-be-identified karrikin-like" (KL) ligand (Morffy et al. 2016). KAI2 proteins are evolutionarily conserved in the plant kingdom and are also found in plants not associated with fire-prone habitats (Ahmad et al. 2022). It has been suggested that the main function of D14L/KAI2

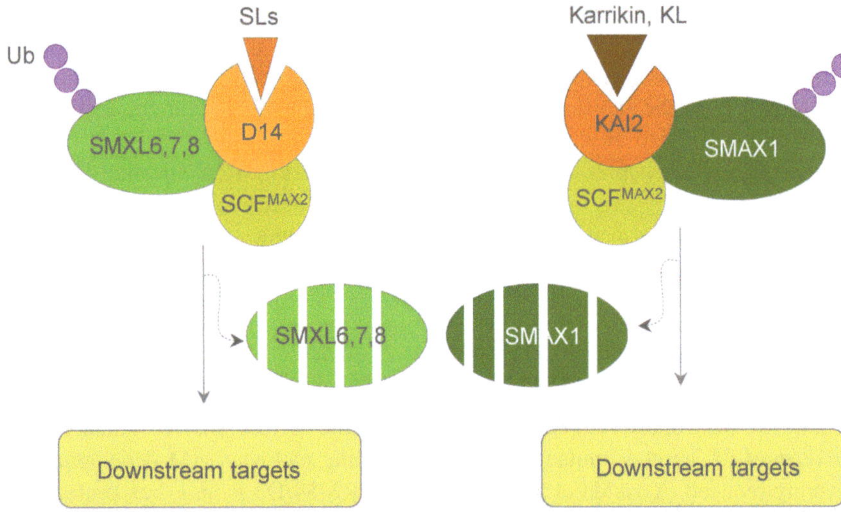

Fig. 4.5 A model of SL signaling and hypothetic KAR signaling. (A) The SL receptor AtD14 binds and hydrolyzes the SL, triggering the formation of a D14–SCF^{MAX2}–SMXL6/7/8 complex, which targets SMXL6/7/8 for ubiquitination and degradation. This leads to the de-repression of unknown TFs and activates the expression of downstream targets. (B) KAR, or a putative KAI2 ligand, is perceived through KAI2. The ligand–receptor interaction triggers the formation of a KAI2–SCF ^{MAX2}–SMAX1 complex to induce the ubiquitination and degradation of SMAX1, which then activates downstream responses

signaling is related to AM presymbiotic signaling, while its effect on post-fire germination is the secondary role (Ho-Plágaro et al. 2021). Consistent with the observations on the disrupted AMF association in SMAX1, LCO responsiveness was eliminated, and CO responsiveness was diminished in *d14l* mutants (see below).

4.3.4 Other Plant Signals

Some other carotenoid-derived metabolites, including blumenols (C13), mycorradicins (C14), and zaxinone, have been shown to be involved in AM association and contribute to the AMF colonization at different stages of this process (Table 4.1).

4.3.5 Plant Hormones

In contrast to a direct role for SLs in the rhizosphere, other phytohormones' role in regulating the structure and function of the AM symbiosis is much more complex. In a mycorrhizal root, both local and systemic responses to the phytohormones occur,

Table 4.1 Some identified (or hypothetical) signaling molecules (other than hormones) from plants with a regulatory role in the AM symbiosis

Potential signaling molecules	Role in AM symbiosis
Blumenols (C_{13} cyclohexenone derivatives)	They are highly accumulated in AM roots; reduced synthesis of blumenols decreases transcript levels of AM markers and increases the number of degenerating arbuscules (Floss et al. 2008a)
Mycorradicins (C14 polyenic dicarboxylic acids)	They are massively accumulated in mycorrhizal roots and are responsible for the characteristic yellow color of strongly colonized roots; mycorradicins contribute to the decay and reemergence of arbuscules, but they are not directly involved in the regulation of the number of active arbuscules (Floss et al. 2008b)
Zaxinone	They are natural apocarotenoid synthesized by a carotenoid cleavage dioxygenase (*zas*) and regulates plant architecture and root growth (Wang et al. 2019). The *zas* mutant displays less colonization than wild-type plants; this phenotype is not rescued by an exogenous supply of zaxinone, while the exogenous synthetic SL analog (GR24) rescues the mycorrhizal phenotype in the *zas* mutant (Votta et al. 2022). *ZAS* orthologs have not been found in genomes of non-host species, including *A. thaliana* (Vallabhaneni et al. 2010). In rice, *OsZAS* is induced during the early stages of the interaction (7 days post-inoculation), and zaxinone content is increased before fungal penetration inside the root (Wang et al. 2019). The *zas* mutant, however, is unable to increase the level of SLs at 7 days post-inoculation with AMF (Votta et al. 2022)
N-acetylglucosamine-like compound) (GlcNAc) (a hypothetical compound)	The plant N-acetylglucosamine exporter, NOPE1 (NO PERCEPTION1), contributes to the excretion of N-acetylglucosamine (GlcNAc)-like compounds, thus influences transcription of genes related to the presymbiotic stage; the root exudates derived from the *nope1* mutant could not induce the transcriptional responses before the physical contact, hyphopodium is not formed, and, subsequently, root penetration by fungal hyphae is inhibited (Nadal et al. 2017)
Coumarins	As a signal in the presymbiotic chemical dialog, it promotes fungal metabolism and stimulates pre-penetration development and metabolism in AMF. Overexpression of genes involved in coumarin production (e.g., scopoletin) and secretion mitigates the incompatibility in the non-host plant, *Arabidopsis* (Cosme et al. 2021)
Flavonoids	These compounds contribute to the signaling, establishment, and regulation of mycorrhizal association in legumes (Singla and Garg 2017), supported by an RNAi silencing study in soybeans (Salloum et al.

(continued)

Table 4.1 (continued)

Potential signaling molecules	Role in AM symbiosis
	2018). Some flavonoids (e.g., apigenin) induce AMF spore germination and hyphal branching (Scervino et al. 2006, 2007). Some flavones (e.g., quercetin and luteolin) improve AMF symbiosis (Steinkellner et al. 2007)
2-hydroxy fatty acids	2-hydroxytetradecanoic acid (2OH-TDA) and 2-hydroxydodecanoic acid (2OH-DDA) induce a hyphal growth response in *Gigaspora gigantean* (Nagahashi and Douds Jr 2011)
Cutin monomers	They are necessary for the formation of hyphopodium and the development of arbuscular because a mutation in the genes responsible for the synthesis of cutin monomers (*ram1* and *ram2*; see below) shows a defect in the formation of these structures (Gobbato et al. 2012; Keymer et al. 2017). Although roots do not contain cutin, the use of cutin compounds as a specific cue for the fungus is likely related to the evolution of AMF in early land plants that have interactions similar to mycorrhizas in their rhizomes as modified stems (Brundrett 2002)

which are necessary to activate the fungus metabolism in the early phase of the interaction but control it and the late stages, i.e., the arbuscule turnover, to guarantee the favorable mutualistic association. As is well-known for other plant developmental processes, phytohormones do not act independently, but a cross-talk between phytohormones regulates AM development and arbuscule formation (Gutjahr 2014; Liao et al. 2018). A summary of information concerning the phytohormones involved in AM symbiosis regulation is provided in Tables 4.2 and 4.3.

4.4 Fungal Signals

SLs have been observed to induce the release of signals by the AM fungus. Due to methodological constraints in applying genetic approaches to AM fungi, the signaling molecules released by AM fungi to establish or regulate symbiosis have not been exhaustively investigated.

4.4.1 Myc-LCOs and COs

The function of lipo-chitooligosaccharide (LCO) as signaling molecules derived from rhizobia (Nod-LCO) in the nodule organogenesis and bacterial colonization has been well documented (Murray 2011). Similarly, branched hyphae of AMF

Table 4.2 The effect of five major phytohormones (auxin, gibberellic acids, cytokinins, abscisic acid, and ethylene) and the mechanisms for their action on the AM establishment and functionality

Major phytohormones	Role in AM symbiosis
Auxin	Auxin acts through numerous mechanisms in regulating AMF symbiosis: (1) Similar to SLs, it regulates early events of the AM association as a part of Pi signaling (Koltai 2015). (2) It controls SLs levels: the *bsh* mutant, with three times less auxin in its roots, shows a significant reduction of SL exudation and low expression level of a key SL synthesis gene (PsCCD8) and defect in AMF colonization; this phenotype is partially restored by the application of GR24 (Foo 2013). (3) It is involved in the post-infection stage of AM symbiosis: exogenous application of auxin analogs stimulates arbuscule formation in the colonized roots (Etemadi et al. 2014); the concentrations of free auxin and auxin conjugates are significantly increased in the mycorrhizal roots (Liao et al. 2015)
Gibberellic acids (GA)	There is evidence of the positive effect of GAs on AM colonization: (1) an increased GA level and upregulation of its biosynthesis and signaling genes is a characteristic of the AM colonized roots (Shaul-Keinan et al. 2002). (2) inhibition of the GA biosynthesis or suppressing of its signaling results in a substantial reduction of hyphal branching and arbuscule formation within the root (Takeda et al. 2015). However, there is evidence showing that GAs act as a negative factor in AM symbiosis: (1) exogenous GAs lead to a substantial reduction of root AM colonization; lower levels of GAs inhibit the arbuscules formation; and higher levels completely inhibit the root colonization (El Ghachtouli et al. 1996). (2) AM colonization and the number of arbuscules are higher in the GA-deficient pea mutant, *na-1* (Foo et al. 2013). (3) Overexpression of the DELLA (Yu et al. 2014) or expression of a non-degradable version of this protein (Floss et al. 2013), a key suppressor in the GA-signaling, results in a significantly increased AM colonization and promotes arbuscule formation. (4) DELLA proteins physically interact with diverse transcriptional regulators in the symbiosis pathway (Pimprikar et al. 2016). (5) GA suppresses Arum-type AM symbiosis but promotes Paris-type AM symbiosis (Tominaga et al. 2020, 2021). These findings suggest the function of mechanisms for precise regulation of the GA biosynthesis and signaling during the establishment of the AM symbiosis
Cytokinins (CK)	Improved levels of CK in both shoots and roots upon mycorrhization have been observed in early studies (Allen et al. 1980), one of the mechanisms for the growth promotion of AM plants. Nevertheless, other studies have observed contradictory results: CK-deficient transgenic tobacco plants (Cosme and Wurst 2013) and CK-overproducing pea mutants (Jones et al. 2015) show higher AM colonization than their wild-type counterparts. Using the root-specific and constitutive expression of CKX (CK oxidase) genes, it has been observed that the shoot CK positively affects AM functioning. At the same time, root CK is responsible for limiting the carbon sink capacity of the fungus to avert fungal parasitism (Cosme et al. 2016)
Abscisic acid (ABA)	ABA treatment induces hyphal branching and promotes fungal spore viability (Mercy et al. 2017). In ABA-deficient mutants, AM colonization, arbuscule formation, and functionality are impaired, suggesting a positive regulation of AM development by ABA (Herrera-Medina et al. 2007). However, ABA seems to modulate the AM symbiosis in a concentration-dependent manner: it stimulates AM colonization at low concentrations. Still, it hampers it at higher levels by impairing the Myc factor-initiated

(continued)

Table 4.2 (continued)

Major phytohormones	Role in AM symbiosis
	symbiotic signaling (Charpentier et al. 2014). In sufficiently AM-colonized roots, the inhibitory effect of ABA observed on the early signaling events could be regarded as a control mechanism for avoiding excess colonization
Ethylene (ET)	ET negatively impacts AM fungal penetration and colonization (Torres Santos et al. 2011). The phenotypes of ET-overproducing and the ET-insensitive mutants of tomato and pea suggest that ET has an inhibitory role in AM colonization (Torres Santos et al. 2011; Foo et al. 2016). However, using mutants with higher sensitivity to ET, it has been shown that ET may mitigate the inhibitory effect of Pi on the AM association (Torres Santos et al. 2016)

Table 4.3 The contribution of three phytohormones (brassinosteroids, salicylic acid, and jasmonic acid) in the AMS

Other phytohormones	Role in AM symbiosis
Brassinosteroids (BRs)	Have a role during AM symbiosis; the tomato biosynthetic mutant (dX) exhibits a lower level of AM colonization (Bitterlich et al. 2014). BR acts likely through inhibition of plant defense or upregulation of invertase (Lin6) (Bitterlich et al. 2014), which is involved in the provision of hexoses to the fungus (Schaarschmidt et al. 2006)
Salicylic acid (SA)	Exogenous application of SA decreases the root colonization at the beginning of the AMF–plant interaction but fails to affect the formation of appressoria (Blilou et al. 2000). The AM-defective (Myc^-) mutants show higher SA accumulation compared with wild (Myc^+) plants (Blilou et al. 1999). The transgenic plant with reduced SA levels is colonized rapidly by AM colonization, while mutants with improved SA levels show a significant delay in AM colonization. However, the ultimate colonization level is not significantly modified (Medina et al. 2003)
Jasmonic acid (JA)	There are reports on both positive and negative effects of JA on AM colonization: JA-deficient tomato *spr2* mutant exhibits a lower colonization rate that could be restored by exogenous application of JA (Tejeda-Sartorius et al. 2008); conversely, an elevated level of AM colonization has been observed in the tomato JA-insensitive mutant *jai*-1 (Herrera-Medina et al. 2008). JA response partially depends on the plant and fungal species (Liao et al. 2018) and is not essential for AM colonization. Still, high levels of JA in the roots have an inhibitory effect on mycorrhization, potentially through the activation of the plant's defense (Gutjahr et al. 2015a, b)

secrete specific signaling molecules, i.e., Myc factors, lipo-chitooligosaccharides (LCOs), and short-chain chitooligosaccharides (CO4/CO5).

Identifying LCOs from AM fungi with a basic structure similar to the Nod-LCO demonstrated that the establishment AMF also implicates LCO-mediated signaling, named Myc-LCOs (Maillet et al. 2011). A generic symbiotic LCO is based on a linear, β(1–4) linked oligomers (tetra- or pentasaccharide) of N-acetyl glucosamine (GlcNAc) with a considerable variation in the N-substitutions (acyl and methyl) and

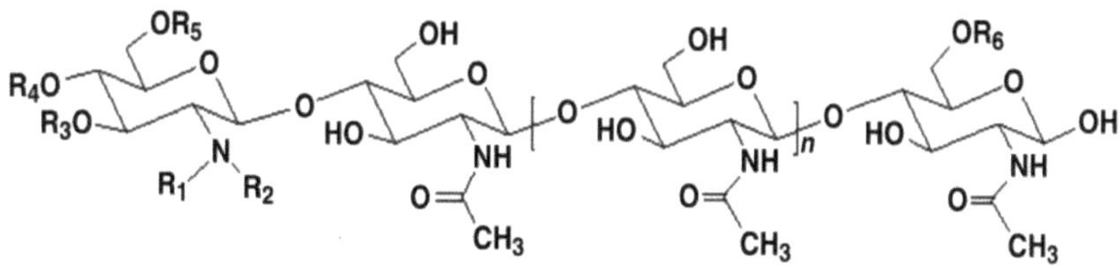

Fungus species	n	R_1	R_2	$R_{3, 4, 5}$	R_6
Rhizophagus irregularis	0, 1, 2	C16:0, C18:0, C18:1	H, Me	H	H, S, Fuc, MeFuc
Rhizophagus intraradices	0, 1, 2	C16:0, C18:0, C18:1	H, Me	H	H, MeFuc
Rhizophagus clarus	0, 1, 2	C16:0, C18:0, C18:1	H	H	H, Fuc, MeFuc
Gigaspora rosea	0, 1, 2	C16:0, C16:1, C18:0, C18:1, C18:2, C20:1	H, Me	H	H, S, Fuc, MeFuc, MeFucS

Fig. 4.6 The generic structure of lipo-chitooligosaccharides (LCOs) shows sites of chemical substitutions. (n) denotes the number of residues of chitin oligomers, (R1) represents the type of fatty acid identified as saturated or unsaturated fatty acid, and (R2–R6) are chemical substitutions: hydrogen (H), acetyl (Ac), carbamoyl (Cb), fucosyl (Fuc), fucosyl sulfate (FucS), methylfucosyl (MeFuc), and sulfate (S)

O-substitutions (methyl, carbamoyl, acetyl, fucosyl, and sulfate) (Gough and Cullimore 2011; Rush et al. 2020) (Fig. 4.6). Exogenously applied Myc-LCOs enhance root colonization and activate Ca^{2+} spiking in the host plants (Sun et al. 2015; Camps et al. 2015). Short-chain chitooligosaccharides (COs) are also able to activate Ca^{2+} spiking, implying that both LCOs and short-chain COs contribute to the recognition of host roots (Genre et al. 2013).

4.4.2 *Plant Perception of the Fungal Signal*

Plant receptors of LCOs released by rhizobia are characterized as lysin motif receptor-like kinases (LysM-RLKs) (Fliegmann et al. 2013). Some Lys-RLKs are necessary for AM colonization and, thus, are the candidate receptors of Myc-COs or Myc-LCOs in various plant species (Buendia et al. 2018; Wu et al. 2022).

CERK1 LysM-RLKs receptors, such as OsCERK1 and OsLYK2 from rice, MtLYK9 from *M. truncatula*, PsLYK9 from *Pisum sativum*, and SlLY10 and SlLYK12 from tomato, are mediated in the COs and LCOs signaling (Ho-Plágaro and García-Garrido 2022). This suggests that the contribution of LysM-RLKs receptors is a conserved feature in AM association of host plants. The rice OsCERK1 (Chitin-elicitor receptor kinase 1) is required for COs-induced responses (Zhang et al. 2015), and the AM association is severely impaired in the *Oscerk1* mutant (Miyata et al. 2014).

CEBiP Chitin is a constituent of the fungal cell wall, a long-chain polymer of GlcNAc. The secreted chitinases from plants break down chitin and release COs (Roberts and Selitrennikoff 1988). Thus, it is important to maintain the immunity response, simultaneous with employing strategies for plant-microbe symbioses. In this process, different classes of LysM receptor kinases (LYKs) or different combinations of single receptors could be used to discriminate different GlcNAc molecules and determine the ultimate response: immunity or symbiosis (Miyata et al. 2014). The LysM protein OsCEBiP (Chitin Elicitor binding protein) collaborates with rice OsCERK1 to regulate chitin signaling in this species (Shimizu et al. 2010). Indeed, in addition to AM signaling, CERK1 is also involved in MAMP-triggered immunity (Miyata et al. 2014). This gene was primarily characterized as a receptor necessary for chitin elicitor signaling, and the *Arabidopsis* knockout mutant of AtCERK1 is unable to respond to chitin (Miya et al. 2007).

Contrary to the impaired mycorrhizal phenotype of rice *Oscerk1* mutant, however, AM response is normal in the *Oscebip* mutant (Miyata et al. 2014), suggesting the involvement of CEBiP only in the plant immune response.

MYR1 In rice, OsCERK1 does not seem to bind to CO4 (Chitotetraose) directly (Kaku et al. 2006), but another lysin motif (LysM)-containing receptor kinase (LYKs), OsMYR1, is involved in the perception of the AM signal, CO4 (He et al. 2019). The *Osmyr1-1/Oslyk2-1* mutant shows a decreased AMF colonization, lower level of Ca^{2+} spiking, and reduced transcription of marker genes of AMS compared to wide-type rice plants upon inoculation with *Rhizophagus irregularis* (Zhang et al. 2021). Evidence suggests that OsMYR1 binds to CO4 from symbiotic fungi and subsequently is associated with OsCERK1. Further dimerization and phosphorylation between OsMYR1 and OsCERK1 trigger the symbiosis signaling pathway (He et al. 2019). Indeed, CERK1 is a common receptor of the AMS and immune response pathway (Zhang et al. 2015; Gibelin-Viala et al. 2019). Such dual function for CERK1 depends on its specific interaction with its coreceptors, OsCERK1 or OsMYR1, in response to either pathogenic or symbiotic signals, respectively (Zhang et al. 2021). The dual function of OsCERK1 homologs in both symbiosis and immunity was also observed in other plant species (Gibelin-Viala et al. 2019; Leppyanen et al. 2017).

Collectively, long-chain COs (CO6, CO7, and CO8) are recognized by OsCEBiP and trigger immunity by the formation of the OsCERK1–OsCEBiP complex, whereas short-chain COs (CO4 or CO5) are sensed by OsMYR1 and trigger symbiotic signaling after the formation of OsCERK1–OsMYR1 complex (Liang et al. 2013) (Fig. 4.7).

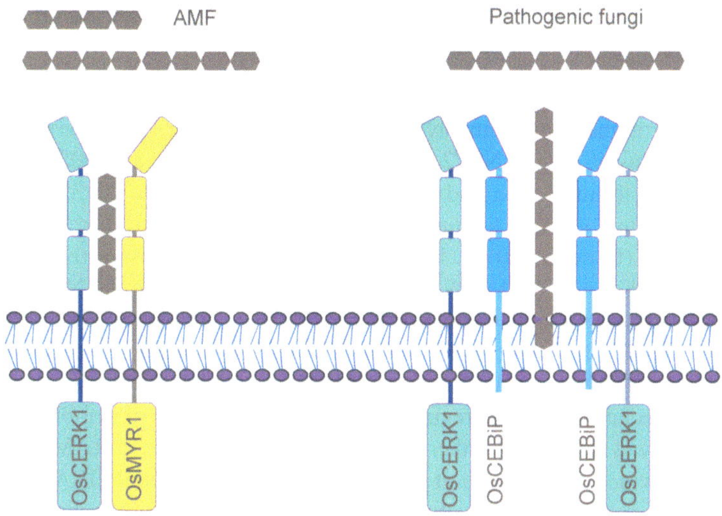

Fig. 4.7 The discrimination of symbiosis and immunity signals through the formation of a specific combination of three different receptors and coreceptors. OsCERK1 acts as a common receptor for both defense and symbiosis pathways that can trigger either of two contrasting signal outputs depending on the context. Symbiotic receptor OsMYR1 binds CO4 (chitotetraose) and subsequently forms a complex with OsCERK1, while long-chain chitins (CO8) bind the MAMP receptor OsCEBiP and then, after the formation of a complex with OsCERK1, triggers immunity response. In the presence of AM fungus, OsCERK1 is mainly allocated to CO4; thus, depletion of OsCERK1 prevents the formation of OsCERK1–OsCEBiP and suppresses immune signaling

4.5 AMF-Activated Genes

After receiving the necessary signals from the fungus and the host plant, the signaling pathway is started, and a wide range of genes are activated as downstream targets to regulate the physiological responses of the plant to the symbiont. A large number of transcription factors and other regulatory proteins are involved in the downstream AMS-specific pathways that are activated in response to CSP induction. Some of the most important molecular players in the AMS are discussed below.

RAM1, RAM2 In searching for molecules specific to the AM signaling pathway, two genes were found, including RAM1 (Reduced Arbuscular Mycorrhization 1) and RAM2 (Wang et al. 2012). Mutations in these genes (*ram1* and *ram2*) are able to form root nodules and also induce hyphal branching in the AM fungi, indicating unaffected strigolactone synthesis in these mutants. However, the root colonization level is severely decreased and associated with a reduction of hyphopodia at the root surface. In addition, *ram2* plants displayed a severe defect in arbuscule formation (Gobbato et al. 2013). Complementation experiments demonstrated that RAM1 encodes a GRAS-type transcription factor responsible for RAM2 expression (Gobbato et al. 2013). RAM2 is a glycerol-3-phosphate acyl transferase (GPAT), which contributes to the synthesis of cutin monomers. The overexpression of RAM2 leads to higher levels of α,ω-dicarboxylic acids and ω-hydroxy fatty acids (Gobbato et al. 2012).

NSP1 and NSP2 In the nodulation pathway, two GRAS-type transcription factors downstream of the Sym pathway have been identified, i.e., NSP1 (Nodulation Signaling Pathway 1) and NSP2 (Kaló et al. 2005; Smit et al. 2005), which are required for both nodulation and mycorrhization (Delaux et al. 2013). Both NSP1 and NSP2 are involved in activating strigolactone biosynthesis through the induction of DWARF27 (Liu et al. 2011). In barley, NSP2 overexpression activates RLK10/NFR5, SYMRK, and CYCLOPS (Li et al. 2022). Interestingly, the NSP2 complex with NSP1 activates strigolactone production (Liu et al. 2011), while the complex of NSP2 with RAM1 leads to the expression of RAM2 responsible for cutin monomer synthesis. Thus, a competition between RAM1 and NSP1 for binding with NSP2 provides a mechanism for the regulation of two different sets of symbiosis-specific genes (Murray et al. 2013). Further evidence on the importance of RAM1 and NSP1 has been provided by Hohnjec et al. (2015), showing that both GRAS-type transcription factors act synergistically in the transduction of diffusible signals and are essential for the presymbiotic transcriptional reprogramming triggered by Myc-LCOs, downstream of the CSP (Hohnjec et al. 2015) (Fig. 4.8).

Given the central role of AMS in P nutrition for plants, a link between Pi deficiency responses and the AMF signaling pathway has been observed in various plant species (Shi et al. 2021; Das et al. 2022). The transcription factors PHR1 and PHR2 are master regulators of the P-starvation response (Sega and Pacak 2019). The PHR2-controlled plant phosphate starvation response is required for pre-contact signaling, gene expression, root colonization, and mycorrhizal phosphate uptake (Shi et al. 2021; Das et al. 2022). Under P-starvation conditions, PHR2 promotes the expression of RAM1 (Shi et al. 2021) and NSP2 (Das et al. 2022), activating the biosynthetic pathway of cutin monomers and SLs, respectively.

4.6 AMF Association and Plant Immune Response

Every organism acts as a non-self-cue and evokes a response in plants. In general, to distinguish these cues and responding appropriately, plants are able to recognize microbe-associated molecular patterns (MAMPs) from pathogen-associated molecular patterns (PAMPs). Upon activation of the specific immunity response for each type of microorganism, the corresponding signaling cascades are initiated and induce the expression of related defense genes leading to the release of chitinases and accumulation of reactive oxygen species (van der Burgh and Joosten 2019). Microorganisms have developed mechanisms to evade these responses, suppress host immunity, and manipulate host cell physiology (Wang et al. 2022). The "effector proteins" promote the colonization of the host by controlling the plant immune system (Plett and Martin 2018). "Pathogen effector proteins" allow successful infection by suppressing the host defense response (Kamoun 2006).

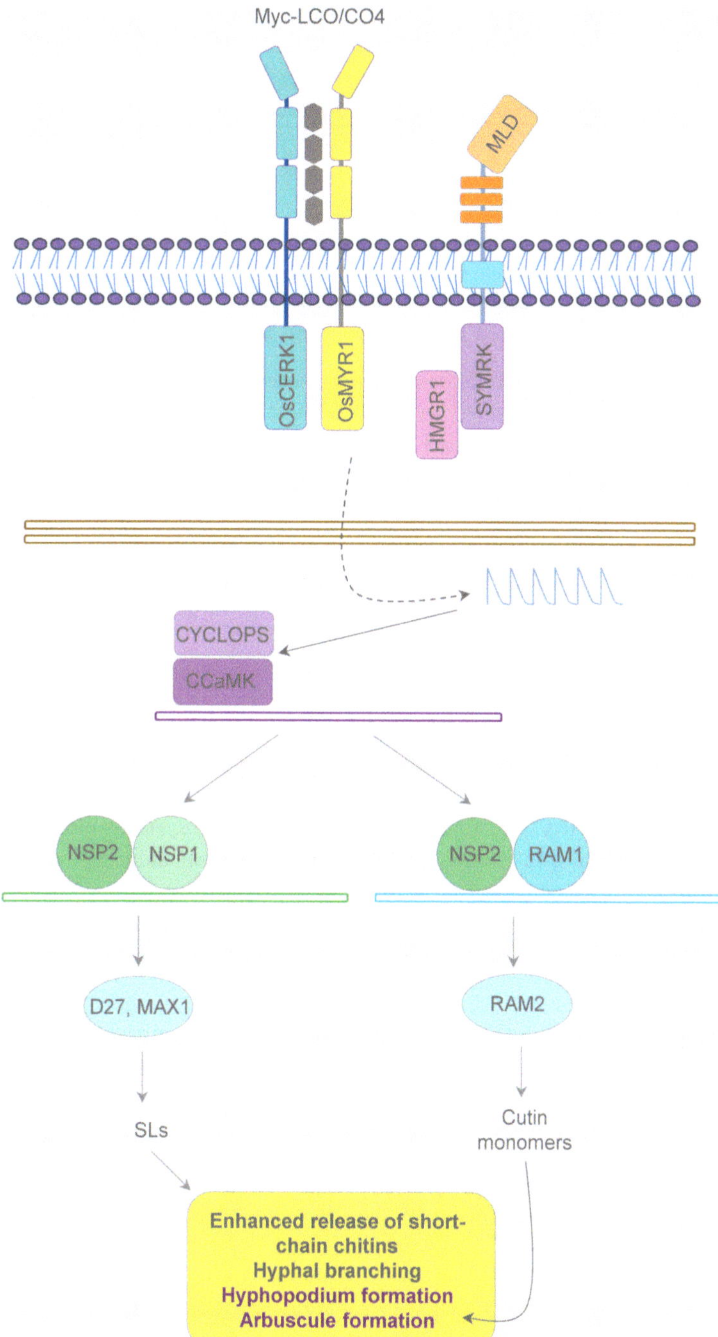

Fig. 4.8 Mycorrhiza-specific signaling. Specific Lys-RLKs detect AM-released Myc factors by forming a heterodimer (OsMYR1/OsCERK1) that leads to the induction of the common symbiosis pathway (CSP). The CSP activates the GRAS transcription factors NSP1, NSP2, and RAM1. NSP1 and NSP2 are involved in elevating strigolactone levels by activating its two biosynthetic genes,

In plant symbiosis, colonization of the host root while avoiding its defense responses is an essential prerequisite for establishing an association. Similar to "pathogen effector proteins," "symbiotic effectors" control the plant immune system, allow successful infection, and promote plant colonization (Kloppholz et al. 2011). Such similarity in the response of plants during infection by symbiotic and pathogenic microorganisms suggests that suppression of plant defense is a conserved feature in plant–microbe interactions.

The secreted effector proteins (SSPs) are involved in subduing the plant defense systems and, thus, facilitating fungal entry into plant cells. Among the predicted SSPs identified in sequencing projects of AM fungi, only a small number of proteins were confirmed to be involved in the AMF–plant interaction.

SP7 SP7 (RP23081 and RP8598) are small, secreted proteins from *Rhizophagus irregularis* that are translocated to the host plant nucleus and interact with the pathogenesis-related transcription factor of plant origin, ETHYLENE-RESPONSIVE FACTOR 19 (ERF19). By counteracting the expression *of ERF19*, SP7 can promote AMF symbiosis (Kloppholz et al. 2011).

SIS1 Another secreted protein, SL-induced putative secreted protein 1 (SIS1) (RP5293), is upregulated in *R. irregularis* and has a role in the colonization of host root (Tsuzuki et al. 2016).

RiCRN1 The third AMF effector, RiCRN1, belongs to the significant pathogen-associated Crinkler (CRN) effector family. CRN effectors are widespread in plant-pathogenic oomycetes and contain an N-terminal motif (LXLFLAK) essential for the effector's intracellular localization. In plant pathogenic oomycetes, CRN enters the plant cell nucleus to exert their function, such as induction of plant cell death (Amaro et al. 2017). RiCRN1 found in *Rhizophagus irregularis* also localizes to the host plant nucleus but, in contrast to other CRN, does not induce plant cell death (Voss et al. 2018). Gene silencing of RiCRN1 (through Host-Induced Gene Silencing, HIGS) results in much smaller arbuscules demonstrating that RiCRN1 expression facilitates AMS and is necessary for symbiosis progression (Voss et al. 2018).

In summary, AMF effector proteins promote symbiosis by impairing the synthesis of plant proteins that are produced upon contact with chitin or its derivatives and are involved in defense, cell death, and immune responses. Despite the lack of host specificity in plant–AMF interactions, there is evidence that fungal SSPs may contribute to host specificity and are likely responsible for variation in symbiosis efficiency among different combinations of AMF species/lines and host plant species/genotypes (Prasad Singh et al. 2019).

Fig. 4.8 (continued) D27 and MAX1. In addition to hyphal branching, the production of short-chain chitin oligomers (CO3-6) by AMF is elevated by SLs leading to the activation of CSP in the initial stages of fungal root colonization. RAM2 expression depends on the formation of a complex between NSP2 and RAM1. Cutin monomers produced by RAM2 promote hyphopodia and arbuscule formation by AMF

4.7 Conclusions

Although our knowledge of signaling pathways in response to AMF has advanced significantly in recent years, several questions remain regarding the nature of the Karrikin-like compound and its biosynthesis, the cause and consequences of the loss of symbiosis ability in non-host plants, and the mechanisms underlying the differential efficiency of specific combinations of plant species/genotypes and AMF species/isolates. However, unlike the legume nodule symbiosis, the diversity of host plants for AMS enables researchers to study and compare plant orders and families for the evolution of signaling pathways and other molecular components, as well as to investigate the diversity and evolutionary changes of symbiotic interactions and their ecological significance during the evolution of terrestrial plants on Earth.

References

Ahmad B, Qadir SU, Dar TA, Alam P, Yousuf PY, Ahmad P (2022) Karrikins: smoke-derived phytohormones from stress alleviation to signaling. J Plant Growth Regul 10:1–3
Akiyama K, Hayashi H (2006) Strigolactones: chemical signals for fungal symbionts and parasitic weeds in plant roots. Ann Bot 97(6):925–931
Akiyama K, Matsuzaki KI, Hayashi H (2005) Plant sesquiterpenes induce hyphal branching in arbuscular mycorrhizal fungi. Nature 435(7043):824–827
Akiyama K, Ogasawara S, Ito S, Hayashi H (2010) Structural requirements of strigolactones for hyphal branching in AM fungi. Plant Cell Physiol 51(7):1104–1117
Allen MF, Moore TS Jr, Christensen M (1980) Phytohormone changes in Bouteloua gracilis infected by vesicular–arbuscular mycorrhizae: I. Cytokinin increases in the host plant. Can J Bot 58(3):371–374
Amaro TM, Thilliez GJ, Motion GB, Huitema E (2017) A perspective on CRN proteins in the genomics age: evolution, classification, delivery and function revisited. Front Plant Sci 8:99
Antolín-Llovera M, Petutsching EK, Ried MK, Lipka V, Nürnberger T, Robatzek S, Parniske M (2014) Knowing your friends and foes–plant receptor-like kinases as initiators of symbiosis or defence. New Phytol 204(4):791–802
Balestrini R, Bonfante P (2005) The interface compartment in arbuscular mycorrhizae: a special type of plant cell wall? Plant Biosyst 139(1):8–15
Bitterlich M, Krügel U, Boldt-Burisch K, Franken P, Kühn C (2014) The sucrose transporter Sl SUT 2 from tomato interacts with brassinosteroid functioning and affects arbuscular mycorrhiza formation. Plant J 78(5):877–889
Blilou I, Ocampo JA, García-Garrido JM (1999) Resistance of pea roots to endomycorrhizal fungus or Rhizobium correlates with enhanced levels of endogenous salicylic acid. J Exp Bot 50(340): 1663–1668
Blilou I, Ocampo JA, García-Garrido JM (2000) Induction of Ltp (lipid transfer protein) and Pal (phenylalanine ammonia-lyase) gene expression in rice roots colonized by the arbuscular mycorrhizal fungus Glomus mosseae. J Exp Bot 51(353):1969–1977
Bonfante P, Genre A (2015) Arbuscular mycorrhizal dialogues: do you speak 'plantish' or 'fungish'? Trends Plant Sci 20(3):150–154
Boyer FD, de Saint Germain A, Pillot JP, Pouvreau JB, Chen VX, Ramos S, Stévenin A, Simier P, Delavault P, Beau JM, Rameau C (2012) Structure-activity relationship studies of strigolactone-

related molecules for branching inhibition in garden pea: molecule design for shoot branching. Plant Physiol 159(4):1524–1544

Brundrett MC (2002) Coevolution of roots and mycorrhizas of land plants. New Phytol 154(2): 275–304

Buendia L, Girardin A, Wang T, Cottret L, Lefebvre B (2018) LysM receptor-like kinase and LysM receptor-like protein families: an update on phylogeny and functional characterization. Front Plant Sci 9:1531

Camps C, Jardinaud MF, Rengel D, Carrère S, Hervé C, Debellé F, Gamas P, Bensmihen S, Gough C (2015) Combined genetic and transcriptomic analysis reveals three major signalling pathways activated by Myc-LCO s in Medicago truncatula. New Phytol 208(1):224–240

Capoen W, Sun J, Wysham D, Otegui MS, Venkateshwaran M, Hirsch S, Miwa H, Downie JA, Morris RJ, Ané JM, Oldroyd GE (2011) Nuclear membranes control symbiotic calcium signaling of legumes. Proc Natl Acad Sci 108(34):14348–14353

Charpentier M (2018) Calcium signals in the plant nucleus: origin and function. J Exp Bot 69(17): 4165–4173

Charpentier M, Bredemeier R, Wanner G, Takeda N, Schleiff E, Parniske M (2008) Lotus japonicus CASTOR and POLLUX are ion channels essential for perinuclear calcium spiking in legume root endosymbiosis. Plant Cell 20(12):3467–3479

Charpentier M, Sun J, Wen J, Mysore KS, Oldroyd GE (2014) Abscisic acid promotion of arbuscular mycorrhizal colonization requires a component of the PROTEIN PHOSPHATASE 2A complex. Plant Physiol 166(4):2077–2090

Charpentier M, Sun J, Martins TV, Radhakrishnan GV, Findlay K, Soumpourou E, Thouin J, Véry AA, Sanders D, Morris RJ, Oldroyd GE (2016) Nuclear-localized cyclic nucleotide–gated channels mediate symbiotic calcium oscillations. Science 352(6289):1102–1105

Chiu CH, Paszkowski U (2020) Receptor-like kinases sustain symbiotic scrutiny. Plant Physiol 182(4):1597–1612

Choi J, Lee T, Cho J, Servante EK, Pucker B, Summers W, Bowden S, Rahimi M, An K, An G, Bouwmeester HJ (2020) The negative regulator SMAX1 controls mycorrhizal symbiosis and strigolactone biosynthesis in rice. Nat Commun 11(1):1–3

Cook CE, Whichard LP, Turner B, Wall ME, Egley GH (1966) Germination of witchweed (Striga lutea Lour.): isolation and properties of a potent stimulant. Science 154(3753):1189–1190

Cosme M, Wurst S (2013) Interactions between arbuscular mycorrhizal fungi, rhizobacteria, soil phosphorus and plant cytokinin deficiency change the root morphology, yield and quality of tobacco. Soil Biol Biochem 57:436–443

Cosme M, Ramireddy E, Franken P, Schmülling T, Wurst S (2016) Shoot-and root-borne cytokinin influences arbuscular mycorrhizal symbiosis. Mycorrhiza 26(7):709–720

Cosme M, Fernández I, Declerck S, van der Heijden MG, Pieterse CM (2021) A coumarin exudation pathway mitigates arbuscular mycorrhizal incompatibility in Arabidopsis thaliana. Plant Mol Biol 106(4):319–334

Das D, Paries M, Hobecker K, Gigl M, Dawid C, Lam HM, Zhang J, Chen M, Gutjahr C (2022) Phosphate starvation response transcription factors enable arbuscular mycorrhiza symbiosis. Nat Commun 13(1):477

De Cuyper C, Struk S, Braem L, Gevaert K, De Jaeger G, Goormachtig S (2017) Strigolactones, karrikins and beyond. Plant Cell Environ 40(9):1691–1703

Delaux PM, Bécard G, Combier JP (2013) NSP 1 is a component of the Myc signaling pathway. New Phytol 199(1):59–65

El Ghachtouli N, Martin-Tanguy J, Paynot M, Gianinazzi S (1996) First-report of the inhibition of arbuscular mycorrhizal infection of Pisum sativum by specific and irreversible inhibition of polyamine biosynthesis or by gibberellic acid treatment. FEBS Lett 385(3):189–192

Etemadi M, Gutjahr C, Couzigou J-M, Zouine M, Lauressergues D, Timmers A, Audran C, Bouzayen M, Bécard G, Combier J-P (2014) Auxin perception is required for arbuscule development in arbuscular mycorrhizal symbiosis. Plant Physiol 166:281–292

Flematti GR, Scaffidi A, Waters MT, Smith SM (2016) Stereospecificity in strigolactone biosynthesis and perception. Planta 243(6):1361–1373
Fliegmann J, Canova S, Lachaud C, Uhlenbroich S, Gasciolli V, Pichereaux C, Rossignol M, Rosenberg C, Cumener M, Pitorre D, Lefebvre B (2013) Lipo-chitooligosaccharidic symbiotic signals are recognized by LysM receptor-like kinase LYR3 in the legume Medicago truncatula. ACS Chem Biol 8(9):1900–1906
Floss DS, Hause B, Lange PR, Küster H, Strack D, Walter MH (2008a) Knock-down of the MEP pathway isogene 1-deoxy-D-xylulose 5-phosphate synthase 2 inhibits formation of arbuscular mycorrhiza-induced apocarotenoids, and abolishes normal expression of mycorrhiza-specific plant marker genes. Plant J 56(1):86–100
Floss DS, Schliemann W, Schmidt J, Strack D, Walter MH (2008b) RNA interference-mediated repression of MtCCD1 in mycorrhizal roots of Medicago truncatula causes accumulation of C27 apocarotenoids, shedding light on the functional role of CCD1. Plant Physiol 148(3):1267–1282
Floss DS, Levy JG, Lévesque-Tremblay V, Pumplin N, Harrison MJ (2013) DELLA proteins regulate arbuscule formation in arbuscular mycorrhizal symbiosis. Proc Natl Acad Sci 110(51): E5025–E5034
Foo E (2013) Auxin influences strigolactones in pea mycorrhizal symbiosis. J Plant Physiol 170: 523–528
Foo E, Ross JJ, Jones WT, Reid JB (2013) Plant hormones in arbuscular mycorrhizal symbioses: an emerging role for gibberellins. Ann Bot 111(5):769–779
Foo E, McAdam EL, Weller JL, Reid JB (2016) Interactions between ethylene, gibberellins, and brassinosteroids in the development of rhizobial and mycorrhizal symbioses of pea. J Exp Bot 67(8):2413–2424
Genre A, Chabaud M, Faccio A, Barker DG, Bonfante P (2008) Prepenetration apparatus assembly precedes and predicts the colonization patterns of arbuscular mycorrhizal fungi within the root cortex of both Medicago truncatula and Daucus carota. Plant Cell 20(5):1407–1420
Genre A, Chabaud M, Balzergue C, Puech-Pagès V, Novero M, Rey T, Fournier J, Rochange S, Bécard G, Bonfante P, Barker DG (2013) Short-chain chitin oligomers from arbuscular mycorrhizal fungi trigger nuclear Ca2+ spiking in M edicago truncatula roots and their production is enhanced by strigolactone. New Phytol 198(1):190–202
Gibelin-Viala C, Amblard E, Puech-Pages V, Bonhomme M, Garcia M, Bascaules-Bedin A, Fliegmann J, Wen J, Mysore KS, le Signor C, Jacquet C (2019) The Medicago truncatula LysM receptor-like kinase LYK9 plays a dual role in immunity and the arbuscular mycorrhizal symbiosis. New Phytol 223(3):1516–1529
Giovannetti M, Mari A, Novero M, Bonfante P (2015) Early Lotus japonicus root transcriptomic responses to symbiotic and pathogenic fungal exudates. Front Plant Sci 6:480
Giuliano G, Al-Babili S, Von Lintig J (2003) Carotenoid oxygenases: cleave it or leave it. Trends Plant Sci 8(4):145–149
Gobbato E, Marsh JF, Vernié T, Wang E, Maillet F, Kim J, Miller JB, Sun J, Bano SA, Ratet P, Mysore KS (2012) A GRAS-type transcription factor with a specific function in mycorrhizal signaling. Curr Biol 22(23):2236–2241
Gobbato E, Wang E, Higgins G, Bano SA, Henry C, Schultze M, Oldroyd GE (2013) RAM1 and RAM2 function and expression during arbuscular mycorrhizal symbiosis and Aphanomyces euteiches colonization. Plant Signal Behav 8(10):e26049
Gough C, Cullimore J (2011) Lipo-chitooligosaccharide signaling in endosymbiotic plant-microbe interactions. Mol Plant-Microbe Interact 24(8):867–878
Groth M, Takeda N, Perry J, Uchida H, Dräxl S, Brachmann A, Sato S, Tabata S, Kawaguchi M, Wang TL, Parniske M (2010) NENA, a Lotus japonicus homolog of Sec13, is required for rhizodermal infection by arbuscular mycorrhiza fungi and rhizobia but dispensable for cortical endosymbiotic development. Plant Cell 22(7):2509–2526
Gutjahr C (2014) Phytohormone signaling in arbuscular mycorrhiza development. Curr Opin Plant Biol 20:26–34

Gutjahr C, Gobbato E, Choi J, Riemann M, Johnston MG, Summers W, Carbonnel S, Mansfield C, Yang SY, Nadal M, Acosta I (2015a) Rice perception of symbiotic arbuscular mycorrhizal fungi requires the karrikin receptor complex. Science 350(6267):1521–1524

Gutjahr C, Siegler H, Haga K, Iino M, Paszkowski U (2015b) Full establishment of arbuscular mycorrhizal symbiosis in rice occurs independently of enzymatic jasmonate biosynthesis. PLoS One 10(4):e0123422

Harrison MJ (2005) Signaling in the arbuscular mycorrhizal symbiosis. Annu Rev Microbiol 59: 19–42

He J, Zhang C, Dai H, Liu H, Zhang X, Yang J, Chen X, Zhu Y, Wang D, Qi X, Li W (2019) A LysM receptor heteromer mediates perception of arbuscular mycorrhizal symbiotic signal in rice. Mol Plant 12(12):1561–1576

Herrera-Medina MJ, Steinkellner S, Vierheilig H, Ocampo Bote JA, Garcia GJ (2007) Abscisic acid determines arbuscule development and functionality in the tomato arbuscular mycorrhiza. New Phytol 175(3):554–564

Herrera-Medina MJ, Tamayo MI, Vierheilig H, Ocampo JA, García-Garrido JM (2008) The jasmonic acid signalling pathway restricts the development of the arbuscular mycorrhizal association in tomato. J Plant Growth Regul 27(3):221–230

Hohnjec N, Czaja-Hasse LF, Hogekamp C, Küster H (2015) Pre-announcement of symbiotic guests: transcriptional reprogramming by mycorrhizal lipochitooligosaccharides shows a strict co-dependency on the GRAS transcription factors NSP1 and RAM1. BMC Genomics 16:1–6

Ho-Plágaro T, García-Garrido JM (2022) Molecular regulation of arbuscular mycorrhizal symbiosis. Int J Mol Sci 23(11):5960

Ho-Plágaro T, Morcillo RJ, Tamayo-Navarrete MI, Huertas R, Molinero-Rosales N, López-Ráez JA, Macho AP, García-Garrido JM (2021) DLK2 regulates arbuscule hyphal branching during arbuscular mycorrhizal symbiosis. New Phytol 229(1):548–562

Hull R, Choi J, Paszkowski U (2021) Conditioning plants for arbuscular mycorrhizal symbiosis through DWARF14-LIKE signalling. Curr Opin Plant Biol 62:102071

Jones JM, Clairmont L, Macdonald ES, Weiner CA, Emery RN, Guinel FC (2015) E151 (sym15), a pleiotropic mutant of pea (Pisum sativum L.), displays low nodule number, enhanced mycorrhizae, delayed lateral root emergence, and high root cytokinin levels. J Exp Bot 66(13): 4047–4059

Kaku H, Nishizawa Y, Ishii-Minami N, Akimoto-Tomiyama C, Dohmae N, Takio K, Minami E, Shibuya N (2006) Plant cells recognize chitin fragments for defense signaling through a plasma membrane receptor. Proc Natl Acad Sci 103(29):11086–11091

Kaló P, Gleason C, Edwards A, Marsh J, Mitra RM, Hirsch S, Jakab J, Sims S, Long SR, Rogers J, Kiss GB (2005) Nodulation signaling in legumes requires NSP2, a member of the GRAS family of transcriptional regulators. Science 308(5729):1786–1789

Kamoun S (2006) A catalogue of the effector secretome of plant pathogenic oomycetes. Annu Rev Phytopathol 44:41–60

Kanamori N, Madsen LH, Radutoiu S, Frantescu M, Quistgaard EM, Miwa H, Downie JA, James EK, Felle HH, Haaning LL, Jensen TH (2006) A nucleoporin is required for induction of Ca2+ spiking in legume nodule development and essential for rhizobial and fungal symbiosis. Proc Natl Acad Sci 103(2):359–364

Keymer A, Pimprikar P, Wewer V, Huber C, Brands M, Bucerius SL, Delaux PM, Klingl V, Röpenack-Lahaye EV, Wang TL, Eisenreich W (2017) Lipid transfer from plants to arbuscular mycorrhiza fungi. elife 6:e29107

Kloppholz S, Kuhn H, Requena N (2011) A secreted fungal effector of Glomus intraradices promotes symbiotic biotrophy. Curr Biol 21(14):1204–1209

Koltai H (2015) Cellular events of strigolactone signalling and their crosstalk with auxin in roots. J Exp Bot 66:4855–4861

Koltai H, LekKala SP, Bhattacharya C, Mayzlish-Gati E, Resnick N, Wininger S, Dor E, Yoneyama K, Yoneyama K, Hershenhorn J, Joel DM (2010) A tomato strigolactone-impaired mutant displays aberrant shoot morphology and plant interactions. J Exp Bot 61(6):1739–1749

Lanfranco L, Fiorilli V, Venice F, Bonfante P (2018) Strigolactones cross the kingdoms: plants, fungi, and bacteria in the arbuscular mycorrhizal symbiosis. J Exp Bot 69(9):2175–2188
Leppyanen IV, Shakhnazarova VY, Shtark OY, Vishnevskaya NA, Tikhonovich IA, Dolgikh EA (2017) Receptor-like kinase LYK9 in Pisum sativum L. is the CERK1-like receptor that controls both plant immunity and AM symbiosis development. Int J Mol Sci 19(1):8
Li XR, Sun J, Albinsky D, Zarrabian D, Hull R, Lee T, Jarratt-Barnham E, Chiu CH, Jacobsen A, Soumpourou E, Albanese A (2022) Nutrient regulation of lipochitooligosaccharide recognition in plants via NSP1 and NSP2. Nat Commun 13(1):6421
Liang Y, Cao Y, Tanaka K, Thibivilliers S, Wan J, Choi J, Kang CH, Qiu J, Stacey G (2013) Nonlegumes respond to rhizobial Nod factors by suppressing the innate immune response. Science 341(6152):1384–1387
Liao D, Chen X, Chen A, Wang H, Liu J, Liu J, Gu M, Sun S, Xu G (2015) The characterization of six auxin-induced tomato GH3 genes uncovers a member, SlGH3.4, strongly responsive to arbuscular mycorrhizal symbiosis. Plant Cell Physiol 56:674–687
Liao D, Wang S, Cui M, Liu J, Chen A, Xu G (2018) Phytohormones regulate the development of arbuscular mycorrhizal symbiosis. Int J Mol Sci 19(10):3146
Liu W, Kohlen W, Lillo A, Op den Camp R, Ivanov S, Hartog M, Limpens E, Jamil M, Smaczniak C, Kaufmann K, Yang WC (2011) Strigolactone biosynthesis in Medicago truncatula and rice requires the symbiotic GRAS-type transcription factors NSP1 and NSP2. Plant Cell 23(10):3853–3865
Lumba S, Holbrook-Smith D, McCourt P (2017) The perception of strigolactones in vascular plants. Nat Chem Biol 13(6):599–606
Maillet F, Poinsot V, André O, Puech-Pagès V, Haouy A, Gueunier M, Cromer L, Giraudet D, Formey D, Niebel A, Martinez EA (2011) Fungal lipochitooligosaccharide symbiotic signals in arbuscular mycorrhiza. Nature 469(7328):58–63
Markmann K, Giczey G, Parniske M (2008) Functional adaptation of a plant receptor-kinase paved the way for the evolution of intracellular root symbioses with bacteria. PLoS Biol 6(3):e68
Mashiguchi K, Seto Y, Yamaguchi S (2021) Strigolactone biosynthesis, transport and perception. Plant J 105(2):335–350
Medina MJ, Gagnon H, Piché Y, Ocampo JA, Garrido JM, Vierheilig H (2003) Root colonization by arbuscular mycorrhizal fungi is affected by the salicylic acid content of the plant. Plant Sci 164(6):993–998
Mercy L, Lucic-Mercy E, Nogales A, Poghosyan A, Schneider C, Arnholdt-Schmitt B (2017) A functional approach towards understanding the role of the mitochondrial respiratory chain in an endomycorrhizal symbiosis. Front Plant Sci 8:417
Miller JB, Pratap A, Miyahara A, Zhou L, Bornemann S, Morris RJ, Oldroyd GE (2013) Calcium/calmodulin-dependent protein kinase is negatively and positively regulated by calcium, providing a mechanism for decoding calcium responses during symbiosis signaling. Plant Cell 25(12):5053–5066
Miya A, Albert P, Shinya T, Desaki Y, Ichimura K, Shirasu K, Narusaka Y, Kawakami N, Kaku H, Shibuya N (2007) CERK1, a LysM receptor kinase, is essential for chitin elicitor signaling in Arabidopsis. Proc Natl Acad Sci 104(49):19613–19618
Miyata K, Kozaki T, Kouzai Y, Ozawa K, Ishii K, Asamizu E, Okabe Y, Umehara Y, Miyamoto A, Kobae Y, Akiyama K (2014) The bifunctional plant receptor, OsCERK1, regulates both chitin-triggered immunity and arbuscular mycorrhizal symbiosis in rice. Plant Cell Physiol 55(11):1864–1872
Morffy N, Faure L, Nelson DC (2016) Smoke and hormone mirrors: action and evolution of karrikin and strigolactone signaling. Trends Genet 32(3):176–188
Mori N, Nomura T, Akiyama K (2020) Identification of two oxygenase genes involved in the respective biosynthetic pathways of canonical and non-canonical strigolactones in Lotus japonicus. Planta 251:40
Moscatiello R, Sello S, Novero M, Negro A, Bonfante P, Navazio L (2014) The intracellular delivery of TAT-aequorin reveals calcium-mediated sensing of environmental and symbiotic

signals by the arbuscular mycorrhizal fungus Gigaspora margarita. New Phytol 203(3): 1012–1020
Murray JD (2011) Invasion by invitation: rhizobial infection in legumes. Mol Plant-Microbe Interact 24(6):631–639
Murray JD, Cousins DR, Jackson KJ, Liu C (2013) Signaling at the root surface: the role of cutin monomers in mycorrhization. Mol Plant 6(5):1381–1383
Nadal M, Sawers R, Naseem S, Bassin B, Kulicke C, Sharman A, An G, An K, Ahern KR, Romag A, Brutnell TP (2017) An N-acetylglucosamine transporter required for arbuscular mycorrhizal symbioses in rice and maize. Nat Plants 3(6):1–7
Nagahashi G, Douds DD Jr (2011) The effects of hydroxy fatty acids on the hyphal branching of germinated spores of AM fungi. Fungal Biol 115(4-5):351–358
Oldroyd GE (2013) Speak, friend, and enter: signalling systems that promote beneficial symbiotic associations in plants. Nat Rev Microbiol 11(4):252–263
Parniske M (2008) Arbuscular mycorrhiza: the mother of plant root endosymbioses. Nat Rev Microbiol 6(10):763–775
Pimprikar P, Carbonnel S, Paries M, Katzer K, Klingl V, Bohmer MJ, Karl L, Floss DS, Harrison MJ, Parniske M, Gutjahr C (2016) A CCaMK-CYCLOPS-DELLA complex activates transcription of RAM1 to regulate arbuscule branching. Curr Biol 26(8):987–998
Plett JM, Martin FM (2018) Know your enemy, embrace your friend: using omics to understand how plants respond differently to pathogenic and mutualistic microorganisms. Plant J 93(4): 729–746
Prasad Singh P, Srivastava D, Jaiswar A, Adholeya A (2019) Effector proteins of Rhizophagus proliferus: conserved protein domains may play a role in host-specific interaction with different plant species. Braz J Microbiol 50:593–601
Rehman NU, Li X, Zeng P, Guo S, Jan S, Liu Y, Huang Y, Xie Q (2021) Harmony but not uniformity: role of strigolactone in plants. Biomol Ther 11(11):1616
Ried MK, Antolín-Llovera M, Parniske M (2014) Spontaneous symbiotic reprogramming of plant roots triggered by receptor-like kinases. elife 3:e03891
Ried MK, Banhara A, Hwu FY, Binder A, Gust AA, Höfle C, Hückelhoven R, Nürnberger T, Parniske M (2019) A set of Arabidopsis genes involved in the accommodation of the downy mildew pathogen Hyaloperonospora arabidopsidis. PLoS Pathog 15(7):e1007747
Roberts WK, Selitrennikoff CP (1988) Plant and bacterial chitinases differ in antifungal activity. Microbiology 134(1):169–176
Rush TA, Puech-Pagès V, Bascaules A, Jargeat P, Maillet F, Haouy A, Maës AQ, Carriel CC, Khokhani D, Keller-Pearson M, Tannous J (2020) Lipo-chitooligosaccharides as regulatory signals of fungal growth and development. Nat Commun 11(1):3897
Saha S, Dutta A, Bhattacharya A, DasGupta M (2014) Intracellular catalytic domain of symbiosis receptor kinase hyperactivates spontaneous nodulation in absence of rhizobia. Plant Physiol 166(4):1699–1708
Saha S, Paul A, Herring L, Dutta A, Bhattacharya A, Samaddar S, Goshe MB, DasGupta M (2016) Gatekeeper tyrosine phosphorylation of SYMRK is essential for synchronizing the epidermal and cortical responses in root nodule symbiosis. Plant Physiol 171(1):71–81
Saito K, Yoshikawa M, Yano K, Miwa H, Uchida H, Asamizu E, Sato S, Tabata S, Imaizumi-Anraku H, Umehara Y, Kouchi H (2007) NUCLEOPORIN85 is required for calcium spiking, fungal and bacterial symbioses, and seed production in Lotus japonicus. Plant Cell 19(2): 610–624
Salloum MS, Menduni MF, Benavides MP, Larrauri M, Luna CM, Silvente S (2018) Polyamines and flavonoids: key compounds in mycorrhizal colonization of improved and unimproved soybean genotypes. Symbiosis 76(3):265–275
Salvioli A, Ghignone S, Novero M, Navazio L, Venice F, Bagnaresi P, Bonfante P (2016) Symbiosis with an endobacterium increases the fitness of a mycorrhizal fungus, raising its bioenergetic potential. ISME J 10(1):130–144

Samaddar S, Dutta A, Sinharoy S, Paul A, Bhattacharya A, Saha S, Chien KY, Goshe MB, DasGupta M (2013) Autophosphorylation of gatekeeper tyrosine by symbiosis receptor kinase. FEBS Lett 587(18):2972–2979

Sanchez E, Artuso E, Lombardi C, Visentin I, Lace B, Saeed W, Lolli ML, Kobauri P, Ali Z, Spyrakis F, Cubas P (2018) Structure–activity relationships of strigolactones via a novel, quantitative in planta bioassay. J Exp Bot 69(9):2333–2343

Scervino JM, Ponce MA, Erra-Bassells R, Bompadre MJ, Vierheilig H, Ocampo JA, Godeas A (2006) Glycosidation of apigenin results in a loss of its activity on different growth parameters of arbuscular mycorrhizal fungi from the genus Glomus and Gigaspora. Soil Biol Biochem 38(9):2919–2922

Scervino JM, Ponce MA, Erra-Bassells R, Bompadre J, Vierheilig H, Ocampo JA, Godeas A (2007) The effect of flavones and flavonols on colonization of tomato plants by arbuscular mycorrhizal fungi of the genera Gigaspora and Glomus. Can J Microbiol 53(6):702–709

Schaarschmidt S, Roitsch T, Hause B (2006) Arbuscular mycorrhiza induces gene expression of the apoplastic invertase LIN6 in tomato (Lycopersicon esculentum) roots. J Exp Bot 57:4015–4023

Sega P, Pacak A (2019) Plant PHR transcription factors: put on a map. Gene 10(12):1018

Shaul-Keinan O, Gadkar V, Ginzberg I, Grünzweig JM, Chet I, Elad Y, Wininger S, Belausov E, Eshed Y, Atzmon N, Ben-Tal Y (2002) Hormone concentrations in tobacco roots change during arbuscular mycorrhizal colonization with Glomus intraradices. New Phytol 154(2):501–507

Shi J, Zhao B, Zheng S, Zhang X, Wang X, Dong W, Xie Q, Wang G, Xiao Y, Chen F, Yu N (2021) A phosphate starvation response-centered network regulates mycorrhizal symbiosis. Cell 184(22):5527–5540

Shimizu T, Nakano T, Takamizawa D, Desaki Y, Ishii-Minami N, Nishizawa Y, Minami E, Okada K, Yamane H, Kaku H, Shibuya N (2010) Two LysM receptor molecules, CEBiP and OsCERK1, cooperatively regulate chitin elicitor signaling in rice. Plant J 64(2):204–214

Shiu SH, Bleecker AB (2003) Expansion of the receptor-like kinase/Pelle gene family and receptor-like proteins in Arabidopsis. Plant Physiol 132(2):530–543

Singla P, Garg N (2017) Plant flavonoids: key players in signaling, establishment, and regulation of rhizobial and mycorrhizal endosymbioses. In: Mycorrhiza-function, diversity, state of the art. Springer, Cham, pp 133–176

Smit P, Raedts J, Portyanko V, Debellé F, Gough C, Bisseling T, Geurts R (2005) NSP1 of the GRAS protein family is essential for rhizobial Nod factor-induced transcription. Science 308(5729):1789–1791

Steinkellner S, Lendzemo V, Langer I, Schweiger P, Khaosaad T, Toussaint JP, Vierheilig H (2007) Flavonoids and strigolactones in root exudates as signals in symbiotic and pathogenic plant-fungus interactions. Molecules 12(7):1290–1306

Sun J, Miller JB, Granqvist E, Wiley-Kalil A, Gobbato E, Maillet F, Cottaz S, Samain E, Venkateshwaran M, Fort S, Morris RJ (2015) Activation of symbiosis signaling by arbuscular mycorrhizal fungi in legumes and rice. Plant Cell 27(3):823–838

Takeda N, Handa Y, Tsuzuki S, Kojima M, Sakakibara H, Kawaguchi M (2015) Gibberellins interfere with symbiosis signaling and gene expression and alter colonization by arbuscular mycorrhizal fungi in Lotus japonicus. Plant Physiol 167(2):545–557

Tamura K, Hara-Nishimura I (2013) The molecular architecture of the plant nuclear pore complex. J Exp Bot 64(4):823–832

Tejeda-Sartorius M, Martinez Vega O, Delano-Frier JP (2008) Jasmonic acid influences mycorrhizal colonization in tomato plants by modifying the expression of genes involved in carbohydrate partitioning. Physiol Plant 133(2):339–353

Tisserant E, Malbreil M, Kuo A, Kohler A, Symeonidi A, Balestrini R, Charron P, Duensing N, Frei Frey N, Gianinazzi-Pearson V, Gilbert LB (2013) Genome of an arbuscular mycorrhizal fungus provides insight into the oldest plant symbiosis. Proc Natl Acad Sci 110(50):20117–20122

Tominaga T, Yamaguchi K, Shigenobu S, Yamato M, Kaminaka H (2020) The effects of gibberellin on the expression of symbiosis-related genes in Paris-type arbuscular mycorrhizal symbiosis in Eustoma grandiflorum. Plant Signal Behav 15(9):1784544

Tominaga T, Miura C, Sumigawa Y, Hirose Y, Yamaguchi K, Shigenobu S, Mine A, Kaminaka H (2021) Conservation and diversity in gibberellin-mediated transcriptional responses among host plants forming distinct arbuscular mycorrhizal morphotypes. Front Plant Sci 2021:2836

Torres Santos R, Vierheilig H, Ocampo JA, García Garrido JM (2011) Altered pattern of arbuscular mycorrhizal formation in tomato ethylene mutants. Plant Signal Behav 6(5):755–758

Torres Santos R, Molinero Rosales N, Ocampo JA, García-Garrido JM (2016) Ethylene alleviates the suppressive effect of phosphate on arbuscular mycorrhiza formation. J Plant Growth Regul 35(3):611–617

Tsuchiya Y, Vidaurre D, Toh S, Hanada A, Nambara E, Kamiya Y, Yamaguchi S, McCourt P (2010) A small-molecule screen identifies new functions for the plant hormone strigolactone. Nat Chem Biol 6(10):741–749

Tsuzuki S, Handa Y, Takeda N, Kawaguchi M (2016) Strigolactone-induced putative secreted protein 1 is required for the establishment of symbiosis by the arbuscular mycorrhizal fungus Rhizophagus irregularis. Mol Plant-Microbe Interact 29(4):277–286

Vallabhaneni R, Bradbury LM, Wurtzel ET (2010) The carotenoid dioxygenase gene family in maize, sorghum, and rice. Arch Biochem Biophys 504(1):104–111

van der Burgh AM, Joosten MH (2019) Plant immunity: thinking outside and inside the box. Trends Plant Sci 24(7):587–601

Venkateshwaran M, Jayaraman D, Chabaud M, Genre A, Balloon AJ, Maeda J, Forshey K, den Os D, Kwiecien NW, Coon JJ, Barker DG (2015) A role for the mevalonate pathway in early plant symbiotic signaling. Proc Natl Acad Sci 112(31):9781–9786

Voss S, Betz R, Heidt S, Corradi N, Requena N (2018) RiCRN1, a crinkler effector from the arbuscular mycorrhizal fungus Rhizophagus irregularis, functions in arbuscule development. Front Microbiol 9:2068

Votta C, Fiorilli V, Haider I, Wang JY, Balestrini R, Petřík I, Tarkowská D, Novák O, Serikbayeva A, Bonfante P, Al-Babili S (2022) Zaxinone synthase controls arbuscular mycorrhizal colonization level in rice. Plant J 111(6):1688–1700

Wakabayashi T, Hamana M, Mori A, Akiyama R, Ueno K, Osakabe K, Suzuki H, Takikawa H, Mizutani M, Sugimoto Y (2019) Direct conversion of carlactonoic acid to orobanchol by cytochrome P450 CYP722C in strigolactone biosynthesis. Sci Adv 5:eaax9067

Wakabayashi T, Shida K, Kitano Y, Takikawa H, Mizutani M, Sugimoto Y (2020) CYP722C from Gossypium arboreum catalyzes the conversion of carlactonoic acid to 5-deoxystrigol. Planta 251:97

Wang E, Schornack S, Marsh JF, Gobbato E, Schwessinger B, Eastmond P, Schultze M, Kamoun S, Oldroyd GE (2012) A common signaling process that promotes mycorrhizal and oomycete colonization of plants. Curr Biol 22(23):2242–2246

Wang JY, Haider I, Jamil M, Fiorilli V, Saito Y, Mi J, Baz L, Kountche BA, Jia KP, Guo X, Balakrishna A (2019) The apocarotenoid metabolite zaxinone regulates growth and strigolactone biosynthesis in rice. Nat Commun 10(1):1–9

Wang Y, Pruitt RN, Nürnberger T, Wang Y (2022) Evasion of plant immunity by microbial pathogens. Nat Rev Microbiol 20(8):449–464

Waters MT, Nelson DC, Scaffidi A, Flematti GR, Sun YK, Dixon KW, Smith SM (2012) Specialisation within the DWARF14 protein family confers distinct responses to karrikins and strigolactones in Arabidopsis. Development 139(7):1285–1295

Waters MT, Gutjahr C, Bennett T, Nelson DC (2017) Strigolactone signaling and evolution. Annu Rev Plant Biol 68:291–322

Wu J, Wang W, Zhu H, Cao Y (2022) Receptor kinases and signal pathway in the arbuscular mycorrhizal symbiosis. In: De Sousa R (ed) Arbuscular mycorrhizal fungi in agriculture - new insights. IntechOpen, London, pp 1–21. https://doi.org/10.5772/intechopen.107261

Yano K, Yoshida S, Müller J, Singh S, Banba M, Vickers K, Markmann K, White C, Schuller B, Sato S, Asamizu E (2008) CYCLOPS, a mediator of symbiotic intracellular accommodation. Proc Natl Acad Sci 105(51):20540–20545

Yoshida S, Kameoka H, Tempo M, Akiyama K, Umehara M, Yamaguchi S, Hayashi H, Kyozuka J, Shirasu K (2012) The D3 F-box protein is a key component in host strigolactone responses essential for arbuscular mycorrhizal symbiosis. New Phytol 196(4):1208–1216
Yu N, Luo D, Zhang X, Liu J, Wang W, Jin Y, Dong W, Liu J, Liu H, Yang W, Zeng L (2014) A DELLA protein complex controls the arbuscular mycorrhizal symbiosis in plants. Cell Res 24(1):130–133
Zhang X, Dong W, Sun J, Feng F, Deng Y, He Z, Oldroyd GE, Wang E (2015) The receptor kinase CERK 1 has dual functions in symbiosis and immunity signalling. Plant J 81(2):258–267
Zhang C, He J, Dai H, Wang G, Zhang X, Wang C, Shi J, Chen X, Wang D, Wang E (2021) Discriminating symbiosis and immunity signals by receptor competition in rice. Proc Natl Acad Sci 118(16):e2023738118
Zhivotovsky B, Orrenius S (2011) Calcium and cell death mechanisms: a perspective from the cell death community. Cell Calcium 50(3):211–221
Zwanenburg B, Pospíšil T (2013) Structure and activity of strigolactones: new plant hormones with a rich future. Mol Plant 6(1):38–62
Zwanenburg B, Nayak SK, Charnikhova TV, Bouwmeester HJ (2013) New strigolactone mimics: structure–activity relationship and mode of action as germinating stimulants for parasitic weeds. Bioorg Med Chem Lett 23(18):5182–5186

Chapter 5
Arbuscular Mycorrhizal Fungi and Plant Secondary Metabolism

Somayeh Rahmat and Zhaleh Soheilikhah

Abstract Plants synthetize and accumulate a various class of organic compounds known as secondary metabolites (SMs). Although SMs do not play a crucial role as primary metabolites in plant growth and development, they are of high ecological significance. Humans use these compounds in the pharmaceutical industry. One of the results of mutual symbiotic association between plants and arbuscular mycorrhizal fungi (AMF) is the reprogramming of metabolic pathways and modulating the range and content of plant SMs such as phenolics, terpenoids, and alkaloids. Many of SMs act as signals for multiple interactions between plants and AMF, from the pre-symbiotic stage to the creation/formation of a functional symbiosis. This chapter briefly reviews the current research status in the field of SM changes under the influence of AMF. The plant association with AMF increases the production and accumulation of SMs directly through improving water and nutrient uptake and enhancing the photosynthetic capacity or indirectly by provoking the biosynthetic pathways of SMs through generation of signaling molecules and changes in the concentration of phytohormones. The extent to which AMF affect plant SMs depends on the plant and fungus species and environmental factors.

Keywords AMF · Secondary metabolites · Phenolics · Flavonoid · Photosynthesis · Stress tolerance

5.1 Introduction

In the long path of evolution, plants have developed a wide range of organic compounds known as secondary metabolites (SMs). SMs have no direct role in plant fundamental processes such as photosynthesis, respiration, and transport of nutrients or in the biosynthetic pathways of carbohydrates, proteins, or lipids. Unlike

S. Rahmat (✉)
Miandoab, Iran

Z. Soheilikhah
Department of Biology, Faculty of Science, Razi University, Kermanshah, Iran

G. J. Ahammed, R. Hajiboland (eds.), *Arbuscular Mycorrhizal Fungi and Higher Plants*, https://doi.org/10.1007/978-981-99-8220-2_5

primary metabolites, SMs have a limited distribution in the plant kingdom; in other words, specific SMs are often found only in one species or plants related to that species (Ghamarnia et al. 2022; Yousefi and Safari 2022). Lack or reduced level of SMs does not cause an immediate death of plants, but it causes disruptions in survival, fertility, and ecological distribution and resistance of plants to natural enemies (Agostini-Costa et al. 2012).

SMs in plants are structurally and chemically diverse groups of products, and despite lack of an essential role for plant growth and development, they are economically important for human well-being in various fields, including pharmaceuticals, nutrients, food additives, and agrochemicals (Chandran et al. 2020). Meanwhile, medicinal plants, which are rich reservoirs of the main active ingredients of numerous medicinal compounds, have attracted special attention (Zhao et al. 2022; Walid et al. 2021).

The most significant and prevalent mutualistic symbiosis of microorganisms with plants is formed between roots and arbuscular mycorrhizal fungi (AMF). Mycorrhiza promotes plant growth; increases the uptake of micro- and macronutrients (Grümberg et al. 2015; Battini et al. 2017; Amani Machiani et al. 2021); improves water relations (Eulenstein et al. 2016; Li et al. 2019; Ostadi et al. 2022), soil fertility, and yield quality (Amani Machiani et al. 2022); enhances tolerance to environmental stresses, such as drought, salinity, and heavy metals; and boosts resistance to pests and pathogens (Moradtalab et al. 2019; Al-Arjani et al. 2020). As the most prevalent symbiotic relationship in nature, AMF symbiosis offers enormous potential for resource utilization in agriculture, especially in low-input and sustainable agriculture, and would cut agricultural expenses and further minimize environmental degradation (Begum et al. 2019; Igiehon and Babalola 2017).

A wide range of environmental (abiotic and biotic stresses) conditions and internal (morphological, developmental, and genetic) factors affect the production and accumulation of SMs (Simões et al. 2016). Moreover, the mutual relationships between plants and soil microorganisms, like AMF, also change the quality and quantity of plant SMs (Kapoor et al. 2017). AMF-mediated increase in the synthesis and accumulation of SMs may be one of the mechanisms for the positive effects of mycorrhization in host plants, particularly those associated with higher tolerance to suboptimal conditions (Kaur and Suseela 2020).

Although the colonization of plant by the fungus takes place morphologically in the root, it affects the metabolism and physiology of the plant systemically (Schweiger and Müller 2015). Due to its dependence on plant carbon (carbohydrates and lipids), AMF are a powerful carbon sink in the roots. They allocate up to 30% of plant's photosynthetic carbon in return for water and nutrient supply (Drigo et al. 2010; Van Der Heijden et al. 2015). Such interaction regulates photosynthesis and other primary metabolic pathways to maintain the balance of carbon in plant. As an inevitable consequence of change in primary metabolism, the profile of plant SMs will change (Kaschuk et al. 2009; Kogel et al. 2010).

A group of SMs act as mediators of AMF-plant interactions, from the pre-symbiotic step to establishing and maintaining a successful symbiotic association. For example, in response to secretion of signaling compounds from roots (strigolactones, flavonoids), AMF release lipocyto-oligosaccharide compounds called "Myc factors" that lead to the induction of symbiotic responses in the host plant (Kaur and Suseela 2020). In addition to the release of molecules necessary for identification and formation of a functional symbiosis, the AMF association leads to reprogramming of the secondary metabolic pathways in the host plants (Pozo et al. 2009; Jung et al. 2012; Song et al. 2013). The host genome is affected by fungal metabolites, and thus, the microbe-host interaction causes changes in the transcriptome, proteome, and ultimately the metabolome of the plant (Mhlongo et al. 2018). This chapter briefly introduces the impacts of AMF on secondary metabolism in host plants and provides insights to achieve higher performance of commercial plants through AMF colonization in sustainable cultivation systems.

5.2 Secondary Metabolism

Unlike primary metabolism, secondary metabolism includes metabolic pathways and their related molecules that are not necessary for plant life (Yang et al. 2018). Products of SMs originate from primary metabolites or their intermediate compounds in the biosynthetic pathways (Piasecka et al. 2015). Methods such as mass spectrometry, gas chromatography, and liquid chromatography have made it possible to detect and identify a wide range of SMs in plants. By examining only 20–30% of all known plant species (Zhu et al. 2023), more than 100,000 different compounds have been isolated and identified (Wink 2008). Despite such a high diversity of SMs, there are limited numbers of corresponding basic biosynthetic pathways (Fig. 5.1). Substrates are usually derived from basic metabolic pathways such as the shikimic acid pathway, Krebs cycle, and glycolysis pathway (Wink 2008). According to the molecular structure and biological functions, three important classes of plant SMs are phenolics, terpenes or terpenoids, and alkaloids (Pandey et al. 2018). Terpenoids and phenolics are synthesized by almost all vascular plants, while alkaloids show a patchy occurrence pattern. So, specific alkaloid compounds are usually considered as indicators of specific taxonomic classes (Wink 2008). Another important group of SMs discussed in this chapter are saponins, a chemically heterogeneous group of triterpenoid and steroidal glycosides that are found in most medicinal plants and have a high molecular weight (Nasseri et al. 2020).

Many SMs show positive effect on numerous aspects of plant growth and development including defense response (Isah 2019), innate immunity (Piasecka et al. 2015), and adaptations to environmental stress (Yang et al. 2018). Moreover, SMs act as signaling compounds for symbiosis between microorganisms and plants, including AMF (Guerrieri et al. 2019).

Apart from their functional significance in plants, the high commercial value of a large group of SMs has caused them to be used as special chemicals such as flavorings, perfumes, insecticides, dyes, and all kinds of anticancer, antidiabetic, antihypertensive, anti-AIDS, heart and blood vessel improvers, memory boosters, and antidepressants (Chandran et al. 2020). Since symbiosis of plants with AMF

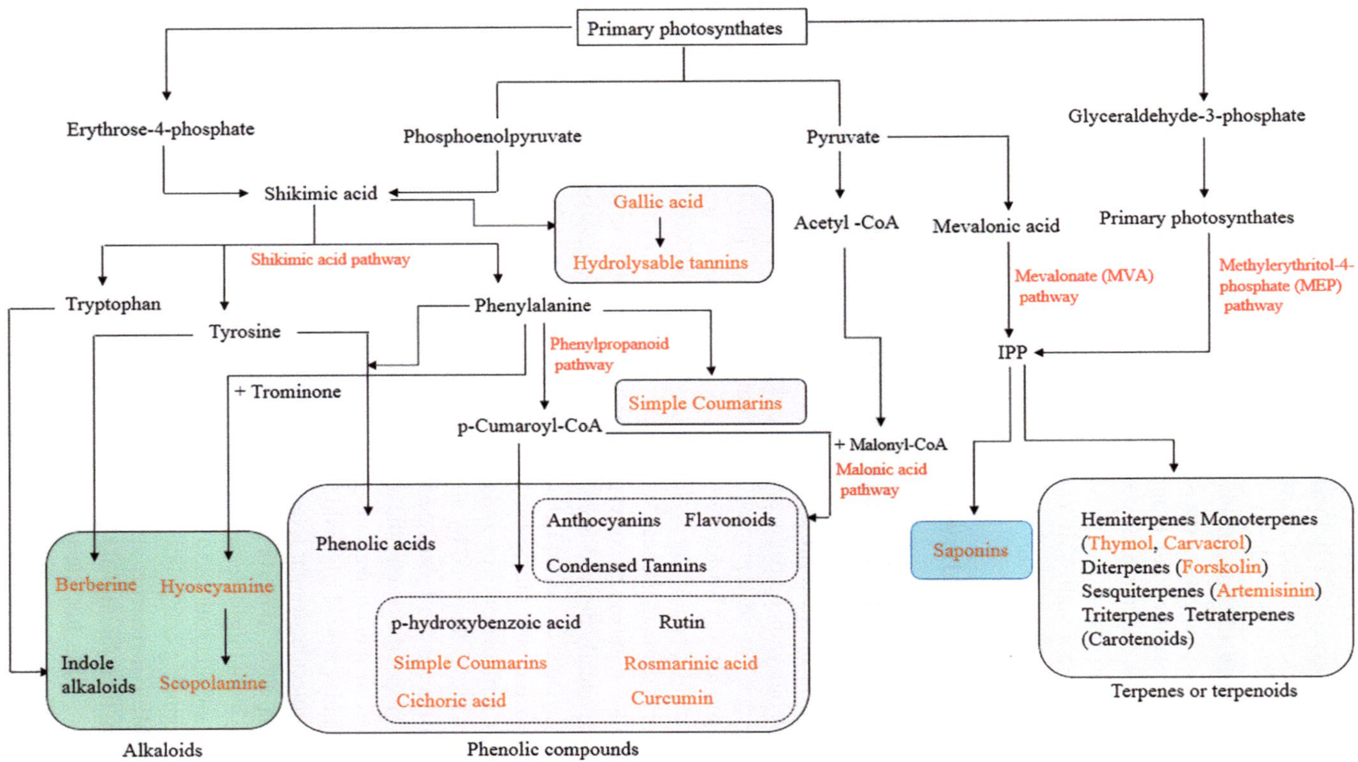

Fig. 5.1 The main metabolic pathways leading to the synthesis of terpenes or terpenoids, phenolics, alkaloids, and saponins in plants (in gray, pink, green, and blue shaded sections, respectively). The brown color shows the category of regulatory compounds in medicinal plants colonized with AMF. Redrawn after Zhao et al. (2022)

significantly modifies production and the pattern of accumulation of SMs (Fig. 5.2) (Tsiokanos et al. 2022), the extent to which the quality and quantity of SMs are affected by AMF association is one of the significant aspects in the production of medicinal and aromatic plants.

5.2.1 Terpenoids

Terpenoids (or isoprenoids) are the most diverse class of plant metabolites, with more than 27,000 compounds (Sacchettini and Poulter 1997; Pandey et al. 2018). Terpenoids contain a ring or chain structure that is obtained from the fusion of isoprene units (C5). Based on the number of isoprene units used in their structure, terpenoids can be classified as hemiterpenes (C5), monoterpenes (C10, geraniol, linalool, limonene, etc.), sesquiterpenes (C15, abscisic acid, humulene, germacrene, bisabolene, etc.), diterpenes (C20, gibberellic acid), triterpenes (C30, ursolic acid, oleanolic acid, betulinic acid, etc.), tetraterpenes (C40, lupeol, diosgenin, stigmasterols, lanosterol, etc.), and polyterpenes ($C > 40$) (Pusztahelyi et al. 2015; Pandey et al. 2018). Terpenoids are synthesized and stored in trichomes and epidermal structures of leaf in plants (Covello et al. 2007). These compounds play an essential role in plant adaptation to adverse environmental conditions, and some of them (e.g., abscisic acid and gibberellic acid) regulate plant growth and stress responses (Pusztahelyi et al. 2015). Some sesquiterpenes (such as caryophyllene) and monoterpenes (such as linalool) are toxic to pathogens and herbivorous insects and can also act as a signaling molecule to attract natural enemies of insects (Kappers et al. 2005; Bakkali et al. 2008; Sharma et al. 2017). In addition, when plants are attacked by pathogens or fungus-eating animals, AMF stimulate the production of some terpenoids (such as catalpol) by the host plant. Catalpol is transferred to soil or external hyphae and acts as a signal or defense substances to protect neighboring plants or symbiotic fungi (Babikova et al. 2013; Duhamel et al. 2013). Despite some reports regarding neutral (Morone-Fortunato and Avato 2008; Nell et al. 2009) and even negative effects (Khaosaad et al. 2006), many studies have shown that the symbiotic relationship between AMF and plants can stimulate the synthesis of terpenoids and affect the composition and content of terpenoids (Karagiannidis et al. 2011; Lermen et al. 2015; Weisany et al. 2015). The studies conducted so far show that the increase in the production of precursors through the methylerythritol 4-phosphate (MEP) pathway (Fig. 5.2) is the reason for the increased concentration of terpenoids after inoculation with AMF (Kapoor et al. 2017). On the other hand, terpenoids, especially volatile mono- and sesquiterpenes with small molecular weight, play an important role in the symbiosis between plants and AMF (Duhamel et al. 2013; Babikova et al. 2013; Sharma et al. 2017). In addition, some terpenoids are bioactive and are the main active component of medicinal plants, such as the sesquiterpene artemisinin found in *Artemisia annua* (Mandal et al. 2015b).

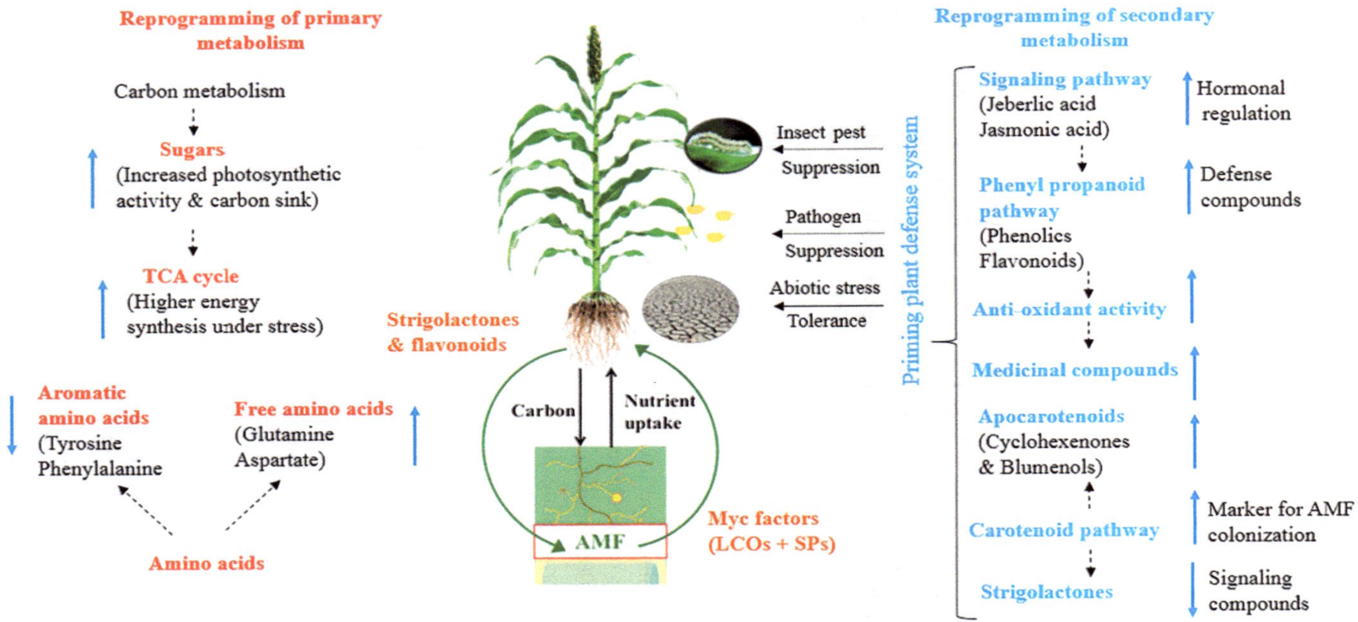

Fig. 5.2 Potential pathways of plant metabolism reprogramming by AMF. Mycorrhization is associated with the reprogramming of primary metabolites and SMs in plants. Reprogramming of SMs stimulates plant defense and increases plant tolerance to abiotic and biotic stresses. Changes in the secondary metabolic pathways contribute to the self-regulation of AMF colonization by modulating the synthesis of signaling compounds. *SA* salicylic acid, *JA* jasmonic acid, *SPs* secreted proteins, *LCOs* lipochitooligosaccharides. Redrawn after Kaur and Suseela (2020)

5.2.1.1 Carotenoid Pathway

Carotenoids are members of the 40-carbon (tetraterpenoids) subfamily of terpenoids, which are made by connecting 8 isoprene units (Li et al. 2020). The important function of the carotenoid pathway is known in various plant processes including signaling, synthesis, and release of hormones, light protection, and photosynthesis (Shumskaya and Wurtzel 2013). Upregulation of the primary enzymes of this pathway, 1-deoxy-D-xylulose 5-phosphate reductoisomerase (DXR) and 1-deoxy-D-xylulose 5-phosphate synthase (DXS) (Fig. 5.3), after inoculation with AMF was associated with the accumulation of the diterpene stevioside and sesquiterpene artemisinin in the leaves of *Stevia rebadiana* and *Artemisia annua*, respectively (Mandal et al. 2015a, b).

A number of molecules derived from the carotenoid pathway (e.g., strigolactones) are released by mycorrhizal roots during the pre-symbiotic step and

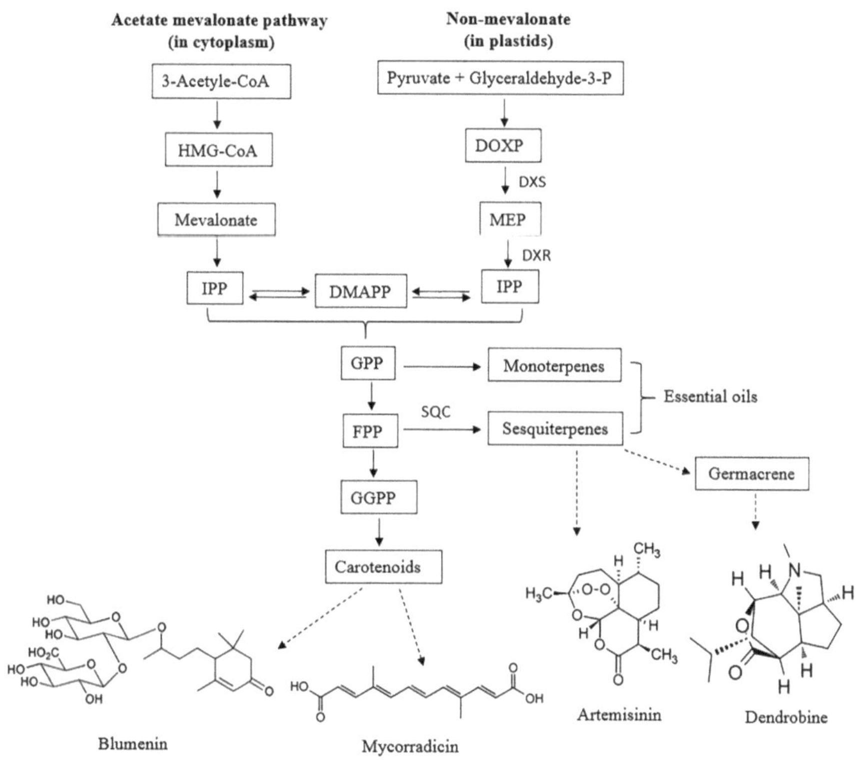

Fig. 5.3 Biosynthesis pathway of DMAPP/IPP promoted or induced by AMF. *GGPP* geranylgeranyl diphosphate, *FPP* farnesyl diphosphate, *SQC* sesquiterpene cyclase, *GPP* geranyl diphosphate, *DMAPP* dimethylallyl diphosphate, *IPP* isopentenyl diphosphate, *DXR* 1-deoxy-D-xylulose 5-phosphate reductoisomerase, *MEP* 2-C-methyl-D-erythritol 4-phosphate, *DXS* 1-deoxy-D-xylulose 5-phosphate synthase, *DOXP* 1-deoxy-D-xylulose 5-phosphate, *HMG-CoA* β-hydroxy β-methylglutaryl-CoA. Redrawn after Zhi-lin et al. (2007)

after colonization (Al-Babili and Bouwmeester 2015). Strigolactones act as signaling compounds and regulate spore germination, hyphal elongation, and hyphopodium formation leading to the start of AMF-plant symbiosis (Al-Babili and Bouwmeester 2015). Significant induction of *PDR1* transporter (an ABC transporter) which is involved in the transport of strigolactones is known in mycorrhizal plants (Kretzschmar et al. 2012). AMF also induce the synthesis and accumulation of some apocarotenoids, such as C14 acyclic polyene mycorradicin and C13 cyclohexenone derivatives (e.g., blumenin) (Fig. 5.3), which are important regulators of the establishment and maintaining of AM symbiosis in colonized roots (Schliemann et al. 2008; Fiorilli et al. 2019). The development of mycorrhization in *Hordeum vulgare* plants was associated with an increased accumulation of cyclohexenone derivatives (Peipp et al. 1997). Blumenols (glycosidic derivatives) are among the most common compounds derived from cyclohexenone that accumulate in the mycorrhizal plants and act as biomarkers of AM symbiosis (Schliemann et al. 2006; Wang et al. 2018). In addition, increased levels of apocarotenoids in AM roots may be important in protecting plants against ROS and pathogens, self-regulation of fungal colonization, and synthesis of signaling molecules (Strack and Fester 2006). There are some reports that the activation of carotenoid biosynthesis is specific to the plant-AMF relationship (Fester et al. 2005; Maier et al. 1997). Furthermore, mycorrhization reduced the emission of volatile sesquiterpenes by herbivory in *Plantago lanceolata*, indicating that terpenoids are also involved in indirectly priming of the host defense system (Fontana et al. 2009).

5.2.1.2 Essential Oils

Plant essential oils (Eos) are aromatic and volatile oil compounds. Complex mixtures of monoterpenes, sesquiterpenes, diterpenes, and compounds derived from phenylpropene are the main bioactive components of Eos (Nurzynska-Wierdak 2013). These compounds are important in the plant growth and development, resistance of plants against adverse environmental conditions, and attraction of pollinating insects. They are used as antiseptic, medicinal, and aromatic compounds in various industries (Chiej 1984; Chukwuka et al. 2011). In addition, studies show that different species of AMF cause different responses in the same plant; specific components of Eos are also affected by AMF in different aromatic plants (Kapoor et al. 2017). Rezaei-Chiyaneh et al. (2021) indicated that colonization by a combination of *Rhizophagus irregularis* and *Funneliformis mosseae* increased the main essential oil compounds such as trans-caryophyllene, borneol, p-cymene, thymol, and geranyl-acetate in *Nigella sativa*. The content of g-terpinene, p-cymene, and thymol in Eos of thyme (*Thymus vulgaris*) also increased after inoculation with *Funneliformis mosseae* (Amani Machiani et al. 2021). Similarly, Eo content was improved in *Mentha requienii* and *Origanum onites* plants inoculated with *Glomus lamellosum* and *G. etunicatum*, respectively (Karagiannidis et al. 2011). Moreover, after inoculation with *Gigaspora rosea*, the accumulation of linalool in *Ocimum basilicum* L. leaves declined, while the accumulation of eugenol and alpha-terpineol enhanced, suggesting that AMF association changes the essential oil composition of plant leaves (Copetta et al. 2006).

5.2.2 *Phenolics*

Phenolics or phenylpropanoids are a large class of SMs produced in plant tissues to protect against predators, parasites, pathogens, and UV radiation (Singla et al. 2019). They are synthesized by the pentose phosphate and shikimate pathways through the phenylpropanoid pathway (Fig. 5.1) (Michalak 2006; Zaynab et al. 2018). The five main groups of phenolics are phenolic acids, flavonoids, coumarins, stilbenes, and monolignols (Deng and Lu 2017). Most of the compounds have antioxidant activity, which, apart from their beneficial effects under stress conditions, give the plant valuable medicinal properties (Krishnaiah et al. 2011). In addition, they are important signaling compounds in microbe-plant relationships and plant development. AMF create changes in the quantity and quality of different SMs by reprogramming the phenylpropanoid pathway, to establish a functional symbiosis (Schweiger and Müller 2015). Two important groups of compounds in the phenylpropanoid pathway are phenolic acids and flavonoids.

5.2.2.1 Flavonoids

Flavonoids form the largest group of phenolics and include flavonols, flavones, isoflavonoids, and anthocyanins (Reyes-Carmona et al. 2005). Flavonoids have high antioxidant activity and contribute to the flavor and color of plants (Mierziak et al. 2014). There are many reports of beneficial effects of AMF on flavonoid concentrations. Inoculation with *Rhizoglomus intraradices* and *Funneliformis mosseae* increased the flavonoid levels in the *Solanum nigrum* L. roots (Muniz et al. 2021). In inoculated corn plants, the appropriate combination of fungus-plant species is essential in determining the result of symbiosis and achieving higher yield. Moreover, the plants inoculated with *Dentiscutata heterogama* and *Claroideoglomus etunicatum* had higher flavonoid levels compared to the plants inoculated with *Acaulospora longula* (Silva et al. 2019; Avio et al. 2020). Furthermore, the levels of phenols and flavonoids in the weed *Aphis gossypii* elevated after inoculation with *Septoglomus constrictum* and *Claroideoglomus etunicatum*, and this was accompanied by a decrease in its herbivory (Wu et al. 2023). These findings indicate the ecological implications of AMF in weed management.

In the AMF-plant relationship, changes in the flavonoid pattern are a function of the developmental stage of the symbiosis. These compounds are essential in initiating and limiting mycorrhization (Kaur and Suseela 2020). Different types of flavonoids affect the inoculation of different AMF species differently. Some flavonoids, including ononin and formononetin, are responsible for autoregulation of AM symbiosis, by limiting the fungal colonization after reaching to a threshold level

(Catford et al. 2006). In the same way, some isoflavonoids, such as malonylononin, ononin, and daidzein, are involved in the later stages of mycorrhizal development (Schliemann et al. 2008). AMF changed flavonoid accumulation in growing roots of *Manihot esculenta Crantz* 6 weeks after inoculation with *Rhizophagus irregularis* (Bag et al. 2022). Comparison of well-colonized roots of *Medicago sativa* and *Medicago truncatula*, and incompletely colonized roots of *Medicago sativa*, showed that the level of flavonoid 4′,7-dihydroxyflavone increased only in the roots with well-developed colonization (Harrison and Dixon 1993). In the same way, the levels of medicarpin, which were increased during initial colonization, were greatly reduced during the advanced stages of colonization. However, the abundance of medicarpin in incompletely colonized roots did not decrease, which means that growth-dependent changes in the flavonoid pattern are necessary for the formation and regulation of symbiotic association (Harrison and Dixon 1993). Moreover, the increase in the concentration of some types of flavonoids, such as coumestrol, daidzein, medicarpin malonyl glucoside, and formononetin malonyl glucoside, is related to the improved resistance of mycorrhizal plants to stressful conditions (Harrison and Dixon 1993).

5.2.2.2 Phenolic Acids

Phenolic acids are considered as important mediators of plant-microorganism interactions (Mandal et al. 2010). Higher level of these compounds in herbal products is an indication of their higher quality (Ghasemzadeh and Ghasemzadeh 2011). There are several reports of improved levels of phenolic acids by inoculation of AMF. AMF colonization enhanced the abundance of phenolic acids in *Arachis hypogaea* roots and leaves (Devi and Reddy 2002). The abundance of phenolic acid derivatives also changed in *Cichorium intybus* (Rozpądek et al. 2014) and *Hordeum vulgare* (Devi and Reddy 2002) plants inoculated with AMF. Hydroxycinnamate amides and caffeic acid derivatives were reported in mycorrhizal roots of chicory, and accumulation of hydroxycinnamate amides in barley mycorrhizal roots. Total phenol content in *Eclipta prostrata* L. plants increased 8 weeks after inoculation with a combination of AMF species (*Acaulospora lacunosa*, *Funneliformis mosseae*, and *Gigaspora albida*) (Duc et al. 2021). Colonization in *Passiflora alata* Curtis plants with *Gigaspora albida* and *Acaulospora longula* increased the total phenol by 94% and 111%, respectively (Muniz et al. 2021). In addition, resistance of *Phoenix dactylifera* against bayoud disease is improved through its colonization with various AMF species. This improvement was related to the increased enzymatic activities of polyphenol oxidases and peroxidases (Jaiti et al. 2007). Therefore, higher concentration of phenolic acids in AMF-inoculated plants implies the improvement of plant defense system against pathogens. In addition, it has been found that artichoke plants inoculated with *Funneliformis mosseae* IMA1 and *Claroideoglomus claroideum* 22W3 had higher phenolic compounds and antioxidant activities than control plants, while *Glomus* sp., *Rhizophagus irregularis*, and other varieties of *Funneliformis mosseae* had no effect on the plant antioxidant activity (Mandal et al. 2010). It has been reported that while ferulic acid level in *Solanum lycopersicum* roots decreased with AMF, the accumulation of caffeic acids increased. Such different response to mycorrhizal association depending on fungus and plant species suggests certain degrees of specificity in the AMF-plant interaction (López-Ráez et al. 2010). Similarly, Aliferis et al. (2015) and Maier et al. (2000) reported increased concentrations of other phenylpropanoid pathway compounds including coumarins and their hydroxyl derivatives, in mycorrhizal *Salix purpurea* L. leaves. These compounds have antioxidant and antimicrobial properties (Maier et al. 2000; Aliferis et al. 2015). Inoculation with *Acaulospora longula* also increased the tannin level in *Libidibia ferrea* fruits (Santos et al. 2020).

5.2.3 Alkaloids

Alkaloids are alkaline-like substances that contain one or more nitrogen atoms in the heterocyclic ring (Zhu et al. 2023). In plants, these metabolites are derived from amino acids, aromatic compounds, as well as terpenes (Herbert 2001). Based on their amino acid precursors, they are classified into pyridine, quinolizidine, indole-quinoline, benzyl-isoquinoline, pyrrolizidine, pyrrolidine, and tropane groups (Song et al. 2014; Dey et al. 2020). So far, 600 bioactive alkaloid compounds have been identified out of a total 20,000 alkaloid compounds discovered (mainly from plants). Due to toxic properties, alkaloids play an important role in strengthening the plant's defense system against biotic stresses (Yeshi et al. 2022). Some types of alkaloids have medicinal properties with a wide application in the pharmaceutical industry. In this case, we can mention morphine, cocaine, nicotine, colchicine, quinine, strychnine, and caffeine (Herbert 2001). Several studies reported the elevated concentrations of alkaloids after the inoculation of plants with AMF. For example, colonization with different species of AMF increased the concentration of alkaloid colchicine in different organs (aerial organ, tuber, and seed) of *Gloriosa superba* L. plant (Pandey et al. 2014). In addition, inoculation with *Gigaspora rosea* was associated with an increase in the content of alkaloid with therapeutic properties, trigonelline in *Prosopis laevigata*. Accumulation of alkaloids trigonelline (Rojas-Andrade et al. 2003), castanospermine (Abu-Zeyad et al. 1999), and camptothecin (Wei and Wang 1989) was also enhanced in plants colonized with different AMF species.

5.2.4 Saponins

An important class of SMs produced in plants are saponins. Their structure is characterized by a carbohydrate part (an oligo or monosaccharide chain) attached to a steroid or triterpene (Hussein and El-Anssary 2019). The various biological functions of saponins explain their wide application in pharmaceutical, cosmetic, and food industries (Isah 2019). An increase in saponin concentration has been reported in mycorrhizal plants. For example, inoculation of *Glycyrrhiza glabra* plants with *Claroideoglomus etunicatum* (Johny et al. 2021) and *Funneliformis mosseae* and/or *Diversispora epigaea* (Liu et al. 2007) was associated with the accumulation of triterpenoid saponin glycyrrhizic acid. Similarly, an accumulation of saponins was observed in the aerial parts of *Anchusa officinalis* L. (Cartabia et al. 2021) and *Passiflora alata* (Muniz et al. 2021) plants inoculated with *Rhizophagus irregularis* and *Acaulospora longula*, respectively.

5.3 Mechanisms of AMF Effects on SM Production in Plants

AM colonization alters the plant SM synthesis pattern in two different ways, nutritional (direct) and non-nutritional (indirect) effects (Fig. 5.4). In the nutritional mode of action, AMF improve water and nutrient uptake (phosphorus, nitrogen, copper, manganese, zinc) (Kapoor et al. 2017) and photosynthetic capacity and growth, which subsequently increase the production of precursor compounds required for the synthesis of SMs (Zhao et al. 2022). Some reports attribute the benefits of AMF on SMs to improved nutrient uptake, especially phosphorus. For example, inoculated and non-inoculated *Foeniculum vulgare* plants, with similar phosphorus levels, showed the similar essential oil content (Kapoor et al. 2017). Improved accumulation of glycyrrhizic acid in *Glycyrrhiza glabra* was also attributed to improved phosphorus uptake in these plants by AMF (Xie et al. 2018a, b). It seems that AMF cause C:N:P (carbon:nitrogen:phosphorus) stoichiometric changes in tissues by increasing phosphorus and nitrogen availability and regulating carbon distribution (Saia et al. 2014; Zhao et al. 2015). Based on the carbon-nutrient balance hypothesis, the synthesis of SMs is regulated by the ratio between carbon and nutrients. Low C:N ratios in plants lead to the synthesis of carbon-based SMs, while high C:N ratios increase the synthesis of nitrogen-based SMs (Gershenzon 1994; Hamilton et al. 2001). A recent study supported this hypothesis. Xie et al. (2018a, b) indicated that AMF-mediated increases in the C:N ratio and decreases in N:P ratio in *Glycyrrhiza glabra* plants were associated with an accumulation of glycyrrhizic acid in the roots. On the other hand, the improved water and nutrient uptake and higher photosynthetic capacity lead to increased division and development in the secretory ducts, glandular trichomes, and Eo channels (Amani Machiani et al. 2021). Moreover, some micronutrients act as cofactors for different enzymes. For example, manganese is an essential cofactor for the function of phytoene synthase, a key enzyme involved in the biosynthesis of carotenoids (tetraterpenoid) (Fraser et al. 2000). Two independent studies have reported increased manganese uptake in *Stevia rebaudiana* plants colonized by *R. fasciculatus* (Mandal et al. 2015b) and *R. intraradices* (Mandal et al. 2013). In addition, improved phosphorus nutrition in AMF plants increases metabolic precursors such as ATP, NADPH,

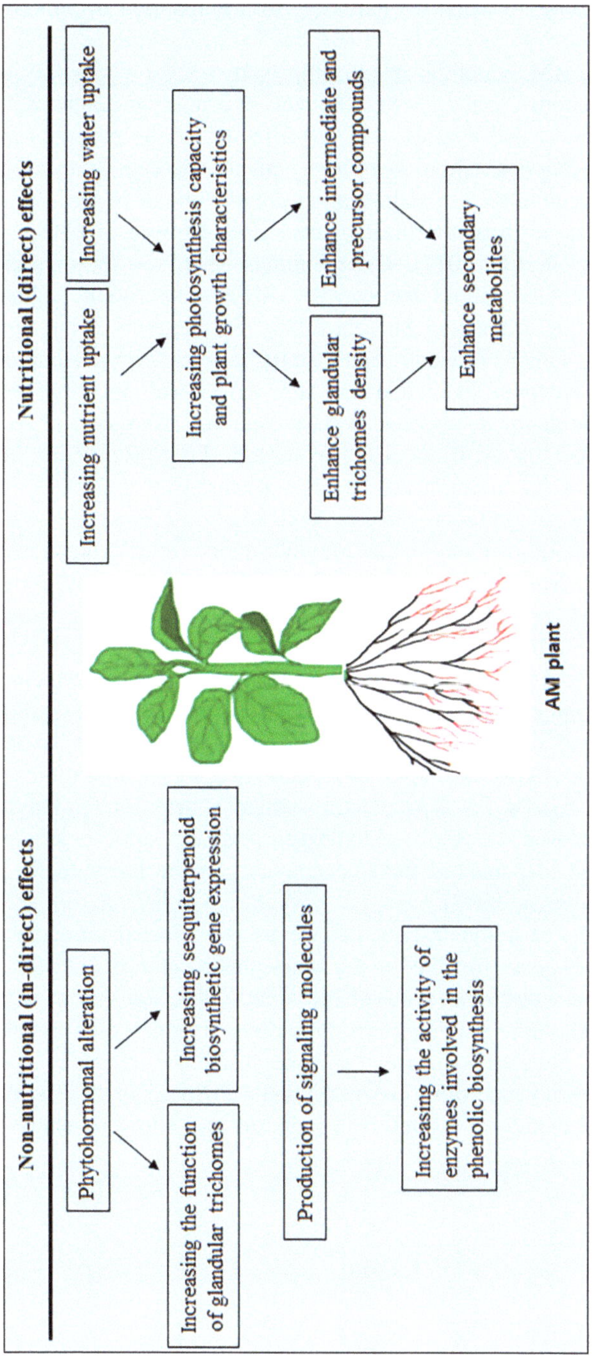

Fig. 5.4 Proposed mechanisms affecting the production of SMs in AMF-colonized plants. Redrawn after Amani Machiani et al. (2022)

phosphoenol-pyruvate, erythrose-4-phosphate, pyruvate, glyceraldehyde phosphate, and acetyl coenzyme A, which are necessary for the synthesis of SMs (Kapoor et al. 2017).

Although the AM symbiotic relationship increases the availability of nutrients, especially phosphorus, leading to higher levels of primary metabolites, the levels of phenols, flavonoids, and tannins were not affected by the phosphorus nutritional status in cebil (*Anadenanthera colubrina*) plants (Pedone-Bonfim et al. 2013). Similarly, the accumulation of anthocyanins and flavonoids in *Medicago truncatula* leaves was observed in mycorrhizal plants in the absence of phosphorus uptake benefits (Adolfsson et al. 2017). These findings emphasize the reprogramming of SMs in the phenylpropanoid pathway by AMF colonization, regardless of its nutritional function (Kaur and Suseela 2020). Furthermore, it has been reported that inoculation with *F. mosseae* significantly increased Eo accumulation in two genotypes of *Origanum* sp., while the same effect was not observed when the non-mycorrhizal plants are supplied with phosphorus (Khaosaad et al. 2006). The authors concluded that AMF-mediated increase in Eo concentration in *Origanum* sp. plants was not due to nutritional benefits (Khaosaad et al. 2006).

In non-nutritional (indirect) effects, AMF affect the biosynthetic pathways of SMs through modifications in the phytohormone levels such as cytokinins, gibberellic acid, and jasmonic acid (Erb and Kliebenstein 2020). For example, Liao et al. (2018) reported that gibberellic acid and jasmonic acid increased the levels of terpenoids by stimulating sesquiterpenoid biosynthetic gene expression and glandular trichome formation. Also, jasmonic acid stimulated two monoterpenes synthesis in *Lycopersicon esculentum* (Van Schie et al. 2007), and sesquiterpenoid biosynthesis gene in *Foeniculum vulgare* (Maes et al. 2011). In addition, the signaling molecules generated upon host plant-AMF interaction influence the levels of SMs in plant tissues. For example, the symbiosis between *Trifolium repens* and *F. mosseae* increased the levels of hydrogen peroxide, nitric oxide, and salicylic acid, which lead to enhanced enzyme activities in the biosynthetic pathway of phenolics (Zhang et al. 2013).

The two modes of action described above are not independent of each other. The expression of key genes involved in the biosynthetic pathways of SMs can also be affected by plant nutritional status (Lillo et al. 2008; Lazzara et al. 2017). For example, the gene expression pattern of various enzymes such as phenylalanine ammonia-lyase, flavonol synthase, chalcone isomerase, chalcone synthase, and cinnamate-4-hydroxylase, from the biosynthetic pathway of flavonoids, has been shown to be affected by the level of phosphorus and nitrogen supply (Lillo et al. 2008).

5.4 The Importance of AMF in Medicinal Plants

Medicinal plants are valuable and perpetual reservoirs of SMs, and in fact, they are the main effective sources of numerous medicinal compounds (Chandran et al. 2020). Due to the lipophilic properties of active oils in several medicinal plants, these substances are capable to entirely disrupt the membrane structure of bacteria, particularly Gram-negative type (Nazzaro et al. 2013). Some other SMs have immune response properties and antibacterial, antioxidant, antiviral, anticancer, and many other capabilities (Chandran et al. 2020).

The widespread approach of the world to the use of herbal medicines has increased the attention of most countries in the world to the identification and therapeutic use of medicinal plants and a return to nature. Chemical drugs as pure compounds, and despite their effectiveness, have several adverse effects, while the bioactive compounds found in medicinal plants are combined with other substances and thus have a state of biological balance. According to the statistics of WHO,[1] more than 80% of the people around the world, especially in developing and poor societies, meet their main medical needs from medicinal plants (Chandran et al. 2020; Zhao et al. 2022).

The growth of human and animal populations particularly in the developing countries, along with the constant increase in food, fodder, and fuel needs, put a huge pressure on the stability of the elements of the agricultural ecosystem. Since the fertility of the land is decreasing, farmers are turning towards the destruction of natural resource lands to obtain more crops. Therefore, if the current trend of population growth continues, in addition to the reduction of forests and rangelands, the production capacity of agricultural lands will also decrease (Shah et al. 2021). Thus, the utilization of phytomicrobiome components is suggested as one of the efficient and necessary solutions for the sustainable growth of crop production (Shah et al. 2021).

5.5 Production of Medicinal Plants

Despite the growing market demand, medicinal plants naturally have a very low performance in producing SMs (Thakur et al. 2019). So, several techniques have been developed aimed at improving the performance of these valuable metabolites in medicinal plants, such as *Agrobacterium*-mediated transformation, metabolic engineering, and mass propagation (Oksman-Caldentey and Inzé 2004; Liu et al. 2005). However, each of these methods is associated with limitations such as low practicality, high cost, and low effectiveness (Oksman-Caldentey and Inzé 2004; El-Sayed and Verpoorte 2007; Kapoor et al. 2017). On the other hand, plant productivity in intensive farming methods has a very close connection with the heavy use of

[1] The World Health Organization.

chemicals including inorganic herbicides, pesticides, and fertilizers. Such widespread use of chemicals, in addition to the risks for human health and environmental issues, reduce the quality and quantity of bioactive molecules in medicinal herbs (Amani Machiani et al. 2022). So, there is a necessity for using environmentally friendly strategies to improve the production of SMs (Li et al. 2023). Symbiosis of plants with AMF can change the pattern of production and accumulation of SMs (Fig. 5.2), depending on the fungus and plant species and the growth environment (Tsiokanos et al. 2022). Therefore, as many studies have confirmed (Chandran et al. 2020; Zhao et al. 2022), inoculation of plants with AMF offers a promising tool and a simpler alternative to overcome these limitations (Kapoor et al. 2017).

5.6 Conclusion

It is well established that AMF symbiosis reprograms plant secondary metabolism. AMF affect the production and accumulation of SMs directly by improving the uptake of water and nutrients, enhancing the photosynthetic capacity, and increasing the biomass, or indirectly by stimulating the biosynthetic pathways of SMs through the production of signaling compounds and changes in the phytohormone levels. Some of these compounds show a characteristic pattern depending on the developmental stage of colonization and act as signals between two partners over the pre-symbiosis stage to the formation of a functional symbiosis and in the maintenance of AM symbiosis. In addition, modifications in secondary metabolism following plant mycorrhization increase the concentration of plant antioxidant compounds. As a result, plant's ability to tolerate stressful conditions and the quality of plant products are improved.

A wide group of SMs have nutritional and medicinal value for human. Wide application and increasing demand for these valuable biological reservoirs make it necessary to invest in research and development to optimize their production. The use of AMF technique is considered as a potential tool and a promising opportunity to improve the quality and yield of plant active ingredients. AMF technique is a potential tool in crop management and plays a prominent role in promoting new environmentally friendly products in low-input and sustainable agricultural systems. In general, three research priorities are suggested to advance future studies: (1) more in-depth studies for understanding the mechanisms through which AMF association changes the SMs quantities and qualities, especially in terms of the species-specific nature of AMF-plant symbiosis; (2) the effects of environmental factors and developmental stages of the symbiosis, which have been greatly underestimated in the current studies, and (3) development of strategies for the implementation of mycorrhiza in the production of aromatic and medicinal compounds for industrial and commercial purposes.

References

Abu-Zeyad R, Khan AG, Khoo C (1999) Occurrence of arbuscular mycorrhiza in Castanospermum australe A. Cunn. & C. Fraser and effects on growth and production of castanospermine. Mycorrhiza 9(2):111–117

Adolfsson L, Nziengui H, Abreu IN, Šimura J, Beebo A, Herdean A, Spetea C (2017) Enhanced secondary-and hormone metabolism in leaves of arbuscular mycorrhizal Medicago truncatula. Plant Physiol 175(1):392–411

Agostini-Costa TDS, Vieira RF, Bizzo HR, Silveira D, Gimenes MA (2012) Secondary metabolites. In: Chromatography and its applications, vol 1. IntechOpen, London, pp 131–164

Al-Arjani ABF, Hashem A, Abd Allah EF (2020) Arbuscular mycorrhizal fungi modulates dynamics tolerance expression to mitigate drought stress in Ephedra foliata Boiss. Saudi J Biol Sci 27(1):380–394

Al-Babili S, Bouwmeester HJ (2015) Strigolactones, a novel carotenoid-derived plant hormone. Annu Rev Plant Biol 66:161–186

Aliferis KA, Chamoun R, Jabaji S (2015) Metabolic responses of willow (Salix purpurea L.) leaves to mycorrhization as revealed by mass spectrometry and 1H NMR spectroscopy metabolite profiling. Front Plant Sci 6:344

Amani Machiani M, Javanmard A, Morshedloo MR, Aghaee A, Maggi F (2021) Funneliformis mosseae inoculation under water deficit stress improves the yield and phytochemical characteristics of thyme in intercropping with soybean. Sci Rep 11(1):15279

Amani Machiani M, Javanmard A, Habibi Machiani R, Sadeghpour A (2022) Arbuscular mycorrhizal fungi and changes in primary and secondary metabolites. Plants 11(17):2183

Avio L, Maggini R, Ujvári G, Incrocci L, Giovannetti M, Turrini A (2020) Phenolics content and antioxidant activity in the leaves of two artichoke cultivars are differentially affected by six mycorrhizal symbionts. Sci Hortic 264:109153

Babikova Z, Gilbert L, Bruce TJ, Birkett M, Caulfield JC, Woodcock C, Johnson D (2013) Underground signals carried through common mycelial networks warn neighbouring plants of aphid attack. Ecol Lett 16(7):835–843

Bag S, Mondal A, Majumder A, Mondal SK, Banik A (2022) Flavonoid mediated selective cross-talk between plants and beneficial soil microbiome. Phytochem Rev 21(5):1739–1760

Bakkali F, Averbeck S, Averbeck D, Idaomar M (2008) Biological effects of essential oils–a review. Food Chem Toxicol 46(2):446–475

Battini F, Grønlund M, Agnolucci M, Giovannetti M, Jakobsen I (2017) Facilitation of phosphorus uptake in maize plants by mycorrhizosphere bacteria. Sci Rep 7(1):4686

Begum N, Qin C, Ahanger MA, Raza S, Khan MI, Ashraf M, Zhang L (2019) Role of arbuscular mycorrhizal fungi in plant growth regulation: implications in abiotic stress tolerance. Front Plant Sci 10:1068

Cartabia A, Tsiokanos E, Tsafantakis N, Lalaymia I, Termentzi A, Miguel M, Declerck S (2021) The arbuscular mycorrhizal fungus Rhizophagus irregularis MUCL 41833 modulates metabolites production of Anchusa officinalis L. under semi-hydroponic cultivation. Front Plant Sci 12: 724352

Catford JG, Staehelin C, Larose G, Piché Y, Vierheilig H (2006) Systemically suppressed isoflavonoids and their stimulating effects on nodulation and mycorrhization in alfalfa split-root systems. Plant Soil 285:257–266

Chandran H, Meena M, Barupal T, Sharma K (2020) Plant tissue culture as a perpetual source for production of industrially important bioactive compounds. Biotechnol Rep 26:e00450

Chiej R (1984) The Macdonald encyclopedia of medicinal plants. Macdonald & Co (Publishers) Ltd., London

Chukwuka KS, Ikheloa JO, Okonko IO, Moody JO, Mankinde TA (2011) The antimicrobial activities of some medicinal plants on Escherichia coli as an agent of diarrhea in livestock. Adv Appl Sci Res 2(4):37–48

Copetta A, Lingua G, Berta G (2006) Effects of three AM fungi on growth, distribution of glandular hairs, and essential oil production in Ocimum basilicum L. var. Genovese. Mycorrhiza 16:485–494
Covello PS, Teoh KH, Polichuk DR, Reed DW, Nowak G (2007) Functional genomics and the biosynthesis of artemisinin. Phytochemistry 68(14):1864–1871
Deng Y, Lu S (2017) Biosynthesis and regulation of phenylpropanoids in plants. Crit Rev Plant Sci 36(4):257–290
Devi MC, Reddy MN (2002) Phenolic acid metabolism of groundnut (Arachis hypogaea L.) plants inoculated with VAM fungus and rhizobium. Plant Growth Regul 37:151–156
Dey P, Kundu A, Kumar A, Gupta M, Lee BM, Bhakta T, Kim HS (2020) Analysis of alkaloids (indole alkaloids, isoquinoline alkaloids, tropane alkaloids). In: Recent advances in natural products analysis. Elsevier, London, pp 505–567
Drigo B, Pijl AS, Duyts H, Kielak AM, Gamper HA, Houtekamer MJ, Kowalchuk GA (2010) Shifting carbon flow from roots into associated microbial communities in response to elevated atmospheric CO2. Proc Natl Acad Sci 107(24):10938–10942
Duc NH, Vo AT, Haddidi I, Daood H, Posta K (2021) Arbuscular mycorrhizal fungi improve tolerance of the medicinal plant Eclipta prostrata (L.) and induce major changes in polyphenol profiles under salt stresses. Front Plant Sci 11:612299
Duhamel M, Pel R, Ooms A, Bücking H, Jansa J, Ellers J, Kiers ET (2013) Do fungivores trigger the transfer of protective metabolites from host plants to arbuscular mycorrhizal hyphae? Ecology 94(9):2019–2029
El-Sayed M, Verpoorte R (2007) Catharanthus terpenoid indole alkaloids: biosynthesis and regulation. Phytochem Rev 6:277–305
Erb M, Kliebenstein DJ (2020) Plant secondary metabolites as defenses, regulators, and primary metabolites: the blurred functional trichotomy. Plant Physiol 184(1):39–52
Eulenstein F, Tauschke M, Behrendt A, Monk J, Schindler U, Lana MA, Monk S (2016) The application of mycorrhizal fungi and organic fertilisers in horticultural potting soils to improve water use efficiency of crops. Horticulturae 3(1):8
Fester T, Wray V, Nimtz M, Strack D (2005) Is stimulation of carotenoid biosynthesis in arbuscular mycorrhizal roots a general phenomenon? Phytochemistry 66(15):1781–1786
Fiorilli V, Wang JY, Bonfante P, Lanfranco L, Al-Babili S (2019) Apocarotenoids: old and new mediators of the arbuscular mycorrhizal symbiosis. Front Plant Sci 10:1186
Fontana A, Reichelt M, Hempel S, Gershenzon J, Unsicker SB (2009) The effects of arbuscular mycorrhizal fungi on direct and indirect defense metabolites of Plantago lanceolata L. J Chem Ecol 35:833–843
Fraser PD, Schuch W, Bramley PM (2000) Phytoene synthase from tomato (Lycopersicon esculentum) chloroplasts–partial purification and biochemical properties. Planta 211:361–369
Gershenzon J (1994) Metabolic costs of terpenoid accumulation in higher plants. J Chem Ecol 20: 1281–1328
Ghamarnia H, Palash M, Dousti B (2022) Camelina zoning for different climate conditions in Kurdistan Province. Agrotechniques Ind Crops 2(1):49–56
Ghasemzadeh A, Ghasemzadeh N (2011) Flavonoids and phenolic acids: role and biochemical activity in plants and human. J Med Plant Res 5(31):6697–6703
Grümberg BC, Urcelay C, Shroeder MA, Vargas-Gil S, Luna CM (2015) The role of inoculum identity in drought stress mitigation by arbuscular mycorrhizal fungi in soybean. Biol Fertil Soils 51:1–10
Guerrieri A, Dong L, Bouwmeester HJ (2019) Role and exploitation of underground chemical signaling in plants. Pest Manag Sci 75(9):2455–2463
Hamilton JG, Zangerl AR, DeLucia EH, Berenbaum MR (2001) The carbon–nutrient balance hypothesis: its rise and fall. Ecol Lett 4(1):86–95
Harrison MJ, Dixon RA (1993) Isoflavonoid accumulation and expression of defense gene transcripts during the establishment of vesicular-arbuscular mycorrhizal associations in roots of Medicago truncatula. Mol Plant-Microbe Interact 6(5):643–654

Herbert RB (2001) The biosynthesis of plant alkaloids and nitrogenous microbial metabolites. Nat Prod Rep 18(1):50–65

Hussein RA, El-Anssary AA (2019) Plants secondary metabolites: the key drivers of the pharmacological actions of medicinal plants. Herb Med 1(3):76139

Igiehon NO, Babalola OO (2017) Biofertilizers and sustainable agriculture: exploring arbuscular mycorrhizal fungi. Appl Microbiol Biotechnol 101:4871–4881

Isah T (2019) Stress and defense responses in plant secondary metabolites production. Biol Res 52: 39

Jaiti F, Meddich A, El Hadrami I (2007) Effectiveness of arbuscular mycorrhizal fungi in the protection of date palm (Phoenix dactylifera L.) against Bayoud disease. Physiol Mol Plant Pathol 71(4-6):166–173

Johny L, Cahill DM, Adholeya A (2021) AMF enhance secondary metabolite production in ashwagandha, licorice, and marigold in a fungi-host specific manner. Rhizosphere 17:100314

Jung SC, Martinez-Medina A, Lopez-Raez JA, Pozo MJ (2012) Mycorrhiza-induced resistance and priming of plant defenses. J Chem Ecol 38:651–664

Kapoor R, Anand G, Gupta P, Mandal S (2017) Insight into the mechanisms of enhanced production of valuable terpenoids by arbuscular mycorrhiza. Phytochem Rev 16:677–692

Kappers IF, Aharoni A, Van Herpen TW, Luckerhoff LL, Dicke M, Bouwmeester HJ (2005) Genetic engineering of terpenoid metabolism attracts bodyguards to Arabidopsis. Science 309(5743):2070–2072

Karagiannidis N, Thomidis T, Lazari D, Panou-Filotheou E, Karagiannidou C (2011) Effect of three Greek arbuscular mycorrhizal fungi in improving the growth, nutrient concentration, and production of essential oils of oregano and mint plants. Sci Hortic 129(2):329–334

Kaschuk G, Kuyper TW, Leffelaar PA, Hungria M, Giller KE (2009) Are the rates of photosynthesis stimulated by the carbon sink strength of rhizobial and arbuscular mycorrhizal symbioses? Soil Biol Biochem 41(6):1233–1244

Kaur S, Suseela V (2020) Unraveling arbuscular mycorrhiza-induced changes in plant primary and secondary metabolome. Metabolites 10(8):335

Khaosaad T, Vierheilig H, Nell M, Zitterl-Eglseer K, Novak J (2006) Arbuscular mycorrhiza alter the concentration of essential oils in oregano (Origanum sp., Lamiaceae). Mycorrhiza 16:443–446

Kogel KH, Voll LM, Schäfer P, Jansen C, Wu Y, Langen G, Sonnewald U (2010) Transcriptome and metabolome profiling of field-grown transgenic barley lack induced differences but show cultivar-specific variances. Proc Natl Acad Sci 107(14):6198–6203

Kretzschmar T, Kohlen W, Sasse J, Borghi L, Schlegel M, Bachelier JB, Martinoia E (2012) A petunia ABC protein controls strigolactone-dependent symbiotic signalling and branching. Nature 483(7389):341–344

Krishnaiah D, Sarbatly R, Nithyanandam R (2011) A review of the antioxidant potential of medicinal plant species. Food Bioprod Process 89(3):217–233

Lazzara S, Militello M, Carrubba A, Napoli E, Saia S (2017) Arbuscular mycorrhizal fungi altered the hypericin, pseudohypericin, and hyperforin content in flowers of Hypericum perforatum grown under contrasting P availability in a highly organic substrate. Mycorrhiza 27(4):345–354

Lermen C, Morelli F, Gazim ZC, da Silva AP, Goncalves JE, Dragunski DC, Alberton O (2015) Essential oil content and chemical composition of Cymbopogon citratus inoculated with arbuscular mycorrhizal fungi under different levels of lead. Ind Crop Prod 76:734–738

Li J, Meng B, Chai H, Yang X, Song W, Li S, Sun W (2019) Arbuscular mycorrhizal fungi alleviate drought stress in C3 (Leymus chinensis) and C4 (Hemarthria altissima) grasses via altering antioxidant enzyme activities and photosynthesis. Front Plant Sci 10:499

Li C, Swofford CA, Sinskey AJ (2020) Modular engineering for microbial production of carotenoids. Metabol Eng Commun 10:e00118

Li Z, Xiong K, Wen W, Li L, Xu D (2023) Functional endophytes regulating plant secondary metabolism: current status, prospects and applications. Int J Mol Sci 24(2):1153

Liao D, Wang S, Cui M, Liu J, Chen A, Xu G (2018) Phytohormones regulate the development of arbuscular mycorrhizal symbiosis. Int J Mol Sci 19(10):3146

Lillo C, Lea US, Ruoff P (2008) Nutrient depletion as a key factor for manipulating gene expression and product formation in different branches of the flavonoid pathway. Plant Cell Environ 31(5): 587–601

Liu Y, Wang H, Ye HC, Li GF (2005) Advances in the plant isoprenoid biosynthesis pathway and its metabolic engineering. J Integr Plant Biol 47(7):769–782

Liu J, Wu L, Wei S, Xiao X, Su C, Jiang P, Yu Z (2007) Effects of arbuscular mycorrhizal fungi on the growth, nutrient uptake and glycyrrhizin production of licorice (Glycyrrhiza uralensis Fisch). Plant Growth Regul 52:29–39

López-Ráez JA, Flors V, García JM, Pozo MJ (2010) AM symbiosis alters phenolic acid content in tomato roots. Plant Signal Behav 5(9):1138–1140

Maes L, Van Nieuwerburgh FC, Zhang Y, Reed DW, Pollier J, Vande Casteele SR, Goossens A (2011) Dissection of the phytohormonal regulation of trichome formation and biosynthesis of the antimalarial compound artemisinin in Artemisia annua plants. New Phytol 189(1):176–189

Maier W, Hammer K, Dammann U, Schulz B, Strack D (1997) Accumulation of sesquiterpenoid cyclohexenone derivatives induced by an arbuscular mycorrhizal fungus in members of the Poaceae. Planta 202:36–42

Maier W, Schmidt J, Nimtz M, Wray V, Strack D (2000) Secondary products in mycorrhizal roots of tobacco and tomato. Phytochemistry 54(5):473–479

Mandal SM, Chakraborty D, Dey S (2010) Phenolic acids act as signaling molecules in plant-microbe symbioses. Plant Signal Behav 5(4):359–368

Mandal S, Evelin H, Giri B, Singh VP, Kapoor R (2013) Arbuscular mycorrhiza enhances the production of stevioside and rebaudioside-A in Stevia rebaudiana via nutritional and non-nutritional mechanisms. Appl Soil Ecol 72:187–194

Mandal S, Upadhyay S, Singh VP, Kapoor R (2015a) Enhanced production of steviol glycosides in mycorrhizal plants: a concerted effect of arbuscular mycorrhizal symbiosis on transcription of biosynthetic genes. Plant Physiol Biochem 89:100–106

Mandal S, Upadhyay S, Wajid S, Ram M, Jain DC, Singh VP, Kapoor R (2015b) Arbuscular mycorrhiza increase artemisinin accumulation in Artemisia annua by higher expression of key biosynthesis genes via enhanced jasmonic acid levels. Mycorrhiza 25:345–357

Mhlongo MI, Piater LA, Madala NE, Labuschagne N, Dubery IA (2018) The chemistry of plant–microbe interactions in the rhizosphere and the potential for metabolomics to reveal signaling related to defense priming and induced systemic resistance. Front Plant Sci 9:112

Michalak A (2006) Phenolic compounds and their antioxidant activity in plants growing under heavy metal stress. Pol J Environ Stud 15(4):523–530

Mierziak J, Kostyn K, Kulma A (2014) Flavonoids as important molecules of plant interactions with the environment. Molecules 19(10):16240–16265

Moradtalab N, Hajiboland R, Aliasgharzad N, Hartmann TE, Neumann G (2019) Silicon and the association with an arbuscular-mycorrhizal fungus (Rhizophagus clarus) mitigate the adverse effects of drought stress on strawberry. Agronomy 9(1):41

Morone-Fortunato I, Avato P (2008) Plant development and synthesis of essential oils in micropropagated and mycorrhiza inoculated plants of Origanum vulgare L. ssp. hirtum (Link) Ietswaart. Plant Cell Tissue Organ Cult 93:139–149

Muniz BC, Falcão EL, de Paula Monteiro R, dos Santos EL, Bastos Filho CJA, da Silva FSB (2021) Acaulospora longula Spain & NC Schenck: a low-cost bioinsumption to optimize phenolics and saponins production in Passiflora alata Curtis. Ind Crop Prod 167:113498

Nasseri MA, Keshtkar H, Kazemnejadi M, Allahresani A (2020) Phytochemical properties and antioxidant activity of Echinops persicus plant extract: green synthesis of carbon quantum dots from the plant extract. SN Appl Sci 2:1–12

Nazzaro F, Fratianni F, De Martino L, Coppola R, De Feo V (2013) Effect of essential oils on pathogenic bacteria. Pharmaceuticals 6(12):1451–1474

Nell M, Voetsch M, Vierheilig H, Steinkellner S, Zitterl-Eglseer K, Franz C, Novak J (2009) Effect of phosphorus uptake on growth and secondary metabolites of garden sage (Salvia officinalis L.). J Sci Food Agric 89(6):1090–1096
Nurzynska-Wierdak R (2013) Does mineral fertilization modify essential oil content and chemical composition in medicinal plants? Acta Sci Pol Hortorum Cultus 12:5
Oksman-Caldentey KM, Inzé D (2004) Plant cell factories in the post-genomic era: new ways to produce designer secondary metabolites. Trends Plant Sci 9(9):433–440
Ostadi A, Javanmard A, Amani Machiani M, Sadeghpour A, Maggi F, Nouraein M, Lorenzo JM (2022) Co-application of TiO2 nanoparticles and arbuscular mycorrhizal fungi improves essential oil quantity and quality of sage (Salvia officinalis L.) in drought stress conditions. Plants 11(13):1659
Pandey DK, Malik T, Dey A, Singh J, Banik RM (2014) Improved growth and colchicine concentration in Gloriosa superba on mycorrhizal inoculation supplemented with phosphorus-fertilizer. Afr J Tradit Complement Altern Med 11(2):439–446
Pandey DK, Kaur P, Dey A (2018) Arbuscular mycorrhizal fungi: effects on secondary metabolite production in medicinal plants. In: Fungi and their role in sustainable development: current perspectives. Springer, Cham, pp 507–538
Pedone-Bonfim MV, Lins MA, Coelho IR, Santana AS, Silva FS, Maia LC (2013) Mycorrhizal technology and phosphorus in the production of primary and secondary metabolites in cebil (Anadenanthera colubrina (Vell.) Brenan) seedlings. J Sci Food Agric 93(6):1479–1484
Peipp H, Maier W, Schmidt J, Wray V, Strack D (1997) Arbuscular mycorrhizal fungus-induced changes in the accumulation of secondary compounds in barley roots. Phytochemistry 44(4): 581–587
Piasecka A, Jedrzejczak-Rey N, Bednarek P (2015) Secondary metabolites in plant innate immunity: conserved function of divergent chemicals. New Phytol 206(3):948–964
Pozo MJ, Verhage A, García-Andrade J, García JM, Azcón-Aguilar C (2009) Priming plant defence against pathogens by arbuscular mycorrhizal fungi. In: Mycorrhizas-functional processes and ecological impact. Springer, Cham, pp 123–135
Pusztahelyi T, Holb IJ, Pócsi I (2015) Secondary metabolites in fungus-plant interactions. Front Plant Sci 6:573
Reyes-Carmona J, Yousef GG, Martínez-Peniche RA, Lila MA (2005) Antioxidant capacity of fruit extracts of blackberry (Rubus sp.) produced in different climatic regions. J Food Sci 70(7):s497–s503
Rezaei-Chiyaneh E, Battaglia ML, Sadeghpour A, Shokrani F, Nasab ADM, Raza MA, von Cossel M (2021) Optimizing intercropping systems of black cumin (Nigella sativa L.) and fenugreek (Trigonella foenum-graecum L.) through inoculation with bacteria and mycorrhizal fungi. Adv Sustainable Syst 5(9):2000269
Rojas-Andrade R, Cerda-García-Rojas C, Frías-Hernández J, Dendooven L, Olalde-Portugal V, Ramos-Valdivia A (2003) Changes in the concentration of trigonelline in a semi-arid leguminous plant (Prosopis laevigata) induced by an arbuscular mycorrhizal fungus during the presymbiotic phase. Mycorrhiza 13(1):49–52
Rozpądek P, Wężowicz K, Stojakowska A, Malarz J, Surówka E, Anielska T, Turnau K (2014) Mycorrhizal fungi modulate phytochemical production and antioxidant activity of Cichorium intybus L.(Asteraceae) under metal toxicity. Chemosphere 112:217–224
Sacchettini JC, Poulter CD (1997) Creating isoprenoid diversity. Science 277(5333):1788–1789
Saia S, Benítez E, García-Garrido JM, Settanni L, Amato G, Giambalvo D (2014) The effect of arbuscular mycorrhizal fungi on total plant nitrogen uptake and nitrogen recovery from soil organic material. J Agric Sci 152(3):370–378
Santos EL, Silva WAV, Ferreira MRA, Soares LA, Sampaio EVSB, Silva FA, Silva FSB (2020) Acaulospora longula increases the content of phenolic compounds and antioxidant activity in fruits of Libidibia ferrea. Open Microbiol J 14(1):132–139

Schliemann W, Schmidt J, Nimtz M, Wray V, Fester T, Strack D (2006) Accumulation of apocarotenoids in mycorrhizal roots of Ornithogalum umbellatum. Phytochemistry 67(12): 1196–1205
Schliemann W, Ammer C, Strack D (2008) Metabolite profiling of mycorrhizal roots of Medicago truncatula. Phytochemistry 69(1):112–146
Schweiger R, Müller C (2015) Leaf metabolome in arbuscular mycorrhizal symbiosis. Curr Opin Plant Biol 26:120–126
Shah A, Nazari M, Antar M, Msimbira LA, Naamala J, Lyu D, Smith DL (2021) PGPR in agriculture: a sustainable approach to increasing climate change resilience. Front Sustainable Food Syst 5:667546
Sharma E, Anand G, Kapoor R (2017) Terpenoids in plant and arbuscular mycorrhiza-reinforced defence against herbivorous insects. Ann Bot 119(5):791–801
Shumskaya M, Wurtzel ET (2013) The carotenoid biosynthetic pathway: thinking in all dimensions. Plant Sci 208:58–63
Silva FA, Maia LC, Silva FS (2019) Arbuscular mycorrhizal fungi as biotechnology alternative to increase concentrate of secondary metabolites in Zea mays L. Rev Bras Bot 42:189–193
Simões CMO, Schenkel EP, de Mello JCP, Mentz LA, Petrovick PR (2016) Farmacognosia: do produto natural ao medicamento. Artmed Editora, Porto Alegre
Singla RK, Dubey AK, Garg A, Sharma RK, Fiorino M, Ameen SM, Al-Hiary M (2019) Natural polyphenols: chemical classification, definition of classes, subcategories, and structures. J AOAC Int 102(5):1397–1400
Song YY, Ye M, Li CY, Wang RL, Wei XC, Luo SM, Zeng RS (2013) Priming of anti-herbivore defense in tomato by arbuscular mycorrhizal fungus and involvement of the jasmonate pathway. J Chem Ecol 39:1036–1044
Song MC, Kim EJ, Kim E, Rathwell K, Nam SJ, Yoon YJ (2014) Microbial biosynthesis of medicinally important plant secondary metabolites. Nat Prod Rep 31(11):1497–1509
Strack D, Fester T (2006) Isoprenoid metabolism and plastid reorganization in arbuscular mycorrhizal roots. New Phytol 172(1):22–34
Thakur M, Bhattacharya S, Khosla PK, Puri S (2019) Improving production of plant secondary metabolites through biotic and abiotic elicitation. J Appl Res Med Aromat Plants 12:1–12
Tsiokanos E, Cartabia A, Tsafantakis N, Lalaymia I, Termentzi A, Miguel M, Fokialakis N (2022) The metabolic profile of Anchusa officinalis L. differs according to its associated arbuscular mycorrhizal fungi. Metabolites 12(7):573
Van Der Heijden MG, Martin FM, Selosse MA, Sanders IR (2015) Mycorrhizal ecology and evolution: the past, the present, and the future. New Phytol 205(4):1406–1423
Van Schie CC, Haring MA, Schuurink RC (2007) Tomato linalool synthase is induced in trichomes by jasmonic acid. Plant Mol Biol 64:251–263
Walid, N., Kassam, R., & Ashfaq, M. A. (2021). Interaction between mycorrhizal and medicinal plants towards enhancement of secondary metabolites. Int. J. Chem. Stud, 9(2), 140–152.
Wang M, Schäfer M, Li D, Halitschke R, Dong C, McGale E, Baldwin IT (2018) Blumenols as shoot markers of root symbiosis with arbuscular mycorrhizal fungi. elife 7:e37093
Wei GT, Wang HG (1989) Effects of VA mycorrhizal fungi on growth, nutrient uptake and effective compounds in Chinese medicinal herb Datura stramonium L. Sci Agric Sin 25(5): 56–61
Weisany W, Raei Y, Pertot I (2015) Changes in the essential oil yield and composition of dill (Anethum graveolens L.) as response to arbuscular mycorrhiza colonization and cropping system. Ind Crop Prod 77:295–306
Wink M (2008) Plant secondary metabolism: diversity, function and its evolution. Nat Prod Commun 3(8):1934578X0800300801
Wu QS, Silva FS, Hijri M, Kapoor R (2023) Arbuscular mycorrhiza-mediated augmentation of plant secondary metabolite production. Front Plant Sci 14:1150900

Xie W, Hao Z, Yu M, Wu Z, Zhao A, Li J, Chen B (2018a) Improved phosphorus nutrition by arbuscular mycorrhizal symbiosis as a key factor facilitating glycyrrhizin and liquiritin accumulation in Glycyrrhiza uralensis. Plant Soil 439:243–257

Xie W, Hao Z, Zhou X, Jiang X, Xu L, Wu S, Chen B (2018b) Arbuscular mycorrhiza facilitates the accumulation of glycyrrhizin and liquiritin in Glycyrrhiza uralensis under drought stress. Mycorrhiza 28:285–300

Yang L, Wen KS, Ruan X, Zhao YX, Wei F, Wang Q (2018) Response of plant secondary metabolites to environmental factors. Molecules 23(4):762

Yeshi K, Crayn D, Ritmejerytė E, Wangchuk P (2022) Plant secondary metabolites produced in response to abiotic stresses has potential application in pharmaceutical product development. Molecules 27(1):313

Yousefi B, Safari H (2022) Evaluation of essential oil, morphological traits and crop yield in Satureja mutica Fisch. & CA Mey. in rainfed cultivation under different row spacing. Agrotechniques Ind Crops 2(4):187–197

Zaynab M, Fatima M, Abbas S, Sharif Y, Umair M, Zafar MH, Bahadar K (2018) Role of secondary metabolites in plant defense against pathogens. Microb Pathog 124:198–202

Zhang RQ, Zhu HH, Zhao HQ, Yao Q (2013) Arbuscular mycorrhizal fungal inoculation increases phenolic synthesis in clover roots via hydrogen peroxide, salicylic acid and nitric oxide signaling pathways. J Plant Physiol 170(1):74–79

Zhao R, Guo W, Bi N, Guo J, Wang L, Zhao J, Zhang J (2015) Arbuscular mycorrhizal fungi affect the growth, nutrient uptake and water status of maize (Zea mays L.) grown in two types of coal mine spoils under drought stress. Appl Soil Ecol 88:41–49

Zhao Y, Cartabia A, Lalaymia I, Declerck S (2022) Arbuscular mycorrhizal fungi and production of secondary metabolites in medicinal plants. Mycorrhiza 32(3-4):221–256

Zhi-lin Y, Chuan-chao D, Lian-qing C (2007) Regulation and accumulation of secondary metabolites in plant-fungus symbiotic system. Afr J Biotechnol 6(11):57436

Zhu F, Wen W, Cheng Y, Alseekh S, Fernie AR (2023) Integrating multiomics data accelerates elucidation of plant primary and secondary metabolic pathways. Abiotech 2023:1–10

Chapter 6
Roles of Arbuscular Mycorrhizal Fungi for Essential Nutrient Acquisition Under Nutrient Deficiency in Plants

Kana Miyata and Mikihisa Umehara

Abstract Plants absorb mineral nutrients for growth and development from the soil though their roots; nutrient acquisition is therefore limited by their root area. To improve it, especially in nutrient-poor conditions, many plant species depend on symbiotic interactions with arbuscular mycorrhizal (AM) fungi, which provide essential nutrients obtained through the network of hyphae to the host plants. When nitrogen, phosphate, or sulfur is deficient, plants produce strigolactones, key signaling molecules, to initiate the interaction with AM fungi. Here, first, we introduce the diversity of AM fungi and their host plants. Second, we summarize the structural features of the symbiotic interaction. Third, we describe strigolactone biosynthesis and the symbiosis signaling pathway. Finally, we describe nutrient exchange system between AM fungi and host plants. Overall, we focus on the roles of AM symbiosis for nutrient acquisition in plants and detail the mechanisms. Understanding how plants adapt to their environment in response to deficiency of mineral nutrients could help to improve sustainable agricultural processes, because the use of AM fungi enables crop production in nutrient-poor environments and allows use of pesticides and fertilizers to be reduced.

Keywords Nitrogen · Nutrient exchange · Phosphate · Strigolactone · Symbiotic signaling pathway

6.1 Introduction

Plants absorb nutrients and minerals for growth from the region accessible by their roots; this region is limited. To expand their nutrient acquisition capacity, plants rely on a community of nearly invisible microorganisms. Mycorrhizal fungi are the most well-known symbiont and they promote uptake of ions, including phosphate and other minerals, and provide various other benefits for the host plants. The word

K. Miyata · M. Umehara (✉)
Department of Applied Biosciences, Toyo University, Itakura, Gunma, Japan
e-mail: umehara@toyo.jp

G. J. Ahammed, R. Hajiboland (eds.), *Arbuscular Mycorrhizal Fungi and Higher Plants*, https://doi.org/10.1007/978-981-99-8220-2_6

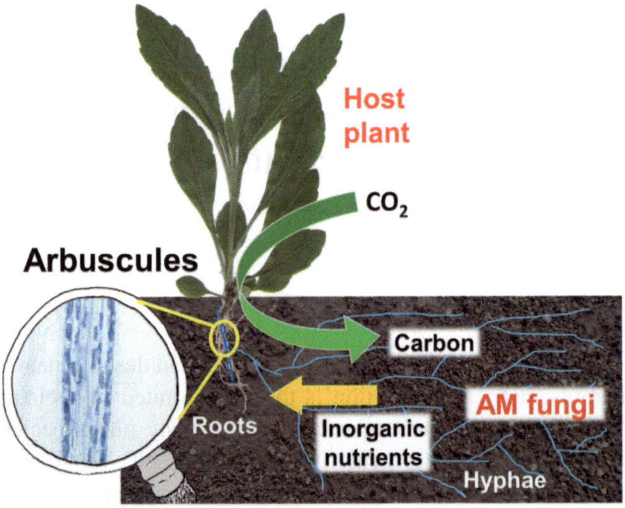

Fig. 6.1 Schematic of AM symbiosis

"mycorrhiza" comes from two Greek words, "myco," meaning "fungus," and "rhiza" meaning "root." Mycorrhizal fungi live within or adjacent to host root cells and provide various benefits for host plants by improving ion uptake and protecting the host plants from various threats (Smith and Read 2008). Mycorrhizal hyphae reach a broad region of soil and transfer the necessary resources to host plants, in exchange of carbohydrates, the product of photosynthesis (Smith and Smith 1990) (Fig. 6.1).

To facilitate resource exchange, mycorrhizal fungi produce unique structures, which can be used to classify mycorrhizal symbiosis into several types: ericoid mycorrhizal, orchid mycorrhizal, ectomycorrhizal (EcM), and arbuscular mycorrhizal (AM) (Brundrett and Tedersoo 2018). For example, EcM symbiosis is defined by the Hartig net, hyphae in the apoplast of cortex cells, and the fungal mantle, a hyphal layer that covers the root surface. Several commercially valuable mushrooms are the products of EcM symbiosis, e.g., matsutake (*Tricholoma matsutake*) and truffles (*Tuber himalayense* and *T. japonicum*). Some host plants of these mycorrhizal symbioses are mycoheterotrophic, meaning they obtain all of their nutrients from their associated fungi instead of from photosynthesis, e.g., species in the tribe Arbuteae. In orchid mycorrhizal symbiosis, host plants decompose the fungal body, so-called peloton, to obtain nutrients for germination and growth.

Mycorrhizal fungi thus establish unique associations with their host plant that can be either mutualistic or non-mutualistic. AM symbiosis is thought to be the most ancient of these associations, and the other types are thought to have evolved from it. The host range of AM symbioses is extremely wide, and it is estimated that more than 70% of terrestrial plants can have interaction with AM fungi (Brundrett and Tedersoo 2018). To initiate the symbiotic relationship, AM fungi and their host plants cooperate to produce arbuscules, specialized structures for resource exchange

(Fig. 6.1) (Smith and Read 2008; Smith and Smith 1997). In addition to improving uptake of mineral ion such as nitrogen and phosphate, AM symbiosis contributes to tolerance of heavy metals, desiccation, osmotic stress, and high temperatures and to pathogen defense responses (Gianinazzi et al. 2010; Lenoir et al. 2016; Thirkell et al. 2017). In the mycorrhizal symbiosis between rice and *Rhizophagus irregularis*, which is the model species of mycorrhizal fungi, the arbuscules have a basal trunk and a bush-like, highly branching structure in each host cell (Arum type, Figs. 6.1 and 6.2). Several other host–fungal pairs produce arbuscular coils instead of arbuscules (Paris type). Differentiation of these types is not clear-cut (Dickson et al. 2007). A single fungal species can form both Arum-type and Paris-type structures, relying on the host species, and both types can be observed in one host plant by different fungal species. Previously, AM symbiosis was mentioned as vesicular arbuscular (VA) mycorrhizal symbiosis. Vesicles, lipid storage structures, are also a structural feature of AM symbiosis, but because several species do not produce vesicles, the term AM symbiosis has been more widely adopted.

The association between AM fungi and plants began in the early Devonian, over 400 million years ago (Taylor et al. 1995). Structures identical to arbuscules were discovered in fossils of *Aglaophyton major*, the early land plant. Ancient AM fungal spores estimated to be from the Ordovician, 460 million years ago (Redecker et al. 2000), have also been found. This timing coincides with the emergence of early land plants, and the ancestors of AM fungi are thought to have supported terrestrialization of ancestral plants, aiding them to adapt to harsh environments by improving uptake of water and mineral nutrients in soil (Simon et al. 1993).

In addition to its evolutionary importance, mycorrhizal symbiosis is increasingly important in agriculture for ensuring a stable food supply. In this book chapter, we describe the fundamental roles of AM symbiosis in nutrient acquisition in plants and describe the details of its mechanisms.

6.2 Diversity of Arbuscular Mycorrhizal Fungi and Host Plants

AM fungi belong to the phylum Glomeromycota, which consists of 5 orders, 14 families, 29 genera, and approximately 230 species (Oehl et al. 2011). The taxonomic position of AM fungi remains debated, but recent molecular phylogenetic analyses indicate that the phylum Glomeromycota belong to the subphylum Mucoromycota (Lanfranco and Young 2012; Spatafora et al. 2016). The species most commonly used as a model for studying AM fungi is *R. irregularis* (previously named as *Glomus intraradices*) (Stockinger et al. 2009). The distribution of *R. irregularis* is global, and *its* spores are commercially available in pure culture (Premier Tech, Quebec, Canada); it is widely used in many types of studies, including molecular biology. Owing to its wide availability, its phenotype and life cycle have been well studied. Although early attempts at a complete genome

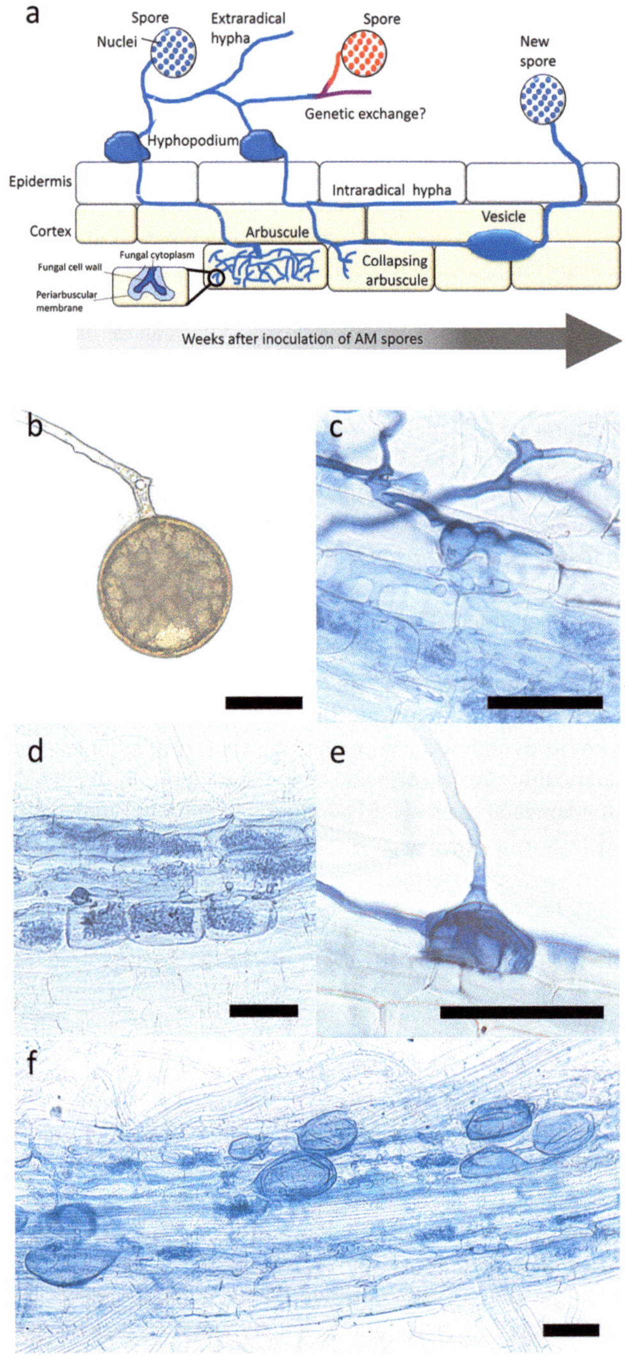

Fig. 6.2 Structural features of AM colonization. (**a**) A model life cycle of AM fungi. (**b**) Spore of *R. irregularis*. (**c**) Hyphopodium of *R. irregularis* in wild-type rice. (**d**) Arbuscules of *R. irregularis* in rice. (**e**) Hyphopodium of *R. irregularis* in *OsSYMRK* mutant. (**f**) Vesicles of *R. irregularis* in

sequence were fragmentary, owing to technical problems, the full sequence of *R. irregularis* DAOM197198 is now available (153 Mb, 23,561 genes) (Tisserant et al. 2012, 2013).

A wide range of terrestrial plants can establish AM symbiosis, including angiosperms, gymnosperms, ferns, and some bryophytes. Among bryophytes, AM colonization has been observed in liverworts (e.g., *Marchantia paleacea*), hornworts (e.g., *Anthoceros agrestis*), and in the moss genus *Takakia* (Grosche et al. 2018; Choi et al. 2018; Delaux et al. 2014). The evolutionary development of AM symbiosis might have proceeded as a stepwise acquisition; part of the main module of symbiotic signaling is present in charophytes, before the water-to-land transition; the other essential symbiosis genes have been acquired gradually (Delaux et al. 2015; Grosche et al. 2018).

Some plants appear to have lost their ability to have AM fungal partnerships during the course of evolution, for example, plants in the family Brassicaceae, including *Arabidopsis thaliana* known as a model plant. The common ancestor of Limnanthaceae, Cleomaceae, Resedaceae, and Brassicaceae families in Brassicales lost the ability to establish AM symbiosis (Delaux et al. 2014). The essential genes for symbiosis were lost in non-host species of brassicas, although they were conserved in peripheral host species like Moringaceae and Caricaceae. Brassica plants prefer phosphate-rich habitats and therefore do not require associations with AM fungi to obtain sufficient nutrients (Lambers and Teste 2013); on the contrary, the growth of brassicas is suppressed by mycorrhizal plants (Veiga et al. 2013). AM fungi can act as pathogens for *Arabidopsis thaliana* and cause negative effects (Lambers and Teste 2013).

In addition to the order Brassicales, several other angiosperms do not host AM fungi and seem to have lost the function to form AM symbiosis, e.g., sugar beet (*Beta vulgaris*, Amaranthaceae), spinach (*Spinacia oleracea*, Amaranthaceae), the obligate parasites *Striga hermonthica*, *Phelipanche aegyptiaca* (Orobanchaceae), *Cuscuta sativa* (Convolvulaceae), and the carnivorous plant *Utricularia gibba* (Lentibulariaceae). Most of these non-AM host plants are habitat or nutrient specialists (Brundrett and Tedersoo 2018), and the phylogenetic distance between them indicates that loss of AM symbiosis occurred independently (Delaux et al. 2014). In some cases, AM symbiosis was lost after a transition to another type of symbiosis. EcM and orchid symbiosis are thought to have developed and supplanted from AM symbiosis. Including transitions, AM symbiosis is estimated to have been lost between 30 and 50 times in total (Brundrett and Tedersoo 2018).

Another well-known but non-fungal symbiosis also originated from AM symbiosis: root nodule (RN) symbiosis of legumes. Nitrogen-fixation bacteria, in the genus *Rhizobium*, produce special structures called nodules in the roots of legumes. Rhizobial bacteria fix N_2 and provide usable N to host legumes in the form of

Fig. 6.2 (continued) wild-type rice at 5 weeks after inoculation. (**c–f**) AM fungi are stained with trypan blue. Bars = 50 μL

ammonium ions. In return, host plants give carbon sources to the bacteria, similar to AM symbiosis. Several genes that are essential for the AM symbiotic signaling pathway are also required for the RN symbiosis; this group of genes is called the common symbiotic pathway (CSP)/common symbiosis signaling pathway (Parniske 2008). Genes in this pathway are thought to have originated about 400 Mya, and they are conserved in most terrestrial plants; RN host plants might have adapted these modules for establishing RN symbiosis (Parniske 2008). Recent research indicates that RN symbiosis may have originated about 110 Mya, followed by the diversification of plants in the nitrogen-fixing clade into four orders: Cucurbitales, Fabales, Fagales, and Rosales (Li et al. 2015; van Velzen et al. 2019). However, most of the plants in this group subsequently lost the ability, except legumes and *Parasponia* in the Cannabaceae. A genetic analysis of *Parasponia* and *Trema*, a non-RN genus of host plants that is closely related to *Parasponia*, showed that several genes that are essential for symbiosis have a frameshift mutation in *Trema* that could explain how the ability for RN symbiosis was lost in *Trema* (Huisman and Geurts 2019; van Velzen et al. 2018; Rutten et al. 2020).

Most legumes can form symbiosis with both AM fungi and rhizobial bacteria. An exception is *Lupinus* (Fabaceae), which has lost its ability for AM symbiosis but can establish RN symbiosis. *Lupinus* developed a different adaptation that allows it to live in low-phosphate conditions: cluster roots that promote phosphate uptake (Harley and Harley 1987; Lambers et al. 2013; Oba et al. 2001).

Thus, plants establish various types of symbiosis depending on their habitat, and most of these seem to originate from AM symbiosis, the most ancient form. Plant–microbe associations are complex and rich in variety, and studying AM symbiosis could provide insight into the mechanisms.

6.3 Morphological Features and Life Cycles of AM Fungi

AM fungi cannot propagate without host plants, and they are associated with host plants at all stages of the life cycle (Fig. 6.2a). This is so-called obligate symbionts. AM fungi disperse to new host plants via spores. The diameter of the spores is very large (≥100 μm; Fig. 6.2b) compared with those of other fungi (1–10 μm) (Golan and Pringle 2017; Marleau et al. 2011). The spores are coenocytic, meaning each spore contains a large number of nuclei (e.g., 20–748 in *G. diaphanum*). Nuclei number in a single spore increases as the size of the spore increases (Marleau et al. 2011). Spores germinate when water and temperature conditions are adequate, uninfluenced by the presence of host plants (Giovannetti et al. 2010). AM hyphae are aseptate, and large numbers of nuclei move slowly and randomly inside (Uetake et al. 2002; Saito et al. 2004).

AM hyphae are attracted to host plants by chemotaxis after spore germination. Strigolactones (SLs) are one of the known chemicals that activate branching of AM fungi hyphae and increase the opportunity to catch the host plants (Akiyama et al. 2005). After the mycorrhizal hyphae come into contact with host root epidermal cells, the fungi form a special structure, the hyphopodium, to attach to host plant roots (Fig. 6.2c). If the host plant permits hyphal entry, the hyphae spread in the cortex of the host root and produce arbuscules in the cells for resource exchange between plant and fungus (Fig. 6.2d). Host plants may recognize compatible symbionts at this stage because hyphae of knockout mutants defective in the symbiotic pathway (e.g., *OsPOLLUX, OsSYMRK, and OsCCaMK* mutants in rice) are blocked at this stage (Fig. 6.2e) (Gutjahr et al. 2008; Banba et al. 2008; Miyata et al. 2023).

Through arbuscule production, the host plants increase the surface area of plasma membrane in colonized cells up to tenfold for the exchange of resources (Smith and Smith 1990). Hyphae outside the root, extraradical hyphae, absorb P from a region of soil broader than that reached by the plant's roots and transfer polyphosphates to the host plant through tubular vacuoles in extraradical and intraradical hyphae (Uetake et al. 2002). Phosphates are exported to the host plant through transporters in the vacuole membrane of AM fungi and in the membrane of host plants at the arbuscules (i.e., periarbuscular membrane) (Ezawa and Saito 2018). Arbuscules collapse shortly after formation, lasting 1 week or less (Bonfante and Genre 2010), and new arbuscules are formed in uncolonized neighbor cells; the region of active AM colonization is dynamic. Interestingly, once colonized cells are not colonized again (Kobae and Fujiwara 2014). Extraradical hyphae spread to a broad area of soil to collect resources and initiate hyphal entry into the host plant at different locations, thereby extending the area of colonization. After a period of production and collapse of arbuscules, with expansion of the colonization region, AM fungi produce a vesicle, a lipid storage organ, in the root cortex cells. In rice, vesicles are observed 1 month after colonization in soil culture (Fig. 6.2f). Some species of AM fungi do not develop vesicles at any point in their life cycle, e.g., *Scutellospora calospora* (van Aarle et al. 2002).

At a later stage, spores are produced by the extraradical hyphae. Multiple nuclei migrate from the hyphae into the spores (Jany and Pawlowska 2010). Spore primordia are small to begin with, then increase in volume and number of nuclei for 60 days after the spore primordia appeared (Marleau et al. 2011). The spores detached from the hypha and then spread to their next habitat.

Little information is available about how AM fungi generate genetic variation in progeny and whether or not they reproduce sexually remains debated. In fact, it remains unclear whether the nuclei in AM fungi are heterokaryotic (multiple genomes per spore) or homokaryotic (one genome per spore). Both explanations have been proposed on the basis of various genomic analyses (Corradi and Brachmann 2017). However, recent findings indicate a new scenario which can explain this contradiction: that AM fungi may fluctuate between homokaryotic and dikaryotic (two genomes in one spore) at different life stages (Corradi and Brachmann 2017; Ropars et al. 2016). According to this hypothesis, spores and hypha are mostly homokaryotic, but sometimes two genetically distinct fungi may induce karyogamy and meiosis and produce genetic variation within one spore. Evidence of genetic exchange has frequently been observed in genetically close lines within the same fungal species (Fig. 6.2a) (Croll et al. 2009; Angelard and Sanders 2011). The dikaryon hypothesis is also supported by the report that two distinct genomes have sometimes been detected in one spore in equal ratios. This finding challenges conventional knowledge about sex-related nuclear inheritance in AM fungi.

Because AM fungi are obligate symbionts, it has presented a large obstacle to molecular studies of AM symbiosis. However, recent technological developments have enabled us to culture them and to produce spores without host plants by adding specific lipids that are lacking in natural mycorrhizal metabolism (Kameoka et al. 2019). These new technologies will enable novel findings and help to improve our understanding of AM symbiosis.

6.4 Strigolactone Biosynthesis in Host Plants

Strigol, one of natural SLs, was first isolated from plant root exudates as germination inducers of a root parasitic plant *Striga lutea* (Cook et al. 1966). Later, another natural SL 5-deoxystrigol (5DS) was identified as stimulators of hyphae branching in AM fungi (Akiyama et al. 2005). More recently, SLs were found to inhibit axillary bud outgrowth by analysis of excess shoot branching mutants (Gomez-Roldan et al. 2008; Umehara et al. 2008). Now SLs are well-known as a group of phytohormones controlling plant growth and development (Al-Babili and Bouwmeester 2015). Various types of SLs are synthesized from carotenoid precursors (Fig. 6.3). In the SL biosynthesis, all-*trans*-β-carotene is first converted to 9-*cis*-β-carotene by DWARF27 (D27), which catalyzes the isomerization of β-carotene in plastid (Alder et al. 2012). Subsequently, two classes of carotenoid cleavage dioxygenases (CCD7 and CCD8) contribute to synthesize an SL intermediate, carlactone (CL), via oxidative cleavage and cyclization reactions (Alder et al. 2012; Seto et al. 2014). And then CL is moved from plastid to endoplasmic reticulum and is further oxidized to carlactonoic acid (CLA) by CYP711A, a cytochrome P450 monooxygenase (Abe et al. 2014). These chemical reactions are common to many plant species, but the pathways downstream of CLA differ among species. Identified natural SLs can be classified into two groups: canonical and non-canonical SLs (Yoneyama et al. 2018b). Canonical SLs contain a tricyclic lactone structure (ABC-ring portion of the SL structure shown in Fig. 6.3) connected to a methyl butenolide (D-ring portion of the SL structure shown in Fig. 6.3) by an enol-ether bridge, whereas non-canonical SLs lack A-, B-, or C-ring structures but still have the D-ring moiety connected to an enol-ether, which is required for shoot branching inhibition (Umehara et al. 2015).

In the biosynthesis of non-canonical SLs, CLA is converted to methyl carlactonoate (MeCLA) by a CLA methyltransferase (CLAMT), which belongs to a member of the SABATH protein family (Mashiguchi et al. 2022). Furthermore, lateral branching oxidoreductase (LBO), which belongs to the 2-oxoglutarate-dependent dioxygenase family, catalyzes the formation of hydroxymethyl carlactonoate (1′-OH-MeCLA) from MeCLA and also catalyzes demethylation of MeCLA to return to CLA (Yoneyama et al. 2020; Brewer et al. 2016).

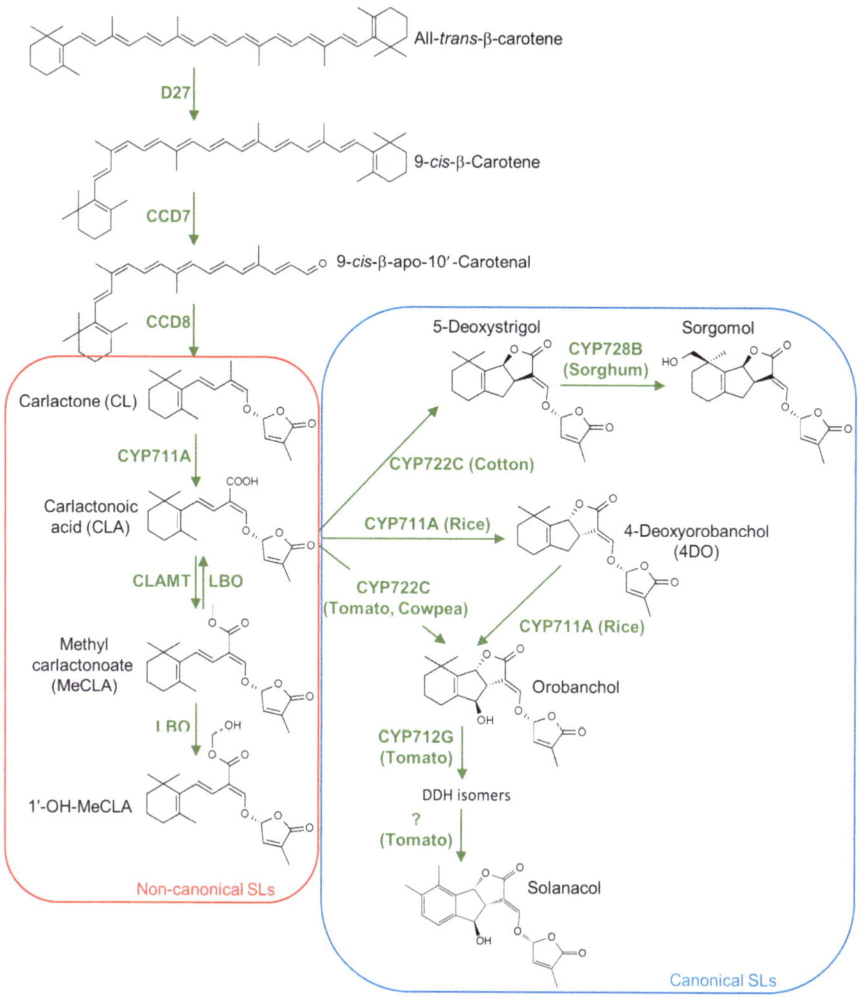

Fig. 6.3 Pathway of biosynthesis of the major strigolactones

In the biosynthesis of canonical SLs in rice, CLA is converted to 4-deoxyorobanchol (4DO) and 4DO is further converted to orobanchol by CYP711A (Zhang et al. 2014; Yoneyama et al. 2018a). In tomato and cowpea, CYP722C contributes to the synthesis of orobanchol from CLA without 4DO synthesis (Wakabayashi et al. 2019). In tomato, CYP712G catalyzes the oxidation of orobanchol for biosynthesis of didehydro-orobanchol isomers and solanacol (Wang et al. 2022). In cotton, CLA is converted to 5-deoxystrigol (5DS) by other cytochrome P450 CYP722C (Wakabayashi et al. 2020); and in sorghum, 5DS is further metabolized to sorgomol by CYP728B (Wakabayashi et al. 2021). Other non-canonical SLs, such as avenaol, heliolactone, zealactone, and lotuslactone, have been respectively identified from the wild oat, sunflower, corn, and *Lotus japonicus*, but the enzymatic reactions that produce them are not yet fully known. To date, more than 30 canonical and non-canonical SLs have been identified from various plants (Yoneyama and Brewer 2021).

SLs are released from host plant roots into the rhizosphere by the active transport through an ATP-binding cassette transporter pleiotropic drug resistance 1 (PDR1), and they result in enhanced symbiotic interaction with AM fungi (Kretzschmar et al. 2012). Structure–activity relationship of SLs in hyphal branching of AM fungi demonstrated that truncation of the A- and AB-rings in the tricyclic lactone of SLs drastically reduced their bioactivity. In addition, connection of the C- and D-rings was shown to have an important role in hyphal branching, but the C–D bridge structure can be replaced with either alkoxy or imino ethers without loss of function (Akiyama et al. 2010).

Levels of some inorganic nutrients in the rhizosphere strongly affect SL biosynthesis and exudation from roots of host plants, likely related to mechanisms for efficient acquisition of inorganic nutrients in host plants. Under phosphate deficiency, many canonical SLs are highly synthesized in roots and released into soil to induce AM symbiosis for phosphate uptake in various plants (Akiyama et al. 2005; Yoneyama et al. 2007b, 2012; Lopez-Raez et al. 2008). Nitrogen deficiency does not have significant effect on SL production in leguminous plants that partner with *Rhizobium*, even though nitrogen fixation by *Rhizobium* is the primary source of nitrogen for these plants. In contrast, non-leguminous plants increase SL levels in roots under nitrogen deficiency as well as phosphate deficiency and they rely on support of AM fungi for acquisition of both phosphate and nitrogen (Yoneyama et al. 2007a, 2012; Sun et al. 2014). Many SL biosynthesis genes are upregulated under phosphate and nitrogen deficiencies in rice and *Medicago truncatula* (Umehara et al. 2010; Bonneau et al. 2013; Sun et al. 2014). Furthermore, sulfur deficiency increases SL levels in roots through upregulation of *D27* in the first step of rice SL biosynthesis (Shindo et al. 2018). In addition, *D27* expression is activated by the transcription factors nodulation signaling pathway 1 (NSP1) and NSP2, which also function in the symbiosis signaling pathway of both rice and *M. truncatula* (Liu et al. 2011). AM fungi can supply sulfate ions in addition to nitrogen and phosphate (Allen and Shachar-Hill 2009). Because *D27* expression is upregulated under sulfur deficiency, the regulation might be associated with effective sulfur uptake via AM fungi. In a split-root experiment, in which one portion of the roots was held in nutrient-rich medium and the other in nitrogen-, phosphate-, or sulfur-deficient medium, the nutrient was supplied from roots in the nutrient-rich medium to roots in the nutrient-deficient roots, resulting in systemic reduction of SL levels in both parts of the roots, indicating that shoot parts, not root parts, perceive nutrient deficiencies for the SL production in roots (Shindo et al. 2021; Yoneyama et al. 2015).

In rice and pea, colonization of AM fungi is lower in SL-biosynthesis mutants (*ccd7* and *ccd8* mutants) than in wild-type plants because the mutants cannot produce the SLs required for the formation of the hyphopodium, the hyphal protrusion that allows AM fungi to penetrate host roots (Gomez-Roldan et al. 2008; Yoshida et al. 2012; Kobae et al. 2018). In contrast, a rice SL signaling mutant, *d14*, exhibited higher AM colonization than the wild type because of excess SL accumulation (Yoshida et al. 2012). The rice *d3* and *kai2* (named *dwarf14-like* in rice) mutants cannot support AM fungi colonization, indicating that the KAI2 signal is critical for AM symbiosis (Yoshida et al. 2012; Gutjahr et al. 2015). In addition, the SL signaling repressor SMAX1 negatively controls expression of both SL-biosynthesis and AM-symbiosis genes, and its degradation initiates SL production and other signaling cascades related to AM symbiosis (Choi et al. 2020).

6.5 Initiation of AM Symbiosis in Host Plants

According to a widely accepted scenario, RN symbiosis developed by repurposing of the genetic signaling pathway for AM symbiosis (Parniske 2008). The earliest molecular studies of AM symbiosis signaling investigated the pathway common to RN and AM symbioses, the CSP, and research into the initiation of AM symbiosis began with studies in legumes, based on the known mechanisms of RN symbiosis. Recently, rice and tomato have also become useful in the study of AM symbiosis. Even though the pathways share many common steps, some parts are specific to AM symbiosis, and much of the process has yet to be elucidated. Current understanding of the signaling pathway for AM symbiosis is summarized in Fig. 6.4.

In RN symbiosis, lipochitooligosaccarides (LCOs), known as Nod ("nodulation") factors, are secreted by rhizobial bacteria and trigger the initiation of RN symbiosis (Dénarié et al. 1996). In AM symbiosis, similar molecules, Myc-LCOs, or Myc factors, which are produced by AM fungi, have been identified as the signal molecules (Maillet et al. 2011). Both Nod and Myc factors induce perinuclear Ca^{2+} oscillation, or Ca^{2+} spiking, in host plants (Sun et al. 2015; Genre et al. 2013; Walker et al. 2000; Oldroyd et al. 2001). Legumes show clear Ca^{2+} spiking in response to Myc factors. In rice, however, Myc factors do not induce Ca^{2+} spiking or expression of downstream genes. Instead, Ca^{2+} spiking is induced by chitin tetramers (CO4), suggesting that CO4 may be the AM signal (Genre et al. 2013; Sun et al. 2013; Miyata et al. 2014). Recent analysis showed that chitin heptamers (CO7) also induce Ca^{2+} spiking in *Medicago* (Feng et al. 2019). CO7 is also known as a typical pathogen-associated molecular pattern (PAMP), and it induces defense responses to pathogen in plants (Desaki et al. 2017). Thus, CO7 is associated with both immunity and symbiosis. To explain this contradiction, it has been proposed that Myc-LCO reduces induction of the defense response caused by CO7 (Feng et al. 2019).

Nod factors are recognized by lysin motif receptor-like kinases (LysM-RLKs), such as LYK3/NFR1 in *Medicago* and NFP/NFR5 in *Lotus* (Radutoiu et al. 2003; Madsen et al. 2003; Amor et al. 2003; Limpens et al. 2003; Arrighi et al. 2006; Broghammer et al. 2012). In rice, OsCERK1, an ortholog of LYK3/NFR1, is involved in AM symbiosis. *OsCERK1* mutants showed severe delay of AM colonization, although the phenotype is unlike that of other CSP mutants (Miyata et al. 2014, 2022). In addition, OsCERK1 is essential for chitin-triggered immunity, a

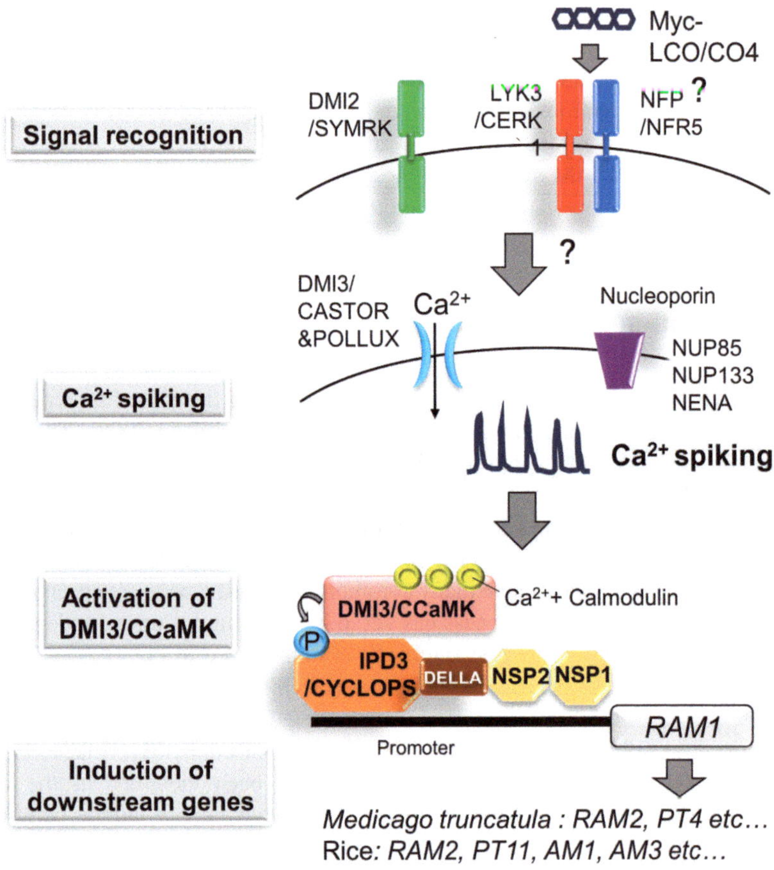

Fig. 6.4 Model of initiation of AM symbiosis

typical PAMP-triggered immune response (Shimizu et al. 2010). Furthermore, CO4-dependent Ca^{2+} spiking is not observed in *OsCERK1* mutants, supporting the involvement of OsCERK1 in AM symbiosis (Carotenuto et al. 2017). Therefore, OsCERK1 has a dual function in symbiosis and defense response (Miyata et al. 2014; Zhang et al. 2015). Similar to OsCERK1, LYK3/NFR1 orthologs are involved in chitin-triggered immunity and AM symbiosis in various plants (Rutten et al. 2020; Liao et al. 2018). Although the mechanisms by which host plants initiate responses to the microbes have yet to be clarified, OsCERK1 is involved in both defense response and symbiosis.

One possible scenario is that downstream response is determined by the partner receptor of OsCERK1 orthologs, such as NFP/NFR5 orthologs, which are orthologs of Nod factor receptors in legumes (He et al. 2019; Zhang et al. 2021). In contrast, Miyata et al. (2014) showed weak mycorrhizal phenotype in *OsNFR5* knockout mutants and no significant difference between the mutants and the wild type. In addition, Ca^{2+} spiking in response to CO4 is observed in *OsNFR5*, although

OsCERK1 mutants lack a response to CO4 (Miyata et al. 2016; Carotenuto et al. 2017). In tomato, lack of the NFP/NFR5 ortholog SILYK10 results in low colonization by AM fungi (Buendia et al. 2015). On the other hand, the AM colonization ratio did not change in *PanNFP1;NFP2* mutants of *Parasponia* (Rutten et al. 2020), although RNAi of PanNFP1 plants gave a low colonization ratio (Op den Camp et al. 2011). The possible influence of NFR5/NFP homologs in AM symbiosis needs further investigation.

Recognition of the symbiosis signal molecule activates downstream signaling in the CSP. The AM symbiosis gene *DMI2/SYMRK* encodes a leucine-rich repeat (LRR) receptor-like kinase (Endre et al. 2002; Stracke et al. 2002). Similar to the LysM receptors involved in the recognition of signal molecules, does not infection (DMI) 2/SYMRK is expressed on the plasma membrane. However, its function remains unclear, including how it regulates AM symbiosis and whether it is involved in the recognition of specific ligands or not. Its function in the regulation of AM/RN symbiosis has been studied in both legumes and non-legumes. In rice, it was shown that OsSYMRK is important for AM symbiosis (Miyata et al. 2023). Most AM host plants have DMI2/SYMRK orthologs (Delaux et al. 2015); thus, a wide range of plant species seem to require DMI2/SYMRK homologs for the initiation of AM symbiosis. Several genes in addition to *SYMRK* have been identified as components of the CSP. For example, *DMI1* in *Medicago* encodes a Ca^{2+}-regulated Ca^{2+} channel on the nuclear membrane (Kim et al. 2019), although it was originally reported to encode potassium-permeable channels (Ané et al. 2004). Similarly, lotus has two DMI1 orthologs, POLLUX and CASTOR (Imaizumi-Anraku et al. 2005). DMI1/POLLUX/CASTOR is essential for both AM and RN symbiosis. DMI1 orthologs in non-legumes also have a key role in AM symbiosis (Banba et al. 2008; Gutjahr et al. 2008). No CO4- or Myc-LCO-dependent Ca^{2+} spiking is observed in *dmi1* and *dmi2* mutants of *Medicago* (Genre et al. 2013; Sun et al. 2015), and *OsSYMRK* mutants of rice also lack Ca^{2+} spiking in response to CO4 (Miyata et al. 2023). The finding indicates that Ca^{2+} spiking lies downstream of DMI2/SYMRK and DMI1/POLLUX/CASTOR.

AM signaling and Ca^{2+} spiking also require components of nucleoporins, NUP85, NUP133, and NENA (Saito et al. 2007; Kanamori et al. 2006; Groth et al. 2010). Although the involvement of Ca^{2+} spiking in AM symbiosis has been discussed, the mechanisms underlying the induction of Ca^{2+} spiking by signal molecules and signaling from the plasma membrane to the nucleus are not yet fully understood.

The Ca^{2+} spiking signal is decoded by the DMI3/CCaMK protein in the nucleus. DMI3/CCaMK is composed of three parts: a serine/threonine kinase domain at the N-terminus, an autoinhibitory domain between the N and C termini, and a visinin-like domain with three EF-hand motifs at the C-terminus (Levy et al. 2004; Mitra et al. 2004). Ca^{2+}-bound calmodulins bind to the EF hands of DMI3/CCaMK and release the kinase domain (Miller et al. 2013). Activated CCaMK directly phosphorylates and activates INTERACTING PROTEIN OF DMI3 (IPD3)/CYCLOPS, which is a DNA-binding transcriptional activator. In RN symbiosis, IPD3/CYCLOPS forms a complex that binds to promoter elements and induces the expression of downstream genes (Yano et al. 2008; Singh et al. 2014). CYCLOPS makes a protein complex with NSP1 and NSP2, which were initially thought to be specific to nodulation but which are also important for AM symbiosis, even though the phenotype of *nsp1* and *nsp2* knockout mutants is not strong (Maillet et al. 2011; Delaux et al. 2013). NSPs and CYCLOPS/IPD3 can interact with DELLA proteins, which are also required for AM symbiosis (Jin et al. 2016; Floss et al. 2013; Liu et al. 2011).

A protein complex of DELLA, NSP1, NSP2, and IPD3/CYCLOPS induces gene expression of *reduced arbuscular mycorrhization* (*RAM*) *1*, which is a transcription factor belonging to the GRAS family, by directly binding the promoter region of *RAM1* in lotus (Gobbato et al. 2012; Pimprikar et al. 2016). RAM1 is required for signaling in AM symbiosis but not in RN symbiosis, and it may represent the branching point of AM symbiosis and RN symbiosis in legumes.

The most well-studied downstream gene of RAM1 is *RAM2*, which encodes a glycerol-3-phosphate acyl transferase. RAM2 is involved in AM symbiosis of *Medicago* and rice (Liu et al. 2022; Gobbato et al. 2012). The lack of RAM1 results in a low AM colonization ratio in *Medicago* and an abnormal distribution of lipid in lotus (Montero and Paszkowski 2022). Besides RAM1 and 2, many AM marker genes have been reported in rice (Gutjahr et al. 2008), along with mycorrhiza-induced subtilase (*SbtM1*) in lotus (Takeda et al. 2009).

Many open questions remain about signaling for the initiation of AM symbiosis, such as what is the missing link of signaling pathway between the cell membrane and the nuclear, how is Ca^{2+} spiking induced, and how are AM-specific responses induced although AM and RN symbioses share the same pathway? Here we focused on host plant symbiosis genes; further research is needed on mutual communication between AM fungi and host plants.

6.6 Nutrient Transport System Between Host Plants and AM Fungi

In colonized root cells, carbon is transported from host plants to AM fungi, and soil mineral nutrients including nitrogen, phosphate ion, and sulfate ion are transferred the other way (Fig. 6.5). Plants absorb carbon dioxide from the atmosphere and convert it into organic compounds by photosynthesis. In AM symbiosis, host plants provide up to approximately 20% of their photosynthetic products for propagation of the AM fungi (Bago et al. 2000). In most plants, sucrose can be transferred from shoots to roots via the phloem. Sucrose is apoplasmically or symplasmically transported from the phloem into AM-colonized cells. Sucrose is also metabolized into glucose (Glc) and fructose (Fru) by invertase in the cell wall or in colonized cells. Glucose is transported into AM-colonized cells through sugar transporters. Gene expression of several sucrose transporters (SUTs) is upregulated in the host root colonized by AM fungi to activate the transport of sucrose (Boldt et al. 2011;

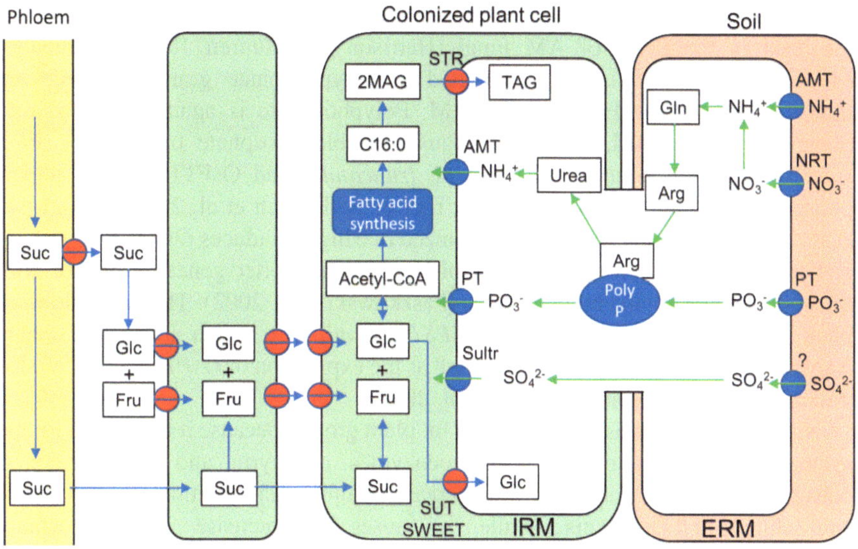

Fig. 6.5 Nutrient exchange system between plant cells and AM fungi

Doidy et al. 2012). The SWEET transporter family exports sucrose and monosaccharides to the symbiotic interface in colonized cells (Manck-Götzenberger and Requena 2016). Carbohydrate transporters import glucose from the symbiotic interface to the fungal intraradical mycelium (IRM) (Schüssle et al. 2006; Helber et al. 2011). Hexose in colonized cells is converted into acetyl-CoA (C16:0) and is produced via fatty acid synthesis and further converted into C16:0 - 2-monoacylglycerol (2MAG) by RAM2 (Luginbuehl et al. 2017). 2MAG is transported into the IRM of AM fungi by STUNTED ARBUSCULE (STR) 1 and STR2, which are ATP-binding cassette transporters, on the periarbuscular membrane (Zhang et al. 2010; Gutjahr et al. 2012; Jiang et al. 2017). 2MAG is finally converted into triacylglycerol in AM fungi. Carbon incorporated by AM fungi is transported from the IRM to the extraradical mycelium (ERM) and used for hyphal extension and spore production in the AM fungi.

As mentioned previously, phosphate, nitrogen, and sulfate deficiencies stimulate the production of SLs that stimulate symbiotic interaction with AM fungi, which support plant growth in nutrient-deficient conditions. Phosphate is essential for plant growth and development because it is required for the synthesis of indispensable compounds for plant growth such as ATP, coenzymes, nucleic acids, and phospholipids. Phosphate deficiency causes the phosphate starvation response: i.e., expression of phosphate transporter genes; secretion of acid phosphatase, ribonuclease, and organic acids from roots into the rhizosphere for the phosphate acquisition; morphological change of root architecture; and anthocyanin biosynthesis (Yuan and Liu 2008). In AM symbiosis, the hyphal network efficiently absorbs small amounts of available phosphorus in soil (Jakobsen et al. 1992a, b). In AM-colonized rice approximately 70% of phosphorus is delivered by the AM fungi (Yang et al.

2012). Phosphate in soil is first absorbed through inorganic phosphate transporters localized in the ERM of AM fungi (Harrison and Buuren 1995; Maldonado-Mendoza et al. 2001). It is converted to polyphosphate granules, which are transported from the ERM to the IRM. Polyphosphate is again converted into phosphate in the IRM, and AM fungus-inducible phosphate transporters (PTs) such as StPT3 in potato, MtPT4 in *M. truncatula*, and OsPT11 in rice import phosphate from the IRM into the host root cells (Rausch et al. 2001; Paszkowski et al. 2002; Javot et al. 2007). AM-colonized rice highly induces *OsPT11* expression and reduces the expression of other phosphate transporter genes associated with direct phosphate uptake by plant roots (Paszkowski et al. 2002). The AM-colonized plant roots reduce the expression of *IPS1* and *miR399*, which are the phosphate starvation response marker genes, as well as the expression of *OsPT2* and *OsPT6* for direct phosphate acquisition (Yang et al. 2012).

Nitrogen is also an essential nutrient for plant growth because it is required for the synthesis of proteins, nucleic acids, coenzymes, porphyrin, and plant hormones. Nitrogen deficiency causes a severe decrease of amino acid (particularly glutamine) and chlorophyll contents, while it causes an increase of anthocyanin, phenylpropanoid, and starch levels (Diaz et al. 2006; Fritz et al. 2006). Inorganic N such as nitrate and ammonium ions is absorbed through nitrogen transporters in the fungal ERM and incorporated into glutamine via the major nitrogen assimilation system GS/GOGAT cycle; glutamine is finally metabolized into arginine (Chalot et al. 1994; Johansen et al. 1996). Positively charged arginine is transferred from the ERM to the IRM by association with negatively charged polyphosphate granules (Parniske 2008). Arginine is degraded into urea and ornithine in the IRM. Furthermore, ammonium ions are produced by urea hydrolyzation and subsequently transported into root cells through plant ammonium transporters (Guether et al. 2009; Kobae et al. 2010; Breuillin-Sessoms et al. 2015).

Sulfur is required for the production of proteins, glutathione, sulfolipids, and coenzymes. In rice, sulfur deficiency causes a severe decrease of the chlorophyll content in young leaves, the number of tillers, and shoot growth (Dobermann and Fairhurst 2001). Expression of *MtSultr1;2* in *Medicago* and *LjSultr1;2* in *Lotus*, which *are* sulfate transporters, are elevated by symbiotic interaction with AM fungi when sulfur is deficient in soil (Giovannetti et al. 2014; Wipf et al. 2014). However, the mechanism of sulfate transport in the ERM of AM fungi is still unknown.

6.7 Conclusion and Perspectives

Here we have discussed the diversity of AM fungi and their host plants, the morphological characteristics of colonization, SL biosynthesis and symbiosis signaling pathways, and the systems for nutrient transport system between AM fungi and host plants. Both partners produce molecules to initiate the symbiosis (fungi produce Myc factors and plants produce SLs), and the production pathways and interactions of these molecules are now better understood. Many plants can make symbiotic associations with AM fungi for improvement of their nutrient absorption from soil. AM fungi provide various types of essential minerals (macronutrients and metal ions) to their host plants. In reciprocation, AM fungi receive sugars and lipids as carbon sources for their propagation from their host plants. Uptake of nitrogen, phosphate, and sulfur is supported by AM fungi when these nutrients are deficient in soil.

Climate change and freshwater scarcity are strongly affecting the environment and reducing nutrient content in soils. Plants, being rooted in place, must adapt to their environment, even in stressful conditions. Understanding how plants do so in response to inorganic nutrient deficiency is important for food security and sustainable agriculture. Although plant–microbe interactions are complex, future work would focus on the molecular mechanisms of nutrient transportation system between host plants and AM fungi under nutrient deficiency. It would be a major advance to understand how sensors of nutrient levels in plants regulate nutrient exchange. In addition, AM colonization increases resistance to disease and drought. Understanding the common signaling pathways in AM colonization may ultimately contribute to increased agricultural and horticultural crop yields. The utilization of AM fungi enables crop production in nutrient-poor environments and reduces the utilization of pesticides and fertilizers, supporting sustainable agriculture.

Acknowledgments This work was in part supported by a grant from the Japan Society for the Promotion of Science (KAKENHI, 20K05776 and 23K05065 to M.U. and 20K15532 to K.M.). We thank Naoto Shibuya, Meiji University, for the helpful advice.

References

Abe S, Sado A, Tanaka K, Kisugi T, Asami K, Ota S, Kim HI, Yoneyama K, Xie X, Ohnishi T, Seto Y, Yamaguchi S, Akiyama K, Yoneyama K, Nomura T (2014) Carlactone is converted to carlactonoic acid by MAX1 in Arabidopsis and its methyl ester can directly interact with AtD14 in vitro. Proc Natl Acad Sci U S A 111:18084–18089

Akiyama K, Matsuzaki K, Hayashi H (2005) Plant sesquiterpenes induce hyphal branching in arbuscular mycorrhizal fungi. Nature 435:824–827

Akiyama K, Ogasawara S, Ito S, Hayashi H (2010) Structural requirements of strigolactones for hyphal branching in AM fungi. Plant Cell Physiol 51:1104–1117

Al-Babili S, Bouwmeester HJ (2015) Strigolactones, a novel carotenoid-derived plant hormone. Annu Rev Plant Biol 66:161–186

Alder A, Jamil M, Marzorati M, Bruno M, Vermathen M, Bigler P, Ghisla S, Bouwmeester H, Beyer P, Al-Babili S (2012) The path from β-carotene to carlactone, a strigolactone-like plant hormone. Science 335:1348–1351

Allen JW, Shachar-Hill Y (2009) Sulfur transfer through an arbuscular mycorrhiza. Plant Physiol 149:549–560

Amor BB, Shaw SL, Oldroyd GE, Maillet F, Penmetsa RV, Cook D, Long SR, Denarie J, Gough C (2003) The NFP locus of Medicago truncatula controls an early step of Nod factor signal transduction upstream of a rapid calcium flux and root hair deformation. Plant J 34:495–506

Ané JM, Kiss GB, Riely BK, Penmetsa RV, Oldroyd GE, Ayax C, Lévy J, Debellé F, Baek JM, Kalo P, Rosenberg C, Roe BA, Long SR, Dénarié J, Cook DR (2004) Medicago truncatula DMI1 required for bacterial and fungal symbioses in legumes. Science 303:1364–1367
Angelard C, Sanders IR (2011) Effect of segregation and genetic exchange on arbuscular mycorrhizal fungi in colonization of roots. New Phytol 189:652–657
Arrighi JF, Barre A, Ben Amor B, Bersoult A, Soriano LC, Mirabella R, de Carvalho-Niebel F, Journet EP, Gherardi M, Huguet T, Geurts R, Denarie J, Rouge P, Gough C (2006) The Medicago truncatula lysin motif-receptor-like kinase gene family includes NFP and new nodule-expressed genes. Plant Physiol 142:265–279
Bago B, Pfeffer PE, Shachar-Hill Y (2000) Carbon metabolism and transport in arbuscular mycorrhizas. Plant Physiol 124:949–958
Banba M, Gutjahr C, Miyao A, Hirochika H, Paszkowski U, Kouchi H, Imaizumi-Anraku H (2008) Divergence of evolutionary ways among common sym genes: CASTOR and CCaMK show functional conservation between two symbiosis systems and constitute the root of a common signaling pathway. Plant Cell Physiol 49:1659–1671
Boldt K, Pörs Y, Haupt B, Bitterlich M, Kühn C, Grimm B, Franken P (2011) Photochemical processes, carbon assimilation and RNA accumulation of sucrose transporter genes in tomato arbuscular mycorrhiza. J Plant Physiol 168:1256–1263
Bonfante P, Genre A (2010) Mechanisms underlying beneficial plant-fungus interactions in mycorrhizal symbiosis. Nat Commun 1:48
Bonneau L, Huguet S, Wipf D, Pauly N, Truong H-N (2013) Combined phosphate and nitrogen limitation generates a nutrient stress transcriptome favorable for arbuscular mycorrhizal symbiosis in Medicago truncatula. New Phytol 199:188–202
Breuillin-Sessoms F, Floss DS, Gomez SK, Pumplin N, Ding Y, Levesque-Tremblay V, Noar RD, Daniels DA, Bravo A, Eaglesham JB, Benedito VA, Udvardi MK, Harrison MJ (2015) Suppression of arbuscule degeneration in Medicago truncatula phosphate transporter4 mutants is dependent on the ammonium transporter 2 family protein AMT2;3. Plant Cell 27:1352–1366
Brewer PB, Yoneyama K, Filardo F, Meyers E, Scaffidi A, Frickey T, Akiyama K, Seto Y, Dun EA, Cremer JE, Kerr SC, Waters MT, Flematti GR, Mason MG, Weiller G, Yamaguchi S, Nomura T, Smith SM, Yoneyama K, Beveridge CA (2016) Lateral branching oxidoreductase acts in the final stages of strigolactone biosynthesis in Arabidopsis. Proc Natl Acad Sci U S A 113:6301–6306
Broghammer A, Krusell L, Blaise M, Sauer J, Sullivan JT, Maolanon N, Vinther M, Lorentzen A, Madsen EB, Jensen KJ, Roepstorff P, Thirup S, Ronson CW, Thygesen MB, Stougaard J (2012) Legume receptors perceive the rhizobial lipochitin oligosaccharide signal molecules by direct binding. Proc Natl Acad Sci U S A 109:13859–13864
Brundrett MC, Tedersoo L (2018) Evolutionary history of mycorrhizal symbioses and global host plant diversity. New Phytol 220:1108–1115
Buendia L, Wang T, Girardin A, Lefebvre B (2015) The LysM receptor-like kinase SlLYK10 regulates the arbuscular mycorrhizal symbiosis in tomato. New Phytol 210(1):184–195
Carotenuto G, Chabaud M, Miyata K, Capozzi M, Takeda N, Kaku H, Shibuya N, Nakagawa T, Barker DG, Genre A (2017) The rice LysM receptor-like kinase OsCERK1 is required for the perception of short-chain chitin oligomers in arbuscular mycorrhizal signaling. New Phytol 214:1440–1446
Chalot M, Brun A, Finlay RD, Söderström B (1994) Metabolism of [14C]glutamate and [14C] glutamine by the ectomycorrhizal fungus Paxillus involutus. Microbiology 140:1641–1649
Choi J, Summers W, Paszkowski U (2018) Mechanisms underlying establishment of arbuscular mycorrhizal symbioses. Annu Rev Phytopathol 56:135–160
Choi J, Lee T, Cho J, Servante EK, Pucker B, Summers W, Bowden S, Rahimi M, An K, An G, Bouwmeester HJ, Wallington EJ, Oldroyd G, Paszkowski U (2020) The negative regulator SMAX1 controls mycorrhizal symbiosis and strigolactone biosynthesis in rice. Nat Commun 11:2114

Cook CE, Whichard LP, Turner B, Wall ME, Egley GH (1966) Germination of witchweed (Striga lutea Lour.): isolation and properties of a potent stimulant. Science 154:1189–1190
Corradi N, Brachmann A (2017) Fungal mating in the most widespread plant symbionts? Trends Plant Sci 22:175–183
Croll D, Giovannetti M, Koch AM, Sbrana C, Ehinger M, Lammers PJ, Sanders IR (2009) Nonself vegetative fusion and genetic exchange in the arbuscular mycorrhizal fungus Glomus intraradices. New Phytol 181:924–937
Delaux PM, Becard G, Combier JP (2013) NSP1 is a component of the Myc signaling pathway. New Phytol 199:59–65
Delaux PM, Varala K, Edger PP, Coruzzi GM, Pires JC, Ane JM (2014) Comparative phylogenomics uncovers the impact of symbiotic associations on host genome evolution. PLoS Genet 10:e1004487
Delaux PM, Radhakrishnan GV, Jayaraman D, Cheema J, Malbreil M, Volkening JD, Sekimoto H, Nishiyama T, Melkonian M, Pokorny L, Rothfels CJ, Sederoff HW, Stevenson DW, Surek B, Zhang Y, Sussman MR, Dunand C, Morris RJ, Roux C, Wong GK, Oldroyd GE, Ané JM (2015) Algal ancestor of land plants was preadapted for symbiosis. Proc Natl Acad Sci U S A 112:13390–13395
Dénarié J, Debellé F, Promé JC (1996) Rhizobium lipo-chitooligosaccharide nodulation factors: signaling molecules mediating recognition and morphogenesis. Annu Rev Biochem 65:503–535
Desaki Y, Miyata K, Suzuki M, Shibuya N, Kaku H (2017) Plant immunity and symbiosis signaling mediated by LysM receptors. Innate Immun 2017:1753425917738885
Diaz C, Saliba-Colombani V, Loudet O, Belluomo P, Moreau L, Daniel-Vedele F, Morot-Gaudry J-F, Masclaux-Daubresse C (2006) Leaf yellowing and anthocyanin accumulation are two genetically independent strategies in response to nitrogen limitation in Arabidopsis thaliana. Plant Cell Physiol 47:74–83
Dickson S, Smith FA, Smith SE (2007) Structural differences in arbuscular mycorrhizal symbioses: more than 100 years after Gallaud, where next? Mycorrhiza 17:375–393
Dobermann A, Fairhurst T (2001) Rice: nutrient disorders & nutrient management. PPI/PPIC, Singapore
Doidy J, van Tuinen D, Lamotte O, Corneillat M, Alcaraz G, Wipf D (2012) The Medicago truncatula sucrose transporter family: characterization and implication of key members in carbon partitioning towards arbuscular mycorrhizal fungi. Mol Plant 5:1346–1358
Endre G, Kereszt A, Kevei Z, Mihacea S, Kalo P, Kiss GB (2002) A receptor kinase gene regulating symbiotic nodule development. Nature 417:962–966
Ezawa T, Saito K (2018) How do arbuscular mycorrhizal fungi handle phosphate? New insight into fine-tuning of phosphate metabolism. New Phytol 220:1116–1121
Feng F, Sun J, Radhakrishnan GV, Lee T, Bozsóki Z, Fort S, Gavrin A, Gysel K, Thygesen MB, Andersen KR, Radutoiu S, Stougaard J, Oldroyd GED (2019) A combination of chitooligosaccharide and lipochitooligosaccharide recognition promotes arbuscular mycorrhizal associations in Medicago truncatula. Nat Commun 10:5047
Floss DS, Levy JG, Lévesque-Tremblay V, Pumplin N, Harrison MJ (2013) DELLA proteins regulate arbuscule formation in arbuscular mycorrhizal symbiosis. Proc Natl Acad Sci U S A 110:5025–5034
Fritz C, Palacios-Rojas N, Feil R, Stitt M (2006) Regulation of secondary metabolism by the carbon–nitrogen status in tobacco: nitrate inhibits large sectors of phenylpropanoid metabolism. Plant J 46:533–548
Genre A, Chabaud M, Balzergue C, Puech-Pages V, Novero M, Rey T, Fournier J, Rochange S, Becard G, Bonfante P, Barker DG (2013) Short-chain chitin oligomers from arbuscular mycorrhizal fungi trigger nuclear Ca2+ spiking in Medicago truncatula roots and their production is enhanced by strigolactone. New Phytol 198:190–202
Gianinazzi S, Gollotte A, Binet MN, van Tuinen D, Redecker D, Wipf D (2010) Agroecology: the key role of arbuscular mycorrhizas in ecosystem services. Mycorrhiza 20:519–530

Giovannetti M, Avio L, Sbrana C (2010) Fungal spore germination and pre-symbiotic mycelial growth – physiological and genetic aspects. Springer, Cham

Giovannetti M, Tolosano M, Volpe V, Kopriva S, Bonfante P (2014) Identification and functional characterization of a sulfate transporter induced by both sulfur starvation and mycorrhiza formation in Lotus japonicus. New Phytol 204:609–619

Gobbato E, Marsh JF, Vernie T, Wang E, Maillet F, Kim J, Miller JB, Sun J, Bano SA, Ratet P, Mysore KS, Denarie J, Schultze M, Oldroyd GE (2012) A GRAS-type transcription factor with a specific function in mycorrhizal signaling. Curr Biol 22:2236–2241

Golan J, Pringle A (2017) Long-distance dispersal of fungi. Microbiol Spectr 5:4

Gomez-Roldan V, Fermas S, Brewer PB, Puech-Pages V, Dun EA, Pillot JP, Letisse F, Matusova R, Danoun S, Portais JC, Bouwmeester H, Becard G, Beveridge CA, Rameau C, Rochange SF (2008) Strigolactone inhibition of shoot branching. Nature 455:189–194

Grosche C, Genau AC, Rensing SA (2018) Evolution of the symbiosis-specific GRAS regulatory network in bryophytes. Front Plant Sci 9:1621

Groth M, Takeda N, Perry J, Uchida H, Dräxl S, Brachmann A, Sato S, Tabata S, Kawaguchi M, Wang TL, Parniske M (2010) NENA, a Lotus japonicus homolog of Sec13, is required for rhizodermal infection by arbuscular mycorrhiza fungi and rhizobia but dispensable for cortical endosymbiotic development. Plant Cell 22:2509–2526

Guether M, Neuhäuser B, Balestrini R, Dynowski M, Ludewig U, Bonfante P (2009) A mycorrhizal-specific ammonium transporter from Lotus japonicus acquires nitrogen released by arbuscular mycorrhizal fungi. Plant Physiol 150:73–83

Gutjahr C, Banba M, Croset V, An K, Miyao A, An G, Hirochika H, Imaizumi-Anraku H, Paszkowski U (2008) Arbuscular mycorrhiza-specific signaling in rice transcends the common symbiosis signaling pathway. Plant Cell 20:2989–3005

Gutjahr C, Radovanovic D, Geoffroy J, Zhang Q, Siegler H, Chiapello M, Casieri L, An K, An G, Guiderdoni E, Kumar CS, Sundaresan V, Harrison MJ, Paszkowski U (2012) The half-size ABC transporters STR1 and STR2 are indispensable for mycorrhizal arbuscule formation in rice. Plant J 69:906–920

Gutjahr C, Gobbato E, Choi J, Riemann M, Johnston MG, Summers W, Carbonnel S, Mansfield C, Yang S-Y, Nadal M, Acosta I, Takano M, Jiao W-B, Schneeberger K, Kelly KA, Paszkowski U (2015) Rice perception of symbiotic arbuscular mycorrhizal fungi requires the karrikin receptor complex. Science 350:1521–1524

Harley JL, Harley EL (1987) A check-list of mycorrhiza in the British flora. New Phytol 105:1–102

Harrison MJ, Buuren MLV (1995) A phosphate transporter from the mycorrhizal fungus Glomus versiforme. Nature 378:626–629

He J, Zhang C, Dai H, Liu H, Zhang X, Yang J, Chen X, Zhu Y, Wang D, Qi X, Li W, Wang Z, An G, Yu N, He Z, Wang YF, Xiao Y, Zhang P, Wang E (2019) A LysM receptor heteromer mediates perception of arbuscular mycorrhizal symbiotic signal in rice. Mol Plant 12:1561–1576

Helber N, Wippel K, Sauer N, Schaarschmidt S, Hause B, Requena N (2011) A versatile monosaccharide transporter that operates in the arbuscular mycorrhizal fungus Glomus sp is crucial for the symbiotic relationship with plants. Plant Cell 23:3812–3823

Huisman R, Geurts R (2019) A roadmap toward engineered nitrogen-fixing nodule symbiosis. Plant Commun 1:100019

Imaizumi-Anraku H, Takeda N, Charpentier M, Perry J, Miwa H, Umehara Y, Kouchi H, Murakami Y, Mulder L, Vickers K, Pike J, Downie JA, Wang T, Sato S, Asamizu E, Tabata S, Yoshikawa M, Murooka Y, Wu GJ, Kawaguchi M, Kawasaki S, Parniske M, Hayashi M (2005) Plastid proteins crucial for symbiotic fungal and bacterial entry into plant roots. Nature 433:527–531

Jakobsen I, Abbott LK, Robson AD (1992a) External hyphae of vesicular-arbuscular mycorrhizal fungi associated with Trifolium subterraneum L. New Phytol 120:371–380

Jakobsen I, Abbott LK, Robson AD (1992b) External hyphae of vesicular-arbuscular mycorrhizal fungi associated with Trifolium subterraneum L. New Phytol 120:509–516

Jany JL, Pawlowska TE (2010) Multinucleate spores contribute to evolutionary longevity of asexual glomeromycota. Am Nat 175:424–435

Javot H, Penmetsa RV, Terzaghi N, Cook DR, Harrison MJ (2007) A Medicago truncatula phosphate transporter indispensable for the arbuscular mycorrhizal symbiosis. Proc Natl Acad Sci U S A 104:1720–1725

Jiang Y, Wang W, Xie Q, Liu N, Liu L, Wang D, Zhang X, Yang C, Chen X, Tang D, Wang E (2017) Plants transfer lipids to sustain colonization by mutualistic mycorrhizal and parasitic fungi. Science 356:1172–1175

Jin Y, Liu H, Luo D, Yu N, Dong W, Wang C, Zhang X, Dai H, Yang J, Wang E (2016) DELLA proteins are common components of symbiotic rhizobial and mycorrhizal signalling pathways. Nat Commun 7:12433

Johansen A, Finlay RD, Olsson PA (1996) Nitrogen metabolism of external hyphae of the arbuscular mycorrhizal fungus Glomus intraradices. New Phytol 133:705–712

Kameoka H, Tsutsui I, Saito K, Kikuchi Y, Handa Y, Ezawa T, Hayashi H, Kawaguchi M, Akiyama K (2019) Stimulation of asymbiotic sporulation in arbuscular mycorrhizal fungi by fatty acids. Nat Microbiol 4:1654–1660

Kanamori N, Madsen LH, Radutoiu S, Frantescu M, Quistgaard EM, Miwa H, Downie JA, James EK, Felle HH, Haaning LL, Jensen TH, Sato S, Nakamura Y, Tabata S, Sandal N, Stougaard J (2006) A nucleoporin is required for induction of Ca2+ spiking in legume nodule development and essential for rhizobial and fungal symbiosis. Proc Natl Acad Sci U S A 103:359–364

Kim S, Zeng W, Bernard S, Liao J, Venkateshwaran M, Ane JM, Jiang Y (2019) Ca(2+)-regulated Ca(2+) channels with an RCK gating ring control plant symbiotic associations. Nat Commun 10:3703

Kobae Y, Fujiwara T (2014) Earliest colonization events of Rhizophagus irregularis in rice roots occur preferentially in previously uncolonized cells. Plant Cell Physiol 55:1497–1510

Kobae Y, Tamura Y, Takai S, Banba M, Hata S (2010) Localized expression of arbuscular mycorrhiza-inducible ammonium transporters in soybean. Plant Cell Physiol 51:1411–1415

Kobae Y, Kameoka H, Sugimura Y, Saito K, Ohtomo R, Fujiwara T, Kyozuka J (2018) Strigolactone biosynthesis genes of rice is required for the punctual entry of arbuscular mycorrhizal fungi into the roots. Plant Cell Physiol 59:544–553

Kretzschmar T, Kohlen W, Sasse J, Borghi L, Schlegel M, Bachelier JB, Reinhardt D, Bours R, Bouwmeester HJ, Martinoia E (2012) A petunia ABC protein controls strigolactone-dependent symbiotic signalling and branching. Nature 483:341–344

Lambers H, Teste FP (2013) Interactions between arbuscular mycorrhizal and non-mycorrhizal plants: do non-mycorrhizal species at both extremes of nutrient availability play the same game? Plant Cell Environ 36:1911–1915

Lambers H, Clements JC, Nelson MN (2013) How a phosphorus-acquisition strategy based on carboxylate exudation powers the success and agronomic potential of lupines (Lupinus, Fabaceae). Am J Bot 100:263–288

Lanfranco L, Young JP (2012) Genetic and genomic glimpses of the elusive arbuscular mycorrhizal fungi. Curr Opin Plant Biol 15:454–461

Lenoir I, Fontaine J, Lounès-Hadj Sahraoui A (2016) Arbuscular mycorrhizal fungal responses to abiotic stresses: a review. Phytochemistry 123:4–15

Levy J, Bres C, Geurts R, Chalhoub B, Kulikova O, Duc G, Journet EP, Ane JM, Lauber E, Bisseling T, Denarie J, Rosenberg C, Debelle F (2004) A putative Ca2+ and calmodulin-dependent protein kinase required for bacterial and fungal symbioses. Science 303:1361–1364

Li H-L, Wang W, Mortimer PE, Li R-Q, Li D-Z, Hyde KD, Xu J-C, Soltis DE, Chen Z-D (2015) Large-scale phylogenetic analyses reveal multiple gains of actinorhizal nitrogen-fixing symbioses in angiosperms associated with climate change. Sci Rep 5:14023

Liao D, Sun X, Wang N, Song F, Liang Y (2018) Tomato LysM receptor-like kinase SlLYK12 is involved in arbuscular mycorrhizal symbiosis. Front Plant Sci 9:1004

Limpens E, Franken C, Smit P, Willemse J, Bisseling T, Geurts R (2003) LysM domain receptor kinases regulating rhizobial Nod factor-induced infection. Science 302:630–633

Liu W, Kohlen W, Lillo A, Op Den Camp R, Ivanov S, Hartog M, Limpens E, Jamil M, Smaczniak C, Kaufmann K, Yang WC, Hooiveld GJ, Charnikhova T, Bouwmeester HJ, Bisseling T, Geurts R (2011) Strigolactone biosynthesis in Medicago truncatula and rice requires the symbiotic GRAS-Type transcription factors NSP1 and NSP2. Plant Cell 23: 3853–3865

Liu YN, Liu CC, Zhu AQ, Niu KX, Guo R, Tian L, Wu YN, Sun B, Wang B (2022) OsRAM2 function in lipid biosynthesis is required for arbuscular mycorrhizal symbiosis in rice. Mol Plant-Microbe Interact 35:187–199

Lopez-Raez JA, Charnikhova T, Gomez-Roldan V, Matusova R, Kohlen W, de Vos R, Verstappen F, Puech-Pages V, Becard G, Mulder P, Bouwmeester H (2008) Tomato strigolactones are derived from carotenoids and their biosynthesis is promoted by phosphate starvation. New Phytol 178:863–874

Luginbuehl LH, Menard GN, Kurup S, Van Erp H, Radhakrishnan GV, Breakspear A, Oldroyd GED, Eastmond PJ (2017) Fatty acids in arbuscular mycorrhizal fungi are synthesized by the host plant. Science 356:1175–1178

Madsen EB, Madsen LH, Radutoiu S, Olbryt M, Rakwalska M, Szczyglowski K, Sato S, Kaneko T, Tabata S, Sandal N, Stougaard J (2003) A receptor kinase gene of the LysM type is involved in legume perception of rhizobial signals. Nature 425:637–640

Maillet F, Poinsot V, Andre O, Puech-Pages V, Haouy A, Gueunier M, Cromer L, Giraudet D, Formey D, Niebel A, Martinez EA, Driguez H, Becard G, Denarie J (2011) Fungal lipochitooligosaccharide symbiotic signals in arbuscular mycorrhiza. Nature 469:58–63

Maldonado-Mendoza IE, Dewbre GR, Harrison MJ (2001) A phosphate transporter gene from the extra-radical mycelium of an arbuscular mycorrhizal fungus Glomus intraradices is regulated in response to phosphate in the environment. Mol Plant Microbe Interact 14:1140–1148

Manck-Götzenberger J, Requena N (2016) Arbuscular mycorrhiza symbiosis induces a major transcriptional reprogramming of the potato SWEET sugar transporter family. Front Plant Sci 7:487

Marleau J, Dalpé Y, St-Arnaud M, Hijri M (2011) Spore development and nuclear inheritance in arbuscular mycorrhizal fungi. BMC Evol Biol 11:51

Mashiguchi K, Seto Y, Onozuka Y, Suzuki S, Takemoto K, Wang Y, Dong L, Asami K, Noda R, Kisugi T, Kitaoka N, Akiyama K, Bouwmeester H, Yamaguchi S (2022) A carlactonoic acid methyltransferase that contributes to the inhibition of shoot branching in arabidopsis. Proc Natl Acad Sci U S A 119:e2111565119

Miller JB, Pratap A, Miyahara A, Zhou L, Bornemann S, Morris RJ, Oldroyd GE (2013) Calcium/calmodulin-dependent protein kinase is negatively and positively regulated by calcium, providing a mechanism for decoding calcium responses during symbiosis signaling. Plant Cell 25: 5053–5066

Mitra RM, Gleason CA, Edwards A, Hadfield J, Downie JA, Oldroyd GE, Long SR (2004) A Ca2+/calmodulin-dependent protein kinase required for symbiotic nodule development: gene identification by transcript-based cloning. Proc Natl Acad Sci U S A 101:4701–4705

Miyata K, Kozaki T, Kouzai Y, Ozawa K, Ishii K, Asamizu E, Okabe Y, Umehara Y, Miyamoto A, Kobae Y, Akiyama K, Kaku H, Nishizawa Y, Shibuya N, Nakagawa T (2014) The bifunctional plant receptor, OsCERK1, regulates both chitin-triggered immunity and arbuscular mycorrhizal symbiosis in rice. Plant Cell Physiol 55:1864–1872

Miyata K, Hayafune M, Kobae Y, Kaku H, Nishizawa Y, Masuda Y, Shibuya N, Nakagawa T (2016) Evaluation of the role of the LysM receptor-like kinase, OsNFR5/OsRLK2 for AM symbiosis in rice. Plant Cell Physiol 57:2283–2290

Miyata K, Hasegawa S, Nakajima E, Nishizawa Y, Kamiya K, Yokogawa H, Shirasaka S, Maruyama S, Shibuya N, Kaku H (2022) OsCERK2/OsRLK10, a homolog of OsCERK1, has a potential role for chitin-triggered immunity and arbuscular mycorrhizal symbiosis in rice. Plant Biotechnol 39:119–128

Miyata K, Hosotani M, Akamatsu A, Takeda N, Jiang W, Sugiyama T, Takaoka R, Matsumoto K, Abe S, Shibuya N, Kaku H (2023) OsSYMRK plays an essential role in AM symbiosis in rice (Oryza sativa). Plant Cell Physiol 64:378–391

Montero H, Paszkowski U (2022) A simple and versatile fluorochrome-based procedure for imaging of lipids in arbuscule-containing cells. Plant J 112:294–301

Oba H, Tawaraya K, Wagatsuma T (2001) Arbuscular mycorrhizal colonization in Lupinus and related genera. Soil Sci Plant Nutr 47:685–694

Oehl F, Sieverding E, Palenzuela J, Ineichen K, da Silva GA (2011) Advances in glomeromycota taxonomy and classification. IMA Fungus 2:191–199

Oldroyd GED, Mitra RM, Wais RJ, Long SR (2001) Evidence for structurally specific negative feedback in the Nod factor signal transduction pathway. Plant J 28:191–199

Op den Camp R, Streng A, de Mita S, Cao Q, Polone E, Liu W, Ammiraju JS, Kudrna D, Wing R, Untergasser A, Bisseling T, Geurts R (2011) LysM-type mycorrhizal receptor recruited for rhizobium symbiosis in nonlegume Parasponia. Science 331:909–912

Parniske M (2008) Arbuscular mycorrhiza: the mother of plant root endosymbioses. Nat Rev Microbiol 6:763–775

Paszkowski U, Kroken S, Roux C, Briggs SP (2002) Rice phosphate transporters include an evolutionarily divergent gene specifically activated in arbuscular mycorrhizal symbiosis. Proc Natl Acad Sci U S A 99:13324–13329

Pimprikar P, Carbonnel S, Paries M, Katzer K, Klingl V, Bohmer MJ, Karl L, Floss DS, Harrison MJ, Parniske M, Gutjahr C (2016) A CCaMK-CYCLOPS-DELLA complex activates transcription of RAM1 to regulate arbuscule branching. Curr Biol 26:987–998

Radutoiu S, Madsen LH, Madsen EB, Felle HH, Umehara Y, Gronlund M, Sato S, Nakamura Y, Tabata S, Sandal N, Stougaard J (2003) Plant recognition of symbiotic bacteria requires two LysM receptor-like kinases. Nature 425:585–592

Rausch C, Daram P, Brunner S, Jansa J, Laloi M, Leggewie G, Amrhein N, Bucher M (2001) A phosphate transporter expressed in arbuscule-containing cells in potato. Nature 414:462–465

Redecker D, Kodner R, Graham LE (2000) Glomalean fungi from the Ordovician. Science 289:1920–1921

Ropars J, Toro KS, Noel J, Pelin A, Charron P, Farinelli L, Marton T, Krüger M, Fuchs J, Brachmann A, Corradi N (2016) Evidence for the sexual origin of heterokaryosis in arbuscular mycorrhizal fungi. Nat Microbiol 1:16033

Rutten L, Miyata K, Roswanjaya YP, Huisman R, Bu F, Hartog M, Linders S, Van Velzen R, van Zeijl A, Bisseling T, Kohlen W, Geurts R (2020) Duplication of symbiotic lysin motif receptors predates the evolution of nitrogen-fixing nodule symbiosis. Plant Physiol 184:1004–1023

Saito K, Kuga-Uetake Y, Saito M (2004) Acidic vesicles in living hyphae of an arbuscular mycorrhizal fungus, Gigaspora margarita. Plant Soil 261:231–237

Saito K, Yoshikawa M, Yano K, Miwa H, Uchida H, Asamizu E, Sato S, Tabata S, Imaizumi-Anraku H, Umehara Y, Kouchi H, Murooka Y, Szczyglowski K, Downie JA, Parniske M, Hayashi M, Kawaguchi M (2007) NUCLEOPORIN85 is required for calcium spiking, fungal and bacterial symbioses, and seed production in Lotus japonicus. Plant Cell 19:610–624

Schüssle A, Martin H, Cohen D, Fitz M, Wipf D (2006) Characterization of a carbohydrate transporter from symbiotic glomeromycotan fungi. Nature 444:933–936

Seto Y, Sado A, Asami K, Hanada A, Umehara M, Akiyama K, Yamaguchi S (2014) Carlactone is an endogenous biosynthetic precursor for strigolactones. Proc Natl Acad Sci U S A 111:1640–1645

Shimizu T, Nakano T, Takamizawa D, Desaki Y, Ishii-Minami N, Nishizawa Y, Minami E, Okada K, Yamane H, Kaku H, Shibuya N (2010) Two LysM receptor molecules, CEBiP and OsCERK1, cooperatively regulate chitin elicitor signaling in rice. Plant J 64:204–214

Shindo M, Shimomura K, Yamaguchi S, Umehara M (2018) Upregulation of DWARF27 is associated with increased strigolactone levels under sulfur deficiency in rice. Plant Direct 2:e00050

Shindo M, Nagasaka S, Kashiwada S, Shimomura K, Umehara M (2021) Shoot has important roles in strigolactone production of rice roots under sulfur deficiency. Plant Signal Behav 16: e1880738
Simon L, Bousquet J, Lévesque RC, Lalonde M (1993) Origin and diversification of endomycorrhizal fungi and coincidence with vascular land plants. Nature 363:67–69
Singh S, Katzer K, Lambert J, Cerri M, Parniske M (2014) CYCLOPS, a DNA-binding transcriptional activator, orchestrates symbiotic root nodule development. Cell Host Microbe 15:139–152
Smith S, Read D (2008) Mycorrhizal symbiosis, 3rd edn. Academic, London
Smith SE, Smith FA (1990) Structure and function of the interfaces in biotrophic symbioses as they relate to nutrient transport. New Phytol 114:1–38
Smith FA, Smith SE (1997) Structural diversity in (vesicular)-arbuscular mycorrhizal symbioses. New Phytol 137:373–388
Spatafora JW, Chang Y, Benny GL, Lazarus K, Smith ME, Berbee ML, Bonito G, Corradi N, Grigoriev I, Gryganskyi A, James TY, O'Donnell K, Roberson RW, Taylor TN, Uehling J, Vilgalys R, White MM, Stajich JE (2016) A phylum-level phylogenetic classification of zygomycete fungi based on genome-scale data. Mycologia 108:1028–1046
Stockinger H, Walker C, Schüssle A (2009) 'Glomus intraradices DAOM197198', a model fungus in arbuscular mycorrhiza research, is not Glomus intraradices. New Phytol 183:1176–1187
Stracke S, Kistner C, Yoshida S, Mulder L, Sato S, Kaneko T, Tabata S, Sandal N, Stougaard J, Szczyglowski K, Parniske M (2002) A plant receptor-like kinase required for both bacterial and fungal symbiosis. Nature 417:959–962
Sun Y, Li L, Macho AP, Han Z, Hu Z, Zipfel C, Zhou JM, Chai J (2013) Structural basis for flg22-induced activation of the Arabidopsis FLS2-BAK1 immune complex. Science 342:624–628
Sun H, Tao J, Liu S, Huang S, Chen S, Xie X, Yoneyama K, Zhang Y, Xu G (2014) Strigolactones are involved in phosphate- and nitrate-deficiency-induced root development and auxin transport in rice. J Exp Bot 65:6735–6746
Sun J, Miller JB, Granqvist E, Wiley-Kalil A, Gobbato E, Maillet F, Cottaz S, Samain E, Venkateshwaran M, Fort S, Morris RJ, Ane JM, Denarie J, Oldroyd GE (2015) Activation of symbiosis signaling by arbuscular mycorrhizal fungi in legumes and rice. Plant Cell 27:823–838
Takeda N, Sato S, Asamizu E, Tabata S, Parniske M (2009) Apoplastic plant subtilases support arbuscular mycorrhiza development in Lotus japonicus. Plant J 58:766–777
Taylor TN, Remy W, Hass H, Kerp H (1995) Fossil arbuscular mycorrhizae from the Early Devonian. Mycologia 87:560–573
Thirkell TJ, Charters MD, Elliott AJ, Sait SM, Field KJ (2017) Are mycorrhizal fungi our sustainable saviours? Considerations for achieving food security. J Ecol 105:921–929
Tisserant E, Kohler A, Dozolme-Seddas P, Balestrini R, Benabdellah K, Colard A, Croll D, Da Silva C, Gomez SK, Koul R, Ferrol N, Fiorilli V, Formey D, Franken P, Helber N, Hijri M, Lanfranco L, Lindquist E, Liu Y, Malbreil M, Morin E, Poulain J, Shapiro H, van Tuinen D, Waschke A, Azcón-Aguilar C, Bécard G, Bonfante P, Harrison MJ, Küster H, Lammers P, Paszkowski U, Requena N, Rensing SA, Roux C, Sanders IR, Shachar-Hill Y, Tuskan G, Young JPW, Gianinazzi-Pearson V, Martin F (2012) The transcriptome of the arbuscular mycorrhizal fungus Glomus intraradices (DAOM 197198) reveals functional tradeoffs in an obligate symbiont. New Phytol 193:755–769
Tisserant E, Malbreil M, Kuo A, Kohler A, Symeonidi A, Balestrini R, Charron P, Duensing N, Frei Dit Frey N, Gianinazzi-Pearson V, Gilbert LB, Handa Y, Herr JR, Hijri M, Koul R, Kawaguchi M, Krajinski F, Lammers PJ, Masclaux FG, Murat C, Morin E, Ndikumana S, Pagni M, Petitpierre D, Requena N, Rosikiewicz P, Riley R, Saito K, San Clemente H, Shapiro H, van Tuinen D, Bécard G, Bonfante P, Paszkowski U, Shachar-Hill YY, Tuskan GA, Young JP, Sanders IR, Henrissat B, Rensing SA, Grigoriev IV, Corradi N, Roux C, Martin F (2013) Genome of an arbuscular mycorrhizal fungus provides insight into the oldest plant symbiosis. Proc Natl Acad Sci U S A 110:20117–20122

Uetake Y, Kojima T, Ezawa T, Saito M (2002) Extensive tubular vacuole system in an arbuscular mycorrhizal fungus, Gigaspora margarita. New Phytol 154:761–768
Umehara M, Hanada A, Yoshida S, Akiyama K, Arite T, Takeda-Kamiya N, Magome H, Kamiya Y, Shirasu K, Yoneyama K, Kyozuka J, Yamaguchi S (2008) Inhibition of shoot branching by new terpenoid plant hormones. Nature 455:195–200
Umehara M, Hanada A, Magome H, Takeda-Kamiya N, Yamaguchi S (2010) Contribution of strigolactones to the inhibition of tiller bud outgrowth under phosphate deficiency in rice. Plant Cell Physiol 51:1118–1126
Umehara M, Mengmeng C, Akiyama K, Akatsu T, Seto Y, Hanada A, Weiqiang L, Takeda-Kamiya N, Morimoto Y, Yamaguchi S (2015) Structural requirements of strigolactones for shoot branching inhibition in rice and Arabidopsis. Plant Cell Physiol 56:1059–1072
van Aarle IM, Olsson PA, Söderström B (2002) Arbuscular mycorrhizal fungi respond to the substrate pH of their extraradical mycelium by altered growth and root colonization. New Phytol 155:173–182
van Velzen R, Holmer R, Bu F, Rutten L, van Zeijl A, Liu W, Santuari L, Cao Q, Sharma T, Shen D, Roswanjaya Y, Wardhani TAK, Kalhor MS, Jansen J, van den Hoogen J, Gungor B, Hartog M, Hontelez J, Verver J, Yang WC, Schijlen E, Repin R, Schilthuizen M, Schranz ME, Heidstra R, Miyata K, Fedorova E, Kohlen W, Bisseling T, Smit S, Geurts R (2018) Comparative genomics of the nonlegume parasponia reveals insights into evolution of nitrogen-fixing rhizobium symbioses. Proc Natl Acad Sci U S A 115:E4700–e4709
van Velzen R, Doyle JJ, Geurts R (2019) A resurrected scenario: single gain and massive loss of nitrogen-fixing nodulation. Trends Plant Sci 24:49–57
Veiga RSL, Faccio A, Genre A, Pieterse CMJ, Bonfante P, van der Heijden MGA (2013) Arbuscular mycorrhizal fungi reduce growth and infect roots of the non-host plant Arabidopsis thaliana. Plant Cell Environ 36:1926–1937
Wakabayashi T, Hamana M, Mori A, Akiyama R, Ueno K, Osakabe K, Osakabe Y, Suzuki H, Takikawa H, Mizutani M, Sugimoto Y (2019) Direct conversion of carlactonoic acid to orobanchol by cytochrome P450 CYP722C in strigolactone biosynthesis. Sci Adv 5:eaax9067
Wakabayashi T, Shida K, Kitano Y, Takikawa H, Mizutani M, Sugimoto Y (2020) CYP722C from Gossypium arboreum catalyzes the conversion of carlactonoic acid to 5-deoxystrigol. Planta 251:97
Wakabayashi T, Ishiwa S, Shida K, Motonami N, Suzuki H, Takikawa H, Mizutani M, Sugimoto Y (2021) Identification and characterization of sorgomol synthase in sorghum strigolactone biosynthesis. Plant Physiol 185:902–913
Walker SA, Viprey V, Downie JA (2000) Dissection of nodulation signaling using pea mutants defective for calcium spiking induced by nod factors and chitin oligomers. Proc Natl Acad Sci U S A 97:13413–13418
Wang Y, Durairaj J, Suárez Duran HG, van Velzen R, Flokova K, Liao CY, Chojnacka A, Macfarlane S, Schranz ME, Medema MH, van Dijk ADJ, Dong L, Bouwmeester HJ (2022) The tomato cytochrome P450 CYP712G1 catalyses the double oxidation of orobanchol en route to the rhizosphere signalling strigolactone, solanacol. New Phytol 235:1884–1899
Wipf D, Mongelard G, van Tuinen D, Gutierrez L, Casieri L (2014) Transcriptional responses of Medicago truncatula upon sulfur deficiency stress and arbuscular mycorrhizal symbiosis. Front Plant Sci 5:680
Yang S-Y, Grønlund M, Jakobsen I, Grotemeyer MS, Rentsch D, Miyao A, Hirochika H, Kumar CS, Sundaresan V, Salamin N, Catausan S, Mattes N, Heuer S, Paszkowski U (2012) Nonredundant regulation of rice arbuscular mycorrhizal symbiosis by two members of the PHOSPHATE TRANSPORTER1 gene family. Plant Cell 24:4236–4251
Yano K, Yoshida S, Müller J, Singh S, Banba M, Vickers K, Markmann K, White C, Schuller B, Sato S, Asamizu E, Tabata S, Murooka Y, Perry J, Wang TL, Kawaguchi M, Imaizumi-Anraku H, Hayashi M, Parniske M (2008) CYCLOPS, a mediator of symbiotic intracellular accommodation. Proc Natl Acad Sci U S A 105:20540–20545

Yoneyama K, Brewer PB (2021) Strigolactones, how are they synthesized to regulate plant growth and development? Curr Opin Plant Biol 63:102072
Yoneyama K, Xie X, Kusumoto D, Sekimoto H, Sugimoto Y, Takeuchi Y, Yoneyama K (2007a) Nitrogen deficiency as well as phosphorus deficiency in sorghum promotes the production and exudation of 5-deoxystrigol, the host recognition signal for arbuscular mycorrhizal fungi and root parasites. Planta 227:125–132
Yoneyama K, Yoneyama K, Takeuchi Y, Sekimoto H (2007b) Phosphorus deficiency in red clover promotes exudation of orobanchol, the signal for mycorrhizal symbionts and germination stimulant for root parasites. Planta 225:1031–1038
Yoneyama K, Xie X, Kim HI, Kisugi T, Nomura T, Sekimoto H, Yokota T, Yoneyama K (2012) How do nitrogen and phosphorus deficiencies affect strigolactone production and exudation? Planta 235:1197–1207
Yoneyama K, Kisugi T, Xie X, Arakawa R, Ezawa T, Nomura T, Yoneyama K (2015) Shoot-derived signals other than auxin are involved in systemic regulation of strigolactone production in roots. Planta 241:687–698
Yoneyama K, Mori N, Sato T, Yoda A, Xie X, Okamoto M, Iwanaga M, Ohnishi T, Nishiwaki H, Asami T, Yokota T, Akiyama K, Yoneyama K, Nomura T (2018a) Conversion of carlactone to carlactonoic acid is a conserved function of MAX1 homologs in strigolactone biosynthesis. New Phytol 218(4):1522–1533
Yoneyama K, Xie X, Yoneyama K, Kisugi T, Nomura T, Nakatani Y, Akiyama K, McErlean CSP (2018b) Which are the major players, canonical or non-canonical strigolactones? J Exp Bot 69: 2231–2239
Yoneyama K, Akiyama K, Brewer PB, Mori N, Kawano-Kawada M, Haruta S, Nishiwaki H, Yamauchi S, Xie X, Umehara M, Beveridge CA, Yoneyama K, Nomura T (2020) Hydroxyl carlactone derivatives are predominant strigolactones in Arabidopsis. Plant Direct 4:e00219
Yoshida S, Kameoka H, Tempo M, Akiyama K, Umehara M, Yamaguchi S, Hayashi H, Kyozuka J, Shirasu K (2012) The D3 F-box protein is a key component in host strigolactone responses essential for arbuscular mycorrhizal symbiosis. New Phytol 196:1208–1216
Yuan H, Liu D (2008) Signaling components involved in plant responses to phosphate starvation. J Integr Plant Biol 50:849–859
Zhang Q, Blaylock LA, Harrison MJ (2010) Two Medicago truncatula half-ABC transporters are essential for arbuscule development in arbuscular mycorrhizal symbiosis. Plant Cell 22:1483–1497
Zhang Y, van Dijk AD, Scaffidi A, Flematti GR, Hofmann M, Charnikhova T, Verstappen F, Hepworth J, van der Krol S, Leyser O, Smith SM, Zwanenburg B, Al-Babili S, Ruyter-Spira C, Bouwmeester HJ (2014) Rice cytochrome P450 MAX1 homologs catalyze distinct steps in strigolactone biosynthesis. Nat Chem Biol 10:1028–1033
Zhang X, Dong W, Sun J, Feng F, Deng Y, He Z, Oldroyd GE, Wang E (2015) The receptor kinase CERK1 has dual functions in symbiosis and immunity signalling. Plant J 81:258–267
Zhang C, He J, Dai H, Wang G, Zhang X, Wang C, Shi J, Chen X, Wang D, Wang E (2021) Discriminating symbiosis and immunity signals by receptor competition in rice. Proc Natl Acad Sci U S A 118:e2023738118

Chapter 7
Mechanisms of Arbuscular Mycorrhizal Fungi-Induced Drought Stress Amelioration in Plants

Vinay Shankar, Amanso Tayang, and Heikham Evelin

Abstract Drought stress is a global problem affecting society, economy, and the environment. The foremost and worst affected sector impacted by drought is the agriculture sector, thus threatening food security. Hence, it is imperative that we adopt methods that impart drought tolerance to plants. One such approach is the inoculation of plants with arbuscular mycorrhizal fungi (AMF) as it is economical, environment friendly, and sustainable. AMF are soil-dwelling symbiotic fungi that impart drought tolerance to many plants. In this article, we have discussed the recent findings of AMF-induced adaptive strategies of plants in overcoming drought stress. These strategies include modifications in the host plant's structure (leaf and root architecture), biochemistry (such as accumulation of osmolytes), and physiology (acquisition of water and nutrients, chlorophyll synthesis, hormone signaling). The article has also identified and presented gaps in research, the unraveling of which can improve our knowledge of mechanisms of drought tolerance brought about by AMF colonization.

Keywords AMF · Water deficit · Stomata · Proline · Osmotic adjustment · Root architecture

7.1 Introduction

Drought is a natural phenomenon characterized by abnormally low precipitation over a long period of time causing water shortage. It is a global concern, and in the last four decades, no other natural hazard has affected human life as much as drought. In addition, climate change is aggravating drought in various parts of the

V. Shankar
Department of Botany, Gaya College, Magadh University, Bodh Gaya, Bihar, India

A. Tayang · H. Evelin (✉)
Department of Botany, Rajiv Gandhi University, Rono Hills, Doimukh, Arunachal Pradesh, India
e-mail: heikham.evelin@rgu.ac.in

G. J. Ahammed, R. Hajiboland (eds.), *Arbuscular Mycorrhizal Fungi and Higher Plants*, https://doi.org/10.1007/978-981-99-8220-2_7

world. Drought has impacted society, economy, and environment and has led to massive shifting of families, degradation of natural resources, and weakened economy. Of all the sectors impacted by drought, agriculture is the first and worst affected sector impacted directly, affecting agricultural produce, thereby threatening food security and rural livelihoods (FAO 2023, accessed 25 June 2023). In the recent drought in Kenya in 2017, an approximate 2.5 million people were left hungry prompting a national emergency (Gichure 2017; He et al. 2019). Unfortunately, drought can neither be halted nor its forecast easy. However, these negative impacts of drought can be mitigated by adopting approaches that target to enhance communities' resilience in coping with drought. In this regard, an environmentally and sustainable technique is the application of beneficial microorganisms like AMF in agriculture.

AMF are soil- dwelling symbiotic fungi of the phylum Glomeromycota. These fungi have been well-documented to ameliorate stresses caused by biotic and abiotic factors like drought, heavy metals, salinity, as well as temperature in plants (Evelin et al. 2019; Shankar and Evelin 2021; Begum et al. 2021; Liu et al. 2023a). These beneficial effects of AMF have been related with their ability to inhabit different types of substratum and colonize 80% of land plants ranging from bryophytes to flowering plants and the ability to cast a wide underground mycelial network for efficient uptake of nutrients and water. The article is aimed at understanding the mechanisms imparted by AMF colonization in enhancing resilience of plants against drought.

7.2 Impacts of Drought on Plants

Morphologically, stress caused by drought makes the plant shorter with fewer and smaller leaves. Anatomically, the outer layer of the epidermal cell develops a thick cuticle, and leaves have lower ratio of spongy:palisade tissue, more cell layers, but less intercellular spaces leading to a decrease in volume (Chartzoulakis et al. 2002). These changes are also reflected on various biochemical as well as physiological functions like synthesis of chlorophyll, photosynthesis, mineral uptake and assimilation, osmotic regulation, and sequestration of reactive oxygen species (Jaleel et al. 2008; Farooq et al. 2009; Li and Liu 2016). Drought also induces the expression of aquaporins, late embryogenesis abundant (LEA) proteins, and dehydrins.

7.3 Plant Response to Drought Stress

In order to overcome stress caused by drought, plants manifest an array of mechanisms that work in unison. Plants undergo adaptive changes in the structure, physiology, and biochemistry in response to drought. They reduce their biomass by cutting down the leaf number as well as leaf size. The leaf also reduces the number of stomata and assumes vertical orientation (Liu et al. 2023a). Besides, it tends to extend its root growth to absorb maximum water and nutrients from the soil resulting in a higher root- to- shoot ratio. These changes greatly reduce loss of water through transpiration (Mohammadi et al. 2019). Biochemical and physiological mechanisms, such as osmoregulation, upped antioxidant system, prevention of injury to the photosynthetic system, and regulation of phytohormone dynamics, contribute to plant drought tolerance (Ilyas et al. 2021).

7.4 Impact of Drought on Diversity and Colonization Potential of AMF

Soil moisture is a key agent that influences the diversity and colonization potential of AMF. *Hordeum vulgare* roots subjected to three soil moisture conditions—flooding, drought, and 15–29% soil moisture regime—showed varying levels of AMF diversity and colonization potential. Roots grown in soil with 15–20% moisture showed maximum AMF diversity followed by roots under drought and flooded soil (Sharma and Kothamasi 2015). Drought stress reduced the colonization potential of AMF in maize (Begum et al. 2019), tobacco (Begum et al. 2021), tea (Liu et al. 2023b), etc. This can be explained by drought-induced inhibition on germination of spore and disruption of hyphal development, thus minimizing the colonization potential of AMF (Giovannetti et al. 2010).

7.5 AMF-Induced Drought Tolerance Mechanisms in Plants

Amelioration of drought stress effects in plants colonized by AMF is brought about by acquiring adaptive changes in the structure, biochemistry, and physiology of the host plants (Fig. 7.1). Recent knowledge of these mechanisms is discussed one by one hereunder.

7.5.1 Improved Plant Growth and the Associated Structural Changes

The ameliorative effects of AM symbioses in plants subjected to drought stress are well-demonstrated. Under water stress conditions, AMF-colonized plants showed better growth and biomass as compared to the non-AMF plants in tobacco (Begum et al. 2021), white clover (Liang et al. 2021), and wheat (Tereucán et al. 2022). Plant growth parameters such as height, leaf number, and biomass production were more

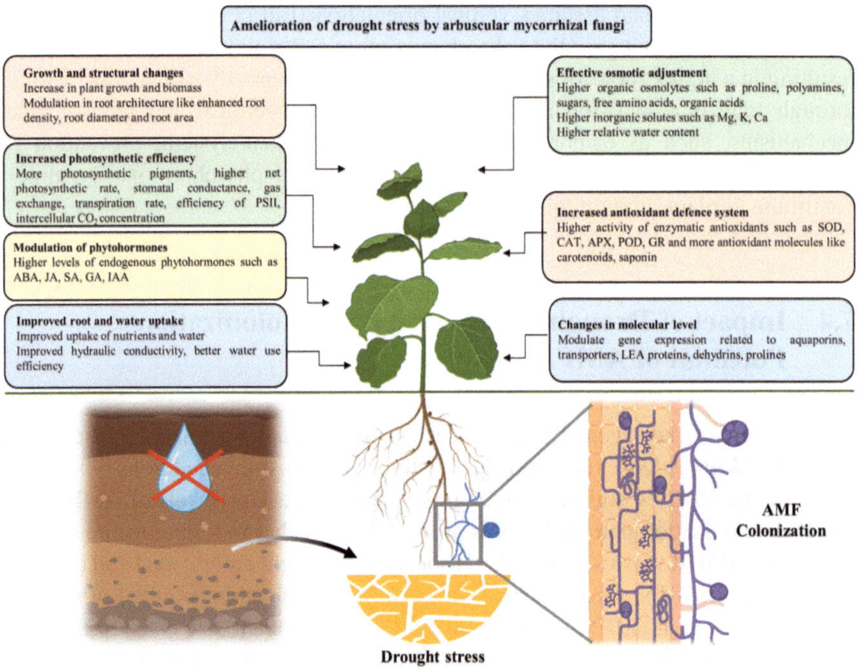

Fig. 7.1 Arbuscular mycorrhizal fungi (AMF)-mediated amelioration of drought stress in plants. Mycorrhizal plants increased resilience to drought by adapting structural, biochemical, and physiological changes induced by AMF

in AMF- colonized plants (Liang et al. 2021; Begum et al. 2021; Tereucán et al. 2022). This can be attributed to the more efficient water and mineral acquisition in the mycorrhizal plants facilitated by adaptive structural modifications in host plants brought about by AMF (Abdel-Salam et al. 2018; Zhang et al. 2018a). AM symbiosis has been reported to induce changes in the root architecture of the host plants (Liu et al. 2016). Roots showed increased biomass as a result of an increase in branching, density, length, and extension of root hairs (Zou et al. 2017; Liu et al. 2018; Zhang et al. 2018a). Additionally, AMF equipped the host plant with an extensive network of AMF mycelium to explore and extract water as well as nutrients from the rhizosphere soil.

7.5.2 *Efficient Acquisition of Water and Nutrient*

In the soil, AMF produce and secrete a super glue, glomalin, which aids in soil aggregation (Chi et al. 2018). AMF hyphae in well- aggregated soil improve the capacity of soil to retain water (Querejeta et al. 2003) and promote better root-soil connect, facilitating better water conductance from soil to root (Augé 2004). Thus,

AMF help in maintaining continued transportation of water from soil to the plant (Bitterlich et al. 2018). In addition, drought stress also promotes higher water absorption rate in AMF hyphae (Zhang et al. 2018a). This is facilitated by the modulation of aquaporin *AQP* gene expression on cell membranes by AMF (He et al. 2019; Cheng et al. 2020). Till date, six aquaporin proteins have been identified in AMF: three each in *Glomus intraradices* (GintAQP1, GintAQPF1, GintAQPF2) and *Rhizophagus clarus* (RcAQP1, RcAQP2, RcAQP3) (Aroca et al. 2009; Li et al. 2013; Kikuchi et al. 2016). Drought stress induced the expression of *GintAQPF1* and *GintAQPF2*, while the expression of *GintAQP1* remained unaltered. Similarly, AMF also modulate *PIP* (plasma membrane intrinsic protein) genes' expression in the host plants under water- deficit condition; it may be upregulated (Zézé et al. 2008) or downregulated (Aroca et al. 2007). These findings suggest that different aquaporins may impart drought tolerance in different plants. Another view is that aquaporin gene expression is upregulated to improve water uptake by plants and downregulated to prevent water loss from plants (Ruiz-Lozano and Aroca 2010; Cheng et al. 2020).

Along with improved water absorption, AMF also ensure optimum nutrient availability to host plants by increasing their acquisition. Wu and Zou (2009) reported increase in the concentrations of Ca, Fe, P, and K in leaves and roots of trifoliate orange plants following *G. versiforme* application under both drought and well-watered conditions. Farias et al. (2014) found similar observations in blueberry in which AMF colonization increased NPK concentrations in leaves. Better nutrient uptake in mycorrhizal plants is due to their access to nutrients present in the micro pores owing to their leaner hyphae as well as the ability to activate genes and enzymes involved in assimilation and transportation of nutrients. For example, AMF improves the activity of ammonium transporter protein and nitrate reductase for efficient N uptake and assimilation (Garg 2013; Balestrini et al. 2019), phosphatase activity for efficient P absorption, and metabolism (Abd-Allah et al. 2015; Egamberdieva et al. 2017). In tomato, AMF induced the expression of phosphate transporter genes such as *LePT4* and *LePT5* to elevate tolerance against drought (Volpe et al. 2018). Balestrini et al. (2019) reported AMF-induced upregulation of peptide transporter, amino acid transporter, and sulfate transporter genes in plants under drought stress. Besides, AMF also up regulated the activities of H^+-ATPase and Ca^{2+}-ATPase under drought conditions. This, in turn, acidifies the soil environment, allowing for efficient uptake of nutrients and exchange of signals between AMF and plants to increase tolerance to drought (Xu et al. 2018a).

7.5.3 Better Photosynthetic Capacity

Photosynthesis, a key primary metabolic process essential for the synthesis of energy in plant, is drastically affected by drought stress (Ilyas et al. 2021). Moderate and severe levels of drought-induced decrease in leaf number and leaf area reduce the concentration of photosynthetic pigments like chlorophylls and carotenoids (Zhang et al. 2018a; Mashabela et al. 2023), which ultimately hampers photosynthesis in plants (Spinoso-Castillo et al. 2023; El-Sawah et al. 2023). Up regulated expression of chlorophyll- degrading enzymes and down regulation of chlorophyll biosynthetic enzymes also contribute to a decline in chlorophyll content (Ilyas et al. 2021; Saxena et al. 2022). However, AMF inoculation can increase chlorophyll content in plants growing under drought conditions. For example, under drought stress (50% field capacity), *Rhizophagus intraradices*-colonized *Ricinus communis* had a significant amount of chlorophylls a and b and total chlorophyll than the non-colonized plants (Zhang et al. 2018a). Similar observations were reported in *Cicer arietinum*, *Zea mays*, *Abelmoschus esculentus*, and *Fagopyrum esculentum* on AMF inoculation (Hashem et al. 2019; Hu et al. 2020; Jabborova et al. 2021; Mohammadi et al. 2022). Recently, Miceli et al. (2023) demonstrated that *Claroideoglomus etunicatum* inoculation increased the level of photosynthetic pigments and chlorophyll a/ chlorophyll b ratio in *Physalis peruviana* under soil moisture- deficit condition, subsequently improving photosynthesis to enhance plant biomass and fruit production.

Drought stress is also known to reduce efficiency of photosynthesis by injuring PSII, and disrupting stomatal conductance, intercellular CO_2 concentration, and transpiration (Chen et al. 2022). However, the inimical effects of drought on photosynthetic machinery can be prevented by AMF inoculation to the host plants. AMF can improve quantum efficiency of PS II, stomatal conductance, transpiration, and intercellular CO_2 concentration (Mathur et al. 2019; Qian et al. 2023). Application of a consortium of AMF consisting of *Rhizoglomus intraradices*, *Funneliformis geosporum*, and *F. mosseae* in *Triticum aestivum* enhanced quantum efficiency of PSII (Mathur et al. 2019). Such findings were also demonstrated in *Zea mays*, *Oryza sativa*, and *Ceratonia siliqua* (Quiroga et al. 2019; Tisarum et al. 2020; Hu et al. 2020; Chareesri et al. 2020; Boutasknit et al. 2020). AM symbiosis promoted transpiration, intercellular CO_2 concentration, and stomatal conductance in *Ricinus communis*, *Olea europaea*, *Phaseolus vulgaris*, and *Paris polyphylla* (Zhang et al. 2018a; Ouledali et al. 2019; Al-Amri 2021; Qian et al. 2023) under different levels of water- deficit conditions. Thus, AM symbiosis facilitates plants to overcome drought stress by synthesizing more photosynthetic pigments and preventing damage to the photosynthetic machinery.

7.5.4 Maintenance of ROS Level

ROS is a term for derivatives of O_2 generated as byproducts of metabolic reactions in chloroplasts, mitochondria and peroxisomes, plasma membrane, and cell wall (Halliwell and Gutteridge 2015; Huang et al. 2019; Mansoor et al. 2022). ROS include singlet oxygen ($^1O_2$), superoxide anions (O_2^-), hydroxyl radicals (OH•), and hydrogen peroxide (H_2O_2) and are involved in maintaining biochemical and cellular processes like cell cycle, programmed cell death, growth and development, hormonal signaling, and biotic and abiotic stress reactions (Conrad et al. 2016; Mansoor et al. 2022). However, ROS production is elevated under drought conditions leading to oxidative stress causing damage to nucleic acid and proteins, ultimately hampering plant's growth and development (Gechev et al. 2006; Raja et al. 2017; Laxa et al. 2019; Khatri and Rathore 2022). Therefore, ROS concentration needs to be maintained at its normal level (McLachlan 2020; Pammi et al. 2023). Antioxidant defense system, which includes molecules and enzymes, is responsible for maintaining ROS at the normal level (Noctor et al. 2018). Antioxidant molecules include alkaloids, flavonoids, ascorbic acid, carotenoids, glutathione, phenolics, and non-protein amino acids, while enzymatic antioxidants include catalase (CAT), ascorbate peroxidase (APXs), glutathione reductase (GRs), glutathione peroxidase (GPXs), and superoxide dismutase (SOD) (Laxa et al. 2019; Mansoor et al. 2022). Antioxidant molecules scavenge ROS through several mechanisms such as electron donation, redox homeostasis, protection of light- harvesting complex, and maintenance of thylakoid membrane and act as chelators and inhibitors of ROS (Hasanuzzaman et al. 2019, 2020; Pammi et al. 2023). Enzymatic antioxidants catalyze dismutation of ROS to prevent hydroxyl formation, conversion of H_2O_2 into H_2O molecules, oxidation for production of phenoxyl radical, etc. (Gill et al. 2015; Pammi et al. 2023).

Many studies have demonstrated higher level of antioxidant molecules as well as higher activity of enzymes in AMF-colonized plants than their non-colonized counterparts under drought stress. Qian et al. (2023) reported higher saponin VI accumulation in *Paris polyphylla* when inoculated with *Rhizophagus irregularis* in water- deficit condition. Similarly, greater carotenoid content was observed in mycorrhizal *Brassica juncea*, *Lactuca sativa*, and *Cupressus arizonica* (Srivastava et al. 2022; Ouhaddou et al. 2022; Aalipour et al. 2023). In addition, *Funneliformis mosseae* colonization improved coumarin, phenolic compounds, and flavonoids in *Olea europaea*, *Poncirus trifoliata*, and *Mentha piperita* (Tekaya et al. 2022; Liu et al. 2022; Ostadi et al. 2023). Similarly, elevated activities of leaf APX, CAT, APX, and GR were reported in *Ephedra foliata* upon colonization by a mixture of *Funneliformis mosseae*, *Claroideoglomus etunicatum*, and *Rhizophagus intraradices* under different levels of water stress (Al-Arjani et al. 2020). Most recently, Liu et al. (2023a) observed that a consortium of *Rhizophagus clarus*, *R. irregularis*, *Glomus lamellosum*, and *Funneliformis mosseae* significantly upped the activities of CAT, APX, SOD, and POD in *Arachis hypogaea* leaf under different abiotic stresses like drought, salinity, and cold. Higher enzymatic activities of these antioxidants can be partially attributed to enhanced nutritional status in AMF-colonized plants as these enzymes are metalloenzymes, and their activities are dependent on the availability of micronutrients like Fe, Cu, and Mn. Besides, other factors like species of AMF and plant and level and duration of stress also determine their activities (Evelin et al. 2019).

7.5.5 Better Osmotic Adjustment

Organic osmolytes are low- molecular- weight, water- soluble organic compounds that normally accumulate during abiotic stresses. Commonly accumulated organic osmolytes include proline, glycine betaines, polyamines, and sugars (Evelin et al. 2019; Shankar and Evelin 2021). Accumulation of these organic solutes predominantly takes place in the cytoplasm; however, the synthesis of specific osmolytes depends on plant tissue, species of plant, and type of stress involved (Yang and Guo 2018). Under drought stress, concentration of these solutes increases to maintain an optimum water potential gradient and facilitate the flow of water from soil to plant roots, consequently preventing cellular dehydration (Evelin et al. 2019; Saxena et al. 2022). Thus, these solutes participate in osmotic adjustment and can be used to efficiently overcome the deleterious influence of osmotic stress (Tang et al. 2022). Besides, these osmolytes exhibit functional diversity to ameliorate stress effects in plants as they maintain membrane integrity and stabilize structure of protein, enhance ROS defense system to prevent oxidative damage, provide carbon and energy to recover from stress, protect photosynthetic apparatus, prevent K efflux from cytoplasm, and function as signaling molecules (Lokhande and Suprasanna 2012; Shankar and Evelin 2021).

Organic osmolyte accumulation in plants under stress conditions is further enhanced upon inoculation with AMF (Evelin et al. 2019; Tang et al. 2022; Liang et al. 2022). It was observed that proline content was significantly increased in *Eucalyptus camaldulensis*, *Valeriana officinalis*, *Citrus aurantium*, and *Saccharum* spp. upon inoculation with *Gigaspora albida*, *Funneliformis mosseae*, and a mixture of *Glomus hoi*, *G. mosseae*, and *G. intraradices*, respectively (Klinsukon et al. 2021; Amanifar and Toghranegar 2020; Hadian-Deljou et al. 2020; Spinoso-Castillo et al. 2023). The enhanced proline content in plants following AMF inoculation can be due to enhanced activity of enzyme P5CS (Δ'-pyrroline-5-carboxylate synthase), higher expression of proline biosynthetic gene encoding *P5CS*, enhanced activity of enzyme, glutamate dehydrogenase, and inactivation of enzyme and proline dehydrogenase (Abo-Doma et al. 2011). Besides osmoregulation, proline also participates in other cellular activities like ROS scavenging and stabilization of membrane, proteins, and DNA. It also acts as a reservoir for excessive NAD^+ and $NADP^+$ for photosynthesis and respiration and sources of carbon, energy, and nitrogen to recover from stress (Kaur and Asthir 2015; Evelin et al. 2019; Shankar and Evelin 2021). In contrast, there are a few studies that report lesser proline content in the AMF-colonized plants as compared to the non-AMF plants under abiotic stresses. For example, *Glycyrrhiza uralensis*, *Myrtus communis*, and *Physalis peruviana* when inoculated with *Rhizophagus irregularis*, *Funneliformis mosseae*, and *Claroideoglomus claroideum*, respectively, had lesser leaf proline content than control plants (Xie et al. 2018; Azizi et al. 2021; Geneva et al. 2023). Proline is a stress marker and lesser proline accumulation in the AMF-colonized plants may indicate less stress in them (Evelin et al. 2019). Moreover, decreased activity of

P5CS could be attributed to lower proline content in *Poncirus trifoliata* inoculated with *Paraglomus occultum* (Zhang et al. 2018b).

Glycine betaine is an amphoteric molecule that acts as an osmoregulator and/or osmoprotectant in plants under drought stress (Saxena et al. 2022). AMF-colonized plants showed more glycine betaine concentration than control plants, resulting in higher osmotic potential than control plants under drought stress (Abd-Elghany et al. 2021). Increase in accumulation of glycine betaine could be attributed to enhanced expression of choline monooxygenase and betaine aldehyde dehydrogenase (Zhang et al. 2009). Higher concentration of glycine betaine is correlated to better cell membrane and photosynthetic apparatus protection, scavenging of ROS, and stabilization of protein and enzyme complexes (Sakamoto and Murata 2000; He et al. 2010; Niu et al. 2021).

Free amino acids provide tolerance to plants against drought stress. AMF colonization can further enhance the level of free amino acids in the plant. AM symbiosis in *Helianthus tuberosus* and *Sesamum indicum* increased the level of free amino acids (Gholinezhad and Darvishzadeh 2021; Nacoon et al. 2022). *Zea mays* had higher accumulation of γ-aminobutyric acid, free amino acids, arginine, and ornithine when inoculated with *Rhizophagus irregularis* as compared to the non-AMF plants growing in water stress conditions (Hu et al. 2020). Similarly, Hu and Chen (2020) observed that *Zea mays* also accumulated higher concentration of glutamine and γ-aminobutyric acid and increased activity of glutamine synthetase when inoculated with a mixed consortium of *Claroideoglomus etunicatum* and *Rhizophagus irregularis* as compared to the control plants.

Sugars are major osmoprotectants that contribute to nearly 50% of a plant's osmotic potential by providing carbon to recover from stress and aiding in osmotic adjustment (Santander et al. 2017). Under stress conditions, total soluble sugar increased in *Malus hupehensis*, *Cinnamomum migao*, and *Fagopyrum esculentum* when inoculated with *Rhizophagus irregularis*, *Glomus lamellosum*, and *G. etunicatum* and a consortium of *Funneliformis mosseae*, *Rhizophagus irregularis*, and *R. fasciculatus*, respectively, as compared to the non-AMF plants (Huang et al. 2020; Liao et al. 2021; Mohammadi et al. 2022). Tekaya et al. (2022) reported higher accumulation of mannitol in *Olea europaea* when colonized by *Rhizophagus irregularis* as compared to the non-mycorrhizal plants. Under osmotic stress, AM symbiosis also improved glucose content in *Oryza sativa* and *Ephedra foliata* (Tisarum et al. 2020; Al-Arjani et al. 2020). The increase in these sugars can be attributed to increased photosynthesis and carbohydrate metabolism in the AMF-colonized plants under abiotic stress conditions (Santander et al. 2017; Tarnabi et al. 2020).

Polyamines are low- molecular-weight polycationic nitrogen-containing compounds of aliphatic nature. They are found in all living organisms. During stress, it takes part in several physiological responses like osmoregulation, stabilization of membranes, ROS scavenging, cellular pH modulation, and photosynthesis (Pang et al. 2007; Chen et al. 2019; Liang et al. 2022). AMF colonization modulated polyamine levels in host plants under drought stress. Putrescine content was enhanced by *Rhizophagus irregularis* in *Zea mays* under drought stress (Hu et al.

2020). Interestingly, Hu and Chen et al. (2020) observed a decrease in putrescine content in *Zea mays* when inoculated with *Rhizophagus irregularis*.

Drought stress also induces dynamism in accumulation of other osmolytes, organic acids. Organic acids prevent the buildup of toxic chloride ions, while providing cellular ion homeostasis, pH regulation, and osmotic adjustment (Guo et al. 2010). AM symbiosis can modulate the accumulation of organic acid in plants. *Zea mays* inoculated with *Rhizophagus irregularis* improved concentration of malic acid and total organic acids under water- deficit conditions (Hu et al. 2020; Hu and Chen 2020). Besides malic acid, Hu and Chen et al. (2020) observed increased activity of GABA transaminase and malic enzymes in AMF-colonized maize plants. AMF could provide protection to enzymes of organic acid biosynthesis as well as initiate closure of stomata (Sheng et al. 2011; Hu and Chen 2020; Hu et al. 2020). Increase in organic acids could positively regulate the Calvin cycle to promote synthesis of sugar for plant's growth and development and confer stress tolerance (Guo et al. 2010). The exact mechanisms behind the alteration in organic acids in the AMF-colonized plants are yet to be established and require further intensive research.

7.5.6 Maintaining Higher Unsaturated Fatty Acid: Saturated Fatty Acid Ratio

Drought tolerance in plants is also associated with a greater unsaturated fatty acid/ saturated fatty acid ratio (Mahnaz et al. 2020). AM symbiosis has been found to regulate metabolism of fatty acids in conferring drought tolerance to the host plants. In inoculation with AMF, *F. mosseae* induced a higher UFA (C18:1, C18:2, and C18:3N3)/SFA (C18:0) ratio in roots of trifoliate orange plants as compared to the non-mycorrhizal plants (Wu et al. 2019; Hu et al. 2020). In sesame plants, inoculation with *F. mosseae* and *Rhizophagus intraradices* imparted drought tolerance by maintaining higher UFA/SFA ratio and better antioxidant defense (Gholinezhad and Darvishzadeh 2021). Higher UFA in mycorrhizal plants may be attributed to the induction of FA desaturase 2 and FA desaturase 6 genes involved in desaturation of fatty acids (Wu et al. 2019). AMF-induced maintenance of higher UFA levels prevents the host plants from drought-induced oxidative damage by maintaining membrane fluidity. Alternatively, AMF also induce sporulation and their growth indirectly by increasing the level of C14:0 levels in the host plants (Sugiura et al. 2020; Meng et al. 2021a).

7.5.7 Modulation of Phytohormones

Phytohormones, also known as plant growth regulators, play a crucial role in growth and development. Drought stress alters endogenous phytohormone levels in plants. However, AMF colonization has been shown to modulate levels of phytohormones in imparting tolerance against drought stress in the host plants (Table 7.1). Abscisic acid (ABA) is an important phytohormone that acts as a key signal molecule in roots for AM colonization. AM symbiosis upregulated biosynthesis of ABA resulting in higher ABA content in *Solanum lypersicum* which further induced closing of stomata, thereby preventing transpiration and water loss (Chitarra et al. 2016). Higher level of ABA is essential for maintaining structural fluidity and efficient regulation of transport ions in plants (Yang et al. 2014). Under drought stress conditions, a mixed consortium of *Funneliformis mosseae*, *Claroideoglomus etunicatum*, and *Rhizophagus intraradices* increased the content of ABA in *Ephedra foliata* (Al-Arjani et al. 2020). More recently, Qian et al. (2023) demonstrated that colonization with *Rhizophagus irregularis* increased ABA content in *Paris polyphylla* var. *yunnanensis* under drought stress. Increase in ABA can be attributed to key genes such as 14-3-3 genes involved in ABA signaling. AM symbiosis induced the expressions of *PcGRF10* and *PcGRF11* in *Populus cathayana* as their expressions were correlated with improved osmoregulation and antioxidant defense system (Han et al. 2022). However, it has also been shown that ABA content in the AMF-colonized plants could reduce under stress conditions, indicating that ABA regulation in the AMF-colonized host plants could vary (Cheng et al. 2021). Reduction of ABA content in *Cichorium intybus* was observed under drought stress upon *Rhizophagus irregularis* inoculation (Langeroodi et al. 2020). Chen et al. (2020) also reported reduced ABA content in *Catalpa bungei* upon *Rhizophagus intraradices* inoculation under drought stress. Similarly, ABA content was lesser in *R. irregularis*- colonized *Olea europaea* than in control plants (Tekaya et al. 2022). The decrease in ABA content in the mycorrhizal plants can be due to the role of ABA in developing fungal mycelium in host plants (Kandowangko et al. 2009; Goicoechea et al. 2010).

AM symbiosis imparts a positive influence on salicylic acid (SA) in improving carbohydrate metabolism and ion homeostasis and formation of fungal structures like arbuscules and vesicles, resulting in increased AM colonization (Liu et al. 2018). Moreover, salicylic acid also regulates root aquaporins to modulate water conductivity and control root water transport (Quiroga et al. 2018). Higher levels of SA were observed in mycorrhizal *Olea europaea* than in control plants under drought stress (Tekaya et al. 2022).

Phytohormone jasmonic acid (JA) also takes part in improving plant's tolerance against abiotic and biotic stresses. Levels of jasmonic acid increased in stress conditions (Gul et al. 2022). Liu et al. (2016) and Zhang et al. (2019) reported that AM symbiosis enhanced the level of methyl jasmonate in trifoliate orange grown under drought conditions. As JA promotes accumulation of osmolytes and enhance

Table 7.1 Role of phytohormones in AMF-induced drought tolerance

Phytohormone involved	AM fungi	Plant	Effect on phytohormone	Plant response under drought stress	References
Indole-3-acetic acid (IAA)	*Rhizophagus irregularis*	*Poncirus trifoliata*	Upregulation of IAA biosynthetic genes such as *PtYUC3* and *PtYUC8* Upregulation of root auxin-species influx carriers such as *PtLAX2* and PtABCB19 Downregulation of root auxin efflux carriers like *PtPIN1* and *PtPIN3* IAA level increased	Improved plant growth and biomass	Liu et al. (2018)
IAA	*Claroideoglomus etunicatum*	*Camellia sinensis*	IAA level increased	Improved root characteristics such as density of root hair, average diameter, volume, and lateral root numbers for better water and nutrient absorption	Liu et al. (2023b)
Abscisic acid (ABA)	*R. irregularis*	*Solanum lycopersicum*	Upregulation in expression of ABA signaling 14-3-3 genes such as *TFT5, TFT7, TFT9,* and *TFT10* ABA level increased	Improved photosynthetic rate and water use efficiency, and decrease in transpiration rate and gas exchange contributed to drought tolerance	Xu et al. (2018b)
ABA	*Funneliformis mosseae*	*Zea mays*	ABA level decreased	Decline in stomatal conductance and net photosynthetic rate postponed, enhanced antioxidant activity, and water use efficiency, reduced proline and malondialdehyde level for better drought acclimatization	Ren et al. (2019)
ABA	*R. irregularis* and exogenous IAA application	*Z. mays*	ABA level increased	Improved root hydraulic conductivity (Lpr) and positively regulated aquaporin protein like PIP2B protein for better water uptake	Quiroga et al. (2020)
Jasmonic acid (JA)	*R. irregularis*	*Z. mays*	JA level increased	Improved Lpr and osmotic root hydraulic conductivity for better root water	Quiroga et al. (2018)

				transport and cell to cell water flow respectively for drought tolerance	
JA	*Rhizoglomus irregulare*	*Trifolium repens*	JA level increased	Increased accumulation of a primary precursor of JA, 12-oxo-phytodienoic acid, that activates defensive signaling response and functions as an anti-transpirant against drought stress	Fresno et al. (2023)
IAA, Indole-3-butyric acid (IBA), ABA	*C. etunicatum, R. intraradices, F. mosseae*	*Ephedra foliata*	Levels of IAA, IBA, ABA increased	Increased plant growth and decreased water loss by stomatal regulation	Al-Arjani et al. (2020)
Strigolactone	*R. irregularis*	*Malus domestica*	Overexpression of strigolactone synthesis gene, *MdIAA24*, resulting in increased levels of strigolactone	Improved mycorrhizal colonization which resulted in increased phosphorus uptake, lower relative electrolytic leakage, high relative water content, photosynthetic parameters and high reactive oxygen species scavenging activity for drought tolerance	Huang et al. (2021)

the antioxidant defense system (Qiu et al. 2014a, b), AM-induced enhancement of JA protects host plants from drought stress.

Strigolactones are a class of plant hormone that participate in the detection of fungal structures and establishment of AM symbiosis in the host plants (Chauhan et al. 2022). They are also involved in regulatory roles against abiotic stress. Under osmotic stress, production of strigolactones was down regulated; however, with the application of AM fungi, levels of strigolactones were improved along with ABA, consequently promoting AM symbiosis and related benefits to counter deleterious effects of osmotic stress (López-Ráez 2016). Higher levels of strigolactones in *Sesbania cannabina* could be as a result of high ABA in the mycorrhizal plants, as ABA induces production of strigolactones that improve the ability of the plant to tolerate abiotic stress (Ren et al. 2018).

There are also other phytohormones whose levels were modulated upon AM symbiosis in abiotic stress conditions. Indole-3-acetic acid (IAA), an auxin, dramatically improved in *Ephedra foliata* and *Camellia sinensis* in the presence of mixture AMF (*Funneliformis mosseae*, *Claroideoglomus etunicatum*, and *Rhizophagus intraradices*) and *C. etunicatum*, respectively, under osmotic stress (Al-Arjani et al. 2020; Liu et al. 2023b). Induction of auxin synthesis is key to enhanced growth of root hairs that will improve plant nutrition and help plant to resist the negative effects of osmotic stress (Liu et al. 2018). Chen et al. (2020) reported an increase in IAA, along with gibberellins, IAA/ABA, and GA_3/ABA in *Catalpa bungei* in the presence of *Rhizophagus intraradices* under drought stress and a decrease in ABA and zeatin. The role of hormones in improving tolerance against osmotic stress under AM symbiosis is not fully understood; therefore, more research should be directed towards studies involving phytohormones and AM fungi under stress conditions.

7.5.8 Influence on LEA Proteins, Dehydrins, and ATPase

LEA proteins are rich in glycine, alanine, and serine and consist of a diverse family of minute hydrophilic polypeptides (10–30 kDa) (Karpinska et al. 2022). They are housed in subcellular compartments like cytosol, mitochondria, and chloroplast and are regulated by phytohormones like ABA (Huang et al. 2016; Artur et al. 2019). Some LEA proteins function as molecular chaperones under abiotic stress and help in the restoration of proteins into their natural conformations, thus enhancing stress tolerance (Kovacs et al. 2008). Expression of genes LEA proteins could contribute to drought tolerance. *Gh_A08G0694*, a LEA3 family gene when overexpressed in *Gossypium hirsutum*, in collaboration with other key stress- tolerant genes such as glyceraldehyde-3-phosphate dehydrogenase and a voltage-dependent anion channel 1 and other regulatory networks could promote drought tolerance (Shiraku et al. 2022). Wang et al. (2023b) reported that *ZmNHL1* protein encoded by a LEA-2 protein family could improve activities of POD and SOD and lower electrolyte leakage and promote drought tolerance in *Zea mays*.

Dehydrins or group II LEA are the LEA proteins which are most widely described. They are low molecular weight, thermostable, and hydrophilic and generally assemble in the later phase of embryogenesis in response to abiotic stress and widely scattered in bryophytes, angiosperms, and gymnosperms (Yu et al. 2018). Response of dehydrins varies with levels of drought stress. For instance, expressions of *ShDHN* and *SiDhn2* reached the maximum level (12-fold change) when subjected to 6 and 3 h of drought conditions, respectively (Qiu et al. 2014a, b; Liu et al. 2015). Under drought stress, *MtCAS31* responds positively to drought by promoting autophagy of *MtPIP2;7* proteins and reducing hydraulic conductivity of roots, consequently minimizing water loss (Li et al. 2020). Overexpression of *CaDHN3* , a dehydrin gene of *Capsicum annum* enhanced tolerance against water-deficit condition in *Arabidopsis thaliana* by promoting phenotypic characteristics and enzymatic antioxidant defense system (Meng et al. 2021b).

The contribution of AM symbiosis in the light of LEA proteins and dehydrins under drought is limited. However, Porcel et al. (2005) observed that AMF can alter the accumulation dehydrin pattern in *Glycine max* and *Lactuca sativa* under drought stress. Expression of LEA genes such as *Islea 1*, *gmlea 8*, and *gmlea 10* was lesser in the mycorrhizal plants than in the non-AMF plants under drought conditions, thus indicating that AM symbiosis does not involve LEA proteins as a drought tolerance mechanism. In contrast, AMF contributed to drought tolerance by increased accumulation of dehydrins (Ruiz-Lozano et al. 2008).

H^+-ATPase is a membrane protein that generates the H^+ electrochemical gradient which regulates solute and nutrient transport through the membranc (Garry et al. 2007). The activity of plasma membrane H^+-ATPase under drought is inhibited causing net H^+ influx which results in depolarization of membrane potential. Therefore, higher activity of H^+-ATPase is crucial to prevent cellular H^+ influx and K^+ influx in order to maintain K^+ homoeostasis (Zhang et al. 2018c). Dong et al. (2013) observed that over expression of *MdVHA-A*, a H^+-ATPase subunit A, improved the activity of vacuole H^+-ATPase to confer drought tolerance in *Malus domestica*, thus making it a potential gene candidate for drought tolerance. It was also shown that the activity of H^+-ATPase in plasma membranes could be promoted by drought- induced augmentation of polyamines like spermine and spermidine in *Triticum aestivum* (Du et al. 2015). Transcriptional study of four H^+-ATPase genes in *Solanum lypersicum* namely, *SlHA1*, *SlHA4*, *SlHA2*, and *SlHA8*, was investigated and it was reported that in the presence of *Rhizophagus irregularis*, *SlHA2* and *SlHA8* in leaf and root were upregulated, while *SlHA1* and *SlHA4* remained unaffected; thus, AM regulation of HA genes depends on HA homologous genes and the host plants (Liu et al. 2016). Further, expression of *SlHA8* was linked to arbuscule development and nitrogen uptake (Liu et al. 2020). Under drought stress, *Funneliformis mosseae* distinctly improved the activity of H^+-ATPase in roots and leaf by 22.61% and 26.06%, respectively, than in control *Poncirus trifoliata* plants. Transcriptome study also revealed that expression of *PtAHA2* in both leaf and roots was significantly improved by 9.50- fold and 20.92-fold, respectively, in the mycorrhizal plants (Cheng et al. 2021).

Drought stress triggers the level of calcium (Ca^{2+}), an important intracellular second messenger, resulting in the aggregation of nucleic acids and proteins, membrane lipid disintegration, and phosphatase precipitation, consequently leading to death of cells (Knight and Knight 2001; Case et al. 2007). The enhanced accumulation of cytosolic Ca^{2+} ions is effluxed out to restore Ca^{2+} homeostasis by antiporters and a high- affinity, low- capacity transporter known as Ca^{2+} ATPases (Bose et al. 2011). Besides, Ca^{2+} ATPases also play a remarkable role in stomatal regulation, pollen development, floral inflorescence architecture, and gibberellin signaling (Huda et al. 2013). However, there is a lack of studies that provide direct data on the role of Ca^{2+} ATPases in stress tolerance. Drought stress triggered overexpression of *OsACA6*, a Ca^{2+} ATPases gene in transgenic *Oryza sativa*, and this modulated scavenging machinery of ROS and synthesis of proline. In addition, overexpressing transgenic lines showed higher content of chlorophyll, photosynthesis, stability of membrane, and reduced loss of water, thus ensuring drought tolerance (Huda et al. 2013). Under drought condition, the activity of Ca^{2+} ATPases along with H^+-ATPase was induced in extraradical hyphae of *R. irregularis* in symbiosis with *Daucus carota.* This caused acidification of rhizosphere soil which promoted plant nutrition and exchange of signals between the plant and the AMF (Xu et al. 2018a). More recently Sheteiwy et al. (2021) also noted improved activitie s of Ca^{2+} ATPases along with H^+-ATPase in soybean plant in the presence of AMF (*Acaulospora laevis*, *Septoglomus deserticola*, and *R. irregularis*) to confer drought tolerance. However, the studies concerning induction of ATPase activities, especially Ca^{2+} ATPases under drought conditions in the presence of AM fungi, are limited. Therefore, extensive research should be directed towards understanding the regulation of ATPase in different plants by different AMF at different levels of drought stress.

7.6 Conclusions

The benefits of AMF in ameliorating drought stress are well documented. AMF symbiosis have proved to be a key tool in bestowing drought tolerance and consequently mitigate drought- induced damages in plants. The ameliorative mechanisms under drought stress include AMF-induced adaptive changes in structure, biochemistry, and physiology of the host plants. However, the current study finds that the following research areas need further investigation to gain full understanding of AMF-induced drought tolerance in host plants:

1. Genomic studies have enabled us to study whole genome sequences in *R. irregularis* and a few stress- related genes encoded by AM fungi (Tisserant et al. 2013). Regulatory mechanisms of these genes in the light of drought stress could provide new information regarding stress tolerance.
2. Molecular studies involving functional genes, transcription factors, and regulatory genes that influence drought tolerance are limited (Wang et al. 2023a, b) and

are insufficient to decipher the mechanisms of drought tolerance at molecular level.

3. Though AM symbiosis upregulates various hormonal signaling pathways in response to drought stress, the information on their regulation is limited.
4. As the response of plant varies with AMF species and drought stress level, successful plant-AM symbiosis relies on the selection of suitable host plant and AM fungi under drought. Therefore, future research should be directed to screen out indigenous and drought- tolerant AMF species to enhance the effectiveness of AM symbiosis.

Acknowledgment AT is thankful to the Ministry of Tribal Affairs for National Fellowship for ST students.

References

Aalipour H, Nikbakht A, Sabzalian MR (2023) Essential oil composition and total phenolic content in *Cupressus arizonica* G. in response to microbial inoculation under water stress conditions. Sci Rep 13(1):1–11. https://doi.org/10.1038/s41598-023-28107

Abd-Allah EF, Abeer H, Alqarawi AA, Hend AA (2015) Alleviation of adverse impact of cadmium stress in sunflower (*Helianthus annuus* L.) by arbuscular mycorrhizal fungi. Pak J Bot 47(2): 785–795

Abd-Elghany SE, Moustafa AA, Gomaa NH, Hamed BEA (2021) Mycorrhizal impact on *Ocimum basilicum* grown under drought stress. BJBAS 10(1):1–13. https://doi.org/10.1186/s43088-021-00166

Abdel-Salam E, Alatar A, El-Sheikh MA (2018) Inoculation with arbuscular mycorrhizal fungi alleviates harmful effects of drought stress on damask rose. Saudi J Biol Sci 25(8):1772–1780. https://doi.org/10.1016/j.sjbs.2017.10.015

Abo-Doma A, Edrees S, Abdel-Aziz SH (2011) The effect of mycorrhiza growth and expression of some genes in barley. EJGC 40:2

Al-Amri SM (2021) Application of bio-fertilizers for enhancing growth and yield of common bean plants grown under water stress conditions. Saudi J Biol Sci 28(7):3901–3908. https://doi.org/10.1016/j.sjbs.2021.03.064

Al-Arjani ABF, Hashem A, Abd Allah EF (2020) Arbuscular mycorrhizal fungi modulates dynamics tolerance expression to mitigate drought stress in *Ephedra foliata* Boiss. Saudi J Biol Sci 27(1):380–394. https://doi.org/10.1016/j.sjbs.2019.10.008

Amanifar S, Toghranegar Z (2020) The efficiency of arbuscular mycorrhiza for improving tolerance of *Valeriana officinalis* L. and enhancing valerenic acid accumulation under salinity stress. Ind Crops Prod 147:112234. https://doi.org/10.1016/j.indcrop.2020.112234

Aroca R, Porcel R, Ruiz-Lozano JM (2007) How does arbuscular mycorrhizal symbiosis regulate root hydraulic properties and plasma membrane aquaporins in *Phaseolus vulgaris* under drought, cold or salinity stresses? New Phytol 173(4):808–816. https://doi.org/10.1111/j.1469-8137.2006.01961.x

Aroca R, Bago A, Sutka M, Paz JA, Cano C, Amodeo G, Ruiz-Lozano JM (2009) Expression analysis of the first arbuscular mycorrhizal fungi aquaporin described reveals concerted gene expression between salt-stressed and nonstressed mycelium. MPMI 22(9):1169–1178. https://doi.org/10.1094/MPMI-22-9-1169

Artur MAS, Zhao T, Ligterink W, Schranz E, Hilhorst HW (2019) Dissecting the genomic diversification of late embryogenesis abundant (LEA) protein gene families in plants. GBE 11(2):459–471. https://doi.org/10.1093/gbe/evy248

Augé RM (2004) Arbuscular mycorrhizae and soil/plant water relations. Canadian J Soil Sci 84(4): 373–381. https://doi.org/10.4141/S04-002

Azizi S, Kouchaksaraei MT, Hadian J, Abad ARFN, Sanavi SAMM, Ammer C, Bader MKF (2021) Dual inoculations of arbuscular mycorrhizal fungi and plant growth-promoting rhizobacteria boost drought resistance and essential oil yield of common myrtle. For Ecol Manag 497:119478. https://doi.org/10.1016/j.foreco.2021.119478

Balestrini R, Rosso LC, Veronico P, Melillo MT, De Luca F, Fanelli E, Pentimone I (2019) Transcriptomic responses to water deficit and nematode infection in mycorrhizal tomato roots. Front Microbiol 10:1807. https://doi.org/10.3389/fmicb.2019.01807

Begum N, Ahanger MA, Su Y, Lei Y, Mustafa NSA, Ahmad P, Zhang L (2019) Improved drought tolerance by AMF inoculation in maize (*Zea mays*) involves physiological and biochemical implications. Plants 8(12):579. https://doi.org/10.3390/plants8120579

Begum N, Akhtar K, Ahanger MA, Iqbal M, Wang P, Mustafa NS, Zhang L (2021) Arbuscular mycorrhizal fungi improve growth, essential oil, secondary metabolism, and yield of tobacco (*Nicotiana tabacum* L.) under drought stress conditions. ESPR 28:45276–45295. https://doi.org/10.1007/s11356-021-13755-3

Bitterlich M, Sandmann M, Graefe J (2018) Arbuscular mycorrhiza alleviates restrictions to substrate water flow and delays transpiration limitation to stronger drought in tomato. Front Plant Sci 9:154. https://doi.org/10.3389/fpls.2018.00154

Bose J, Pottosin II, Shabala SS, Palmgren MG, Shabala S (2011) Calcium efflux systems in stress signalling and adaptation in plants. Front Plant Sci 85:1–16. https://doi.org/10.3389/fpls.2011.00085

Boutasknit A, Baslam M, Ait-El-Mokhtar M, Anli M, Ben-Laouane R, Douira A, El Modafar C, Mitsui T, Wahbi S, Meddich A (2020) Arbuscular mycorrhizal fungi mediate drought tolerance and recovery in two contrasting carob (*Ceratonia siliqua* L.) ecotypes by regulating stomatal, water relations, and (in) organic adjustments. Plants 9(1):80. https://doi.org/10.3390/plants9010080

Case RM, Eisner D, Gurney A, Jones O, Muallem S, Verkhratsky A (2007) Evolution of calcium homeostasis: from birth of the first cell to an omnipresent signalling system. Cell Calcium 42(4-5):345–350. https://doi.org/10.1016/j.ceca.2007.05.001

Chareesri A, De Deyn GB, Sergeeva L, Polthanee A, Kuyper TW (2020) Increased arbuscular mycorrhizal fungal colonization reduces yield loss of rice (*Oryza sativa* L.) under drought. Mycorrhiza 30:315–328. https://doi.org/10.1007/s00572-020-00953-z

Chartzoulakis K, Patakas A, Kofidis G, Bosabalidis A, Nastou A (2002) Water stress affects leaf anatomy, gas exchange, water relations and growth of two avocado cultivars. Sci Hortic 95(1-2): 39–50. https://doi.org/10.1016/S0304-4238(02)00016-X

Chauhan S, Mahawar S, Jain D, Udpadhay SK, Mohanty SR, Singh A, Maharjan E (2022) Boosting sustainable agriculture by arbuscular mycorrhiza under stress condition: mechanism and future prospective. BioMed Research International 2022:5275449. https://doi.org/10.1155/2022/5275449

Chen D, Shao Q, Yin L, Younis A, Zheng B (2019) Polyamine function in plants: metabolism, regulation on development, and roles in abiotic stress responses. Front Plant Sci 9:1945. https://doi.org/10.3389/fpls.2018.01945

Chen W, Meng P, Feng H, Wang C (2020) Effects of arbuscular mycorrhizal fungi on growth and physiological performance of *Catalpa bungei* CA Mey. under drought stress. Forests 11(10): 1117. https://doi.org/10.3390/f11101117

Chen Z, Liu Z, Han S, Jiang H, Xu S, Zhao H, Ren S (2022) Using the diurnal variation characteristics of effective quantum yield of PSII photochemistry for drought stress detection in maize. Ecol Indic 138:108842. https://doi.org/10.1016/j.ecolind.2022.108842

Cheng HQ, Ding YE, Shu B, Zou YN, Wu QS, Kuča K (2020) Plant aquaporin responses to mycorrhizal symbiosis under abiotic stress. Int J Agric Biol 23:786–794. https://doi.org/10.17957/IJAB/15.1353

Cheng HQ, Zou YN, Kuča K (2021) Arbuscular mycorrhizal fungi alleviate drought stress in trifoliate orange by regulating H+-ATPase activity and gene expression. Front Plant Sci 12: 659694. https://doi.org/10.3389/fpls.2021.659694

Chi GG, Srivastava AK, Wu QS (2018) Exogenous easily extractable glomalin-related soil protein improves drought tolerance of trifoliate orange. Arc Agron Soil Sci 64(10):1341–1350. https://doi.org/10.1080/03650340.2018.1432854

Chitarra W, Pagliarani C, Maserti B, Lumini E, Siciliano I, Cascone P, Schubert A, Gambino G, Balestrini R, Guerrieri E (2016) Insights on the impact of arbuscular mycorrhizal symbiosis on tomato tolerance to water stress. Plant Physiol 171(2):1009–1023. https://doi.org/10.1104/pp.16.00307

Conrad M, Angeli JPF, Vandenabeele P, Stockwell BR (2016) Regulated necrosis: disease relevance and therapeutic opportunities. Nat Rev Drug Discov 15(5):348–366. https://doi.org/10.1038/nrd.2015.6

Dong QL, Wang CR, Liu DD, Hu DG, Fang MJ, You CX, Hao YJ (2013) MdVHA-A encodes an apple subunit A of vacuolar H^+-ATPase and enhances drought tolerance in transgenic tobacco seedlings. J Plant Physiol 170(6):601–609. https://doi.org/10.1016/j.jplph.2012.12.014

Du H, Zhou X, Yang Q, Liu H, Kurtenbach R (2015) Changes in H^+-ATPase activity and conjugated polyamine contents in plasma membrane purified from developing wheat embryos under short-time drought stress. Plant Growth Regul 75:1–10. https://doi.org/10.1007/s10725-014-9925-9

Egamberdieva D, Wirth SJ, Alqarawi AA, Abd-Allah EF, Hashem A (2017) Phytohormones and beneficial microbes: essential components for plants to balance stress and fitness. Front Microbiol 8:2104. https://doi.org/10.3389/fmicb.2017.02104

El-Sawah AM, Abdel-Fattah GG, Holford P, Korany SM, Alsherif EA, AbdElgawad H, Ulhassan Z, Josko L, Ali B, Sheteiwy MS (2023) *Funneliformis constrictum* modulates polyamine metabolism to enhance tolerance of *Zea mays* L. to salinity. Microbiol Res 266: 127254. https://doi.org/10.1016/j.micres.2022.127254

Evelin H, Devi TS, Gupta S, Kapoor R (2019) Mitigation of salinity stress in plants by arbuscular mycorrhizal symbiosis: current understanding and new challenges. Front Plant Sci 10:470. https://doi.org/10.3389/fpls.2019.00470

FAO (2023). http://www.fao.org. Accessed on 25 June 2023

Farias DDH, Pinto MAB, Carra B, Schuch MW, Souza PVDD (2014) Development of seedlings of blueberry inoculated arbuscular mycorrhizal fungi. Rev Bras Frutic 36:655–663. https://doi.org/10.1590/0100-2945-128/13

Farooq M, Wahid A, Kobayashi NSMA, Fujita DBSMA, Basra SMA (2009) Plant drought stress: effects, mechanisms and management. In: Lichtfouse E, Navarrete M, Debaeke P, Véronique S, Alberola C (eds) Sustainable agriculture. Springer, Dordrecht, pp 153–188. https://doi.org/10.1007/978-90-481-2666-8_12

Fresno DH, Solé-Corbatón H, Munné-Bosch S (2023) Water stress protection by the arbuscular mycorrhizal fungus *Rhizoglomus irregulare* involves physiological and hormonal responses in an organ-specific manner. Physiol Plant 175(1):e13854. https://doi.org/10.1111/ppl.13854

Garg SK (2013) Role and hormonal regulation of nitrate reductase activity in higher plants: a review. Plant Sci Feed 3(1):13–20. https://doi.org/10.1007/s11356-021-18232-5

Garry M, Rosewarne F, Andrew Smith DP, Schachtman SE (2007) Localization of proton-ATPase genes expressed in arbuscular mycorrhizal tomato plants. Mycorrhiza 17:249–258. https://doi.org/10.1007/s00572-006-0101-6

Gechev TS, Van Breusegem F, Stone JM, Denev I, Laloi C (2006) Reactive oxygen species as signals that modulate plant stress responses and programmed cell death. Bioessays 11:1091–1101. https://doi.org/10.1002/bies.20493

Geneva M, Hristozkova M, Kirova E, Sichanova M, Stancheva I (2023) Response to drought stress of in vitro and in vivo propagated *Physalis peruviana* L. plants inoculated with arbuscular mycorrhizal fungi. Agriculture 13(2):472. https://doi.org/10.3390/agriculture13020472

Gholinezhad E, Darvishzadeh R (2021) Influence of arbuscular mycorrhiza fungi and drought stress on fatty acids profile of sesame (*Sesamum indicum* L.). Field Crops Res 262:108035. https://doi.org/10.1016/j.fcr.2020.108035

Gichure RW (2017) Effects of drought on crop production and coping mechanisms undertaken by small scale farmers: a case of Makueni County, Kenya. Doctoral Dissertation, University of Nairobi.

Gill SS, Anjum NA, Gill R, Yadav S, Hasanuzzaman M, Fujita M, Mishra P, Sabat SC, Tuteja N (2015) Superoxide dismutase—mentor of abiotic stress tolerance in crop plants. ESPR 22: 10375–10394. https://doi.org/10.1007/s11356-015-4532-5

Giovannetti M, Avio L, Sbrana C (2010) Fungal spore germination and pre-symbiotic mycelial growth -physiological and genetic aspects. In: Koltai H, Kapulnik Y (eds) Arbuscular mycorrhizas: physiology and function. Springer, Dordrecht, pp 3–32. https://doi.org/10.1007/978-90-481-9489-6_1

Goicoechea N, Garmendia I, Sanchez-Diaz M, Aguirreolea J (2010) Arbuscular mycorrhizal fungi (AMF) as bioprotector agents against wilt induced by *Verticillium* spp. in pepper: a review. SJAR 8:25–42. https://doi.org/10.5424/sjar/201008S1-5300

Gul Z, Tang ZH, Arif M, Ye Z (2022) An insight into abiotic stress and influx tolerance mechanisms in plants to cope in saline environments. Biology 11(4):597. https://doi.org/10.3390/biology11040597

Guo LQ, Shi DC, Wang DL (2010) The key physiological response to alkali stress by the alkali-resistant halophyte *Puccinellia tenuiflora* is the accumulation of large quantities of organic acids and into the rhyzosphere. J Agron Crop Sci 196(2):123–135. https://doi.org/10.1111/j.1439-037X.2009.00397.x

Hadian-Deljou M, Esna-Ashari M, Mirzaie-asl A (2020) Alleviation of salt stress and expression of stress-responsive gene through the symbiosis of arbuscular mycorrhizal fungi with sour orange seedlings. Sci Hortic 268:109373. https://doi.org/10.1016/j.scienta.2020.109373

Halliwell B, Gutteridge JM (2015) Free radicals in biology and medicine. Oxford University Press, Oxford

Han Y, Lou X, Zhang W, Xu T, Tang M (2022) Arbuscular mycorrhizal fungi enhanced drought resistance of *Populus cathayana* by regulating the 14-3-3 family protein genes. Microbiol Spectrum 10(3):e02456. https://doi.org/10.1128/spectrum.02456-21

Hasanuzzaman M, Bhuyan MB, Anee TI, Parvin K, Nahar K, Mahmud JA, Fujita M (2019) Regulation of ascorbate-glutathione pathway in mitigating oxidative damage in plants under abiotic stress. Antioxidants 8(9):384. https://doi.org/10.3390/antiox8090384

Hasanuzzaman M, Bhuyan MB, Zulfiqar F, Raza A, Mohsin SM, Mahmud JA, Fujita M, Fotopoulos V (2020) Reactive oxygen species and antioxidant defense in plants under abiotic stress: revisiting the crucial role of a universal defense regulator. Antioxidants 9(8):681. https://doi.org/10.3390/antiox9080681

Hashem A, Kumar A, Al-Dbass AM, Alqarawi AA, Al-Arjani ABF, Singh G, Farooq M, Abd Allah EF (2019) Arbuscular mycorrhizal fungi and biochar improves drought tolerance in chickpea. Saudi J Biol Sci 26(3):614–624. https://doi.org/10.1016/j.sjbs.2018.11.005

He C, Yang A, Zhang W, Gao Q, Zhang J (2010) Improved salt tolerance of transgenic wheat by introducing betA gene for glycine betaine synthesis. PCTOC 101:65–78. https://doi.org/10.1007/s11240-009-9665-0

He JD, Dong T, Wu HH, Ying-Ning Z, Qiang-Sheng W, Kamil K (2019) Mycorrhizas induce diverse responses of root TIP aquaporin gene expression to drought stress in trifoliate orange. Sci Hortic 243:64–69. https://doi.org/10.1016/j.scienta.2018.08.010

Hu Y, Chen B (2020) Arbuscular mycorrhiza induced putrescine degradation into γ-aminobutyric acid, malic acid accumulation, and improvement of nitrogen assimilation in roots of water-stressed maize plants. Mycorrhiza 30:329–339. https://doi.org/10.1007/s00572-020-00952-0

Hu Y, Xie W, Chen B (2020) Arbuscular mycorrhiza improved drought tolerance of maize seedlings by altering photosystem II efficiency and the levels of key metabolites. Chem Biol Technol Agric 7:1–14. https://doi.org/10.1186/s40538-020-00186-4

Huang Z, Jin SH, Guo HD, Zhong XJ, He J, Li X, Jiang MY, Yu XF, Long H, Ma MD, Chen QB (2016) Genome-wide identification and characterization of TIFY family genes in Moso Bamboo (*Phyllostachys edulis*) and expression profiling analysis under dehydration and cold stresses. PeerJ 4:e2620. https://doi.org/10.7717/peerj.2620

Huang H, Ullah F, Zhou DX, Yi M, Zhao Y (2019) Mechanisms of ROS regulation of plant development and stress responses. Front Plant Sci 10:800. https://doi.org/10.3389/fpls.2019.00800

Huang D, Ma M, Wang Q, Zhang M, Jing G, Li C, Ma F (2020) Arbuscular mycorrhizal fungi enhanced drought resistance in apple by regulating genes in the MAPK pathway. Plant Physiol Biochem 149:245–255. https://doi.org/10.1016/j.plaphy.2020.02.020

Huang D, Wang Q, Jing G, Ma M, Li C, Ma F (2021) Overexpression of MdIAA24 improves apple drought resistance by positively regulating strigolactone biosynthesis and mycorrhization. Tree Physiol 41(1):134–146. https://doi.org/10.1093/treephys/tpaa109

Huda KM, Banu MS, Garg B, Tula S, Tuteja R, Tuteja N (2013) OsACA6, a P-type IIB Ca 2+ ATPase promotes salinity and drought stress tolerance in tobacco by ROS scavenging and enhancing the expression of stress-responsive genes. Plant J 76(6):997–1015. https://doi.org/10.1111/tpj.12352

Ilyas M, Nisar M, Khan N, Hazrat A, Khan AH, Hayat K, Fahad S, Khan A, Ullah A (2021) Drought tolerance strategies in plants: a mechanistic approach. J Plant Growth Regul 40:926–944. https://doi.org/10.1007/s00344-020-10174-5

Jabborova D, Annapurna K, Al-Sadi AM, Alharbi SA, Datta R, Zuan ATK (2021) Biochar and Arbuscular mycorrhizal fungi mediated enhanced drought tolerance in Okra (*Abelmoschus esculentus*) plant growth, root morphological traits and physiological properties. Saudi J Biol Sci 28(10):5490–5499. https://doi.org/10.1016/j.sjbs.2021.08.016

Jaleel CA, Manivannan, Lakshmanan GMA, Gomathinayagam M, Panneerselvam R (2008) Alterations in morphological parameters and photosynthetic pigment responses of *Catharanthus roseus* under soil water deficits. Colloids Surf B 61(2):298–303. https://doi.org/10.1016/j.colsurfb.2007.09.008

Kandowangko NY, Suryatmana GIAT, Nurlaeny N, Simanungkalit RDM (2009) Proline and abscisic acid content in droughted corn plant inoculated with *Azospirillum* sp. and Arbuscular mycorrhizae fungi. HAYATI J Biosci 1:15–20. https://doi.org/10.4308/hjb.16.1.15

Karpinska B, Razak N, Shaw DS, Plumb W, Van De Slijke E, Stephens J, Jaeger GD, Murcha MW, Foyer CH (2022) Late embryogenesis abundant (LEA) 5 regulates translation in mitochondria and chloroplasts to enhance growth and stress tolerance. Front Plant Sci 13:875799. https://doi.org/10.3389/fpls.2022.875799

Kaur G, Asthir B (2015) Proline: a key player in plant abiotic stress tolerance. Biol Plant 59:609–619. https://doi.org/10.1007/s10535-015-0549-3

Khatri K, Rathore MS (2022) Salt and osmotic stress-induced changes in physio-chemical responses, PSII photochemistry and chlorophyll a fluorescence in peanut. Plant Stress 3:100063. https://doi.org/10.1016/j.stress.2022.100063

Kikuchi Y, Hijikata N, Ohtomo R, Handa Y, Kawaguchi M, Saito K, Masuta C, Ezawa T (2016) Aquaporin-mediated long-distance polyphosphate translocation directed towards the host in arbuscular mycorrhizal symbiosis: application of virus-induced gene silencing. New Phytol 211(4):1202–1208. https://doi.org/10.1111/nph.14016

Klinsukon C, Lumyong S, Kuyper TW, Boonlue S (2021) Colonization by arbuscular mycorrhizal fungi improves salinity tolerance of eucalyptus (*Eucalyptus camaldulensis*) seedlings. Sci Rep 11(1):1–10. https://doi.org/10.1038/s41598-021-84002-5

Knight H, Knight MR (2001) Abiotic stress signalling pathways: specificity and crosstalk. Trends Plant Sci 6(6):262–267. https://doi.org/10.1016/S1360-1385(01)01946-X

Kovacs D, Agoston B, Tompa P (2008) Disordered plant LEA proteins as molecular chaperones. Plant Signal Behav *3*(9):710–713. https://doi.org/10.4161/psb.3.9.6434

Langeroodi ARS, Osipitan OA, Radicetti E, Mancinelli R (2020) To what extent arbuscular mycorrhiza can protect chicory (*Cichorium intybus* L.) against drought stress. Sci Hortic 263: 109109. https://doi.org/10.1016/j.scienta.2019.109109

Laxa M, Liebthal M, Telman W, Chibani K, Dietz KJ (2019) The role of the plant antioxidant system in drought tolerance. Antioxidants 8(4):94. https://doi.org/10.3390/antiox8040094

Li X, Liu F (2016) Drought stress memory and drought stress tolerance in plants: biochemical and molecular basis. In: Hossain M, Wani S, Bhattacharjee S, Burritt D, Tran LS (eds) Drought stress tolerance in plants, vol 1. Springer, Cham. https://doi.org/10.1007/978-3-319-28899-4_2

Li T, Hu YJ, Hao ZP, Li H, Wang YS, Chen BD (2013) First cloning and characterization of two functional aquaporin genes from an arbuscular mycorrhizal fungus *Glomus intraradices*. New Phytol 197(2):617–630. https://doi.org/10.1111/nph.12011

Li X, Liu Q, Feng H, Deng J, Zhang R, Wen J, Dong J, Wang T (2020) Dehydrin MtCAS31 promotes autophagic degradation under drought stress. Autophagy 16(5):862–877. https://doi.org/10.1080/15548627.2019.1643656

Liang SM, Jiang DJ, Miao-Miao XIE, Ying-Ning ZOU, Qiang-Sheng WU, Kamil KUČA (2021) Physiological responses of mycorrhizal symbiosis to drought stress in white clover. Not Bot Horti Agrobot 49(1):12209–12209. https://doi.org/10.15835/nbha49112209

Liang SM, Zheng FL, Wu QS (2022) Elucidating the dialogue between arbuscular mycorrhizal fungi and polyamines in plants. World J Microbiol Biotechnol 38(9):159. https://doi.org/10.1007/s11274-022-03336-y

Liao X, Chen J, Guan R, Liu J, Sun Q (2021) Two arbuscular mycorrhizal fungi alleviate drought stress and improves plant growth in *Cinnamomum migao* seedlings. Mycobiology 49(4): 396–405. https://doi.org/10.1080/12298093.2021.1938803

Liu H, Yu C, Li H, Ouyang B, Wang T, Zhang J, Wang X, Ye Z (2015) Overexpression of ShDHN, a dehydrin gene from *Solanum habrochaites* enhances tolerance to multiple abiotic stresses in tomato. Plant Sci 231:198–211. https://doi.org/10.1016/j.plantsci.2014.12.006

Liu J, Guo C, Chen ZL, He JD, Zou YN (2016) Mycorrhizal inoculation modulates root morphology and root phytohormone responses in trifoliate orange under drought stress. Emirates J Food Agric 28(4):251–256. https://doi.org/10.9755/ejfa.2015-11-1044

Liu CY, Zhang F, Zhang DJ, Srivastava AK, Wu QS, Zou YN (2018) Mycorrhiza stimulates root-hair growth and IAA synthesis and transport in trifoliate orange under drought stress. Sci Rep 8(1):1–9. https://doi.org/10.1038/s41598-018-20456-4

Liu J, Chen J, Xie K, Tian Y, Yan A, Liu J, Xu G (2020) A mycorrhiza-specific H+-ATPase is essential for arbuscule development and symbiotic phosphate and nitrogen uptake. Plant Cell Environ 43(4):1069–1083. https://doi.org/10.1111/pce.13714

Liu Z, Cheng S, Liu XQ, Kuča K, Hashem A, Al-Arjani ABF, Almutairi KF, Abd-Allah EF, Wu QS, Zou YN (2022) Cloning of a *CHS* gene of *Poncirus trifoliata* and its expression in response to soil water deficit and arbuscular mycorrhizal fungi. Front Plant Sci 13:1101212. https://doi.org/10.3389/fpls.2022.1101212

Liu Y, Lu J, Cui L, Tang Z, Ci D, Zou X, Si T (2023a) The multifaceted roles of arbuscular mycorrhizal fungi in peanut responses to salt, drought, and cold stress. BMC Plant Biol 23(1): 1–19. https://doi.org/10.1186/s12870-023-04053-w

Liu CY, Hao Y, Wu XL, Dai FJ, Abd-Allah EF, Wu QS, Liu SR (2023b) Arbuscular mycorrhizal fungi improve drought tolerance of tea plants via modulating root architecture and hormones. Plant Growth Regul 2023:1–10. https://doi.org/10.1007/s10725-023-00972-8

Lokhande VH, Suprasanna P (2012) Prospects of halophytes in understanding and managing abiotic stress tolerance. In: Environmental adaptations and stress tolerance of plants in the era of climate change. Springer, Cham, pp 29–56. https://doi.org/10.1007/978-1-4614-0815-4_2

López-Ráez JA (2016) How drought and salinity affect arbuscular mycorrhizal symbiosis and strigolactone biosynthesis? Planta 243:1375–1385. https://doi.org/10.1007/s00425-015-2435-9

Mahnaz Z, Ali E, Mohammad S, Sodabe J (2020) Alleviating effect of 24-epibrassinolide on seed oil content and fatty acid composition under drought stress in safflower. J Food Compos Analy 92:103544. https://doi.org/10.1016/j.jfca.2020.103544

Mansoor S, Ali Wani O, Lone JK, Manhas S, Kour N, Alam P, Ahmad A, Ahmad P (2022) Reactive oxygen species in plants: from source to sink. Antioxidants 11(2):225. https://doi.org/10.3390/antiox11020225

Mashabela MD, Masamba P, Kappo AP (2023) Applications of metabolomics for the elucidation of abiotic stress tolerance in plants: a special focus on osmotic stress and heavy metal toxicity. Plants 12(2):269. https://doi.org/10.3390/plants12020269

Mathur S, Tomar RS, Jajoo A (2019) Arbuscular mycorrhizal fungi (AMF) protects photosynthetic apparatus of wheat under drought stress. Photosynth Res 139:227–238. https://doi.org/10.1007/s11120-018-0538-4

McLachlan DH (2020) Systemic signalling, and the synchronization of stomatal response. New Phytol 225(1):5–6. https://doi.org/10.1111/nph.16253

Meng L, Srivastava AK, Kuča K, Giri B, Rahman MM, Wu Q (2021a) Interaction between earthworms and arbuscular mycorrhizal fungi in plants: a review. Phyton 90(3):687–699. https://doi.org/10.32604/phyton.2021.015427

Meng YC, Zhang HF, Pan XX, Chen N, Hu HF, Haq SU, Khan A, Chen RG (2021b) CaDHN3, a pepper (*Capsicum annuum* L.) dehydrin gene enhances the tolerance against salt and drought stresses by reducing ROS accumulation. Int J Mol Sci 22(6):3205. https://doi.org/10.3390/ijms22063205

Miceli A, Vetrano F, Torta L, Esposito A, Moncada A (2023) Effect of mycorrhizal inoculation on melon plants under deficit irrigation regimes. Agronomy 13(2):440. https://doi.org/10.3390/agronomy13020440

Mohammadi R, Etminan A, Shoshtari L (2019) Agro-physiological characterization of durum wheat genotypes under drought conditions. Exp Agric 55(3):484–499. https://doi.org/10.1017/S0014479718000133

Mohammadi E, Fattahi M, Barin M, Ashrafi-Saeidlou S (2022) Arbuscular mycorrhiza and vermicompost alleviate drought stress and enhance yield, total flavonoid, rutin content, and antioxidant activity of buckwheat (*Fagopyrum esculentum* Moench). S Afr J Bot 148:588–600. https://doi.org/10.1016/j.sajb.2022.05.020

Nacoon S, Seemakram W, Ekprasert J, Jogloy S, Kuyper TW, Mongkolthanaruk W, Boonlue S (2022) Promoting growth and production of sunchoke (*Helianthus tuberosus*) by co-inoculation with phosphate solubilizing bacteria and arbuscular mycorrhizal fungi under drought. Front Plant Sci 13:1022319. https://doi.org/10.3389/fpls.2022.1022319

Niu T, Zhang T, Qiao Y, Wen P, Zhai G, Liu E, Al-Bakre DA, AlHarbi MS, Gao X, Yang X (2021) Glycinebetaine mitigates drought stress-induced oxidative damage in pears. PLOS ONE 16: e0251389. https://doi.org/10.1371/journal.pone.0251389

Noctor G, Reichheld JP, Foyer CH (2018) ROS-related redox regulation and signaling in plants. In: Seminars in cell & developmental biology, vol 80. Academic Press, London, pp 3–12. https://doi.org/10.1016/j.semcdb.2017.07.013

Ostadi A, Javanmard A, Amani Machiani M, Kakaei K (2023) Optimizing antioxidant activity and phytochemical properties of peppermint (*Mentha piperita* L.) by integrative application of biofertilizer and stress-modulating nanoparticles under drought stress conditions. Plants 12(1): 151. https://doi.org/10.3390/plants12010151

Ouhaddou R, Ben-Laouane R, Lahlali R, Anli M, Ikan C, Boutasknit A, Meddich A (2022) Application of indigenous rhizospheric microorganisms and local compost as enhancers of lettuce growth, development, and salt stress tolerance. Microorganisms 10(8):1625. https://doi.org/10.3390/microorganisms10081625

Ouledali S, Ennajeh M, Ferrandino A, Khemira H, Schubert A, Secchi F (2019) Influence of arbuscular mycorrhizal fungi inoculation on the control of stomata functioning by abscisic acid (ABA) in drought-stressed olive plants. S Afr J Bot 121:152–158. https://doi.org/10.1016/j.sajb.2018.10.024

Pammi SS, Suresh B, Giri A (2023) Antioxidant potential of medicinal plants. JCSB 26(1):13–26. https://doi.org/10.1007/s12892-022-00159-z

Pang XM, Zhang ZY, Wen XP, Ban Y, Moriguchi T (2007) Polyamines, all-purpose players in response to environment stresses in plants. Plant Stress 1(2):173–188

Porcel R, Azcon R, Ruiz-Lozano JM (2005) Evaluation of the role of genes encoding for dehydrin proteins (LEA D-11) during drought stress in arbuscular mycorrhizal *Glycine max* and *Lactuca sativa* plants. JXB 56(417):1933–1942. https://doi.org/10.1093/jxb/eri188

Qian S, Zhang Q, Li S, Shi R, He X, Zi S, Liu T (2023) Arbuscular mycorrhiza and plant growth promoting endophytes facilitates accumulation of saponin under moderate drought stress. Chin Herb Med. https://doi.org/10.1016/j.chmed.2022.11.004

Qiu Z, Guo J, Zhu A, Zhang L, Zhang M (2014a) Exogenous jasmonic acid can enhance tolerance of wheat seedlings to salt stress. EES 104:202–208. https://doi.org/10.1016/j.ecoenv.2014.03.014

Qiu H, Zhang L, Liu C, He L, Wang A, Liu HL, Zhu JB (2014b) Cloning and characterization of a novel dehydrin gene, SiDhn2, from *Saussurea involucrata* Kar. et Kir. Plant Mol Biol 84:707–718. https://doi.org/10.1007/s11103-013-0164-7

Querejeta J, Egerton-Warburton LM, Allen MF (2003) Direct nocturnal water transfer from oaks to their mycorrhizal symbionts during severe soil drying. Oecologia 134:55–64. https://doi.org/10.1007/s00442-002-1078-2

Quiroga G, Erice G, Aroca R, Zamarreño ÁM, García-Mina JM, Ruiz-Lozano JM (2018) Arbuscular mycorrhizal symbiosis and salicylic acid regulate aquaporins and root hydraulic properties in maize plants subjected to drought. Agric Water Manag 202:271–284. https://doi.org/10.1016/j.agwat.2017.12.012

Quiroga G, Erice G, Ding L, Chaumont F, Aroca R, Ruiz-Lozano JM (2019) The arbuscular mycorrhizal symbiosis regulates aquaporins activity and improves root cell water permeability in maize plants subjected to water stress. Plant Cell Environ 42(7):2274–2290. https://doi.org/10.1111/pce.13551

Quiroga G, Erice G, Aroca R, Zamarreño ÁM, García-Min JM, Ruiz-Lozano JM (2020) Radial water transport in arbuscular mycorrhizal maize plants under drought stress conditions is affected by indole-acetic acid (IAA) application. J Plant Physiol 246:153115. https://doi.org/10.1016/j.jplph.2020.153115

Raja V, Majeed U, Kang H, Andrabi KI, John R (2017) Abiotic stress: Interplay between ROS, hormones and MAPKs. EEB 137:142–157. https://doi.org/10.1016/j.envexpbot.2017.02.010

Ren CG, Kong CC, Xie ZH (2018) Role of abscisic acid in strigolactone-induced salt stress tolerance in arbuscular mycorrhizal *Sesbania cannabina* seedlings. BMC Plant Biol 18(1):1–10. https://doi.org/10.1186/s12870-018-1292-7

Ren AT, Zhu Y, Chen YL, Ren HX, Li JY, Abbott LK, Xiong YC (2019) Arbuscular mycorrhizal fungus alters root-sourced signal (abscisic acid) for better drought acclimation in *Zea mays* L. seedlings. EEB 167:103824. https://doi.org/10.1016/j.envexpbot.2019.103824

Ruiz-Lozano JM, Aroca R (2010) Modulation of aquaporin genes by the arbuscular mycorrhizal symbiosis in relation to osmotic stress tolerance. In: Seckbach J, Grube M (eds) Symbioses and stress. Cellular origin, life in extreme habitats and astrobiology, vol 17. Springer, Dordrecht, pp 357–375. https://doi.org/10.1007/978-90-481-9449-0_17

Ruiz-Lozano JM, Porcel R, Aroca R (2008) Evaluation of the possible participation of drought-induced genes in the enhanced tolerance of arbuscular mycorrhizal plants to water deficit. In: Varma A (ed) Mycorrhiza. Springer, Berlin, pp 185–205. https://doi.org/10.1007/978-3-540-78826-3_10

Sakamoto A, Murata N (2000) Genetic engineering of glycinebetaine synthesis in plants: current status and implications for enhancement of stress tolerance. JXB 51(342):81–88. https://doi.org/10.1093/jexbot/51.342.81

Santander C, Aroca R, Ruiz-Lozano JM, Olave J, Cartes P, Borie F, Cornejo P (2017) Arbuscular mycorrhiza effects on plant performance under osmotic stress. Mycorrhiza 27:639–657

Saxena B, Sharma K, Kapoor R, Wu QS, Giri B (2022) Insights into the molecular aspects of salt stress tolerance in mycorrhizal plants. World J Microbiol Biotechnol 38(12):253. https://doi.org/10.1007/s11274-022-03440-z

Shankar V, Evelin H (2021) Contribution of organic solutes in amelioration of salt stress in plants. In: Organic solutes, oxidative stress, and antioxidant enzymes under abiotic stressors. CRC Press, Boca Raton, pp 35–60

Sharma D, Kothamasi D (2015) Soil moisture—a regulator of arbuscular mycorrhizal fungal community assembly and symbiotic phosphorus uptake. Mycorrhiza 25:67–75. https://doi.org/10.1007/s00572-014-0596-1

Sheng M, Tang M, Zhang F, Huang Y (2011) Influence of arbuscular mycorrhiza on organic solutes in maize leaves under salt stress. Mycorrhiza 21:423–430. https://doi.org/10.1007/s00572-010-0353-z

Sheteiwy MS, Abd Elgawad H, Xiong YC, Macovei A, Brestic M, Skalicky M, Shaghaleh H, Hamoud YA, El-Sawah AM (2021) Inoculation with *Bacillus amyloliquefaciens* and mycorrhiza confers tolerance to drought stress and improve seed yield and quality of soybean plant. Physiol Plant 172(4):2153–2169. https://doi.org/10.1111/ppl.13454

Shiraku ML, Magwanga RO, Zhang Y, Hou Y, Kirungu JN, Mehari TG, Liu F (2022) Late embryogenesis abundant gene LEA3 (Gh_A08G0694) enhances drought and salt stress tolerance in cotton. Int J Biol Macromol 207:700–714. https://doi.org/10.1016/j.ijbiomac.2022.03.110

Spinoso-Castillo JL, Moreno-Hernández MDR, Mancilla-Álvarez E, Sánchez-Segura L, Sánchez-Páez R, Bello-Bello JJ (2023) Arbuscular Mycorrhizal Symbiosis improves ex vitro acclimatization of sugarcane plantlets (*Saccharum* Spp.) under drought stress conditions. Plants 12(3): 687. https://doi.org/10.3390/plants12030687

Srivastava A, Sharma VK, Kaushik P, El-Sheikh MA, Qadir S, Mansoor S (2022) Effect of silicon application with mycorrhizal inoculation on *Brassica juncea* cultivated under water stress. Plos one 17(4):e0261569. https://doi.org/10.1371/journal.pone.0261569

Sugiura Y, Akiyama R, Tanaka S, Yano K, Kameoka H, Marui S, Saito M, Kawaguchi M, Akiyama K, Saito K (2020) Myristate can be used as a carbon and energy source for the asymbiotic growth of arbuscular mycorrhizal fungi. Proc Natl Acad Sci 117:25779–25788. https://doi.org/10.1073/pnas.2006948117

Tang H, Hassan MU, Feng L, Nawaz M, Shah AN, Qari SH, Miao J (2022) The critical role of arbuscular mycorrhizal fungi to improve drought tolerance and nitrogen use efficiency in crops. Front Plant Sci 13:919166. https://doi.org/10.3389/fpls.2022.919166

Tarnabi MZ, Iranbakhsh A, Mehregan I, Ahmadvand R (2020) Impact of arbuscular mycorrhizal fungi (AMF) on gene expression of some cell wall and membrane elements of wheat (*Triticum aestivum* L.) under water deficit using transcriptome analysis. Physiol Mol Biol Plants 26:143–162. https://doi.org/10.1007/s12298-019-00727-8

Tekaya M, Dabbaghi O, Guesmi A, Attia F, Chehab H, Khezami L, Algathami FK, Hamadi NF, Hammami M, Prinsen E, Mechri B (2022) Arbuscular mycorrhizas modulate carbohydrate, phenolic compounds and hormonal metabolism to enhance water deficit tolerance of olive trees (*Olea europaea*). Agric Water Manag 274:107947. https://doi.org/10.1016/j.agwat.2022.107947

Tereucán G, Ruiz A, Nahuelcura J, Oyarzún P, Santander C, Winterhalter P, Cornejo P (2022) Shifts in biochemical and physiological responses by the inoculation of arbuscular mycorrhizal fungi in *Triticum aestivum* growing under drought conditions. J Sci Food Agric 102(5): 1927–1938. https://doi.org/10.1002/jsfa.11530

Tisarum R, Theerawitaya C, Samphumphuang T, Polispitak K, Thongpoem P, Singh HP, Cha-Um S (2020) Alleviation of salt stress in upland rice (*Oryza sativa* L. ssp. indica cv. Leum Pua) using arbuscular mycorrhizal fungi inoculation. Front Plant Sci 11:348. https://doi.org/10.3389/fpls.2020.00348

Tisserant E, Malbreil M, Kuo A, Kohler A, Symeonidi A, Balestrini R, Charron P, Duensing N, Frey NF et al (2013) Genome of an arbuscular mycorrhizal fungus provides insight into the oldest plant symbiosis. Proceedings of the National Academy of Sciences 110(50): 20117–20122. https://doi.org/10.1073/pnas.1313452110

Volpe V, Chitarra W, Cascone P, Volpe MG, Bartolini P, Moneti G, Pieraccini G, Serio CD, Maserti B, Guerrieri E, Balestrini R (2018) The association with two different arbuscular mycorrhizal fungi differently affects water stress tolerance in tomato. Front Plant Sci 9:1480. https://doi.org/10.3389/fpls.2018.01480

Wang G, Su H, Abou-Elwafa SF, Zhang P, Cao L, Fu J, Wei L (2023a) Functional analysis of a late embryogenesis abundant protein *ZmNHL1* in maize under drought stress. J Plant Physiol 280: 153883. https://doi.org/10.1016/j.jplph.2022.153883

Wang Y, Zou YN, Shu B, Wu QS (2023b) Deciphering molecular mechanisms regarding enhanced drought tolerance in plants by arbuscular mycorrhizal fungi. Sci Hortic 308:111591. https://doi.org/10.1016/j.scienta.2022.111591

Wu QS, Zou YN (2009) Mycorrhiza has a direct effect on reactive oxygen metabolism of drought-stressed citrus. Plant Soil Environ 55(10):436–442

Wu QS, He JD, Srivastava AK, Zou YN, Kuča K (2019) Mycorrhizas enhance drought tolerance of citrus by altering root fatty acid compositions and their saturation levels. Tree Physiol 39(7): 1149–1158. https://doi.org/10.1093/treephys/tpz039

Xie W, Hao Z, Zhou X, Jiang X, Xu L, Wu S, Zhao A, Zhang X, Chen B (2018) Arbuscular mycorrhiza facilitates the accumulation of glycyrrhizin and liquiritin in *Glycyrrhiza uralensis* under drought stress. Mycorrhiza 28:285–300. https://doi.org/10.1007/s00572-018-0827-y

Xu LJ, Hao ZP, Xie W, Li F, Chen BD (2018a) Transmembrane H^+ and Ca^{2+} fluxes through extraradical hyphae of arbuscular mycorrhizal fungi in response to drought stress. Chin J Plant Ecol 42:764–773. https://doi.org/10.17521/cjpe.2018.0089

Xu L, Li T, Wu Z, Feng H, Yu M, Zhang X, Chen B (2018b) Arbuscular mycorrhiza enhances drought tolerance of tomato plants by regulating the 14-3-3 genes in the ABA signaling pathway. Appl Soil Ecol 125:213–221. https://doi.org/10.1016/j.apsoil.2018.01.012

Yang Y, Guo Y (2018) Elucidating the molecular mechanisms mediating plant salt-stress responses. New Phytol 217(2):523–539. https://doi.org/10.1111/nph.14920

Yang R, Yang T, Zhang H, Qi Y, Xing Y, Zhang N, Li R, Weeda S, Ren S, Ouyang B, Guo YD (2014) Hormone profiling and transcription analysis reveal a major role of ABA in tomato salt tolerance. Plant Physiol Biochem 77:23–34. https://doi.org/10.1016/j.plaphy.2014.01.015

Yu Z, Wang X, Zhang L (2018) Structural and functional dynamics of dehydrins: a plant protector protein under abiotic stress. Int J Mol Sci 19(11):3420. https://doi.org/10.3390/ijms19113420

Zézé A, Brou YC, Meddich A, Marty F (2008) Molecular identification of *MIP* genes expressed in the roots of an arbuscular mycorrhizal *Trifolium alexandrium* L. under water stress. Afr J Agric Res 3(1):78–83

Zhang H, Dong H, Li W, Sun Y, Chen S, Kong X (2009) Increased glycine betaine synthesis and salinity tolerance in *AhCMO* transgenic cotton lines. Mol Breed 23:289–298. https://doi.org/10.1007/s11032-008-9233-z

Zhang T, Hu Y, Zhang K, Tian C, Guo J (2018a) Arbuscular mycorrhizal fungi improve plant growth of *Ricinus communis* by altering photosynthetic properties and increasing pigments under drought and salt stress. Ind Crops Prod 117:13–19. https://doi.org/10.1016/j.indcrop.2018.02.087

Zhang F, Jia-Dong HE, Qiu-Dan NI, Qiang-Sheng WU, Ying-Ning ZOU (2018b) Enhancement of drought tolerance in trifoliate orange by mycorrhiza: changes in root sucrose and proline metabolisms. Not Bot Horti Agrobot 46(1):270–276. https://doi.org/10.15835/nbha46110983

Zhang X, Wu H, Chen L, Liu L, Wan X (2018c) Maintenance of mesophyll potassium and regulation of plasma membrane H+-ATPase are associated with physiological responses of tea plants to drought and subsequent rehydration. Crop J 6(6):611–620. https://doi.org/10.1016/j.cj.2018.06.001

Zhang F, Wang P, Zou YN, Wu QS, Kuča K (2019) Effects of mycorrhizal fungi on root-hair growth and hormone levels of taproot and lateral roots in trifoliate orange under drought stress. Arch Agron Soil Sci 65(9):1316–1330. https://doi.org/10.1080/03650340.2018.1563780

Zou YN, Wang P, Liu CY, Ni QD, Zhang DJ, Wu QS (2017) Mycorrhizal trifoliate orange has greater root adaptation of morphology and phytohormones in response to drought stress. Sci Rep 7(1):41134. https://doi.org/10.1038/srep41134

Chapter 8
Arbuscular Mycorrhizal Fungi in Plant Tolerance to Organic Pollutants and Associated Food Safety

Hamideh Bakhshayeshan-Agdam, Mahdieh Houshani, and Seyed Yahya Salehi-Lisar

Abstract Industrialization and the modern world have led to environmental contamination by organic pollutants, many of which are persistent and could engender hazardous disorders in ecosystems and living organisms including humans. Organic pollutants are phytotoxic compounds and affect the plant biochemical and physiological processes. However, plants can uptake these compounds leading to inclusion of pollutants into food chains. Nowadays, several studies have been carried out to explore and develop novel strategies that could lessen or eradicate pollution while posing no deleterious effects on the environment. Bioremediation (including phytoremediation) is green technology for the remediation of contaminated soils and water. The majority of plant species form symbiosis with mycorrhizal fungi which usually leads to improved plant resistance to organic contaminations. Therefore, plant root inoculation with mycorrhizal fungi could be a biological solution for the problem of organic pollutants.

Keywords Organic pollutants · Arbuscular mycorrhizal fungi (AMF) · Bioremediation · Crops

8.1 Introduction

Large amounts of mineral and organic pollutants are released into the environment because of human activities such as agriculture and mining, which causes soil pollution and ultimately the accumulation of these pollutants in agricultural products (Wang et al. 2019). For example, pesticides as contaminations are widely used to secure crops from pathogens and weeds. Besides, other organic contaminations, such as polychlorinated biphenyls (PCBs), polycyclic aromatic hydrocarbons (PAHs), phthalates (PAEs), and petroleum hydrocarbons, generally have long

H. Bakhshayeshan-Agdam · M. Houshani · S. Y. Salehi-Lisar (✉)
Department of Plant, Cell and Molecular Biology, Faculty of Natural Sciences, University of Tabriz, Tabriz, East Azerbaijan, Iran
e-mail: y_salehi@tabrizu.ac.ir

G. J. Ahammed, R. Hajiboland (eds.), *Arbuscular Mycorrhizal Fungi and Higher Plants*, https://doi.org/10.1007/978-981-99-8220-2_8

persistence in agricultural soils. A lot of these organic pollutants are mutagens, carcinogens, and teratogens (Wang et al. 2019; Casida 2017). Biological mechanisms are commonly used for the elimination of these toxic compounds from the environment (Fecih and Baoune 2019). Symbiotic interactions, mycorrhizal associations, in particular, are valuable tools for the bioremediation of polluted ecosystems (Whitacre 2015). Mycorrhizoremediation, the use of mycorrhizae in bioremediation, is one of the biological ways that could alleviate the matter of persistent organic pollutants (POPs) that results in removing the pollutants from soils and increases the host plant growth and development in contaminated soils (Fecih and Baoune 2019). In the current chapter, the arbuscular mycorrhizal fungi (AMF)-induced plant response to organic pollutants, the role of AMF in plant-based remediation of organic pollutants, the positive effect of AMF inoculation on host plants grown in contaminated environments, as well as their application for ensuring food safety are discussed.

8.2 Mycorrhiza: Definition and Types

Mycorrhiza is a unique symbiotic relationship between beneficial fungi and host plant roots. This symbiosis can be divided into two types: ectomycorrhiza and endomycorrhiza.

8.2.1 Ectomycorrhiza

Ectomycorrhizae are formed by basidiomycetes and ascomycetes with the roots of plants mainly from cold temperate regions, such as beech, birch, pine, willow, oak, spruce, and fir (Smith and Read 2008; Ganugi et al. 2019). Ectomycorrhizal symbiosis is featured in intercellular fungi hyphae net called as the "Hartig Net." The Hartig Net includes vastly branched hyphae attaching the root epidermal and cortical cells. The fungus creates a covering of hyphae around the root and completely covers the tip of the root, but never enters the cells (Ganugi et al. 2019; Dighton and Allen 1991).

8.2.2 Endomycorrhiza

Endomycorrhizae are found in over 80% of vascular plants. Endomycorrhiza is distinct from ectomycorrhiza in formation and function. The fungus hyphae grow inside the plant roots and even transpire to the root cell walls. They have more invasive relationship than ectomycorrhiza (Weber 2005; Ganugi et al. 2019). Endomycorrhizae are divided into five types including arbuscular mycorrhizae, ericaceous mycorrhizae, arbutoid mycorrhizae, monotropoid, and orchidaceous mycorrhizae (Ganugi et al. 2019). Arbuscular mycorrhizae are the most widespread and important relationship in the plant kingdom.

8.2.2.1 Arbuscular Mycorrhizae Fungi (AMFs)

Arbuscular mycorrhiza symbiosis is the largest association between over 100,000 plant species and a few 100 AMF morphotypes (Chen et al. 2018). AMFs are obligate biotrophs and indeed need plant photosynthetic products which are necessary for their metabolism (Begum et al. 2019).

The penetration of the fungus into the root can be done in various ways including (1) appressorium generation from which intracellular hyphae originate, (2) penetration through a radical hair, and (3) entering across the cells of external layers of roots that are often dead (Ganugi et al. 2019).

8.3 Benefits of AMFs for Plants

Based on the obtained results from fossil and molecular data, the symbiosis of mycorrhizal fungi and plants dates back to the time of the first plant appearance on land, which is about 400–450 million years ago (Smith and Read 2008; Selosse et al. 2015). The symbiosis between mycorrhizal fungi and plants has many advantages for plants:

8.3.1 Effects on Plant Growth and Nutrition

Among the symbiotic microorganisms, AMFs are widely distributed and establish symbiotic relationships with most crops (Posta and Hong Duc 2020). In many studied plant species, AMFs improve growth (Bona et al. 2017; Gamalero et al. 2004; Gogoi and Singh 2011; Nakmee et al. 2016; Parihar and Bora 2018) and uptake of mineral elements like nitrogen and phosphorus (Jansa et al. 2019; Song et al. 2020). AMFs enhance the mineral absorption in the plants as a result of an escalation in the absorption network of nutrients around the roots, which makes the roots available to a large amount of soil (Smith and Smith 2011; Begum et al. 2019). Besides, the fungal hyphae are very tiny compared with the roots of plants, so they can enter the tiny pores and subsequently absorb more nutrients. The direct transfer of nutrients to the inner layers of the root tissue has also been proposed as a mechanism to increase the uptake of elements. Molecular data have shown that during the coexistence of plants with AMFs, the expression of transporter genes such as aquaporins, phosphate, ammonium, nitrate, sulfur, and zinc transporters increases (Diagne et al. 2020). Experimental trials have shown that inoculation with AMFs leads to increased plant leaf area and higher contents of essential elements such as phosphorus, nitrogen, potassium, and calcium and finally boosts plant growth (Begum et al. 2019). In fact, AMFs support plants nutritionally, especially in stressful situations.

8.3.2 *Alleviation of Stress in Plants*

Many studies were carried out to examine the mitigating effects of AM fungi on plants under different stresses. The results revealed that these fungi could improve the tolerance of inoculated host plants to a wide range of stress including biotic and abiotic stress (Diagne et al. 2020).

Drought is a devastating abiotic stress affecting plant growth and development (Salehi-Lisar and Bakhshayeshan-Agdam 2016; Ganugi et al. 2019). Inoculation of plants with AMFs makes them resistant to drought and increases the stability of plants in stressful conditions (Diagne et al. 2020). The increase of water absorption by AMFs is done mainly through the development of more absorption surface by fungal hyphae (Augé 2001), as well as increasing the access to fine soil pores and improving the apoplastic flow of water (Diagne et al. 2020).

Heat and cold stress are also considered as important stress factors affecting plants worldwide (Begum et al. 2019; Hasanuzzaman et al. 2013). Plant inoculation with AMFs increases the resistance of plants to temperature stress by improving the water and mineral element absorption, improving the efficiency and capacity of photosynthesis, alleviation of oxidative stress, and inducing osmotic adaptation (Caradonia et al. 2019; Zhu et al. 2017).

AMF could be a remedy for chemical pesticides in agriculture, leading to higher sustainability and eliminating public health hazards. Indeed, it is well-known that AMF could lessen the soil-borne pathogens that caused damage in plants (Ganugi et al. 2019).

The soil structure is adversely affected by intensive agricultural practices. The hyphae of AMFs are one of the chief binding factors involved in stabilizing soil aggregates (Miller and Jastrow 2000).

8.4 Environmental Pollutants: Types and Problems

The pollution of the environment because of the spreading of industrial and urban wastes produced through anthropogenic acts is one of the significant environmental worries in the world (Jacob and Cherian 2013). Environmental contaminants polluting the water and land resources result in the reduction of food quality and health risks in living organisms (Dean et al. 2001; Houshani et al. 2021). The common pollutants exist abundantly in the environments such as PAHs, pesticides, PCBs, and explosives as organic compounds and inorganic compounds such as metals, metalloids, and radionuclides (Testiati et al. 2013; Vane et al. 2014). Their long-lived toxicity, persistence in the environment, and their large amount require a lot of attention (Aina et al. 2006; Parrish et al. 2005). Organic pollutants can disrupt different plant segments and activities like membranes, metabolism, water and nutrient uptake, and other disorders. In the same manner, these organic contaminations have the ability to cause toxicity and have xenobiotic and carcinogenic properties (Zeb et al. 2022). Human and wildlife exhibition to such steady and toxic compounds leads to their accumulation and movement into plants and also natural and agricultural food chains (Yan et al. 2004).

8.4.1 Persistent Organic Pollutants (POPs)

Organic pollutants are toxic and can harmfully affect living organisms. There are different sources for releasing these pollutants including accidental release, industrial functions (such as chemical and petrochemical plants), agriculture (such as pesticides), and military activities (such as explosives, chemical weapons) (Ensley and Raskin 1999; Ashraf 2017). Further, organic contaminations include various types of compounds like solvents (such as trichloroethylene), explosives (such as trinitrotoluene (TNT) and cyclotrimethylenetrinitramine), polycyclic aromatic hydrocarbons (PAHs), petroleum products (such as benzene, toluene, ethylbenzene, and xylene: BTEX), polychlorinated biphenyles (PCBs), and pesticides (such as atrazine, chlorpyrifos, 2, 4-D) (Singh and Jain 2003; Zeb et al. 2022). As a consequence of standing against biological, chemical, and photolytic transformation processes, they could accumulate in the environment and could be moved by wind and water and thus affect living organisms far from where they are released (Cheng et al. 2017).

8.5 Organic Pollutant Effects on Plants

Many organic pollutants are plant toxic compounds and influence numerous biochemical and physiological activities in plants. These organic pollutants can make acute and chronic damages to plants based on the concentrations and duration of exposure (Kmentova 2003; Cheng et al. 2017). The long-lived effect of toxic organic pollutants can impact the sustainability and performance of ecosystems, reduce biomass production and oxygen release through photosynthesis, and finally lead to the destruction of the environment (Vanova 2009). In addition, the severity of the effects of organic pollutants on plants depends on the ontogenetic phase of growth, the physiological condition of the organisms, the morphological and size features of the leaf surface, cuticle enterability, transpiration rate and leaf length, and the enzymes involved in organic pollutants metabolism such as peroxidases, hydrolases, and cytochrome P450 causing the reactivity of plants to these pollutants (Greipsson 2011). These pollutants cause morphological, cytological, genetic, and metabolic disorganization; for example, through the induction of oxidative stress, they inhibit and reduce seed germination and disrupt the functioning of the photosynthetic apparatus (Reynoso-Cuevas et al. 2008; Tomar and Jajoo 2014).

8.5.1 Seed Germinations

Seed germination and root system formation are vital stages of plant growth influenced by various factors as external factors. Accordingly, as one of the outside factors affecting plants, pollutants can influence all phases of plant life cycle from germination to reproduction (Maila and Cloete 2002; Reynoso-Cuevas et al. 2008; Houshani et al. 2021). Investigators have demonstrated that the reactivity of maize, barley, millet, and lettuce seed germination to PAHs as an organic pollutant is more dependent on the length of the root (Kmentova 2003; Kummerova et al. 2012, Houshani et al. 2019). Salehi-Lisar and Deljoo (2015) reported that organic pollutants had a damaging effect on the germination speed of alfalfa, sunflower, and wheat and caused a decrease in their growth.

8.5.2 Metabolism and Photosynthesis

Recent research has shown that organic pollutants have a profound effect on plant metabolism and mainly have a negative and destructive effect on photosynthesis and synthesis of lipids, amino acids, proteins, nucleic acids, and hormones. For instance, PCBs and herbicides as an organic pollutants have been found to be detrimental to cell metabolism (Brain et al. 2010; Cheng et al. 2017). In parallel, it was found that these pollutants can inhibit photosynthesis in plants through the prevention of photosynthetic pigment biosynthesis, making green leaves and stems white and translucent (Liu et al. 2010; Kaspary et al. 2014; Cheng et al. 2017). Organic contamination can restrain the photosynthesis process. In addition, these pollutants can perforate chloroplasts and affect the initial ways of photosynthesis (Zobiole et al. 2009; Amaia et al. 2013). Correspondingly, the stability and activity of the photosynthetic structure can be affected by organic pollutant contamination and also they can interrupt the electron transport chain by producing quinone compounds and binding to plastoquinones. In addition, these contaminants can inhibit photosystem II (PSII) on both sides (Kummerova et al. 2006; Houshani & Salehi-Lisar 2020).

8.5.3 Oxidative Stress

Plants have enzymatic and non-enzymatic defense techniques to mitigate oxidative stress induced by organic pollutants (Yoshiyama et al. 2013). The oxidative stress induced by these pollutants and their connection with morphological disorders has been previously stated in different plant species such as *Arabidopsis thaliana* (Alkio et al. 2005). Tarigholizadeh et al. (2021) reported that PAHs could cause oxidative stress in *Panicum miliaceum*. Treated plants have shown a remarkable increase in antioxidant enzyme activity along with a substantial increase in non-enzymatic antioxidant levels potentially in order to detoxify reactive oxygen species.

8.6 Plant Resistance to POPs: General Mechanisms

Plants have the potential to uptake contaminations from the environment by the root system or aerial parts. Also, they can remove or reduce pollutant concentration in the environment by different strategies like rhizoremediation, phytoextraction, phytostabilization, phytovolatilization, phytodegradation, and rhizodegradation. These strategies form the basis of phytoremediation technology (Etim 2012). Accordingly, phytoremediation is named rhizodegradation when degradation occurs in the rhizosphere, and it is also named phytostimulation when it includes microbes. In some samples, the pollutants are deactivated in the root zone called phytostabilization. The blended processes of the rhizosphere within the root promote phytoremediation called rhizoremediation. Phytoextraction refers to assembling contaminations in collective plant tissues—uniquely shoot tissues. This approach is notably utilized for phytoremediation of inorganic pollutants. In some cases, plants can break down contaminations inside their tissues; this process is named phytodegradation and is mainly appropriate for organic contaminations. Some pollutants can also be volatilized by plants and this process is called phytovolatilization (Etim 2012; Kafle et al. 2022). The organic pollutants are chiefly degraded as plant responses to organic pollutants by two mechanisms: phytodegradation or phytotransformation and rhizoremediation (Arthur et al. 2005).

8.6.1 Transformation of POPs

The transformation of organic contaminations in plants is done by enzymes (Vanova 2009). It is a characteristic of a process that is species-specified and may be specific to that plant and occur only in a particular tissue or organ or even at a particular growth step (Kmentova 2003; Vanova 2009). The transformation of pollutants in plants is divided into three stages, which include conversion, conjugation, and elimination reactions (Fig. 8.1).

In the primary stage, oxidation, reduction, and hydrolysis reactions are created by enzymes such as cytochrome P450 or carboxylesterases. The reduction reactions are of little importance in transformation, but oxidation is the most involved reaction in detoxification of the organic pollutants. The important reaction in the stage of oxidation is monooxygenation (hydroxylation) where the addition of oxygen to

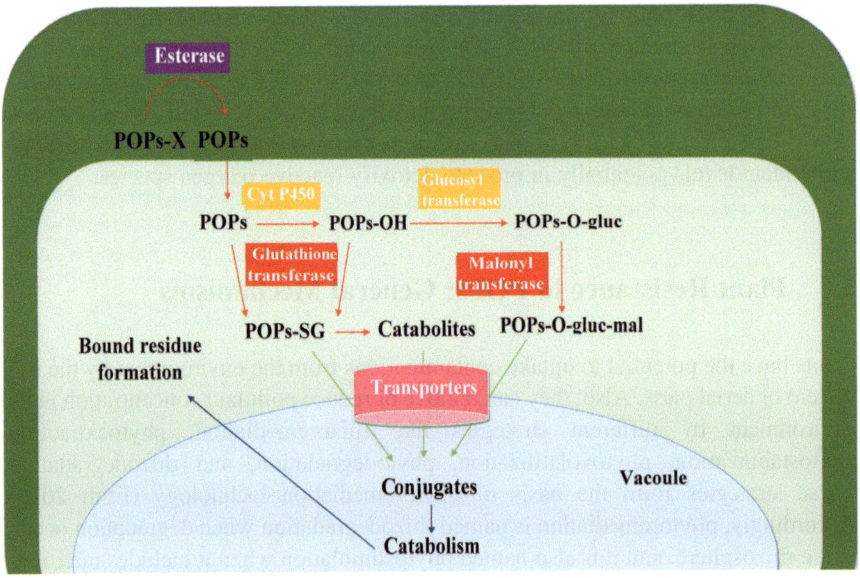

Fig. 8.1 Various stages of metabolism of organic pollutants in plant cell

toxicant molecule occurs (Korte et al. 2000). It is mentioned that the top-said stage is placed within the endoplasmatic reticulum or mitochondria and these reactions may adjust combinations and able to conjugate with amines, alcohols, and acids in the next stage of detoxification (Korte et al. 2000; Syed et al. 2010).

In the subsequent stage (conjugation), the conjugated forms of pollutants are created by adding compounds like sugars, amino acids, organic acids, and peptides. Reactions are carried out by enzymes such as transferases. For example, glucosyltransferase and glutathione transferase enzymes add sugars and glutathione to pollutants, respectively. Then, malonyltransferase enzyme adds malonic acid after adding glucose (Knejzlik et al. 2000; Collins 2011).

The third stage is called detoxification. The conjugated compounds constructed in the former stage are transported to the vacuole by ABC transporters. On the other hand, the metabolites accumulated in the vacuole are catalyzed by hydrolytic reactions inside the vacuole and then again transferred to the cytoplasm for other catabolism (Brazier-Hicks et al. 2008). Also, plants can store soluble conjugated compounds in the vacuole and transport insoluble conjugated compounds to the cell wall for further catabolism (Trapp and Mcfarlane 1995). Moreover, numerous organic contaminations that possess carboxyl, sulfhydryl, and hydroxyl or amine groups are located into lignin or to other cell wall components like hemicellulose, cellulose, or protein (Vanova 2009; Kang et al. 2010).

8.6.2 Phytodegradation

This concept is considered a process using plants to remove or degrade or reduce the toxicity of environmental pollutants (Singh et al. 2017; Kafle et al. 2022). Plant degradation includes the direct destruction of organic pollutants, and this action occurs through the discharge of enzymes from the roots or through metabolic activities in plant tissues. In phytodegradation, organic pollutants are absorbed by roots and metabolized into smaller toxic substances in plant tissues. Some studies have reported that different leguminous and grass plant species have the ability to break down organic contaminants (Adam and Duncan 1999). Some tropical plants significantly have the potential to degrade organic pollutants, and this potential is due to the inherent characteristics of the plant itself including deep fibrous root systems and high hydrocarbon tolerance (Dzantor et al. 2000). Plant intercropping has the ability to increase the speed of breakdown of organic contaminations in comparison with the cultivation of plant individually in soil (Chouychai et al. 2009; Cheema et al. 2010). For example, plants such as alfalfa (*Medicago sativa*), corn (*Zea mays*), rice (*Oryza sativa*), rapeseed (*Brassica napus*), and *Festuca arundinacea* can increase the capability of degradation in polluted environments (Cheema et al. 2008; Du et al. 2011).

8.7 AMFs and Plant Resistance to POPs

AMFs could modify soil structure and plant health and decline the contaminant accumulation in the plants as well. AMFs improve the plant resistance to organic pollutants and enhance plant growth, resulting in higher biomass production in contaminated soils. AMFs reduce organic pollutant residues in the plant shoots and increase the accumulation of pollutants (POPs in particular) in plant roots. AMFs could improve the resistance of plants to POPs in various ways including (1) biomass increase in plants via improved mineral nutrition and water availability, (2) enhanced activities of contaminant-degrading enzymes, (3) alleviation of the harmful effects of oxidative stress induced by pollutants, (4) accumulation of pollutants by AMF structures, (5) reduction of pesticide and herbicide use in agricultural fields, and (6) stimulation of growth of microorganisms involved in the degradation of contaminants in the soil (Wang et al. 2019; Fecih and Baoune 2019). It was observed that AM causes significant degradation of PAHs as one type of organic pollutant. The demotion of pollutants has been observed in plants with mycorrhizal treatments much more and faster (Joner and Leyval 2003). Similarly, Gao et al. (2011) reported greater degradation of pyrene and phenanthrene as PAHs in plant soils with AM cultivation than in soils without AM cultivation. Ectomycorrhizal fungi in pure culture are able to secrete oxidative enzymes such as polyphenol oxidases, laccases, tyrosinases, or peroxidases and these enzymes can degrade PAHs in spike soils under controlled environmental conditions (Gramss and Rudeschko 1998; Genney et al. 2004).

8.8 Use of AMFs for Management of Polluted Environment and Agriculture

One of the most successful bioremediation techniques is the AMF-assisted remediation. In this technique, improvement of soil construction and enhancement of nutrient uptake by plants and the efficiency of POP bioremediation increase considerably. In order to improve the bioremediation process, the adapted strains of AMFs must be selected and used for plant inoculation in polluted ecosystems (Fecih and Baoune 2019). Gao et al. (2010) reported that AMF hyphae have a great potential to uptake PAHs. Combined use of AMFs and rhizobia can be effective in enhancing the phytoremediation efficiency of PAHs such as phenanthrene and pyrene by plants (Ren et al. 2017). Teng et al. (2010) suggested that synergistic interactions between AMFs and rhizobium may have great potential to enhance the phytoremediation of agricultural soil contaminated with weathered PCBs.

8.9 AMFs and Food Safety

Food supply and safety is one of the most important issues for a growing global population. Application of excess chemicals (such as fertilizers, herbicides, pesticides, etc.) in agriculture is not a sustainable approach for the future because it causes severe environmental challenges such as negative impacts on soils, human and ecosystem health, accumulation of biocide residues in the environments, and finally affecting plant productivity and product safety. Therefore, chemicalized agriculture is an important challenge for food safety in the future (George and Ray 2023). In recent years, enough and safe food production to feed the world population has become a serious problem worldwide. Recently there have been many reasons for the extensive production of AMF inocula. Favorable effects on health, plant growth and development, ecosystem restoration, disease management, and phytoremediation improvement are the most important reasons among them (Siddiqui et al. 2008). Organic farming is developing day by day and its general principles are the elimination of most synthetic materials, soil management through the addition of organic materials, and the use of crop rotation (Gosling et al. 2006). Indeed, AMFs are capable not only of relieving lower fertilizer inputs (especially phosphorus) in organic systems but also of helping to phytoremediation of POPs (Fecih and Baoune 2019). Figure 8.2 briefly illustrates the role of AMF in improving food safety.

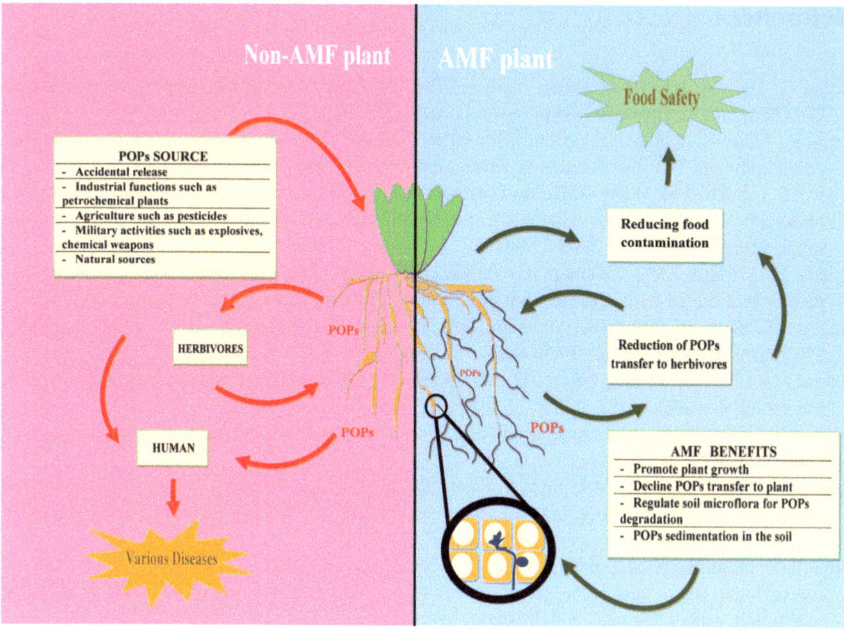

Fig. 8.2 Sources of POP entry into plants and roles of AMF in reducing food chain contamination and improving food safety

8.10 Conclusions

Environmental pollution influences soil characteristics and reduces plant growth and yield along with its impact on food safety and human health. Production of sufficient and safe food for the growing population of the globe is one of the most important concerns in many countries. In many regions farmers cultivate crops in polluted soils or use polluted water for irrigation, leading to contamination of foods directly or indirectly. So, the management of contaminated lands for crop cultivation is an urgent need. Many strategies can be used to improve the cultivation of crops and to decrease the risk of food contamination in polluted soils. Remediation of the environment by different techniques is the most effective strategy for preventing the entry of contaminants into the food chain. Mycorrhizal fungi (especially AMFs) have great potential in the degradation of some organic pollutants. Therefore, colonization with AM fungi helps to protect plants and decrease contaminant entry into crop plants and the food chain. More biochemical, physiological, and molecular studies are needed to determine the role of AMF in removing organic pollutants and the involved mechanisms. These researches could help to progress in the usage of AMF as candidates to assist phytoremediation and restoration of the environments polluted with organic contaminants.

References

Adam G, Duncan HJ (1999) Effect of diesel fuel on growth of selected plant species. Environ Geochem Health 21(4):353–357

Aina R, Palin L, Citterio S (2006) Molecular evidence for benzo[a]pyrene and naphthalene genotoxicity in Trifolium repens L. Chemosphere 65(4):666–673

Alkio M, Tabuchi TM, Wang X, Colón-Carmona A (2005) Stress responses to polycyclic aromatic hydrocarbons in Arabidopsis include growth inhibition and hypersensitive response-like symptoms. J Exp Bot 56(421):2983–2994

Amaia Z, Miriam GM, Villamor JG, Zabalza A, Hoorn RA, Royuela M (2013) Proteolytic pathways induced by herbicides that inhibit amino acid biosynthesis. PLoS One 8:e73847

Arthur EL, Rice PJ, Rice PJ, Anderson TA, Baladi SM, Henderson KLD, Coats JR (2005) Phytoremediation. Crit Rev Plant Sci 24(2):109–122

Ashraf MA (2017) Persistent organic pollutants (POPs): a global issue, a global challenge. Environ Sci Pollut Res 24(5):4223–4227

Augé RM (2001) Water relations, drought and vesicular-arbuscular mycorrhizal symbiosis. Mycorrhiza 11:3–42

Begum N, Qin C, Ahanger MA, Raza S, Khan MI, Ashraf M, Ahmed N, Zhang L (2019) Role of arbuscular mycorrhizal fungi in plant growth regulation: implications in abiotic stress tolerance. Front Plant Sci 10:1–10

Bona E, Cantamessa S, Massa N, Manassero P, Marsano F, Copetta A, Berta G (2017) Arbuscular mycorrhizal fungi and plant growth-promoting pseudomonads improve yield, quality and nutritional value of tomato: a field study. Mycorrhiza 27(1):1–11

Brain RA, Solomon KR, Brooks BW (2010) Targets, effects and risks in aquatic plants exposed to veterinary antibiotics. ACS Symp Ser 1018:169–189

Brazier-Hicks M, Evans KM, Cunningham OD, Hodgson DRW, Steel PG, Edwards R (2008) Catabolism of glutathione conjugates in Arabidopsis thaliana. J Biol Chem 283(30): 21102–21112

Caradonia F, Francia E, Morcia C, Ghizzoni R, Moulin L, Terzi V, Ronga D (2019) Arbuscular mycorrhizal fungi and plant growth promoting rhizobacteria avoid processing tomato leaf damage during chilling stress. Agronomy 9:1–22

Casida JE (2017) Organophosphorus xenobiotic toxicology. Annu Rev Pharmacol Toxicol 57(1): 309–327

Cheema SA, Khan MI, Tang X, Zhang C, Shen C, Malik Z, Ali S, Yang J, Shen K, Chen X (2008) Enhancement of phenanthrene and pyrene degradation in rhizosphere of tall fescue (*Festuca arundinacea*). J Hazard Mater 166(2–3):1226–1231

Cheema SA, Imran Khan M, Shen C, Tang X, Farooq M, Chen L, Zhang C, Chen Y (2010) Degradation of phenanthrene and pyrene in spiked soils by single and combined plants cultivation. J Hazard Mater 177(1–3):384–389

Chen M, Arato M, Borghi L, Nouri E, Reinhardt D (2018) Beneficial Services of Arbuscular Mycorrhizal Fungi—from ecology to application. Front Plant Sci 9:1–14

Cheng ZH, Yao F, Yuan-wang L, Hui-qing CH, Zhao-jun L, Jian-ming X (2017) Uptake and translocation of organic pollutants in plants: a review. J Integr Agric 16(8):1659–1668

Chouychai W, Thongkukiatkul A, Upatham S, Lee H, Pokethitiyook P, Kruatrachue M (2009) Plant-enhanced phenanthrene and pyrene biodegradation in acidic soil. J Environ Biol 30:139–144

Collins CD (2011) Organic xenobiotics and plants. Springer Netherlands, Dordrecht, p 307

Dean LL, Harjono I, Zylstra GJ, Häggblom MM (2001) Isolation and characterization of polycyclic aromatic hydrocarbon-degrading bacteria associated with the rhizosphere of salt marsh plants. Appl Environ Microbio 67 (6): 2683-2691.

Diagne N, Ngom M, Djighaly PI, Fall D, Hocher V, Svistoonoff S (2020) Roles of arbuscular mycorrhizal fungi on plant growth and performance: importance in biotic and abiotic stressed regulation. Diversity 12:1–25

Dighton J, Allen MF (1991) The ecology of Mycorrhizae. J Ecol 79(3):867
Du W, Sun Y, Cao L, Huang J, Ji R, Wang X, Wu J, Zhu J, Guo H (2011) Environmental fate of phenanthrene in lysimeter planted with wheat and rice in rotation. J Hazard Mater 188(1–3): 408–413
Dzantor EK, Chekol T, Vough LR (2000) Feasibility of using forage grasses and legumes for phytoremediation of organic pollutants. J Environ Sci Health 35(9):1645–1661
Ensley BD, Raskin I (1999) Rationale for use of phytoremediation. In: Phytoremediation of toxic metals: using plants to clean up the environment. Wiley, New York
Etim EE (2012) Phytoremediation and its mechanisms: a review. Int J Environ Bioenerg 2(3): 120–136
Fecih T, Baoune H (2019) Arbuscular mycorrhizal fungi remediation potential of organic and inorganic compounds. In: Microbial technology. pp 247–257
Gamalero E, Trotta A, Massa N, Copetta A, Martinotti MG, Berta G (2004) Impact of two fluorescent pseudomonads and an arbuscular mycorrhizal fungus on tomato plant growth, root architecture and P acquisition. Mycorrhiza 14(3):185–192
Ganugi P, Masoni A, Pietramellara G, Benedettelli S (2019) A review of studies from the last twenty years on plant–arbuscular mycorrhizal fungi associations and their uses for wheat crops. Agronomy 9(12):840
Gao Y, Cheng Z, Ling W, Huang J (2010) Arbuscular mycorrhizal fungal hyphae contribute to the uptake of polycyclic aromatic hydrocarbons by plant roots. Bioresour Technol 101(18): 6895–6901
Gao Y, Li Q, Ling W, Zhu X (2011) Arbuscular mycorrhizal phytoremediation of soils contaminated with phenanthrene and pyrene. J Hazard Mater 185(3):703–709
Genney DR, Alexander IJ, Killham K, Meharg AA (2004) Degradation of the polycyclic aromatic hydrocarbon (PAH) fluorene is retarded in a Scots pine Ectomycorrhizosphere. New Phytol 163(3):641–649
George NP, Ray JG (2023) The inevitability of arbuscular mycorrhiza for sustainability in organic agriculture-a critical review. Front Sustain Food Syst 7:1–23
Gogoi P, Singh RK (2011) Differential effect of some arbuscular mycorrhizal fungi on growth of *Piper longum* L. (Piperaceae). Indian J Sci Technol 4(2):119–125
Gosling P, Hodge A, Goodlass G, Bending GD (2006) Arbuscular mycorrhizal fungi and organic farming. Agric Ecosyst Environ 113:17–35
Gramss G, Rudeschko O (1998) Activities of oxidoreductase enzymes in tissue extracts and sterile root exudates of three crop plants, and some properties of the peroxidase component. New Phytol 138(3):401–409
Greipsson, S (2011) Phytoremediation. Nature Education Knowledge 3(10):7
Hasanuzzaman M, Gill SS, Fujita M (2013) Physiological role of nitric oxide in plants grown under adverse environmental conditions. In: Plant acclimation to environmental stress. pp 269–322
Houshani M, Salehi-Lisar SY (2020) Agronomic crop responses and tolerance to polycyclic aromatic hydrocarbon toxicity. In: Agronomic crops. pp 265–283
Houshani M, Tarigholizadeh S, Rajput VD, Jatav HS (2021) The degradation of organic and inorganic pollutants. In: Basic concepts in environmental biotechnology. pp 133–148
Houshani M, Salehi-Lisar SY, Motafakkerazad R, Movafeghi A (2019) Uptake and distribution of phenanthrene and pyrene in roots and shoots of maize (Zea mays L.). Environ Sci Pollut Res 1–7.
Jacob J, Cherian J (2013) Review of environmental and human exposure to persistent organic pollutants. Asian Soc Sci 9(11):107–120
Jansa J, Forczek ST, Rozmoš M, Püschel D, Bukovská P, Hršelová H (2019) Arbuscular mycorrhiza and soil organic nitrogen: network of players and interactions. Chem Biol Technol Agric 6:10
Joner EJ, Leyval C (2003) Rhizosphere Gradients of Polycyclic Aromatic Hydrocarbon (PAH) dissipation in two industrial soils and the impact of arbuscular mycorrhiza. Environ Sci Technol 37(11):2371–2375

Kafle A, Timilsina A, Gautam A, Adhikari K, Bhattarai A, Aryal N (2022) Phytoremediation: mechanisms, plant selection and enhancement by natural and synthetic agents. Environ Adv 8: 100203

Kang F, Chen D, Gao Y, Zhang Y (2010) Distribution of polycyclic aromatic hydrocarbons in subcellular root tissues of ryegrass (*Lolium multiflorum Lam*). BMC Plant Biol 10(1):210–215

Kaspary TE, Lamego FP, Cutti L, Aguiar ACM, Bellas C (2014) Determination of photosynthetic pigments in fleabane biotypes susceptible and resistant to the herbicide glyphosate. Planta Daninha 32:417–426

Kmentova E (2003) Response of plant to fluoranthene in environment. Ph.D. thesis, Masaryk University, Brno, Czech Republic, p 145

Knejzlik Z, Kas J, Ruml T (2000) Mechanismus vstupu xenobiotik do organismu a jejich detoxikace. Chem Listy 94:913–918

Korte F, Kvesitadze G, Ugrekhelidze D, Gordeziani M, Khatisashvili G, Buadze O, Zaalishvili G, Coulston F (2000) Organic toxicants and plants. Ecotoxicol Environ Saf 47:1–26

Kummerova M, Bartak M, Triska J, Zubrova E, Zezulka S (2006) Inhibitory effect of fluoranthene on photosynthetic processes in lichens detected by chlorophyll fluorescence. Ecotoxicology 15: 121–131

Kummerova M, Zezulka S, Vanova L, Fiserova H (2012) Effect of organic pollutant treatment on the growth of pea and maize seedlings. Cent Eur J Biol 7(1):159–166

Liu H, Weisman D, Ye Y, Cui B, Huang Y, Colón-Carmona A, Wang Z (2010) An oxidative stress response to polycyclic aromatic hydrocarbon exposure is rapid and complex in Arabidopsis thaliana. Plant Sci 176(3):375–382

Maila MP, Cloete TE (2002) Germination of Lepidium sativum as a method to evaluate polycyclic aromatic hydrocarbons (PAHs) removal from contaminated soil. Int Biodeterior Biodegradation 50(2):107–113

Miller RM, Jastrow JD (2000) Mycorrhizal fungi influence soil structure. In: Arbuscular mycorrhizas: physiology and function. pp 3–18

Nakmee PS, Techapinyawat S, Ngamprasit S (2016) Comparative potentials of native arbuscular mycorrhizal fungi to improve nutrient uptake and biomass of Sorghum bicolor Linn. Agric Nat Resour 50(3):173–178

Parihar P, Bora M (2018) Effect of mycorrhiza (*Glomus mosseae*) on morphological and biochemical properties of Ashwagandha (*Withania somnifera*) (L.) Dunal. J Appl Nat Sci 10(4): 1115–1123

Parrish ZD, Banks MK, Schwab AP (2005) Assessment of contaminant ability during phytoremediation of polycyclic aromatic hydrocarbon impacted soil. Environ Pollut 137(2): 187–197

Posta K, Hong Duc N (2020) Benefits of arbuscular mycorrhizal fungi application to crop production under water scarcity. In: Drought—detection and solutions. pp 1–26

Ren CG, Kong CC, Bian B, Liu W, Li Y, Luo YM, Xie ZH (2017) Enhanced phytoremediation of soils contaminated with PAHs by arbuscular mycorrhiza and rhizobium. Int J Phytoremediation 19(9):789–797

Reynoso-Cuevas L, Gallegos-Martinez M, Cruz-Sosa F, Gutierrez-Rojas M (2008) In vitro evaluation of germination and growth of five plant species on medium supplemented with hydrocarbons associated with contaminated soils. Bioresour Technol 99(14):6379–6385

Salehi-Lisar SY, Bakhshayeshan-Agdam H (2016) Drought stress in plants: causes, consequences, and tolerance. In: Drought stress tolerance in plants, vol 1. pp 1–16

Salehi-Lisar SY, Deljoo S (2015) Physiological effect of phenanthrene on *Triticum aestivum*, *Helianthus annus* and *Medicago sativa*. Eur J BioSci 29–37

Selosse MA, Strullu-Derrien C, Martin FM, Kamoun S, Kenrick P (2015) Plants, fungi and oomycetes: a 400-million years affair that shapes the biosphere. New Phytol 206:501–506

Siddiqui ZA, Akhtar MS, Futai K (eds) (2008) Mycorrhizae: sustainable agriculture and forestry. Springer, Dordrecht

Singh OV, Jain RK (2003) Phytoremediation of toxic aromatic pollutants from soil. Appl Microbiol Biotechnol 63(2):128–135
Singh H, Verma A, Kumar M, Sharma R, Gupta R, Kaur M, Negi M, Sharma SK (2017) Phytoremediation: a green technology to clean up the sites with low and moderate level of heavy metals. Austin Biochem 2(2):1012
Smith S, Read D (2008) Mycorrhizal symbiosis. In: Mycorrhizal symbiosis
Smith SE, Smith FA (2011) Roles of arbuscular mycorrhizas in plant nutrition and growth: new paradigms from cellular to ecosystem scales. Annu Rev Plant Biol 62:227–250
Song Z, Bi Y, Zhang J, Gong Y, Yang H (2020) Arbuscular mycorrhizal fungi promote the growth of plants in the mining associated clay. Sci Rep 10(1):2663
Syed K, Doddapaneni H, Subramanian V, Lam YW, Yadav JS (2010) Genome-to-function characterization of novel fungal P450 monooxygenases oxidizing polycyclic aromatic hydrocarbons (PAHs). Biochem Biophys Res Commun 399(4):492–497
Tarigholizadeh S, Motafakkerazad R, Salehi-Lisar SY, Mohagel-Kazmi E (2021) High resistance of *Panicum miliaceum* L. to phenanthrene toxicity based on growth response and antioxidant system assessment. Acta Agric Slov 117(2):1–13
Teng Y, Luo Y, Sun X, Tu C, Xu L, Liu W, Li Z, Christie P (2010) Influence of arbuscular mycorrhiza and rhizobium on phytoremediation by alfalfa of an agricultural soil contaminated with weathered PCBs: a field study. Int J Phytoremediation 12(5):516–533
Testiati E, Parinet J, Massiani C, Laffont-Schwob I, Rabier J, Pfeifer HR, Lenoble V, Masotti V, Prudent P (2013) Trace metal and metalloid contamination levels in soils and in two native plant species of a former industrial site: evaluation of the phytostabilization potential. J Hazard Mater 249:131–141
Tomar RS, Jajoo A (2014) Fluoranthene, a polycyclic aromatic hydrocarbon, inhibits light as well as dark reactions of photosynthesis in wheat (*Triticum aestivum*). Ecotoxicol Environ Saf 109: 110–115
Trapp S, McFarlane JC (1995) Plant contamination, modeling and simulation of organic processes. Lewis, Tokyo, p 254
Vane CH, Kim AW, Beriro DJ, Cave MR, Knights K, Moss-Hayes V, Nathanail PC (2014) Polycyclic aromatic hydrocarbons (PAH) and polychlorinated biphenyls (PCB) in urban soils of Greater London, UK. Appl Geochem 51:303–314
Vanova L (2009) The use of in vitro cultures for effect assessment of persistent organic pollutants on plants. PhD thesis, Faculty of Science, Masaryk University, Czech Republic
Wang F, Adams CA, Yang W, Sun Y, Shi Z (2019) Benefits of arbuscular mycorrhizal fungi in reducing organic contaminant residues in crops: implications for cleaner agricultural production. Crit Rev Environ Sci Technol 50(15):1580–1612
Weber RWS (2005) Mycorrhizas: anatomy and cell biology. Mycology 19(3):133
Whitacre DM (ed) (2015) Reviews of environmental contamination and toxicology. Springer
Yan J, Wang L, Fu PP, Yu H (2004) Photomutagenicity of 16 polycyclic aromatic hydrocarbons from the US EPA priority pollutant list. Mutat Res Genet Toxicol Environ Mutagen 557(1): 99–108
Yoshiyama L, Seal CE, Kranner I, Odjaova M (2013) A central role for thiols in plant tolerance to abiotic stress. Int J Mol Sci 14:7405–7432
Zeb BS, Hayat MT, Zeb T, Khan FY, Abbasi HZ, Nawaz I, Ebadi A (2022) Uptake of organic pollutants and the effects on plants. In: Sustainable plant nutrition under contaminated environments. pp 209–234
Zhu X, Song F, Liu F (2017) Arbuscular mycorrhizal fungi and tolerance of temperature stress in plants. In: Arbuscular mycorrhizas and stress tolerance of plants. pp 163–194
Zobiole LH, Kremer RJ, Oliveira RSD, Constantin J (2009) Glyphosate effects on photosynthesis, nutrient accumulation, and nodulation in glyphosate-resistant soybean. J Wuhan Univ Technol 175:319–330

Chapter 9
Arbuscular Mycorrhizal Fungi (AMF)-Mediated Control of Foliar Fungal Diseases

Priyakshi Kashyap, Indrani Sharma, Sampurna Kashyap, and Niraj Agarwala

Abstract Plants witness a variety of disease incidences throughout their life, ultimately resulting in reduced plant growth and productivity. Climate change or human interventions have aggravated the incidences of various plant diseases, among which foliar fungal diseases are serious threats. Arbuscular mycorrhizal fungi (AMF) are a mutualistic group of organisms that play a significant role in enhancing plant growth and resilience under varied environmental circumstances. Moreover, it is well established that AMF confers tolerance against several foliar fungal diseases. This chapter highlights how fungal foliar diseases affect plant health and the various roles of AMF in providing resistance to different crop plants. In addition, AMF-mediated alterations in the root system architecture (RSA), modulation of reactive oxygen species (ROS), and reinforcement of thc physical barrier that prevents pathogen invasion and establishment have been discussed in detail. Furthermore, the intricate cross talk between AMF and phytohormones or plant metabolites has also been explored. Overall, harnessing the potential of AMF in imparting tolerance against foliar fungal diseases might reduce the reliance on chemical fungicides, thereby introducing an environment-friendly approach for plant protection.

Keywords Arbuscular mycorrhizal fungi · Foliar fungal diseases · Plant growth-promoting rhizobacteria · Defense responses · Root exudation · Rhizosphere niche

---

P. Kashyap
Chaiduar College, Gohpur, Gohpur, Assam, India

I. Sharma · S. Kashyap · N. Agarwala (✉)
Department of Botany, Gauhati University, Guwahati, Assam, India
e-mail: niraj_botany@gauhati.ac.in

G. J. Ahammed, R. Hajiboland (eds.), *Arbuscular Mycorrhizal Fungi and Higher Plants*, https://doi.org/10.1007/978-981-99-8220-2_9

9.1 Introduction

The outbreak of plant pathogens in newer areas and their extension to multiple plant hosts along with the emergence of new pathogenic strains has caused a major threat to global food security and sustainable crop production. Alterations in weather patterns of a geographical location might create an environment more conducive for insects, pests, or pathogens (Pawlowski and Hartman 2020), which directly affect plant life and ultimately cause a threat to mankind. Among various plant diseases, fungal diseases have caused major crop loss around the world. For example, in the 1840s, potato blight disease caused by the fungus *Phytophthora infestans*, had created havoc in Ireland, where millions of people lost their lives and livelihoods (Badri and Vivanco 2009). Foliar fungal diseases are one of the key constraints affecting the productivity of crop plants. Blight, wilt, powdery mildew, anthracnose, leaf spot, etc. are some of the well-known foliar fungal diseases, which infect majority of the crop plants causing a heavy loss in crop yield. With the rise in plant disease incidences, agronomists have searched for several alternative ways to achieve disease-free sustainable agriculture. Among various plant disease management practices, the use of AMF, a group of obligate fungal symbionts, in the agricultural sector has gained momentum in recent times. AMF belonging to the group Glomeromycotina are the most comprehensive plant root symbionts, which supply soil nutrients like phosphorous (P) and nitrogen (N) to the plant roots via extra- and intra-radical mycorrhizal hyphae in exchange for lipid and sugar molecules (Marro et al. 2022).

AMF are protective symbionts that help the host plants to survive well in any diseased condition by providing the plants with defensive benefits and thereby shaping balanced ecological dynamics (Eck et al. 2022). About 80% of terrestrial plants live in direct association with different groups of AM fungi, which determines the overall health and fitness of the host plant (van der Heijden et al. 2015; Weng et al. 2022). Mutualistic interactions among the host plants and mycorrhizal partner determine the belowground ecosystem as well as regulate nutrient cycling (Sharma et al. 2023). However, being ubiquitous in nature, the potential of AMF in governing the fate of a host plant undergoing pathogen infection under natural conditions is not yet fully understood.

The distribution, abundance, and diversity of plant species in a natural ecosystem are influenced by the negative effect of any biotic or abiotic stress. Biotic stress-like disease conditions lead to the alteration in the normal physio-biochemical process of the plants, ultimately hampering growth and development. AMF provides nutritional and defensive support to the associated plant and also improves plant primary and secondary metabolism and the mobilization of less mobile elements, thus enhancing plant protection from pathogens (Delavaux et al. 2017; Kaur and Suseela 2020; Dey and Ghosh 2022). AMF association provides the host plant with an augmented state of defense, called defense priming, where they regulate the expression of host defense genes, hence enhancing the efficiency of plant defense mechanisms (Goddard et al. 2021; Eck et al. 2022). Besides, enhancing the plant survivability under any pathogen attack, AMF also help the plants to tolerate abiotic stresses like drought, high temperature, salinity, metal toxicity, etc. (Begum et al. 2019). The rational use of AMF with other microorganisms maximizes plant tolerance to diseases by competing with harmful pathogens (Devi et al. 2022). Studies have also revealed that the combined application of AMF and other beneficial plant microorganisms like plant growth-promoting rhizobacteria (PGPRs) impacts plant root exudation, thereby altering the rhizosphere niche (Ji et al. 2022). Furthermore, deciphering the far-reaching changes in plant metabolome, microbiome, and molecular reprogramming under the influence of AMF in diseased and non-diseased conditions will provide better outcomes in AMF-mediated plant disease management.

This chapter highlights the role of AMF in controlling foliar fungal diseases and the mechanisms associated with AMF-mediated disease resistance. The chapter also aims to describe the benefits of AMF in plant adaptation and provides a glimpse of how AMF along with other plant-beneficial microbes enhances plant overall health.

9.2 The Impact of Foliar Diseases on Plants

The changes in climatic conditions have a significant effect on the occurrence and progression of crop plant diseases as well as the distribution of disease-causing pathogens. Additionally, the changes in agricultural practices and the use of high-yielding crop plants may have a profound role in shifting both the pathogen diversity and the threat they possess (Launay et al. 2014). For instance, survey data on the severity index of *Phytophthora cinnamomi*, a fungal pathogen affecting oak trees in relation to winter temperature (Marĉais et al. 1996), reveals winter survival of *P. cinnamomi* plays a crucial role in limiting canker disease in oaks (Bergot et al. 2004).

Plant-microbe interactions undoubtedly brought various changes in plant adaptation to challenging environmental circumstances. These interactions might negatively impact the plants, causing several diseases. Foliar diseases have significant effects on plant morphology and physiology where the leaf architecture, color, and the process of photosynthesis get negatively afflicted. They often manifest in visible signs such as leaf spots, wilting, and discoloration, which help in identifying specific foliar diseases like late blight of potato, Fusarium or Verticillium wilt, anthracnose, and rust in wheat (Paliwal and Joshi 2022). Fungal pathogens are the primary cause of the majority of foliar diseases, with around 8000 fungal species known to cause plant diseases (Agrios 2009). Wilt diseases, observed in crops like eggplant (Bletsos et al. 2003), melon (Ros et al. 2005), cotton (Cai et al. 2009), tomato (Bidellaoui et al. 2019), and others, are primarily caused by the fungi belonging to *Fusarium* and *Verticillium* spp. Similarly, rice plants are susceptible to blast disease caused by *Magnaporthe oryzae*, which leads to the formation of elliptical reddish spots on the leaf sheath (Younas et al. 2023).

Fungal infection in the plant tissues begins with spore germination followed by germ tube elongation and mycelial penetration inside the host tissues. The mycelial growth phase is considered as the determining phase in triggering fungal crop disease epidemics (Magarey et al. 2007); it is also the phase that is most sensitive to variations in climatic conditions. Subsequent epidemics take place in suitable weather conditions where temperature and water availability play an important role in determining the intensity of infections (Launay et al. 2014). Plant pathogenic fungi utilize the natural openings, injured cells, and tissues to enter the host plants using varied mechanical and chemical methods (Knogge 1998). Based on these strategies, fungal pathogens are classified as biotrophs, hemibiotrophs, and necrotrophs, where the biotrophs directly feed on the living host tissues using a specialized structure known as haustoria; necrotrophs on the other hand, rely on the dead tissues and infected areas of the host plant. On the contrary, hemibiotrophic pathogens can eventually switch from a biotrophic to a necrotrophic mode (Stotz et al. 2014). For example, *Puccinia graminis*, which causes rust on wheat, is a biotrophic fungus, while *Botrytis cinerea*, causing mold disease in various plant species, is a necrotrophic fungus (De Silva et al. 2017), and *P. infestans*, causing potato late blight, is an example of hemibiotrophic fungus (Stotz et al. 2014).

Meanwhile, plants are found to evolve a broad range of direct and indirect resistance mechanisms against varied foliar pathogens. Infection by pathogen triggers several signaling pathways like extracellular-signal-regulated kinase (ERK) and mitogen-activated protein kinases (MAPKs) which leads to hypersensitive response (HR) and prevents the spreading of disease (Hammond-Kosack and Rudd 2008). Plants utilize effector-triggered defense (ETD) where receptor-like proteins (RLPs) engage receptor-like kinase SOBIR1 against the response of pathogenic fungi. The host cell death is stimulated typically by ETD after a period of endophytic pathogen growth (Stotz et al. 2014). On the contrary, plant response toward pathogenic fungi also leads to several changes in the metabolic profile of the host plants. Xiao et al. (2022) reported the enhancement of calcium signaling pathway-related genes such as *CMLs*, *CNGCs*, *CPKs*, and *RBOHDs* in strawberry resistant cultivar against *B. cinerea*. Furthermore, a network involving six different phytohormones, especially auxin (IAA) and jasmonic acid (JA) and genes associated with amino acid and phenylpropanoid biosynthesis, contributed to the strawberry flower's resistance to *B. cenerea*. Similarly, ascorbic acid, citric acid, glucose, galactose, total phenolics, and flavonoids were found to be positively correlated with *B. cinerea* resistance (Zhao et al. 2022). Furthermore, plant-released chemical signaling molecules play a pivotal role in the recruitment of microbial symbionts which promotes plant growth and development under stressed circumstances (Sharma et al. 2023). AMF make close associations with most of the terrestrial plants which facilitates the bidirectional exchange of nutrients, i.e., carbon-based photosynthates and soil minerals, between plants and AMF. These symbiotic associations play an important role in plant ecosystem functioning (Sharma et al. 2017).

9.3 Benefits of AMF to Plants

Symbiotic relationships between AMF and the plants depend upon varied environmental factors where the soil characteristics and the host plant defense also play an important role (Poveda et al. 2022). Plant roots colonized by AMF are known to influence host plant morphology, physiology, and expression patterns of several host genes associated with the terpenoid biosynthesis pathways. AMF-induced changes in the level of secondary metabolites, phytohormones, proteins, and external hyphal networks have a profound effect on plant defense strategy (Sharma et al. 2017; Weng et al. 2022). AMF-colonized plants show an increase in shoot and root length and also the biomass of the host plant (Han et al. 2023). Furthermore, soil chemical and physical properties along with rhizospheric microbial communities are greatly influenced by the roots colonized by AMF, and such changes play a significant role in alleviating numerous plant abiotic and biotic stresses (Miozzi et al. 2019; Weng et al. 2022). AM fungi show diverse functional activities ranging from the mobilization of soil nutrients, fixation of atmospheric nitrogen, and production of phytohormones, siderophores, and antibiotics. AMF-colonized durum wheat cultivars, Saragolla and Odisseo, showed an enhancement in the abundance of Bacteroidetes and Actinobacteria in the host plants (Agnolucci et al. 2019); particularly, AM fungi *Funneliformis mosseae* synergistically regulated the abundance of *Rhodococcus* sp. in Saragolla cultivar, while an increase in *Microbacterium* and *Streptomyces* spp. was observed in both the durum wheat cultivars (Ujvári et al. 2021). Similarly, Han et al. (2023) reported a shift in endophytic microbial diversity in AMF-colonized host plant (Lettuce) where *Sphingomonadales*, *Betaproteobacteria*, and *Burkholderiales* showed a positive correlation with AMF colonization.

Besides nutrition, mycorrhizal association provides several other benefits to the plants. AMF is reported to act as a biofertilizer which aids in the uptake of nutrients and thus increases plant growth. AMF association alters the physical and chemical environment of the plant rhizosphere, thereby modulating the rhizospheric microbiota (Dowarah et al. 2022; Sharma et al. 2023). Changes in rhizospheric microflora might synergistically affect the plant-beneficial microorganisms, thus inhibiting the outbreak of pathogenic strains (Xavier and Boyetchko 2004). Maize plants inoculated with AM fungus *Claroideoglomus etunicatum* altered the bacterial and fungal communities of rhizospheric soil (Hao et al. 2022). The symbiosis between AMF and *Artemisia annua* roots modulates the use of various carbon sources by rhizomicrobiota affecting the survivability and functionality of rhizospheric microbes, which in return improves nutrient acquisition by *A. annua* (Ma et al. 2021). Furthermore, AMF-induced changes in root exudation directly impact the rhizospheric microorganisms. The enzymatic activities of rhizospheric soil as well as the abundance of the bacterial community were found to be increased in the maize rhizosphere when the plants were in association with AMF. However, the AMF association negatively regulated the rhizospheric bacterial diversity, but a reduction in soil-borne diseases of maize was observed (Ma et al. 2022). Hence,

AMF-mediated alteration in the rhizospheric microbial communities and root exudation profiles of host plant confers tolerance to various bacterial and fungal diseases.

Genes for phosphate transporter (*LePT3*, *LePT4*, *LePT5*) in tomato plants were found to be upregulated when the roots were colonized by AM fungi (Nagy et al. 2005). AM fungi form a network of fungal hyphae which helps in enhancing the water-holding capacity (Songachan 2023). Mycorrhizal plants can combat water stress conditions as the water-absorbing surface is increased by the fungal hyphal network which helps in increased absorption of water even from the soil with low water potential (Augé et al. 2001). AMF in the soil increase the production of glomalin or glomalin-related soil protein (GRSP), which helps in soil binding and thus aids in overcoming erosion of soil and water infiltration (Sharma et al. 2017). AMF can reduce heavy metal toxicity in the host plant and the soil by increasing the expression of metal transporters (Ahammed et al. 2023). These transporters help AM fungi to take up heavy metals from the soil and thus lead to environmental cleanup (Göhre and Paszkowski 2006). A Zn transporter, *GintZnT1*, is reported to be upregulated in *Glomus intraradices* when exposed to high Zn concentration (González-Guerrero et al. 2005). As per evidence put forward by Lazcano et al. (2014), it is also found that tomato plants with mycorrhizal association have the potential to reduce N_2O emissions during nitrogen fixation, which is a potent greenhouse gas. Further research involving the use of AMF and selected rhizospheric microbes along with other non-microbial substrates in field conditions might improve plant resistance to various diseases.

AMF-associated plants also acquire additional benefits in developing resilience capacity towards abiotic and biotic stresses (Delavaux et al. 2017). Meanwhile, AMF associations and their benefits vary among species, taxonomic groups, and types of stresses encountered by the host plant. For example, a global meta-analysis report on the effects of AMF species and taxonomic groups on stressed and non-stressed plants reveals positive effect sizes of *F. mosseae*, *Rhizophagus intraradices*, and *Diversispora versiformis* under drought-stressed conditions; *Septoglomus deserticola*, etc. under heavy metal stress; *D. versiformis*, *R. fasciculatus*, etc. under salinity stress; and *Gigaspora margarita* and *R. fasciculatus* under biotic stress conditions of the host plant species (Marro et al. 2022). In a study conducted by Begum et al. (2021), inoculation of AMF was reported to significantly increase the tolerance of tobacco plants against drought stress. It also leads to the increased production of antioxidants which protects the plant from oxidative damage during drought stress. Similarly, Orine et al. (2022), through an experimental study, revealed the importance of plant metabolites such as rutin and alpha-tomatine, accumulated in AMF-associated tomato plant leaves. The accumulated metabolites play a key role in the host plant contributing toward more plasticity and greater phenotypical space in terms of growth and defense against drought-stressed conditions. Under salt stress, AMF association induces increased expression of the chloroplast gene which as a result increases the photosynthesis efficiency in the host plant (Chandrasekaran et al. 2019). Simultaneously, mycorrhizal fungi associated with the host plants act as a protective shield, preventing the host from damage caused by plant pathogens. Considerably, endless studies reported the beneficial role of AMF association in reducing plant diseases caused by pathogenic bacteria, nematodes, and fungi under varied environmental circumstances (Weng et al. 2022). For example, AMF, *G. mossae* inoculation in the soybean plant results in the suppression of *Pseudomonas syringae* infection (Miransari 2011). Mycorrhizal fungi reduce varied nematocidal diseases of soybeans, cucumbers, cotton, alfalfa, peach, tomatoes, kidney beans, and citrus to various extents by parasitizing the cysts of nematodes (de Sá and Campos 2020; Rodrigues e Silva et al. 2021). Whether symbiotic or pathogenic interactions, various fungi have developed diverse strategies to infect the host plants, where pathogen infection leads to plant death in most cases. Phytopathogenic fungi are the most dominant group of organisms causing ecological and economic threats to the crop plants (Doehlemann et al. 2017). For example, Ceylon (Sri Lanka) was one of the great coffee producers until the 1870s. Coffee rust caused by *Hemileia vastatrix* in Ceylon during 1875 declined its production by 95% by completely destroying the coffee plantations (Avelino et al. 2015). Thus, fungal diseases pose a major risk to world crop production, and formulation of immediate effective measures seeks concentration from the scientific community. Here, in this chapter, we are going to discuss various strategies employed by AMF in controlling foliar diseases caused by pathogenic fungi which might cause serious threats to crop plantations.

9.4 Role of AMF in Controlling Fungal Foliar Diseases

AMF forms hyphal interactions with the roots of most agriculturally important crop plants. This interaction promotes great effectiveness in the uptake of soil nutrients confining low mobility like P, resulting in an enhancement in the crop nutrient availability and growth. Due to their ability in promoting health benefits to the associated crop plants, they possess potential economic concerns in agricultural fields (Abarca et al. 2023). AMF-associated plants show significant resistance toward biotic and abiotic stress factors. This protective nature of AMF toward environmental factors might be due to varied mechanisms associated with an increment in nutrient acquisition (Mustafa et al. 2016) and induction of systemic acquired resistance (SAR) (Jung et al. 2012) in the host plant. Foliar fungal diseases develop a wide range of symptoms that differ to a great extent depending on the pathogen type and the host plant. Some of the major fungal foliar diseases and their control measures using AMF are discussed below and enlisted in Table 9.1.

9.4.1 Blight

The symptomatic occurrence of yellowish to brownish spots ranging in diameter from 2 to 30 mm on the leaves results in blight, which later on spreads over a large

Table 9.1 List of fungal foliar diseases which can be controlled by various species of AMF

Sl no.	Disease name	Causal fungus	Host plant	Effective AMF species for disease control	References
1	Early blight	*Alternaria solani*	Tomato (*Solanum lycopersicum* L.)	*G. intraradices*	Fritz et al. (2006)
				G. versiforme	Song et al. (2011b)
				G. clarum	Joseph and Olumide (2022)
2	*Fusarium* blight	*Fusarium vasinfectum*	Cotton (*Gossypium herbaceum* L.)	*G. mosseae*	Zhengjia and Xiangdong (1991)
3	Phytophthora blight	*Phytophthora capsici*	Peeper (*Capsicum annuum* L.)	*G. intraradices*	Hu-zhe et al. (2005)
			Pepper (*Capsicum annuum* L. cv. Charliston Bagci)	*G. fasciculatum*, *G. etunicatum*, *G. mosseae*, and *Gigaspora margarita*	Ozgonen and Erkilic (2007)
4	Sheath blight	*Rhizoctonia solani*	Rice (*Oryza sativa var. japonica* cv. Koshihikari)	*Ceratobasidium* spp.	Mosquera-Espinosa et al. (2013)
5	Leaf blight	*Cylindrocladium quinqueseptatum*	Gum tree (*Eucalyptus* sp.)	*Gigaspora margarita* KKU-BH-01	Klinsukon et al. (2021)
6	*Ascochyta* blight	*Ascochyta rabiei*	Chickpea (*Cicer arietinum* var. *Bivanij*)	*R. irregularis*, *G. versiform*, *F. mosseae*, *G. fasciculatum*, and *Gigaspora margarita*	Moarrefzadeh et al. (2022)
			Chickpea (*Cicer arietinum* var. *Saral*)	*R. irregularis*, *G. versiform*, *F. mosseae*, and *Gigaspora margarita*	
7	Fusarium wilt	*Fusarium oxysporum* f. sp. *cucumerinum*	Cucumber (*Cucumis sativus L.* cv. Jinlv 3)	*G. etunicatum* BEG168	Hao et al. (2005)
		Fusarium oxysporum	Persian violet (*Cyclamen persicum* Mill., cv. Pastel)	*G. fasciculatum*	Maya and Matsubara (2013)

		Fusarium oxysporum f. sp. niveum	Watermelon {*Citrullus lanatus* (Trunb.) Matsum and Nakai cv. Zaojia 84–24}	G. mosseae	Ren et al. (2015)
		Fusarium oxysporum	Tomato (*Lycopersicum esculentum* var. PKM-1)	*Glomus* sp.	Kumari and Prabina (2019)
8	Powdery mildew	*Sphaerotheca macularis*	Strawberry (*Fragaria vesca* cv. Elvira)	*G. caledonium*, *G. fasiculatum*, and *G. mossae*	Lowe et al. (2012)
		Blumeria graminis f. sp. *hordei*	Barley (*Hordeum vulgare*)	*Piriformospora indica*	Molitor et al. (2011)
		Blumeria graminis f. sp. *tritici*	Wheat (*Triticum aestivum* cv. Orvantis and *T. aestivum* cv. Lord)	*R. irregularis* and *R. irregularis* + *G. mosseae*	Mustafa et al. (2013)
			Wheat (*Triticum aestivum* cv. Orvantis)	*F. mosseae*	Mustafa et al. (2014)
			Wheat (*Triticum aestivum* cv. Orvantis and *T. aestivum* cv. Lord)	*F. mosseae*, Solrize®, and *R. irregularis*	Mustafa et al. (2016)
		Podosphaera leucotricha	Apple (*Malus domestica*)	*Glomus* sp.	Yousefi et al. (2018)
9	Anthracnose	*Colletotrichum gloeosporioides*	Persian violet (*Cyclamen persicum* Mill., cv. Pastel)	*G. fasciculatum*	Maya and Matsubara (2013)
		Colletotrichum camelliae	Tea {*Camellia sinensis* (L.) Kuntze var. *Pingyangtezao*}	*R. intraradices* BGC JX04	Chen et al. (2023b)
10	Rust	*Puccinia graminis* Pers. f. sp. *tritici*	Wheat (*Triticum aestivum* L.)	AMF, AMF + *T. harzianum* HL1 and AMF + *T. viride* HL5	El-Sharkawy et al. (2018)
11	Leaf spot	*Bipolaris sorokiniana*	Perennial ryegrass (*Lolium perenne*)	AMF, AMF+ *Epichloe*	Guo et al. (2019)
		Phoma medicaginis	Alfalfa (*Medicago sativa*)	*R. intraradices*	Li et al. (2021)
12	Black Sigatoka	*Mycosphaerella fijiensis*	Banana (*Musa* spp.)	*R. irregularis* MUCL 41833	Anda et al. (2020)
13	Brown spot	*Bipolaris oryzae*	Rice (*Oryza sativa*)	AMF+ *T. viride*; AMF+ *P. fluorescens* + *T. viride*	Saleh and El-Akshar (2020)

(continued)

Table 9.1 (continued)

Sl no.	Disease name	Causal fungus	Host plant	Effective AMF species for disease control	References
		Physoderma maydis	Maize (*Zea mays* L. var. *CMS9015* and *Z. mays* L. var. CMS8704)	*Glomus* sp. and *Gigaspora* sp.	Adamou et al. (2023)

portion of the foliage (Klinsukon et al. 2021). A study carried out by Zhengjia and Xiangdong (1991), reveal the importance of vesicular arbuscular mycorrhizal (VAM) fungi *G. mosseae* against Fusarium blight caused by *F. vasinfectum*. Cotton plants inoculated with *G. mosseae* result in a significant decrease in the severity of disease symptoms like vascular discoloration caused by *F. vasinfectum*. Hu-zhe et al. (2005) reported that microbial infection can reduce the incidence of blight disease in pepper plants caused by *P. capsici*. Pepper plants infected by *P. capsici* along with the mycorrhizal fungus *G. intraradices* show reduced mortality and infection in comparison to those plants which were infected by the pathogen alone. Similarly, AM fungi *G. fasciculatum*, *G. etunicatum*, *G. mosseae*, and *G. margarita* showed enhancement in the growth and development as well as reduction in the blight of pepper plant caused by *P. capsici* (Ozgonen and Erkilic 2007). Tomato plants infected by the fungus *Alternaria solani* cause early blight, which results in damping off of seedlings, collar rot, and destruction of leaves and fruits (Rotem 1994). *G. intraradices* colonization significantly reduces the symptoms caused by *A. solani* in tomato plants when compared with non-mycorrhizal plants (Fritz et al. 2006). Potato plants inoculated with *Glomus* sp. induce two pathogenesis-related genes, *PR1* and *PR2* in the leaves of the host plant. The expression of *PR* genes ultimately shows systemic resistance against the pathogenic fungus *P. infestans* causing potato late blight disease (Gallou et al. 2011). In a similar study conducted by Song et al. (2011b) on the effect of AMF, *G. versiforme* upon tomato plants infected by *A. solani*, it was revealed that *G. versiforme* inoculation showed an enhancement in the expression of three defense-related genes, namely, *PR1*, *PR2*, and *PR3*, in the leaves of the host plant. Mosquera-Espinosa et al. (2013), in their study, depicted the role of mycorrhizal fungi *Ceratobasidium* spp., which were isolated from orchid and evaluated for their ability to control the disease severity caused by *Rhizoctonia solani* in rice plants. A significant reduction in the symptoms of sheath blight disease was observed in pre-inoculated host plants compared to the control. AM fungus *G. margarita* KKU-BH-01, in association with the roots of eucalyptus plants, was reported to reduce leaf blight disease caused by *Cylindrocladium quinqueseptatum*. Moreover, plants pre-inoculated with AMF, prior to pathogen infection, showed 33.4% reduction in the disease (Klinsukon et al. 2021). Moarrefzadeh et al. (2022) successfully inoculated two varieties of chickpea, Saral and Bivanij, with five strains of AM fungi *R. irregularis*, *G. versiform*, *F. mosseae*, *G. fasciculatum*, and *G. margarita* and observed their disease control ability against *Ascochyta* blight caused by *Ascochyta rabiei*. All four strains of AMF except *G. fasciculatum* in the Saral variety significantly reduced the disease extremity. Furthermore, *R. irregularis* and *G. versiform* showed the highest disease suppression capacity, i.e., 46.15% and 42.30%, respectively, on the variety Saral and *G. fasciculatum* by 40% on the variety Bivanij. In another recent experiment with tomato plants, AM fungi *G. clarum* was identified as having an important role in reducing the incidence of early blight disease caused by *A. solani* (Joseph and Olumide 2022).

9.4.2 Wilt

Wilt is caused by soil-borne pathogenic fungi that infect the host plant via roots, ultimately clogging the vascular system and resulting in yellowing, wilting, and death of the infected plants (Noman et al. 2023). *F. oxysporum* and *V. dahliae* are found to cause the majority of wilt diseases in a variety of plant species (Usharani et al. 2022). AM fungus *G. etunicatum* BEG168 inoculation on cucumber seeds at the germination stage showed resistance toward *F. oxysporum* f. sp. *cucumerinum*. Mycorrhizal colonization increases the production of plant secondary metabolites (Hao et al. 2005), which might result in plant tolerance toward Fusarium wilt disease. AMF-associated cyclamen plants exhibited higher antioxidant contents and enhanced plant defense against Fusarium wilt, thereby reducing disease incidences (Maya and Matsubara 2013). Ren et al. (2015), in an experimental study, revealed the importance of *p*-coumaric acid and malic acid secreted in the rhizospheric region of the host plant watermelon (*Citrullus lanatus*), co-inoculated with AMF *G. mosseae* and pathogenic fungus *F. oxysporum* f. sp. *niveum* (FON). The alteration in the composition of the exudate resulted in the stimulation of defense signaling molecules in the host plant ultimately enhancing plant disease tolerance capacity. A similar study was conducted in in vitro conditions, using tomato as a host plant and AM fungus *Glomus* sp. as an antagonist against *F. oxysporum*. The study depicted the potential role of *Glomus* sp. to control the pathogen. Furthermore, the secretion of various exudates around the roots has resulted in a decrease in disease incidences, thereby improving the plant growth. (Kumari and Prabina 2019).

9.4.3 Powdery Mildew

Symptoms of powdery mildew are easily recognized by the presence of a white powder-like substance on the leaf surface of the host plants. Host plants may undergo leaf chlorosis and an overall reduction in plant growth and development due to powdery mildew. Lowe et al. (2012) have reported that AM fungi (*G. caledonium*, *G. fasiculatum*, and *G. mossae*) in combination with chitosan spray significantly reduces the effect of powdery mildew, resulting in higher fruit yield and runner production in strawberry plants. Barley roots colonized with *Piriformospora indica*, a basidiomycetous fungus, upregulate the expression of heat-shock proteins (*Hsp70*) and defense genes *PR5*, *PR2*, and *PR1* against *Blumeria graminis* f. sp. *hordei*, a leaf pathogen causing powdery mildew. The pathogen results in the induction of a distinct set of genes, LRR receptor kinases, and WRKY transcription factors after 12 h of postinoculation (hpi) and vesicle-localized gene products after 24 hpi, which were altered by the root-associated fungus *P. indica*. *P. indica* results shift in the metabolic profile of the host plant, making it unsuitable for pathogen colonization (Molitor et al. 2011). Wheat plants inoculated with AM fungi *R. irregularis* singly and in combination with *G. mosseae* showed resistance of 25% and 43%, respectively, toward a fungal pathogen *B. graminis* (Mustafa et al. 2013). In a similar study conducted by Mustafa et al. (2014), AM fungus *F. mosseae* was reported to decrease powdery mildew infection by 73% in wheat plants caused by *B. graminis* f. sp. *tritici*. Reduction in the disease symptoms results in the accumulation of defense enzymes such as lipoxygenase and peroxidase and induces plant systemic acquired resistance (SAR). Additionally, on testing three different AM fungi, viz., *F. mosseae*, and Solrize®, two commercially available mycorrhizal inocula and *R. irregularis*, a laboratory sample, against the powdery mildew pathogen *B. graminis* f. sp. *tritici* in wheat plants, it was found that all the three inocula exhibit biocontrol activity against *B. graminis* with different extents, i.e., 74%, 58%, and 34%, respectively. Hence, from the study, it can be concluded that the extent of protection against *B. graminis* is determined by the type of AM fungi involved in the mycorrhizal association (Mustafa et al. 2016). In an experiment carried out by Yousefi et al. (2018), apple seedlings treated with powdery mildew-causing fungal pathogen *Podosphaera leucotricha* were given different treatments with two fungicides (Flint and Stroby) and AM fungus *Glomus* sp. After 6 weeks, seedlings in association with AM fungi are reported to show higher resistance to the pathogen in comparison to the fungicide treatments.

9.4.4 Other Fungal Foliar Diseases

Besides the aforementioned fungal diseases, mycorrhizal fungi associations with the host plants can also effectively control several other fungal foliar diseases. Anthracnose disease caused by *Colletotrichum gloeosporioides* reduces the growth and biomass production of the host plant. Additionally, *C. gloeosporioides* infection shows a significant reduction in the plant antioxidant contents like ascorbic acid, ascorbate peroxidase (APX), superoxide dismutase (SOD), and polyphenols. AM fungus *G. fasciculatum*, inoculation induces drastic changes in growth parameters and in the activity of antioxidants in cyclamen, infected with *C. gloeosporioides* (Maya and Matsubara 2013). A similar study conducted by Chen et al. (2023b) reported an increase in the activity of antioxidants {(SOD, catalase (CAT) and peroxidase (POD)} in tea plants inoculated with AM fungi *R. intraradices* BGC JX04B. Tea plants inoculated with AM fungi significantly reduce the anthracnose disease severity caused by *Colletotrichum camelliae*. AM fungi alone or in a combined form with PGPR play a significant role in controlling plant fungal diseases. For example, rust disease in wheat caused by *Puccinia graminis* Pers. f. sp. *tritici* can be reduced by mycorrhizal treatment alone or in combination with *Trichoderma harzianum* HL1 and *T. viride* HL5 (El-Sharkawy et al. 2018).

AM fungi with grass fungal endophyte (*Epichloe*) suppressed the leaf spot disease incidence of perennial ryegrass caused by *Bipolaris sorokiniana*. AMF alone or in association with (*Epichloe*) can increase the lignin contents of the host plant infected with the fungal pathogen *B. sorokiniana* (Guo et al. 2019). Similarly, AMF with two PGPR, viz., *P. fluorescens* (Pf) and *T. viride* (Tv), results in highest fungal disease control efficiency of 65.60% (AMF + Tv) and 48.40% (AMF + Pf + Tv) in rice plant against *B. oryzae*, a fungal pathogen causing brown spot disease (Saleh and El-Akshar 2020). Black sigatoka disease caused by *Mycosphaerella fijiensis*, a fungal pathogen in banana, was found to be controlled by AM fungus, *R. irregularis* MUCL 41833. Four banana cultivars (Batard, C292, CRBP39, and F568) inoculated with *R. irregularis* were grown in a greenhouse and were observed for the effects of the pathogen on the plantlets. After 35 days of growth, it was observed that AM fungi inoculated CRBP39 and F568 showed more resistance than Batard and C292 cultivars toward *M. fijiensis*. The overall study signifies that AMF inoculation resists the development of host plant infectious symptoms by *M. fijiensis* at an early age of infection (Anda et al. 2020). Symbiotic association of AM fungi *R. intraradices*, with alfalfa (*Medicago sativa*), decreases the incidence of leaf spot disease caused by the fungus *Phoma medicaginis* by 39.48% and also reduces its disease index by 56.18% (Li et al. 2021). A soil-borne fungus *Physoderma maydis* causing brown spot disease in *Zea mays* was effectively controlled by AM fungi (*Glomus* sp. and *Gigaspora* sp.). Two different varieties of maize, CMS9015 and CMS8704, were inoculated with AM fungi in combination and were observed for their potential to control *P. maydis* and found that the incidence and severity of the disease is reduced by 33.7% and 35%, respectively (Adamou et al. 2023).

9.5 AMF-Mediated Mechanisms Associated with Plant Disease Management

The use of biocontrol agents in ameliorating the havoc of plant diseases has gained momentum in recent years. With the advances in the use of biological control agents, the application of AMF as a plant disease management tool has also got the attention of the scientific community. AMF association induces resistance in the infected plants through various mechanisms involving structural and chemical changes, improving nutrient uptake by the plants, changing the rhizospheric niche, competing with the pathogenic microbes, and activating the host plant defense system (Fig. 9.1) (Weng et al. 2022).

9.5.1 AMF-Induced Structural Changes in Plants

AMF association induces structural/morphological changes in the roots of the host plants ultimately creating a barrier for the entry of pathogens. Plant roots undergo changes like thickening, lignification, excessive branching, increase in the cell layers, etc. when they are symbiotically associated with the AMF, thus slowing

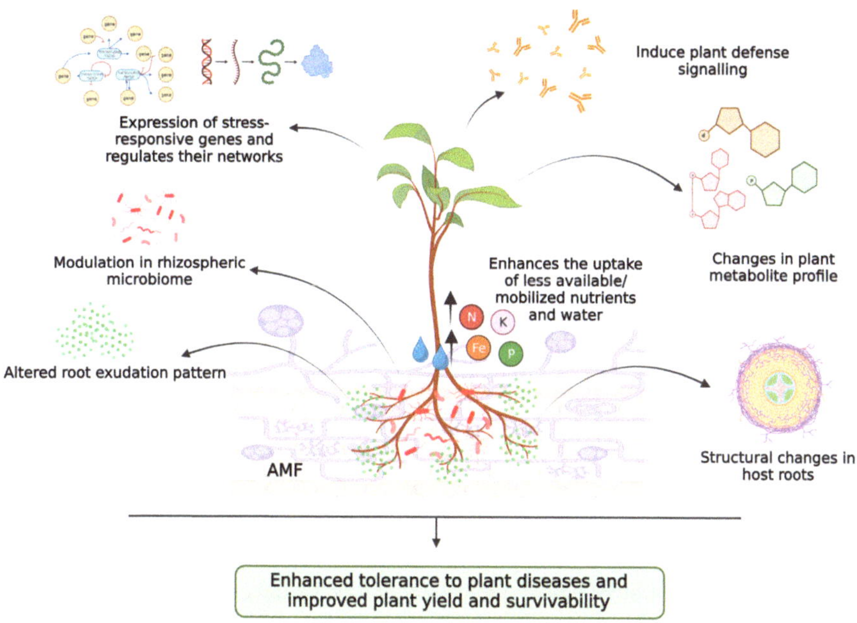

Fig. 9.1 Diagrammatic representation of various AMF-induced mechanisms in host plants associated with improved plant disease tolerance. Created with BioRender.com

down the pathogen from entering the plant tissue (Gao et al. 2018; Basyal and Emery 2021). Moreover, increased root growth and branching enhance the mineral nutrition of the plants ultimately strengthening the plant defense system (Matsubara et al. 1995). Similarly, the deposition of lignin at various plant parts like leaves, roots, and xylem tissue provides mechanical strength to the plants, thus inhibiting pathogen invasion (Goicoechea et al. 2010; Sanmartín et al. 2020). Tomato plants in association with AMF (*G. mosseae* and *G. intraradices*) were found to have altered root anatomy, eventually hindering *Phytophthora parasitica* infection kinetics (Pozo et al. 2002). Yanan et al. (2015) in their study demonstrated that strawberry plants treated with AMF resulted in a reduction in wilt disease caused by *F. oxysporum*. Plants in association with AMF had shown an increase in hydroxyproline-rich glycoprotein (HRGP) and lignin contents in the root tissue of strawberry plants leading to cell wall thickening and contributing to host defense. Similarly, wheat plants inoculated with AM fungus *F. mosseae* showed biocontrol activity against the foliar pathogen *B. graminis* f. sp. *tritici*. Inoculation of *F. mosseae* induced systemic resistance in the leaves by forming papillae at the *B. graminis* penetration site, thus enhancing the wheat plant's resistance to powdery mildew (Mustafa et al. 2017). Furthermore, HPRGs are sugar proteins that protect the cell wall from degradation by various enzymes secreted by pathogens like proteases, cellulases, etc. *Medicago sativa* in symbiosis with AM fungi was found to produce more amount of HPRGs, which act as a defense molecule against *Phoma medicaginis* (Gao et al. 2018).

The functionality of plant roots gets altered during symbiosis with AMF (de Vries et al. 2021). Analysis of mycorrhizal and non-mycorrhizal roots of *Camellia sinensis* L. revealed that AMF (*R. intraradices*) association amends the RSA and increased the sugar content of tea roots (Chen et al. 2021), indicating that change in RSA and root sugar content might modulate the rhizomicrobiome of *C. sinensis*, which may directly or indirectly hamper the pathogen infection in plants. Thus, AMF-mediated structural changes in root and leaf morphology or anatomy protect the host plant from any type of pathogen interventions and minimize disease incidences.

9.5.2 AMF-Induced Enhancement in Plant Nutrition

Plants exchange their photosynthetic outcomes with AMF and in return acquire nutrients and water from the mycorrhizal partner (Smith and Read 2008). AMF helps the plants in uptaking nutrients like N, P, and other essential micronutrients by inducing a series of chemical signaling between the AMF and host plant (Dowarah et al. 2022), which assists plant growth and development. The mycelia formed by AMF develop into a huge branching system that helps in better absorption of water and minerals from the soil. Being an inevitable element for plant functions, P is required by plants during various developmental stages. The utilization of inorganic P (Pi) from the soil is not an easy task for many plant species due to its low mobility and solubility in the soil (Wang et al. 2023). Since plants alone cannot absorb sufficient Pi for their utilization, they seek help from the mycorrhizal fungal partner and fulfill their need (Ferrol et al. 2019). Likewise, AMF also helps in the uptake of both organic and inorganic N by the plants, but the mechanisms of action involved in AMF-induced N acquisition are not yet fully understood (Savolainen and Kytöviita 2022). AMF colonization enhances plant nutrient uptake as their extraradical hyphae explore beneath the soil by expanding the host plant rhizosphere (Schaefer et al. 2021) through enzymatic or physical actions. Furthermore, mycorrhizal hyphae can penetrate deep into the soil due to their thinner appearance, thus improving the efficiency of nutrient acquisition by the host plant (Ebbisa 2022).

Cucumber plants symbiotically associated with AMF had shown increased uptake of inorganic nutrients such as N, P, and potassium (K) from the soil which eventually enhanced the defense system of cucumber plants aiding in resisting Fusarium wilt disease (Wang et al. 2012). A tripartite interaction involving AMF, leguminous plants, and *Rhizobium* sp. helps in N uptake by the host plant. In an intercropping system consisting of soybean and maize plants, when inoculated with AM fungus, *G. mosseae* and *Rhizobium* strain SH212 exhibited an enhanced N fixation rate in soybean plants. Furthermore, improved N fixation efficacy stimulated the transfer of N from soybean to maize plants, thus improving the productivity of both plants (Meng et al. 2015). AM fungi *F. mosseae* BGC HEB02 and *R. intraradices* BGC HEB07D in interaction with wheat plants increased the levels of P and zinc (Zn); especially wheat plants in association with *R. intraradices* significantly increased the Zn concentration 1.13–2.76 times than that of non-inoculated plants (Ma et al. 2019). Similarly, in tomato plants, a maximum uptake of N, P, and K was observed when inoculated with AMF (Kumari and Prabina 2019). High-throughput phenotyping of various sorghum genotypes unveiled that colonization of sorghum roots by the AM fungus *R. irregularis* boosts P, iron (Fe), and Zn concentration in the maize grains (Watts-Williams et al. 2022). However, *Artemisia ordosica* plants inoculated with *F. mosseae*, an AM fungus, lowered the availability of leaf P, indicating that AMF association may not always enhance plant nutrient levels. Moreover, a lowering in leaf P content due to the symbiosis between *A. ordosica* and *F. mosseae* negatively regulated the population of a foliar herbivore *Chrysolina aeruginosa*, emphasizing that AMF-mediated reduction in P concentration of *A. ordosica* leaves might have a positive role in limiting the damages caused by *C. aeruginosa* (Wang et al. 2023). In addition to the enhancement in plant nutrient uptake, AMF association also contributes to the sequestration of heavy metals (HMs) present in HM-polluted soil (Khan et al. 2022), suggesting that AMF interaction may have the potential to alleviate the toxic effect of HM stress. The further implication of AMF with other microbial or nonmicrobial substances might help the plants to survive under challenging environmental conditions.

9.5.3 AMF-Induced Changes in Plant Metabolites

About 4–20% of the photosynthetic output of the host plant either in sugar or lipid form is utilized by the AMF for their proper nutrition (Kaur and Suseela 2020). Acquisition of photosynthetically fixed carbon by the mycorrhizal partner in return generates a carbon sink in host plant roots and activates the photosynthetic machinery (Bago et al. 2000; Kaschuk et al. 2009), thereby forming a positive feedback cycle. Symbiosis by AM fungi stimulates the production of various types of plant metabolites, which play a significant role in plant defense against foliar fungal pathogens. AMF-induced modulation in the root metabolite profile synergistically alters the metabolites of aboveground plant parts (Schweiger and Müller 2015; Wang et al. 2018), thus helping the plant from various harmful phyllospheric and rhizospheric microbes. Moreover, AMF association initiates the shikimic acid pathway in the host plant, which produces several plant secondary metabolites like phytoalexins, phenolics, antimicrobial compounds, and various other enzymes. Several studies have also revealed that mycorrhizal plants' primary and secondary metabolite profiles differ from those of non-mycorrhizal plants (Kaur and Suseela 2020).

Pepper plants pre-inoculated with AM fungus *G. mosseae* have enhanced tolerance against the pepper blight pathogen *P. capsica* in both greenhouse and field conditions. Furthermore, the study also revealed that inoculation of *G. mosseae* enhanced the production of plant secondary metabolites, phytoalexin, and capsidiol which can have a role in defense against *P. capsica* (Ozgonen and Erkilic 2007). Corn variety Gaoyou-115, susceptible to *R. solani*, a fungus responsible for sheath blight disease, when inoculated with the AM fungus *G. mosseae*, a significant reduction in disease outbreak, was observed. Furthermore, the interaction of *G. mosseae* upregulated the production of 2,4-dihydroxy-7-methoxy-2 H-1,4-benzoxazin-3(4H)-one (DIMBOA) in the leaves and roots of corns, helping the plants to cope with sheath blight disease (Song et al. 2011a). Previous studies involving the role of DIMBOA in controlling sheath blight disease also reflected that an increase in DIMBOA concentration in the host plant inhibits the hyphal growth of *R. solani* (Song et al. 2011a). Similarly, cereal crops in association with AMF had shown an increase in the concentration of secondary metabolites like flavones, phenolic acid, and phytic acid (Ryan and Angus 2003; Pepe et al. 2018), which are well known for their key regulatory role in plant stress alleviation. Additionally, the quantity of plant secondary metabolites like alkaloids, flavonoids, organic acids, and terpenoids was also found to be increased in plants interacting with various AM fungi (Card et al. 2016; Latz et al. 2018; Kaddes et al. 2019). AMF-induced changes in plant primary and secondary metabolic pathways and changes in hormonal cross-talk can induce the photosynthetic machinery of the host plants, thereby increasing the photosynthetic output (Khan et al. 2022).

AMF symbiosis also increases antioxidant enzyme synthesis which reduces oxidative damage caused by various fungal pathogens to the plants. Tomato plants infected with vascular wilt disease, when inoculated with a consortium of AMF comprising *Claroideoglomus etunicatum*, *F. mosseae*, and *R. intraradices*, an induction in plant systemic defense, were observed involving the production of fusaric acid, which helped in reducing wilting symptoms in tomato plants. Moreover, inoculation of the AMF consortium also enhanced the concentration of antioxidative enzymes involving CAT, glutathione reductase (GR), etc. and reduced the toxic effect of ROS generated due to *F. oxysporum* f. sp. *lycopersici* (Hashem et al. 2021). Similarly, evaluating the effect of AMF in tomato plants infected with *V. dahliae* revealed that AMF interaction enhanced the intensity of antioxidants like CAT, SOD, and polyphenol oxidase (PPO) in tomato roots and leaves and finally improved tomato tolerance against *V. dahliae* (Ait Rahou et al. 2021). In addition to that, AM fungus *G. viscosum* showed effective biocontrol ability against verticillium wilt disease in artichoke plants by modulating the activity of antioxidant enzymes like SOD, APX, and monodehydroascorbate reductase (MDHAR). Further, the association of *G. viscosum* also increased the concentration of glutathione (GSH) and ascorbate (ASC) in artichoke plants, while a reduction in hydrogen peroxidase (H_2O_2) and lipid peroxidation levels was observed (Villani et al. 2021). Deciphering the effect of AMF on tea anthracnose incidences unraveled that mycorrhizal inoculation significantly increases the content of CAT, POD, and superoxide ions in tea plants and inhibited the proliferation of *C. camelliae*, thus protecting the tea plants from damages caused by anthracnose disease (Chen et al. 2023b). AMF-induced changes in plant metabolic profile directly or indirectly influence plant tolerance to any pathogenic fungal challenges. Furthermore, AMF also modulates the expression of defense signaling molecules in plants and also alters the composition of plant-associated microbiota, ultimately improving plant disease resistance.

9.5.4 AMF-Induced Hormonal Signaling

Plant response to environmental fluctuations or pathogen attack is also regulated by the modulation in plant defense hormonal signaling. Plant hormones are required in trace amounts and play an important role in plant development. Various plant hormones like IAA, abscisic acid (ABA), brassinosteroid (BR), cytokinin (CK), ethylene (ET), gibberellic acid (GA), JA, and salicylic acid (SA) play a crucial role in the establishment of AMF-mediated plant defense (Schmitz and Harrison 2014). From various studies, it has been established that SA resists the infection of biotrophic pathogens and induces SAR in plants, while JA and ET inhibit saprotrophic pathogens and activate induced systemic resistance (ISR) in plants (Hause et al. 2007). The combined application of *G. mosseae* and JA/SA elicitors on Fusarium wilt undergoing tomato plants resulted in a significant reduction in disease incidences. Furthermore, the level of phytohormones like GA, IAA, zeatin, and zeatin-riboside was found to be significantly enhanced in diseased tomato plants after the application of AMF and JA/SA inducers, indicating that changes in hormonal levels might have a positive correlation with the disease resistance (El-Khallal 2007). Likewise, in tomato plants infected with *B. cinerea* when inoculated with the mycorrhizal fungus *G. mosseae*, an upregulation in *Pin II* (a JA marker gene) was observed. Mycorrhiza-induced enhancement in the *Pin II* expression conferred ISR in infected tomato plants suggesting that modulation in JA concentration might have primed the plants against *B. cinerea* (Jung et al. 2009).

The real-time qPCR analysis on the effect of *Glomus* sp. MUCL41833 on late-blight infected potato plants unveiled that AMF induction enhanced the expression of defense genes associated with JA, SA, and JA/ET pathways (Gallou et al. 2011). Analyzing the cross-talk of *G. mosseae*, *G. intraradices*, and the beneficial non-mycorrhizal fungus *T. harzianum* on melon plants showed induction in 1-aminocyclopropane-1-carboxylic acid (ACC), IAA, ABA, JA, SA, and zeatin levels. In addition to the modified hormonal profile, the treated melon plants also exhibited an enhanced tolerance against the Fusarium wilt disease (Martínez-Medina et al. 2011). In a recent study, it was found that *Salvia miltiorrhiza* plants when pre-inoculated with AM fungus *G. versiforme*, a significant reduction in the wilt incidence was observed. Further investigation also revealed that the inoculation of mycorrhizal fungus also upregulated the JA and SA signaling pathways, thereby enhancing the host tolerance to Fusarium wilt (Pu et al. 2022). Further studies involving the modulation in phytohormone levels due to AMF association and tolerance to fungal foliar disease need attention from the scientific community.

9.6 Association of AMF with Other Microbes for Enhancing the Efficiency of Plant Stress Tolerance

The stress tolerance potential of AMF can be amplified by using various strategies. AMF in combination with other microorganisms shows enhanced efficiency in controlling various plant diseases and other abiotic stresses. To minimize the use of chemical fertilizers, the combined application of AMF and PGPR in crop fields has gained momentum in recent years. Tomato plants inoculated with AM fungi *G. mosseae* and *G. versiforme* along with PGPR, and *Bacillus polymyxa* provided control against *Meloidogyne incognita*. Furthermore, the study revealed that the AMF association increased the population of PGPR in the rhizospheric region, whereas PGPR enhanced the colonization of AMF in tomato roots (Liu et al. 2012). The study indicates that inoculation of plants with compatible AMF and PGPR enhances plant tolerance against any biotic stress.

A consortium of AMF and two PGP bacteria *B. subtilis* and *P. fluorescens* resulted in a significant reduction in the infection of *Sclerotium rolfsi* on common bean plants growing in P-deficient soil. Moreover, the synergistic effect of AMF and PGP bacteria enhanced P and Fe uptake by the diseased plants, which ultimately improved plant development under biotic stress (Mohamed et al. 2019). *Triticum durum* L. and *Vicia faba* L. are the two most important economic crops of the Mediterranean semiarid region. Research involving the inoculation of a consortium of AMF-PGPR-rhizobia revealed that combined inoculation of all three enhances soil fertility and wheat/faba bean productivity. Furthermore, inoculation of AMF-PGPR-rhizobia improved the content of nutrients like N, P, calcium (Ca), potassium (K), and sodium (Na) in the shoot tissue of both crops along with an increase in shoot protein and sugar content (Raklami et al. 2019). Similarly, tomato plants inoculated with AM fungus *F. mosseae* and PGPR strain *Paraburkholderia graminis* C4D1M and enhanced tomato seedling growth and the efficacy of PSII by reducing the leakage of electrolyte (Caradonia et al. 2019).

Summer squash inoculated with AMF (*G. etunicatu*, *G. harzianum*, *G. intraradices*, and *G. monosporum*) and PGPR strains *B. subtilis* and *T. harzianum* resulted in an elevated level of free amino acids, total proteins, free phenolics, chlorophyll, and carotenoid contents in the host plant. Furthermore, the application of AMF with PGPRs lowered the disease severity index of Fusarium wilt in the summer squash plants (El-Sharkawy and Abdelrazik 2022). Evaluating the effect of the combined application of AMF and PGPR revealed that the mixture of both microbial substrates enhanced rice plants' nutritional content and yield. In addition to that, the consortium of AMF and PGPR significantly enhanced the levels of inorganic nutrients like boron (B), N, P, and K in rice field soils, thus providing the plants with sufficient nutrients for proper growth and yield (Chen et al. 2023a). Similarly, inoculation of AMF and PGPR consortium on date palms undergoing Fusarium wilt decreased the disease incidences. Further investigation also revealed that the reduction in disease progression was due to the activation of the phenylpropanoid pathway in date palms, induced by the synergistic effect of AMF and PGPR (Ziane et al. 2023). Hence, the combined application of AMF and other plant-beneficial microorganisms plays crucial functions in improving plant-soil characteristics and also ameliorates the ability of host plants to cope with any environmental fluctuations. Further, the large-scale application of AMF and their consortia in field conditions will help in achieving climate-smart crops for a sustainable future.

9.7 Field Application of AMF: constraints and possible measures

Application of AMF in conventional agriculture practices still remains an unaccomplished task due to various hindrances. Excessive use of chemicals, crop rotation using non-mycorrhizal plants, and intensive tillage practices hampers the symbiotic association and efficiency of AMF. Thus, the natural soil is the main reservoir for a diversity of AM fungi, but due to anthropogenic interventions, the population and diversity of AMF are much lower in the agricultural field (Mbuthia et al. 2015). For instance, high doses of P-fertilization inhibit AM symbiosis with the crop plants (Balzergue et al. 2013). Overdose of P in agricultural fields allows the plants to utilize plenty of P from the soil without exchanging the soluble sugars (Garcia-Caparros et al. 2021), thereby limiting the mycorrhizal appressoria formation. Thus, the requirement of P by the host plant and its availability in the agricultural soil determines the extent of mycorrhization. Similar to P, the application of a disproportionate amount of N in the soil negatively impacts AM colonization (Lin et al. 2020). Furthermore, the use of fungicides and pesticides in the crop plants also degrades the AMF association (Channabasava et al. 2015; Kuila and Ghosh 2022), thereby making it more difficult for the AMF to function properly in the field conditions. Hence, minimal use of chemical substances might enhance the field applicability of AMF in conferring plant disease resistance.

One of the wide practices in conventional cultivation is the tillage of agricultural land. It has been estimated from various studies that tillage can crash the AMF spores and cause physical damage to them by disrupting the hyphal networks (Brito et al. 2012). Because of the tillage system, the soil layers get altered in the fields, ultimately making the soil environment suppressive for AMF. In addition to that, no-till activity for a longer duration also reduces soil AMF population (Schlüter et al. 2018). Thus, minimizing the tillage practices in agricultural land may help in the proper proliferation of AM fungi. Moreover, crop rotation is another factor that hampers the proper growth and functioning of AMF. The use of non-mycorrhizal partner plants during crop rotation adversely affects AMF development (Kuila and Ghosh 2022). Therefore, the rational use of various agricultural practices will help in overcoming constraints associated with AMF application. Furthermore, appropriate preparation of mycorrhizal inoculum and their compatibility with the host and soil samples will also determine the effectiveness of AMF in plant disease management.

9.8 Conclusion and Future Prospects

Control of plant disease by means of chemical substances hampers the ecosystem processes and poses a threat to mankind. To avoid/minimize such problems in the agricultural sector, alternative means have been taken into consideration in the recent decade. Due to the widespread availability and eco-friendly nature, the use of AMF in controlling plant diseases has gained popularity nowadays. AMF association enhances plant growth and development and also provides tolerance to various biotic and abiotic stresses. Besides having so many advantages in the application of AMF for agricultural purposes, a few limitations also exist for which the AMF-based solutions are unable to replace/reduce the use of chemicals. Furthermore, the tripartite interaction involving AMF, host plants, and indigenous microbes associated with the host may not always act synergistically to improve plant functionality in field conditions. Thus, depending upon the colonization, survival rate, and potency of plant disease control, a careful selection of the AMF species is very necessary. Furthermore, the interaction between the plant and microbes must also be precisely screened by understanding the mechanism of biocontrol. The application of AMF individually or in consortium with other microbes can provide an eco-friendly approach to plant disease management. However, understanding the in-depth mechanisms and other environmental/edaphic factors involved in AMF-mediated plant disease control will help to devise more effective future strategies.

Acknowledgement The authors are thankful to DST, Govt. of India for providing DST-FIST Support to Department of Botany, Gauhati University and DST-PURSE support to Gauhati University, where this work was carried out.

References

Abarca C, Simón MR, Esquisabel E, Velázquez MS (2023) Effect of spontaneous arbuscular mycorrhizal colonization in bread wheat varieties on the incidence of foliar diseases and grain yield. J Biosci 48(2):12. https://doi.org/10.1007/s12038-023-00335-5

Adamou S, Heu A, Mboussi SB, Kone Nsangou AN, Sinama P, Sesseumaga E, Amawissa ZT, Dooh JPN, Kosma P, Ambang Z (2023) Effect of Arbuscular Mycorrhizal Fungi (AMF) on development of brown spot disease of corn (Zea mays L.) due to Physoderma maydis in Far North Cameroon. MYCOPATH 20(1)

Agnolucci M, Palla M, Cristani C, Cavallo N, Giovannetti M, De Angelis M, Gobbetti M, Minervini F (2019) Beneficial plant microorganisms affect the endophytic bacterial approaches. Front Microbiol 10:2500. https://doi.org/10.3389/fmicb.2019.02500

Agrios GN (2009) Plant pathogens and disease: general introduction. https://api.semanticscholar.org/CorpusID:81630977

Ahammed GJ, Shamsy R, Liu A, Chen S (2023) Arbuscular mycorrhizal fungi-induced tolerance to chromium stress in plants. Environ Pollut 327:121597. https://doi.org/10.1016/j.envpol.2023.121597

Ait Rahou Y, Ait-El-Mokhtar M, Anli M, Boutasknit A, Ben-Laouane R, Douira A, Benkirane R, El Modafar C, Meddich A (2021) Use of mycorrhizal fungi and compost for improving the growth and yield of tomato and its resistance to Verticillium dahliae. Arch Phytopathol Plant Protect 54(13–14):665–690. https://doi.org/10.1080/03235408.2020.1854938

Anda CCO, Ndong AN, Ndoutoumou PN, Loubana PM (2020) Impact of arbuscular mycorrhizal fungus (Rhizophagus irregularis) on disease symptoms caused by the ascomycete fungus (Mycosphaerella fijiensis M.) in Black Sigatoka-resistant banana plantain. Int J Biol Chem Sci 14(2):306–316. https://doi.org/10.4314/ijbcs.v14i2.1

Augé RM, Kubikova E, Moore JL (2001) Foliar dehydration tolerance of mycorrhizal cowpea, soybean and bush bean. New Phytol 151(2):535–541. https://doi.org/10.1046/j.0028-646x.2001.00187.x

Avelino J, Cristancho M, Georgiou S, Imbach P, Aguilar L, Bornemann G, Peter L, Anzueto F, Hruska AJ, Morales C (2015) The coffee rust crises in Colombia and Central America (2008–2013): impacts, plausible causes and proposed solutions. Food Secur 7:303–321. https://doi.org/10.1007/s12571-015-0446-9

Badri DV, Vivanco JM (2009) Regulation and function of root exudates. Plant Cell Environ 32(6): 666–681. https://doi.org/10.1111/j.1365-3040.2008.01926.x

Bago B, Pfeffer PE, Shachar-Hill Y (2000) Carbon metabolism and transport in arbuscular mycorrhizas. Plant Physiol 124(3):949–958. https://doi.org/10.1104/pp.124.3.949

Balzergue C, Chabaud M, Barker DG, Bécard G, Rochange SF (2013) High phosphate reduces host ability to develop arbuscular mycorrhizal symbiosis without affecting root calcium spiking responses to the fungus. Front Plant Sci 4:426. https://doi.org/10.3389/fpls.2013.00426

Basyal B, Emery SM (2021) An arbuscular mycorrhizal fungus alters switchgrass growth, root architecture, and cell wall chemistry across a soil moisture gradient. Mycorrhiza 31(2):251–258. https://doi.org/10.1007/s00572-020-00992-6

Begum N, Qin C, Ahanger MA, Raza S, Khan MI, Ashraf M, Ahmed N, Zhang L (2019) Role of arbuscular mycorrhizal fungi in plant growth regulation: implications in abiotic stress tolerance. Front Plant Sci 10:1068. https://doi.org/10.3389/fpls.2019.01068

Begum N, Akhtar K, Ahanger MA, Iqbal M, Wang P, Mustafa NS, Zhang L (2021) Arbuscular mycorrhizal fungi improve growth, essential oil, secondary metabolism, and yield of tobacco (Nicotiana tabacum L.) under drought stress conditions. Environ Sci Pollut Res 28:45276–45295. https://doi.org/10.1007/s11356-021-13755-3

Bergot M, Cloppet E, Pérarnaud V, Déqué M, Marçais B, Desprez-Loustau ML (2004) Simulation of potential range expansion of oak disease caused by Phytophthora cinnamomi under climate change. Glob Chang Biol 10(9):1539–1552. https://doi.org/10.1111/j.1365-2486.2004.00824.x

Bidellaoui B, Segarra G, Hakkou A, Isabel Trillas M (2019) Beneficial effects of Rhizophagus irregularis and Trichoderma asperellum strain T34 on growth and fusarium wilt in tomato plants. J Plant Pathol 101:121–127. https://doi.org/10.1007/s42161-018-0159-y

Bletsos F, Thanassoulopoulos C, Roupakias D (2003) Effect of grafting on growth, yield, and Verticillium wilt of eggplant. HortScience 38(2):183–186. https://doi.org/10.21273/HORTSCI.38.2.183

Brito I, Goss MJ, Carvalho M (2012) Effect of tillage and crop on arbuscular mycorrhiza colonisation of winter wheat and triticale under Mediterranean conditions. Soil Use Manag 28:201–208. http://hdl.handle.net/10174/6829

Cai Y, Xiaohong H, Mo J, Sun Q, Yang J, Liu J (2009) Molecular research and genetic engineering of resistance to Verticillium wilt in cotton: a review. Afr J Biotechnol 8(25). http://www.academicjournals.org/AJB

Caradonia F, Francia E, Morcia C, Ghizzoni R, Moulin L, Terzi V, Ronga D (2019) Arbuscular mycorrhizal fungi and plant growth promoting rhizobacteria avoid processing tomato leaf damage during chilling stress. Agronomy 9(6):299. https://doi.org/10.3390/agronomy9060299

Card S, Johnson L, Teasdale S, Caradus J (2016) Deciphering endophyte behaviour: the link between endophyte biology and efficacious biological control agents. FEMS Microbiol Ecol 92(8):fiw114. https://doi.org/10.1093/femsec/fiw114

Chandrasekaran M, Chanratana M, Kim K, Seshadri S, Sa T (2019) Impact of arbuscular mycorrhizal fungi on photosynthesis, water status, and gas exchange of plants under salt stress—a meta-analysis. Front Plant Sci 10:457. https://doi.org/10.3389/fpls.2019.00457

Channabasava A, Lakshman HC, Jorquera MA (2015) Effect of fungicides on association of arbuscular mycorrhiza fungus Rhizophagus fasciculatus and growth of Proso millet (Panicum miliaceum L.). J Soil Sci Plant Nutr 15(1):35–45. https://doi.org/10.4067/S0718-95162015005000004

Chen W, Ye T, Sun Q, Niu T, Zhang J (2021) Arbuscular mycorrhizal fungus alters root system architecture in *Camellia sinensis* L. as revealed by RNA-Seq analysis. Front Plant Sci 12: 777357. https://doi.org/10.3389/fpls.2021.777357

Chen D, Saeed M, Ali MNHA, Raheel M, Ashraf W, Hassan Z, Hassan MZ, Farooq U, Hakim MF, Rao MJ, Naqvi SAH, Moustafa M, Al-Shehri M, Negm S (2023a) Plant growth promoting rhizobacteria (PGPR) and arbuscular mycorrhizal fungi combined application reveals enhanced soil fertility and rice production. Agronomy 13(2):550. https://doi.org/10.3390/agronomy13020550

Chen W, Ye T, Sun Q, Niu T, Zhang J (2023b) Arbuscular mycorrhizal fungus alleviates anthracnose disease in tea seedlings. Front Plant Sci 13:1058092. https://doi.org/10.3389/fpls.2022.1058092

de Sá CSB, Campos MAS (2020) Arbuscular mycorrhizal fungi decrease *Meloidogyne enterolobii* infection of Guava seedlings. J Helminthol 94:e183. https://doi.org/10.1017/S0022149X20000668

De Silva DD, Crous PW, Ades PK, Hyde KD, Taylor PW (2017) Life styles of Colletotrichum species and implications for plant biosecurity. Fungal Biol Rev 31(3):155–168. https://doi.org/10.1016/j.fbr.2017.05.001

de Vries J, Evers JB, Kuyper TW, van Ruijven J, Mommer L (2021) Mycorrhizal associations change root functionality: a 3D modelling study on competitive interactions between plants for light and nutrients. New Phytol 231(3):1171–1182. https://doi.org/10.1111/nph.17435

Delavaux CS, Smith-Ramesh LM, Kuebbing SE (2017) Beyond nutrients: a meta-analysis of the diverse effects of arbuscular mycorrhizal fungi on plants and soils. Ecology 98(8):2111–2119. https://doi.org/10.1002/ecy.1892

Devi NO, Tombisana Devi RK, Debbarma M, Hajong M, Thokchom S (2022) Effect of endophytic Bacillus and arbuscular mycorrhiza fungi (AMF) against Fusarium wilt of tomato caused by Fusarium oxysporum f. sp. lycopersici. *Egyptian Journal of Biological.* Pest Control 32(1): 1–14. https://doi.org/10.1186/s41938-021-00499-y

Dey M, Ghosh S (2022) Arbuscular mycorrhizae in plant immunity and crop pathogen control. Rhizosphere 22:100524. https://doi.org/10.1016/j.rhisph.2022.100524

Doehlemann G, Ökmen B, Zhu W, Sharon A (2017) Plant pathogenic fungi. Microbiol Spectr 5(1): 5–1. https://doi.org/10.1128/microbiolspec.funk-0023-2016

Dowarah B, Gill SS, Agarwala N (2022) Arbuscular mycorrhizal fungi in conferring tolerance to biotic stresses in plants. J Plant Growth Regul 41:1429–1444. https://doi.org/10.1007/s00344-021-10392-5

Ebbisa A (2022) Arbuscular mycorrhizal fungi (AMF) in optimizing nutrient bioavailability and reducing agrochemicals for maintaining sustainable agroecosystems. In: Mycorrhiza—new insights. IntechOpen. https://doi.org/10.5772/intechopen.106995

Eck JL, Kytöviita MM, Laine AL (2022) Arbuscular mycorrhizal fungi influence host infection during epidemics in a wild plant pathosystem. New Phytol 236(5):1922–1935. https://doi.org/10.1111/nph.18481

El-Khallal SM (2007) Induction and modulation of resistance in tomato plants against Fusarium wilt disease by bioagent fungi (arbuscular mycorrhiza) and/or hormonal elicitors (jasmonic acid & salicylic acid): 1-Changes in growth, some metabolic activities and endogenous hormones related to defence mechanism. Aust J Basic Appl Sci 1(4):691–705

El-Sharkawy EES, Abdelrazik E (2022) Biocontrol of *Fusarium* root rot in squash using mycorrhizal fungi and antagonistic microorganisms. Egypt J Biol Pest Control 32:13. https://doi.org/10.1186/s41938-022-00513-x

El-Sharkawy HH, Rashad YM, Ibrahim SA (2018) Biocontrol of stem rust disease of wheat using arbuscular mycorrhizal fungi and Trichoderma spp. Physiol Mol Plant Pathol 103:84–91. https://doi.org/10.1016/j.pmpp.2018.05.002

Ferrol N, Azcón-Aguilar C, Pérez-Tienda J (2019) Review: arbuscular mycorrhizas as key players in sustainable plant phosphorus acquisition: an overview on the mechanisms involved. Plant Sci 280:441–447. https://doi.org/10.1016/j.plantsci.2018.11.011

Fritz M, Jakobsen I, Lyngkjær MF, Thordal-Christensen H, Pons-Kühnemann J (2006) Arbuscular mycorrhiza reduces susceptibility of tomato to Alternaria solani. Mycorrhiza 16:413–419. https://doi.org/10.1007/s00572-006-0051-z

Gallou A, Mosquera HPL, Cranenbrouck S, Suárez JP, Declerck S (2011) Mycorrhiza induced resistance in potato plantlets challenged by Phytophthora infestans. Physiol Mol Plant Pathol 76(1):20–26. https://doi.org/10.1016/j.pmpp.2011.06.005

Gao P, Li Y, Guo Y, Duan T (2018) Co-inoculation of lucerne (Medicago sativa) with an AM fungus and a rhizobium reduces occurrence of spring black stem and leaf spot caused by Phoma medicaginis. Crop Pasture Sci 69(9):933–943. https://doi.org/10.1071/CP18135

Garcia-Caparros P, Lao MT, Preciado-Rangel P, Sanchez E (2021) Phosphorus and carbohydrate metabolism in green bean plants subjected to increasing phosphorus concentration in the nutrient solution. Agronomy 11(2):245. https://doi.org/10.3390/agronomy11020245

Goddard ML, Belval L, Martin IR, Roth L, Laloue H, Deglène-Benbrahim L, Valat L, Bertsch C, Chong J (2021) Arbuscular mycorrhizal symbiosis triggers major changes in primary metabolism together with modification of Defense responses and signaling in both roots and leaves of *Vitis vinifera*. Front Plant Sci 12:721614. https://doi.org/10.3389/fpls.2021.721614

Göhre V, Paszkowski U (2006) Contribution of the arbuscular mycorrhizal symbiosis to heavy metal phytoremediation. Planta 223:1115–1122. https://doi.org/10.1007/s00425-006-0225-0

Goicoechea N, Garmendia I, Sanchez-Diaz M, Aguirreolea J (2010) Arbuscular mycorrhizal fungi (AMF) as bioprotector agents against wilt induced by Verticillium spp. in pepper: a review. Span J Agric Res 8:25–42. https://doi.org/10.5424/sjar/201008S1-5300

González-Guerrero M, Azcón-Aguilar C, Mooney M, Valderas A, MacDiarmid CW, Eide DJ, Ferrol N (2005) Characterization of a Glomus intraradices gene encoding a putative Zn transporter of the cation diffusion facilitator family. Fungal Genet Biol 42(2):130–140. https://doi.org/10.1016/j.fgb.2004.10.007

Guo Y, Gao P, Li F, Duan T (2019) Effects of AM fungi and grass endophytes on perennial ryegrass Bipolaris sorokiniana leaf spot disease under limited soil nutrients. Eur J Plant Pathol 154:659–671. https://doi.org/10.1007/s10658-019-01689-z

Hammond-Kosack KE, Rudd JJ (2008) Plant resistance signalling hijacked by a necrotrophic fungal pathogen. Plant Signal Behav 3(11):993–995. https://doi.org/10.4161/psb.6292

Han Z, Zhang Z, Li Y, Wang B, Xiao Q, Li Z, Li G, Xiaowan L, Keqing H, Tao L, Xiaoyu, Chen J (2023) Effect of Arbuscular mycorrhizal fungi (AMF) inoculation on endophytic bacteria of lettuce. Physiol Mol Plant Pathol 126:102036. https://doi.org/10.1016/j.pmpp.2023.102036

Hao Z, Christie P, Qin L, Wang C, Li X (2005) Control of fusarium wilt of cucumber seedlings by inoculation with an arbuscular mycorrhical fungus. J Plant Nutr 28(11):1961–1974. https://doi.org/10.1080/01904160500310997

Hao B, Zhang Z, Bao Z, Hao L, Diao F, Li FY, Guo W (2022) Claroideoglomus etunicatum affects the structural and functional genes of the rhizosphere microbial community to help maize resist Cd and La stresses. Environ Pollut (Barking, Essex: 1987) 307:119559. https://doi.org/10.1016/j.envpol.2022.119559

Hashem A, Akhter A, Alqarawi AA, Singh G, Almutairi KF, Abd Allah EF (2021) Mycorrhizal fungi induced activation of tomato defense system mitigates Fusarium wilt stress. Saudi J Biol Sci 28(10):5442–5450. https://doi.org/10.1016/j.sjbs.2021.07.025

Hause B, Mrosk C, Isayenkov S, Strack D (2007) Jasmonates in arbuscular mycorrhizal interactions. Phytochemistry 68(1):101–110. https://doi.org/10.1016/j.phytochem.2006.09.025

Hu-zhe Z, Chun-lan CUI, Yu-ting Z, Dan WANG, Yu JING, Yong KK (2005) Active changes of lignification-related enzymes in pepper response to Glomus intraradices and/or Phytophthora capsici. J Zhejiang Univ Sci B 6(8):778–786. https://doi.org/10.1007/BF02842437

Ji C, Chen Z, Kong X, Xin Z, Sun F, Xing J, Li C, Li K, Liang Z, Cao H (2022) Biocontrol and plant growth promotion by combined *Bacillus* spp. inoculation affecting pathogen and AMF communities in the wheat rhizosphere at low salt stress conditions. Front Plant Sci 13:1043171. https://doi.org/10.3389/fpls.2022.1043171

Joseph OO, Olumide OV (2022) The effect of amf (glomus clarum) on tomato resistance to early leaf blight (alternaria solani) on tomato. Glob J Agric Sci 21(1):85–90. https://doi.org/10.4314/gjass.v21i1.11

Jung SC, García-Andrade J, Verhage A, Fernández I, García JM, Azcón-Aguilar C, Pozo MJ (2009) Arbuscular mycorrhiza confers systemic resistance against Botrytis cinerea in tomato through priming of JA-dependent defense responses. In: Induced resistance: chances and limits, IOBC/wprs Bulletin, Working Group "Induced resistance in plants against insects and diseases." Proceedings of the meeting at Granada (Spain), pp 8–16

Jung SC, Martinez-Medina A, Lopez-Raez JA, Pozo MJ (2012) Mycorrhiza-induced resistance and priming of plant defenses. J Chem Ecol 38(6):651–664. https://doi.org/10.1007/s10886-012-0134-6

Kaddes A, Fauconnier ML, Sassi K, Nasraoui B, Jijakli MH (2019) Endophytic fungal volatile compounds as solution for sustainable agriculture. Molecules (Basel, Switzerland) 24(6):1065. https://doi.org/10.3390/molecules24061065

Kaschuk G, Kuyper TW, Leffelaar PA, Hungria M, Giller KE (2009) Are the rates of photosynthesis stimulated by the carbon sink strength of rhizobial and arbuscular mycorrhizal symbioses? Soil Biol Biochem 41(6):1233–1244. https://doi.org/10.1016/j.soilbio.2009.03.005

Kaur S, Suseela V (2020) Unraveling arbuscular mycorrhiza-induced changes in plant primary and secondary metabolome. Metabolites 10(8):335. https://doi.org/10.3390/metabo10080335

Khan Y, Shah S, Tian H (2022) The roles of arbuscular mycorrhizal fungi in influencing plant nutrients, photosynthesis, and metabolites of cereal crops—a review. Agronomy 12(9):2191. https://doi.org/10.3390/agronomy12092191

Klinsukon C, Ekprasert J, Boonlue S (2021) Using arbuscular mycorrhizal fungi (Gigaspora margarita) as a growth promoter and biocontrol of leaf blight disease in eucalyptus seedlings caused by Cylindrocladium quinqueseptatum. Rhizosphere 20:100450. https://doi.org/10.1016/j.rhisph.2021.100450

Knogge W (1998) Fungal pathogenicity. Curr Opin Plant Biol 1(4):324–328. https://doi.org/10.1016/1369-5266(88)80054-2

Kuila D, Ghosh S (2022) Aspects, problems and utilization of Arbuscular Mycorrhizal (AM) application as bio-fertilizer in sustainable agriculture. Curr Res Microb Sci 3:100107. https://doi.org/10.1016/j.crmicr.2022.100107

Kumari SMP, Prabina BJ (2019) Protection of tomato, Lycopersicon esculentum from wilt pathogen, Fusarium oxysporum f. sp. lycopersici by arbuscular mycorrhizal fungi, Glomus sp. Int J Curr Microbiol Appl Sci 8:1368–1378. https://doi.org/10.20546/ijcmas.2019.804.159

Latz MA, Jensen B, Collinge DB, Jørgensen HJ (2018) Endophytic fungi as biocontrol agents: elucidating mechanisms in disease suppression. Plant Ecol Divers 11(5–6):555–567. https://doi.org/10.1080/17550874.2018.1534146

Launay M, Caubel J, Bourgeois G, Huard F, de Cortazar-Atauri IG, Bancal MO, Brisson N (2014) Climatic indicators for crop infection risk: application to climate change impacts on five major foliar fungal diseases in Northern France. Agric Ecosyst Environ 197:147–158. https://doi.org/10.1016/j.agee.2014.07.020

Lazcano C, Barrios-Masias FH, Jackson LE (2014) Arbuscular mycorrhizal effects on plant water relations and soil greenhouse gas emissions under changing moisture regimes. Soil Biol Biochem 74:184–192. https://doi.org/10.1016/j.soilbio.2014.03.010

Li Y, Duan T, Nan Z, Li Y (2021) Arbuscular mycorrhizal fungus alleviates alfalfa leaf spots caused by Phoma medicaginis revealed by RNA-seq analysis. J Appl Microbiol 130(2):547–560. https://doi.org/10.1111/jam.14387

Lin C, Wang Y, Liu M, Li Q, Xiao W, Song X (2020) Effects of nitrogen deposition and phosphorus addition on arbuscular mycorrhizal fungi of Chinese fir (Cunninghamia lanceolata). Sci Rep 10(1):12260. https://doi.org/10.1038/s41598-020-69213-6

Liu R, Dai M, Wu X et al (2012) Suppression of the root-knot nematode [*Meloidogyne incognita* (Kofoid & White) Chitwood] on tomato by dual inoculation with arbuscular mycorrhizal fungi and plant growth-promoting rhizobacteria. Mycorrhiza 22:289–296. https://doi.org/10.1007/s00572-011-0397-8

Lowe A, Rafferty-McArdle SM, Cassells AC (2012) Effects of AMF-and PGPR-root inoculation and a foliar chitosan spray in single and combined treatments on powdery mildew disease in strawberry. Agric Food Sci 21(1):28–38. https://doi.org/10.23986/afsci.4997

Ma X, Luo W, Li J, Wu F (2019) Arbuscular mycorrhizal fungi increase both concentrations and bioavilability of Zn in wheat (*Triticum aestivum* L) grain on Zn-spiked soils. Appl Soil Ecol 135:91–97. https://doi.org/10.1016/j.apsoil.2018.11.007

Ma J, Ma Y, Wei Z, Wu J, Sun C, Yang J et al (2021) Effects of arbuscular mycorrhizal fungi symbiosis on microbial diversity and enzyme activities in the rhizosphere soil of Artemisia annua. Soil Sci Soc Am J 85(3):703–716. https://doi.org/10.1002/saj2.20229

Ma J, Xie Y, Yang Y, Jing C, You X, Yang J, Sun C, Qin S, Chen J, Cao K, Huang J, Li Y (2022) AMF colonization affects allelopathic effects of *Zea mays* L. root exudates and community structure of rhizosphere bacteria. Front Plant Sci 13:1050104. https://doi.org/10.3389/fpls.2022.1050104

Magarey RD, Fowler GA, Borchert DM, Sutton TB, Colunga-Garcia M, Simpson JA (2007) NAPPFAST: an Internet system for the weather-based mapping of plant pathogens. Plant Dis 91(4):336–345. https://doi.org/10.1094/PDIS-91-4-0336

Marĉais B, Dupuis F, Desprez-Loustau ML (1996) Susceptibility of the Quercus rubra root system to Phytophthora cinnamomi; comparison with chestnut and other oak species. Eur J For Pathol 26(3):133–143. https://doi.org/10.1111/j.1439-0329.1996.tb00718.x

Marro N, Grilli G, Soteras F, Caccia M, Longo S, Cofré N, Borda V, Burni M, Janoušková M, Urcelay C (2022) The effects of arbuscular mycorrhizal fungal species and taxonomic groups on stressed and unstressed plants: a global meta-analysis. New Phytol 235(1):320–332. https://doi.org/10.1111/nph.18102

Martínez-Medina A, Roldán A, Albacete A, Pascual JA (2011) The interaction with arbuscular mycorrhizal fungi or Trichoderma harzianum alters the shoot hormonal profile in melon plants. Phytochemistry 72(2–3):223–229. https://doi.org/10.1016/j.phytochem.2010.11.008

Matsubara YI, Tamura H, Harada T (1995) Growth enhancement and Verticillium wilt control by vesicular-arbuscular mycorrhizal fungus inoculation in eggplant. J Jpn Soc Hortic Sci 64(3): 555–561. https://doi.org/10.2503/jjshs.64.555

Maya MA, Matsubara YI (2013) Tolerance to Fusarium wilt and anthracnose diseases and changes of antioxidative activity in mycorrhizal cyclamen. Crop Prot 47:41–48. https://doi.org/10.1016/j.cropro.2013.01.007

Mbuthia LW, Acosta-Martínez V, DeBruyn J, Schaeffer S, Tyler D, Odoi E, Mpheshea M, Walker F, Eash N (2015) Long term tillage, cover crop, and fertilization effects on microbial community structure, activity: implications for soil quality. Soil Biol Biochem 89:24–34. https://doi.org/10.1016/j.soilbio.2015.06.016

Meng L, Zhang A, Wang F, Han X, Wang D, Li S (2015) Arbuscular mycorrhizal fungi and rhizobium facilitate nitrogen uptake and transfer in soybean/maize intercropping system. Front Plant Sci 6:339. https://doi.org/10.3389/fpls.2015.00339

Miozzi L, Vaira AM, Catoni M, Fiorilli V, Accotto GP, Lanfranco L (2019) Arbuscular mycorrhizal symbiosis: plant friend or foe in the fight against viruses. Front Microbiol 10:1238. https://doi.org/10.3389/fmicb.2019.01238

Miransari M (2011) Interactions between arbuscular mycorrhizal fungi and soil bacteria. Appl Microbiol Biotechnol 89:917–930. https://doi.org/10.1007/s00253-010-3004-6

Moarrefzadeh N, Khateri H, Sharifi R (2022) The effects of some arbuscular mycorrhizal fungi on plant growth and biocontrol of Ascochyta blight in two chickpea varieties. Biol J Microorg 11(44):23–39. https://doi.org/10.22108/bjm.2021.130031.1410

Mohamed I, Eid KE, Abbas MHH, Salem AA, Ahmed N, Ali M, Shah GM, Fang C (2019) Use of plant growth promoting Rhizobacteria (PGPR) and mycorrhizae to improve the growth and nutrient utilization of common bean in a soil infected with white rot fungi. Ecotoxicol Environ Saf 171:539–548. https://doi.org/10.1016/j.ecoenv.2018.12.100

Molitor A, Zajic D, Voll LM, Pons-Kühnemann J, Samans B, Kogel KH, Waller F (2011) Barley leaf transcriptome and metabolite analysis reveals new aspects of compatibility and Piriformospora indica–mediated systemic induced resistance to powdery mildew. Mol Plant-Microbe Interact 24(12):1427–1439. https://doi.org/10.1094/MPMI-06-11-0177

Mosquera-Espinosa AT, Bayman P, Prado GA, Gómez-Carabalí A, Otero JT (2013) The double life of Ceratobasidium: orchid mycorrhizal fungi and their potential for biocontrol of Rhizoctonia solani sheath blight of rice. Mycologia 105(1):141–150. https://doi.org/10.3852/12-079

Mustafa G, Tisserant B, Randoux B, Fontaine J, Reignault P, Sahraoui AL (2013) Towards a more sustainable agriculture: wheat mycorrhization to protect against powdery mildew. Commun Agric Appl Biol Sci 78(3):467–478

Mustafa G, Tisserant B, Randoux B, Fontaine J, Sahraoui AL, Reignault P (2014) Mechanisms involved in mycorrhizal wheat protection against powdery mildew. Commun Agric Appl Biol Sci 79:403–410

Mustafa G, Randoux B, Tisserant B, Fontaine J, Magnin-Robert M, Lounès-Hadj Sahraoui A, Reignault PH (2016) Phosphorus supply, arbuscular mycorrhizal fungal species, and plant genotype impact on the protective efficacy of mycorrhizal inoculation against wheat powdery mildew. Mycorrhiza 26:685–697. https://doi.org/10.1007/s00572-016-0698-z

Mustafa G, Khong NG, Tisserant B, Randoux B, Fontaine J, Magnin-Robert M, Reignault P, Sahraoui AL (2017) Defence mechanisms associated with mycorrhiza-induced resistance in wheat against powdery mildew. Funct Plant Biol 44(4):443–454. https://doi.org/10.1071/FP16206

Nagy R, Karandashov V, Chague V, Kalinkevich K, Tamasloukht M, Xu G, Jakobsen I, Levy AA, Amrhein N, Bucher M (2005) The characterization of novel mycorrhiza-specific phosphate transporters from Lycopersicon esculentum and Solanum tuberosum uncovers functional redundancy in symbiotic phosphate transport in solanaceous species. Plant J Cell Mol Biol 42(2): 236–250. https://doi.org/10.1111/j.1365-313X.2005.02364.x

Noman M, Ahmed T, Shahid M, Mudassir Nazir M, Azizullah, Li D, Song F (2023) Salicylic acid-doped iron nano-biostimulants potentiate defense responses and suppress Fusarium wilt in watermelon. J Adv Res S2090-1232(23)00174-1. Advance online publication. https://doi.org/10.1016/j.jare.2023.06.011

Orine D, Defossez E, Vergara F, Uthe H, van Dam NM, Rasmann S (2022) Arbuscular mycorrhizal fungi prevent the negative effect of drought and modulate the growth-defence trade-off in tomato plants. J Sustain Agric Environ 1(3):177–190. https://doi.org/10.1002/sae2.12018

Ozgonen H, Erkilic A (2007) Growth enhancement and Phytophthora blight (Phytophthora capsici Leonian) control by arbuscular mycorrhizal fungal inoculation in pepper. Crop Prot 26(11): 1682–1688. https://doi.org/10.1016/j.cropro.2007.02.010

Paliwal J, Joshi S (2022) An overview of deep learning models for foliar disease detection in maize crop. J Artif Intell Syst 4(1):1–21. https://doi.org/10.33969/AIS.2022040101

Pawlowski ML, Hartman GL (2020) Reduction of sudden death syndrome foliar symptoms and *Fusarium virguliforme* DNA in roots inoculated with *Rhizophagus intraradices*. Plant Dis 104(5):1415–1420. https://doi.org/10.1094/PDIS-07-19-1500-RE

Pepe A, Giovannetti M, Sbrana C (2018) Lifespan and functionality of mycorrhizal fungal mycelium are uncoupled from host plant lifespan. Sci Rep 8:10235. https://doi.org/10.1038/s41598-018-28354-5

Poveda J, Díaz-González S, Díaz-Urbano M, Velasco P, Sacristán S (2022) Fungal endophytes of Brassicaceae: molecular interactions and crop benefits. Front Plant Sci 13:932288. https://doi.org/10.3389/fpls.2022.932288

Pozo MJ, Cordier C, Dumas-Gaudot E, Gianinazzi S, Barea JM, Azcón-Aguilar C (2002) Localized versus systemic effect of arbuscular mycorrhizal fungi on defence responses to Phytophthora infection in tomato plants. J Exp Bot 53(368):525–534. https://doi.org/10.1093/jexbot/53.368.525

Pu C, Ge Y, Yang G, Zheng H, Guan W, Chao Z, Shen Y, Liu S, Chen M, Huang L (2022) Arbuscular mycorrhizal fungi enhance disease resistance of *Salvia miltiorrhiza* to Fusarium wilt. Front Plant Sci 13:975558. https://doi.org/10.3389/fpls.2022.975558

Raklami A, Bechtaoui N, Tahiri AI, Anli M, Meddich A, Oufdou K (2019) Use of rhizobacteria and mycorrhizae consortium in the open field as a strategy for improving crop nutrition, productivity and soil fertility. Front Microbiol 10:1106. https://doi.org/10.3389/fmicb.2019.01106

Ren L, Zhang N, Wu P, Huo H, Xu G, Wu G (2015) Arbuscular mycorrhizal colonization alleviates Fusarium wilt in watermelon and modulates the composition of root exudates. Plant Growth Regul 77:77–85. https://doi.org/10.1007/s10725-015-0038-x

Rodrigues e Silva MT, Calandrelli A, Miamoto A, Rinaldi LK, Pereira Moreno B, da Silva C, Dias-Arieira CR (2021) Pre-inoculation with arbuscular mycorrhizal fungi affects essential oil quality and the reproduction of root lesion nematode in Cymbopogon citratus. Mycorrhiza 31:613–623. https://doi.org/10.1007/s00572-021-01045-2

Ros M, Hernandez MT, Garcia C, Bernal A, Pascual JA (2005) Biopesticide effect of green compost against Fusarium wilt on melon plants. J Appl Microbiol 98(4):845–854. https://doi.org/10.1111/j.1365-2672.2004.02508.x

Rotem J (1994) The genus Alternaria: biology, epidemiology, and pathogenicity. American Phytopathological Society

Ryan M, Angus J (2003) Arbuscular mycorrhizae in wheat and field pea crops on a low P soil: increased Zn-uptake but no increase in P-uptake or yield. Plant Soil 250:225–239. https://doi.org/10.1023/A:1022839930134

Saleh MM, El-Akshar YS (2020) Integration between arbuscular mycorrhizal fungi, bacterial and fungal bioagents for controlling rice brown spot disease. Egypt J Phytopathol 48(1):81–93. https://doi.org/10.21608/ejp.2020.47059.1011

Sanmartín N, Pastor V, Pastor-Fernández J, Flors V, Pozo MJ, Sánchez-Bel P (2020) Role and mechanisms of callose priming in mycorrhiza-induced resistance. J Exp Bot 71(9):2769–2781. https://doi.org/10.1093/jxb/eraa030

Savolainen T, Kytöviita MM (2022) Mycorrhizal symbiosis changes host nitrogen source use. Plant Soil 471:643–654. https://doi.org/10.1007/s11104-021-05257-5

Schaefer DA, Gui H, Mortimer PE, Xu J (2021) Arbuscular mycorrhiza and sustainable agriculture. Circ Agric Syst 1(1):1–7. https://doi.org/10.48130/CAS-2021-0006

Schlüter S, Großmann C, Diel J, Wu GM, Tischer S, Deubel A, Rücknagel J (2018) Long-term effects of conventional and reduced tillage on soil structure, soil ecological and soil hydraulic properties. Geoderma 332:10–19. https://doi.org/10.1016/j.geoderma.2018.07.001

Schmitz AM, Harrison MJ (2014) Signaling events during initiation of arbuscular mycorrhizal symbiosis. J Integr Plant Biol 56(3):250–261. https://doi.org/10.1111/jipb.12155

Schweiger R, Müller C (2015) Leaf metabolome in arbuscular mycorrhizal symbiosis. Curr Opin Plant Biol 26:120–126. https://doi.org/10.1016/j.pbi.2015.06.009

Sharma S, Prasad R, Varma A, Sharma AK (2017) Glycoprotein associated with Funneliformis coronatum, Gigaspora margarita and Acaulospora scrobiculata suppress the plant pathogens in vitro. Asian J Plant Pathol 11(4):199–202. https://doi.org/10.3923/ajppaj.2017.199.202

Sharma I, Kashyap S, Agarwala N (2023) Biotic stress-induced changes in root exudation confer plant stress tolerance by altering rhizospheric microbial community. Front Plant Sci 14:1132824. https://doi.org/10.3389/fpls.2023.1132824

Smith SE, Read DJ (2008) Mycorrhizal symbiosis, 3rd edn. Academic, New York

Song YY, Cao M, Xie LJ, Liang XT, Zeng RS, Su YJ, Huang JH, Wang RL, Luo SM (2011a) Induction of DIMBOA accumulation and systemic defense responses as a mechanism of enhanced resistance of mycorrhizal corn (Zea mays L.) to sheath blight. Mycorrhiza 21(8): 721–731. https://doi.org/10.1007/s00572-011-0380-4

Song YY, Wang RL, Wei XC, Lu YJ, Tang ZY, Wu GZ, Zeng RS (2011b) Mechanism of tomato plants enhanced disease resistance against early blight primed by arbuscular mycorrhizal fungus Glomus versiforme. Ying Yong Sheng Tai Xue Bao 22(9):2316–2324

Songachan LS (2023) The usage of arbuscular mycorrhizal fungi (Amf) as a biofertilizer. https://doi.org/10.21203/rs.3.rs-2559546/v1

Stotz HU, Mitrousia GK, de Wit PJ, Fitt BD (2014) Effector-triggered defence against apoplastic fungal pathogens. Trends Plant Sci 19(8):491–500. https://doi.org/10.1016/j.tplants.2014.04.009

Ujvári G, Turrini A, Avio L, Agnolucci M (2021) Possible role of arbuscular mycorrhizal fungi and associated bacteria in the recruitment of endophytic bacterial communities by plant roots. Mycorrhiza 31(5):527–544. https://doi.org/10.1007/s00572-021-01040-7

Usharani TR, Ganachari M, Sowmya HD (2022) Omics in vegetable crops under the family Solanaceae. In: Omics in horticultural crops. Academic, pp 303–332. https://doi.org/10.1016/B978-0-323-89905-5.00022-7

van der Heijden MGA, Martin FM, Selosse MA, Sanders IR (2015) Mycorrhizal ecology and evolution: the past, the present, and the future. New Phytol 205(4):1406–1423. https://doi.org/10.1111/nph.13288

Villani A, Tommasi F, Paciolla C (2021) The arbuscular mycorrhizal fungus *Glomus viscosum* improves the tolerance to verticillium wilt in artichoke by modulating the antioxidant defense systems. Cells 10(8):1944. https://doi.org/10.3390/cells10081944

Wang C, Li X, Song F (2012) Protecting cucumber from Fusarium wilt with arbuscular mycorrhizal fungi. Commun Soil Sci Plant Anal 43(22):2851–2864. https://doi.org/10.1080/00103624.2012.728263

Wang M, Schäfer M, Li D, Halitschke R, Dong C, McGale E, Paetz C, Song Y, Li S, Dong J, Heiling S, Groten K, Franken P, Bitterlich M, Harrison MJ, Paszkowski U, Baldwin IT (2018) Blumenols as shoot markers of root symbiosis with arbuscular mycorrhizal fungi. eLife 7: e37093. https://doi.org/10.7554/eLife.37093

Wang M, Wang Z, Guo M, Qu L, Biere A (2023) Effects of arbuscular mycorrhizal fungi on plant growth and herbivore infestation depend on availability of soil water and nutrients. Front Plant Sci 14:1101932. https://doi.org/10.3389/fpls.2023.1101932

Watts-Williams SJ, Gill AR, Jewell N, Brien CJ, Berger B, Tran BT et al (2022) Enhancement of sorghum grain yield and nutrition: a role for arbuscular mycorrhizal fungi regardless of soil phosphorus availability. Plants, People, Planet 4(2):143–156. https://doi.org/10.1002/ppp3.10224

Weng W, Yan J, Zhou M, Yao X, Gao A, Ma C, Cheng J, Ruan J (2022) Roles of *Arbuscular mycorrhizal* fungi as a biocontrol agent in the control of plant diseases. Microorganisms 10(7): 1266. https://doi.org/10.3390/microorganisms10071266

Xavier LJ, Boyetchko SM (2004) Arbuscular mycorrhizal fungi in plant disease control. Mycol Ser 21:183–194

Xiao G, Zhang Q, Zeng X, Chen X, Liu S, Han Y (2022) Deciphering the molecular signatures associated with resistance to *Botrytis cinerea* in strawberry flower by comparative and dynamic transcriptome analysis. Front Plant Sci 13:888939. https://doi.org/10.3389/fpls.2022.888939

Yanan W, Xusheng Z, Baozhong Y, Wenchao Z, Jintang G (2015) Biochemical defenses induced by mycorrhizae fungi Glomus mosseae in controlling strawberry Fusarium wilt. Open Biomed Eng J 9:301–304. https://doi.org/10.2174/1874120701509010301

Younas MU, Wang G, Du H, Zhang Y, Ahmad I, Rajput N, Li M, Feng Z, Hu K, Khan NU, Xie W, Qasim M, Chen Z, Zuo S (2023) Approaches to reduce rice blast disease using knowledge from host resistance and pathogen pathogenicity. Int J Mol Sci 24(5):4985. https://doi.org/10.3390/ijms24054985

Yousefi Z, Zanganeh S, Riahi H, Kaya Y (2018) Controlling powdery mildew; Use of Arbuscular mycorrhizal fungi as biocontrol agent instead of chemical fungicides J Life Sci Biomed, 8(5): 77–83

Zhao Y, Vlasselaer L, Ribeiro B, Terzoudis K, Van den Ende W, Hertog M, Nicolaï B, De Coninck B (2022) Constitutive defense mechanisms have a major role in the resistance of woodland strawberry leaves against *Botrytis cinerea*. Front Plant Sci 13:912667. https://doi.org/10.3389/fpls.2022.912667

Zhengjia H, Xiangdong G (1991) Pretransplant inoculation with VA mycorrhizal fungi and Fusarium blight of cotton. Soil Biol Biochem 23(2):201–203. https://doi.org/10.1016/0038-0717(91)90135-7

Ziane SO, Talibi ZEA, Douira A, Amir S, Meddich A, El Modafar C (2023) Synergistic effects of arbuscular mycorrhizal fungi associated to plant growth-promoting rhizobacteria in suppression of soil-borne Fusarium wilt of date palm. Biocatal Agric Biotechnol 102753. https://doi.org/10.1016/j.bcab.2023.102753

Chapter 10
Applications of Arbuscular Mycorrhizal Fungi in Controlling Root-Knot Nematodes

Maryluce Albuquerque da Silva Campos

Abstract *Meloidogyne* nematodes cause diseases in economically important plants. These sedentary endoparasites modify plant roots, creating feeding sites and leading to the formation of root galls. Arbuscular mycorrhizal fungi (AMF) form mutualistic associations with many plants, enhancing nutrient uptake and overall plant health. AMF can also provide protection against pathogens, making them valuable for biocontrol. Studies have shown that AMF can reduce the number of *Meloidogyne* galls and eggs while improving plant growth and nutrient absorption, potentially outperforming chemical pesticides. AMF affect *Meloidogyne* infection at various stages, such as making roots less attractive to nematodes and reducing giant cell formation in galls. There is an increase in the production of protective molecules, compounds, and defense genes in mycorrhizal plants infected by *Meloidogyne*, standing out phenolic compounds and defense enzymes like peroxidase and polyphenol oxidase. The activation of defense genes and pathways is suggested to play a role in the tolerance of mycorrhizal plants to *Meloidogyne*. However, there is still a need for further research to understand the physiological and genetic modifications that occur in plants infected by *Meloidogyne* and associated with AMF.

Keywords *Meloidogyne* · AMF · Mycorrhiza · Bioprotection · Sustainability

10.1 Introduction

Among phytonematodes, which are nematodes causing diseases in plants, those of the genus *Meloidogyne* stand out for causing diseases in many economically important plants, and in some of these, they are quite aggressive and challenging to control. These nematodes are classified as sedentary endoparasites, settling close to the central cylinder of the plant root, modifying the root cells by creating their own feeding site. In the places where these feeding sites are located, the root becomes

M. A. da Silva Campos (✉)
University of Pernambuco (UPE) Campus Petrolina, Petrolina, Pernambuco, Brazil
e-mail: maryluce.campos@upe.br

G. J. Ahammed, R. Hajiboland (eds.), *Arbuscular Mycorrhizal Fungi and Higher Plants*, https://doi.org/10.1007/978-981-99-8220-2_10

swollen, forming the so-called galls, a term commonly used to name these nematodes, known as root-knot nematodes (Perry and Wright 1998).

Arbuscular mycorrhizal fungi (AMF), found in the soil, form a mutualistic symbiotic association called mycorrhiza with 80% of plants. They absorb nutrients from the soil, which are provided to the plants in exchange for carbohydrates and lipids. This association is quite ancient, and it is believed to have been essential for the migration of plants from aquatic environments to terrestrial ones. AMF belong to the phylum Glomeromycota (Tedersoo et al. 2018), which comprises approximately 300 species that do not exhibit specificity toward their host; in other words, the same species can associate with any plant species. However, preferences exist, considered compatibility between the fungus and the plant, making the association established between the AMF species and the plant species a promising one, with the robust development of both symbionts and better adaptation to the conditions to which they are exposed (Silva et al. 2004; Püschel et al. 2021; Amanifar et al. 2019; Coninx et al. 2017).

Considering the benefits arising from mycorrhization with AMF, tolerance to biotic stresses, such as pathogens, is well documented in the literature. In this way, these fungi can be used in biocontrol, reducing the number of pathogenic individuals, while allowing better plant development and growth and increased nutrient absorption (Hajra et al. 2015).

Numerous articles address the topic of biocontrol using AMF in plants infected by *Meloidogyne*. In most articles, with experiments conducted in greenhouses or in the field, the presence of AMF reduces the number of galls and parasite eggs, while the plant develops well and produces flowers and fruits normally, in addition to having a higher content of nutrients such as P, N, and K, so the plant is in a situation of tolerance to the pathogen in the presence of AMF (Campos 2020). This scenario depends on the species and isolate of AMF used, as well as the plant species and cultivar, and the factors that can influence this association, such as the substrate.

Some articles highlight that the use of AMF in controlling *Meloidogyne* is equal to or greater than the use of chemical pesticides (Saad et al. 2012; Odeyemi et al. 2013; Abo-Korah 2022). Thus, the use of AMF in controlling *Meloidogyne* would be as efficient as the use of chemical products that are harmful to nature, accumulating in the soil and contaminating the plants, while AMF are natural soil organisms that do not cause any damage to nature, on the contrary, they contribute to the maintenance of ecosystems and cultivation areas.

It is worth noting that the addition of other beneficial microorganisms, as well as natural substrates used in plant production, such as organic fertilizers, plant extracts with nematicidal effects, can positively contribute to the beneficial effect of AMF.

Although there are numerous articles dealing with the use of AMF for *Meloidogyne* control, most only highlight the best results according to the tests performed, providing specific information about plant growth, production, and the reduction of *Meloidogyne* quantity. Few articles address possible physiological and genetic modifications, making this topic still in its early stages and in need of elucidation for a better understanding of the action of AMF in infected plants. Therefore, this chapter aims to provide a brief description of the main causes related

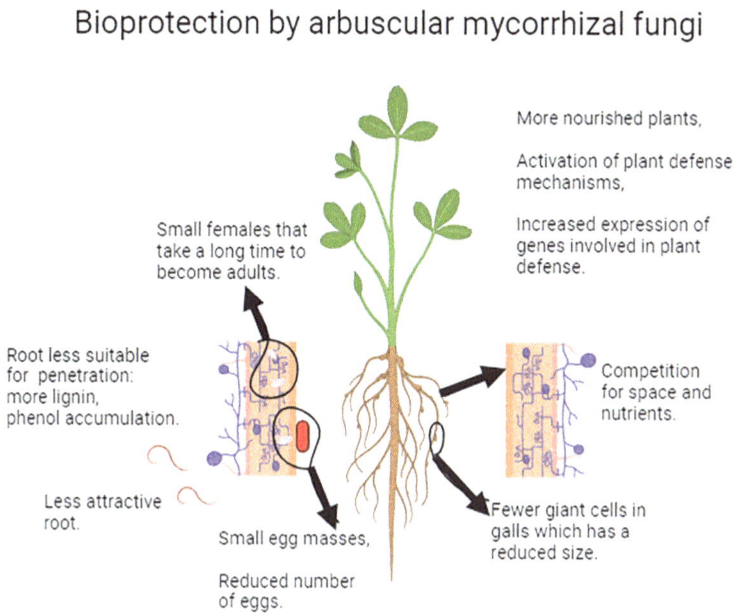

Fig. 10.1 Mechanisms provided by arbuscular mycorrhizal fungi in the protection of plants infected by *Meloidogyne*. Created in BioRender.com

to the bioprotection conferred by AMF on plants infected by *Meloidogyne* and address the latest developments on this topic, citing some possible strategies (Fig. 10.1).

10.2 Key Known Causes of Mycorrhizal Plant Tolerance to *Meloidogyne*

The mechanisms involved in reducing the quantity of *Meloidogyne* in plants associated with AMF range from competition for space and nutrients to the activation of plant defense mechanisms, with an increase in the number of genes involved in this process, as well as possible protective molecules (Sá and Campos 2020; Campos 2020; Ahamad et al. 2023; Malviya et al. 2023). However, studies regarding the biochemical, physiological, and genetic description of protection against *Meloidogyne* modulated by AMF are scarce; thus, there are still many gaps regarding this topic.

It is important to highlight that plants associated with AMF are more robust, as these fungi enhance nutrient and water absorption, making these plants better nourished and vigorous and thus better equipped to withstand the presence of the pathogen (Kolawole et al. 2018; Campos 2020). Table 10.1 summarizes the effects on AMF, *Meloidogyne*, and plant parameters in experiments with AMF and *Meloidogyne*. Furthermore, the presence of AMF associating with plants can positively influence soil microbiota by stimulating other beneficial organisms (Tiwari et al. 2017; Campos 2020).

A highly enlightening aspect related to the tolerance of mycorrhizal plants to *Meloidogyne* is the negative influence during the preinfection, infection, and postinfection phases exerted by AMF. In the preinfection phase, it was observed that the presence of AMF in plants rendered the roots less attractive for nematode infection. This could be attributed to a reduction in the quantity of root exudates or alteration of these exudates. Another possibility could be the accumulation of phenols, such as phytoalexins, flavonoids, and isoflavonoids, which render the root less suitable for pathogen penetration (Morandi 1996). In addition to an increase in the amount of lignin in the roots (Malviya et al. 2023).

In regard to infection, a lower quantity of giant cells in galls was observed in the presence of FMA (Siddiqui and Mahmood 1998), in addition to the reduced size of these cells (Hajra et al. 2015). Another characteristic observed in roots associated with FMA is smaller nematodes, which take longer to develop into adults (Vos et al. 2012a). These modifications result in lower egg production by *Meloidogyne*, reducing the reproductive rate, a characteristic associated with the postinfection phase, as observed in most articles related to this topic (Campos 2020).

The causes mentioned above require more detailed studies that can elucidate how each one occurs, which molecules are involved, and which mechanisms could be employed, in addition to the steps involved.

10.3 Protective Molecules, Compounds and Defense Genes in Mycorrhizal Plants Infected by *Meloidogyne*

The most innovative study regarding the utilization of FMA in the control of *Meloidogyne* involves the qualification and quantification of protective molecules, defense compounds, as well as defense-related genes. Several molecules and compounds produced by the plant are considered important for plant defense. While there are reports of many different types of defense molecules and compounds, few are known in relation to mycorrhizal plants infected by *Meloidogyne*. Most articles quantify these defense compounds, most of which belong to the phenols group. When it comes to molecules, which are also quantified and/or tested for activity, defense enzymes stand out in the studies.

Positive correlations of AMF with plant growth, phenolic content, photosynthetic pigments, and defense enzyme activity were observed by Ahamad et al. (2023) in mycorrhizal and *Meloidogyne incognita*-infected carrot plants (*Daucus carota* L.). In the same article, the authors observed that the activity of peroxidase and polyphenol oxidase, defense enzymes, increased in the presence of AMF and *Meloidogyne*.

Table 10.1 Effects of the interaction between arbuscular mycorrhizal fungi (AMF) and *Meloidogyne* on different plant species (scientific articles published in the last 5 years)

Plant species	AMF species	*Meloidogyne* species	Plant	AMF	*Meloidogyne*	References
			Parameters			
			Growth	MC SD	G E J	
Guava (*Psidium guajava* L.)	*Gigaspora albida* *Acaulospora longula* *Claroideoglomus etunicatum*	*M. enterolobii*	↑ ↑ ↑	↑ NO ↑ NO ↑ NO	NO ↓ NO NO ↓ NO NO ↓ NO	Sá and Campos (2020)
Sweet pepper (*Capsicum annuum L.)*	*Funneliformis geosporum* *Claroideoglomus claroideum* *Glomus ambisporum*	*M. incognita*	↑ ↑ ↑	NO NO NO NO NO NO	↓ ↓ ↓ ↓ ↓ ↓ ↓ ↓ ↓	Herrera-Parra et al. (2021)
Sweet pepper	*Funneliformis mosseae Septoglomus deserticola* *Gigaspora gigantea*	*M. incognita*	↑ ↑ ↑	NE NO NE NO NE NO	↓ NO NO ↓ NO NO ↓ NO NO	Udo et al. (2022)
Pepper var. California Wonder	*Rhizoglomus fasciculatum*	*M. incognita*	NE	↑ NO	↓ ↓ NO	Giri et al. (2022)
Rice (*Oryza sativa* L.)	*F. mosseae* *R. fasciculatus* *R. intraradices*	*M. graminicola*	↑ ↑ ↑	NO NO NO NO NO NO	NO ↓ ↓ NO ↓ NO NO ↓ NO	Malviya et al. (2023)

(continued)

Table 10.1 (continued)

Plant species	AMF species	*Meloidogyne* species	Plant	AMF	*Meloidogyne*	References
			Parameters			
			Growth	MC SD	G E J	
Faba bean (*Vicia faba* L.)	*Mix of genera Glomus, Gigaspora and Acaulospora.*	*M. incognita*	↑	↓ NO	↓ NO ↓	El-Shafeey et al. (2019)
Coffee (*Coffea arabica* L.)	Arbuscular mycorrhizal fungi consortia	*Meloidogyne* spp.	↑	NE NO	NO NO NO	Vallejos-Torres et al. (2020)
Basil (*Ocimum basilicum* L.)	*C. etunicatum* *Rhizophagus clarus*	*M. javanica*	↑ ↑	NO NO	NO ↓ ↓ NO ↓ ↓	Silva et al. (2021)
Soybean (*Glycine max* L.)	*C. etunicatum* *R. clarus*	*M. incognita*	NE NE	NO NO NO NO	NO ↓ ↓ NO ↓ ↓	Messa et al. (2020)
Tomato (*Solanum lycopersicum* L.)	*F. mosseae*	*M. javanica*	↑	NO NO	↓ ↓ ↓	Sohrabi et al. (2020)
Tomato “Saint Pierre”	*Rhizophagus irregularis*	*M. incognita*	↑	NO NO	NO ↓ NO NO ↓ NO	Rodriguez-Heredia et al. (2020)
Tomato cv. super strain P	Mixed with arbuscular mycorrhizal fungi	*M. incognita*	↑	NO NO	↓ ↓ ↓	Ismaiel and Abdelaziz (2019)
Tomato cv. Pritchard	Mixed with arbuscular mycorrhizal fungi	*M. incognita*	↑	↑ NO	↓ NO ↓	Alamri et al. (2022)
Eggplant (*Solanum melongena* L.)	*Glomus* sp.	*M. incognita*	↑	NO NO	↓ NO NO	Khalid et al. (2021)
Chickpea (*Cicer arietinum* L.)	*Glomus* sp.	*M. javanica*	↑	NO NO	↓ NO NO	Aljuboori et al. (2022)

MC mycorrhizal colonization, *SD* spore density, *G* number of galls, *E* number of eggs, *J* number of juveniles, ↑ positive effect, ↓ negative effect, *NE* no effect, *NO* not observed

For various phenolic compounds, negative effects on nematode chemotaxis, motility, or survival have been reported (Wuyts et al. 2006). The quantification of phenolic compounds in the root was higher in the treatment with a mixture of AMF in the presence of *M. incognita* in *Impatiens balsamina* L. plants and fertilized soil. The same result was observed regarding the quantification of antioxidants, although in the latter case, the result extended to the aerial part of the plant (Banuelos et al. 2014). In the article, both compounds are considered plant defense compounds. There was a higher concentration of phenolics in cassava (*Manihot esculenta* Crantz.) plants inoculated with AMF in the presence of *Meloidogyne* spp., both in treatments where AMF and nematodes were inoculated simultaneously and in treatments where nematodes were inoculated 30 days after AMF inoculation (Sery et al. 2016).

In chamomile plants (*Matricaria recutita* L.) infected with M. incognita, the inoculation with the AMF *Glomus intraradices*, as well as the bacteria *Bacillus megaterium* and the filamentous fungus *Trichoderma harzianum*, resulted in the best treatment, exhibiting higher content of flavonoids and phenols, along with greater antioxidant capacity (Gupta et al. 2017).

Sharma and Sharma (2017a) in a study exploring the quantification of defense enzymes in tomato plants (*Solanum lycopersicum* L.) infected with *M. incognita* and inoculated with the beneficial bacteria *Pseudomonas jessenii* and the AMF *Rhizophagus irregularis*, the researchers observed that the treatment containing all microorganisms led to an increase in the quantity of polyphenol oxidase, peroxidase, malondialdehyde, hydrogen peroxide, superoxide dismutase, as well as an increase in the quantity of phenols. In another study by the same authors, they observed higher activity of the enzymes polyphenol oxidase, peroxidase, and superoxide dismutase, while there was a reduction in the content of malondialdehyde and hydrogen peroxide compounds in tomato plants inoculated with *R. irregularis* and infected by *M. incognita* (Sharma and Sharma 2017b).

Regarding your statement, in tomato plants infected with *Meloidogyne javanica* and inoculated with *Rhizophagus clarum*, the quantity of peroxidase, as well as phenylalanine ammonia lyase, increased. This increase intensified over the course of several days, reaching its maximum level at day 11 (Silva et al. 2022). In rice (*Oryza sativa* L.) infected by *Meloidogyne graminicola*, treatment using three different species of AMF (*Funneliformis mosseae*, *Rhizophagus fasciculatus*, *Rhizophagus intraradices*) stood out in the production and activity of defense enzymes: peroxidase, ascorbate peroxidase, polyphenol oxidase, superoxide dismutase, and catalase, both in susceptible and resistant cultivars (Malviya et al. 2023). In pea plants cv Arkil (*Pisum sativum* L.) infected by *M. incognita*, treatments using *F. mosseae* or *Aspergillus awamori* in combination with rhizobacteria (*Pseudomonas putida*, *Pseudomonas alcaligenes*, *Paenibacillus polymyxa*) increased the activity of the enzymes: catalase and peroxidase (Akhtar and Panwar 2012).

The inoculation of AMF in *Thymus vulgaris* L. increased the thymol content (phenolic monoterpene) in these plants; this compound has a proven effect in the reduction of phytonematodes in vitro and in the soil. A leaf extract was made from these mycorrhizal plants, and the effect of this extract on *M. incognita* and

M. javanica was tested in vitro, and it was observed that there was a reduction in the number of eggs and juveniles of these nematodes in the treatments containing extracts of mycorrhizal *T. vulgaris* leaves (Sasanelli et al. 2009).

Another factor that has been studied is exudation, in an attempt to understand the factors that influence *Meloidogyne* infection in mycorrhizal plants. To penetrate a root, nematodes first have to be able to orient themselves to a suitable host and site of infection, for which they depend on exudation from the root (Curtis et al. 2009). The addition of root exudates from mycorrhizal tomato plants decreased the root penetration of *M. incognita*; in addition, second-stage juveniles of this nematode were temporarily paralyzed when exposed in vitro to root exudate from mycorrhizal tomato plants (Vos et al. 2012b).

Considering the possible genes involved in tolerance to *Meloidogyne* in plants inoculated with AMF, Vos et al. (2013) studying tomato plants inoculated with *F. mosseae* and infected with *M. incognita* grouped the genes found into categories: defense, protein synthesis, and modification 17%, signal translation 16%, genes involved in secondary metabolism and hormone 9%, primary metabolism 6%, DNA and RNA metabolism 6%, and cell wall and cell organization 5%. Among the genes found related to defense are induced arachidonic acid (DEA 1), lipid transfer family protein, MYB family transcription factor, heat shock protein 81-1, heat shock protein 70, responsive family protein dry, ML as protein 1, WRKY family transcription factor, glutamate cysteine ligase, NADH ubiquinone oxidoreductase subunit, and NADH ubiquinone oxidoreductase. In the same work, the relative abundance of transcription of some of these defense genes mentioned above was evaluated. The authors observed an increase in the transcription of the genes: glutamate cysteine ligase, transcription factor from the MYB family, protein from the lipid transfer family, and transcription factor from the WRKY family when both organisms were present in the roots.

Balestrini et al. (2019) in a study on tomato associated with *R. intraradices* and infected with *M. incognita* observed that transcripts that encode proteins involved in the biosynthesis and modification of the cell wall were negatively regulated in colonized galls. These authors suggest that AMF colonization could induce changes in the nematode feeding site, neutralizing cell expansion. While in grapevine (*Vitis amurensis* Rupr.), transcription of the VCH3 class III chitinase gene was elevated in mycorrhizal roots with *Glomus versiforme* and resulted in increased tolerance against *M. incognita* (Li et al. 2006).

Malviya et al. (2023), studying rice infected by *M. graminicola*, observed, in the treatment with AMF mix, an increase in the expression of four gene groups: key genes involved in the signaling process, genes that codify proteins related to pathogenesis, key genes involved in the phenylpropanoid pathway, and key genes involved in lignin and callose biosynthesis, highlighting that the presence of AMF can improve the defense mechanism.

According to Pozo and Azcón-Aguilar (2007), resistance induced by mycorrhizae would be related to the activation of plant defense genes such as the jasmonate pattern; however, this option has not yet been studied in the presence of

Meloidogyne, and it has not been proven that such this situation occurs in roots infected by the nematode.

As shown above, several articles bring the evaluation of molecules, compounds, and defense genes, with positive results in the presence of AMF in plants infected by *Meloidogyne*, indicating that there is indeed a defense mechanism that is activated in the presence of AMF; however, how each step of the defense process occurs is still unclear, requiring further studies detailing the minutiae of this process.

10.4 Applications of AMF in Controlling Root-Knot Nematodes

Arbuscular mycorrhizal fungi have been recognized for their potential application in controlling root-knot nematodes that can damage the roots of plants and reduce crop yields. Here are some applications of AM fungi in controlling root-knot nematodes:

1. Biological control: Arbuscular mycorrhizal fungi can act as a biological control agent against root-knot nematodes. These fungi provide multiple benefits to the host plant, including enhanced nutrient uptake and improved plant growth. Studies have shown that AM fungi can also suppress the population and activity of root-knot nematodes, therefore reducing their impact on crops.
2. Induced systemic resistance: Arbuscular mycorrhizal fungi have been found to induce systemic resistance in plants, making them more resistant to pathogenic nematodes like root-knot nematodes. Through their interactions with the plants' root systems, AM fungi activate plant defense mechanisms, which can help protect against nematode infection and reduce the severity of root-knot nematode damage.
3. Soil improvement: Arbuscular mycorrhizal fungi play a crucial role in improving soil health and structure. They form a network of hyphae that create channels in the soil, enhancing its water-holding capacity and nutrient availability. This improved soil structure can indirectly affect root-knot nematodes by creating an unfavorable environment for their survival and reproduction.
4. Competition for resources: Arbuscular mycorrhizal fungi compete with root-knot nematodes for nutrients. By efficiently acquiring and transporting nutrients to the plant, AM fungi can limit the availability of resources for root-knot nematodes, potentially reducing their population and impact on plant health.
5. Biocontrol agent carriers: Arbuscular mycorrhizal fungi can also serve as carriers for other biocontrol agents, such as bacteria or fungi, that have been found to suppress root-knot nematode populations.

It is important to note that the effectiveness of using arbuscular mycorrhizal fungi for nematode control can vary depending on various factors such as the specific species of AM fungi, the crop being grown, and the nematode species involved.

Nevertheless, these fungi hold promise as a sustainable and environmentally friendly approach to managing root-knot nematode infestations in agricultural systems.

10.5 Conclusions and Future Perspectives

AMF are important microorganisms in the biocontrol of *Meloidogyne*, and their positive effect has been proven in numerous papers, allowing the development of plants while reducing the amount of the pathogen. The possible causes involved in this protection are well highlighted in the articles, although some were not necessarily tested.

Some combinations of AMF and plant species do not show positive results, so combinations should be tested in order to suggest the best AMF isolates in different cultures against *Meloidogyne* control.

Additionally, little by little some molecules and defense compounds in mycorrhizal plants infected by *Meloidogyne* have been studied, mainly enzymes and phenols, showing that phenols increase in quantity and enzymes increase their activity, which would be related to the reduction of the amount pathogen, thus providing protection to mycorrhizal plants.

In addition, genes involved in defense have also been studied. These genes have their expression increased, contributing to a better defense of plants associated with AMF against *Meloidogyne*. However, the mechanisms involved in how this bioprotection occurs have not yet been elucidated, as observed for other pathogens that already have some descriptions of the possible mechanisms. There are probably several protection formats that will have to be described little by little, being an important gap when considering this topic.

Despite the lack of knowledge of these detailed aspects, it is a fact that the use of AMF can help in plant growth and the control of *Meloidogyne*. As this fungus is naturally present in the soil and can benefit any plant without damaging the environment, its use is appropriate in any location and plant, aiming at better crop performance and reduction of *Meloidogyne*. AMF can be considered a natural, sustainable, and important biofertilizer for ecosystems, thus ensuring healthy food security, free of toxic inputs.

References

Abo-Korah MS (2022) Effectiveness of certain safety commercial nematicides compare with a chemical nematicide Carbofuran for controlling *M. incognita* infected pomegranate plants. Int J Sci Res Sustain Dev 5:1–12

Ahamad L, Bhat AH, Kumar H, Rana A, Hasan MN, Ahmed I, Ahmed S, Machado RAR, Ameen F (2023) From soil to plant: strengthening carrot defenses against *Meloidogyne incognita* with vermicompost and arbuscular mycorrhizal fungi biofertilizers. Front Microbiol 14:1206217

Akhtar MS, Panwar J (2012) Efficacy of root-associated fungi and PGPR on the growth of *Pisum sativum* (cv. Arkil) and reproduction of the root-knot nematode *Meloidogyne incognita*. J Basic Microbiol 52:1–9

Alamri S, Nafady NA, El-Sagheer AM, El-Aal MA, Mostafa YS, Hashem M, Hassan EA (2022) Current utility of arbuscular mycorrhizal fungi and hydroxyapatite nanoparticles in suppression of tomato root-knot nematode. Agronomy 12:1–16

Aljuboori FK, Ibrahim BY, Mohamed AH (2022) Biological control of the complex disease of Rhizoctonia solani and root-knot nematode Meloidogyne javanica on chickpea by Glomus spp. and Pseudonomas sp. Iraqi J Agric Sci 53:669–676

Amanifar S, Khodabandeloo M, Fard EM, Askari MS, Ashrafi M (2019) Alleviation of salt stress and changes in glycyrrhizin accumulation by arbuscular mycorrhiza in liquorice (*Glycyrrhiza glabra*) grown under salinity stress. Environ Exp Bot 160:25–34

Balestrini R, Rosso LC, Veronico P, Melillo MT, de Luca F, Fanelli E, Colagiero M, Fossalunga AS, Ciancio A, Pentimone I (2019) Transcriptomic responses to water déficit and nematode infection in mycorrhizal tomato roots. Front Microbiol 10:1–17

Banuelos J, Alarcon A, Larsen J, Cruz-Sanchez S, Trejo D (2014) Interactions between arbuscular mycorrhizal fungi and *Meloidogyne incognita* in the ornamental plant *Impatiens balsamina*. J Soil Sci Plant Nutr 14:63–74

Campos MAS (2020) Bioprotection by arbuscular mycorrhizal fungi in plants infected with *Meloidogyne* nematodes: a sustainable alternative. Crop Prot 135:105203

Coninx L, Martinova V, Rineau F (2017) Mycorrhiza-assisted phytoremediation. In: Cuypers A, Vangronsveld J (eds) Advances in botanical research, 1st edn. Elsevier, pp 127–188

Curtis RHC, Robinson AF, Perry RN (2009) Hatch and host location. In: Perry RN, Moens M, Starr JL (eds) Root-knot nematodes. CAB International, Wallingford, pp 139–162

El-Shafeey EI, Abd-El-Hadi MA, Hagag ES, Abu El-Naga GS (2019) Utilization of organic and bio fertilizers against root-knot nematode (*Meloidogyne incognita*) infecting faba bean (*Vicia faba* L.). Bulg J Agric Sci 25:506–513

Giri B, Rawat R, Saxena G, Manchanda P, Wu KS, Sharma A (2022) Effect of *Rhizoglomus fasciculatum* and *Paecilomyces lilacinus* in the biocontrol of root-knot nematode, *Meloidogyne incognita* in *Capsicum annuum* L. Commun Integr Biol 15:75–78

Gupta R, Saikia SK, Pandey R (2017) Bioconsortia augments antioxidant and yield in *Matricaria recutia* L. against *Meloidogyne incognita* (Kofoid and White) Chitwood infestation. Proc Natl Acad Sci India 87:335–342

Hajra N, Shahina F, Firoza K, Maria R (2015) Damage induced by root-knot nematodes and its alleviation by vesicular arbuscular mycorrhizal fungi in roots of *Luffa cylindrica*. Pak J Nematol 33:71–78

Herrera-Parra E, Ramos-Zapata J, Basto-Pool C, Cristóbal-Alejo J (2021) Sweet pepper (Capsicum annuum) response to the inoculation of native arbuscular mycorrhizal fungi and the parasitism of root-knot Meloidogyne incognita. Rev Biocienc 8:e982

Ismaiel G, Abdelaziz S (2019) Effect of arbuscular mycorrhizal fungi and some plant growth promoting rhizobacteria in controlling root-knot nematode (Meloidogyne incognita) on tomato under greenhouse conditions. Desert Res Center 69:131–150

Khalid H, Aminuzzaman FM, Amit K, Faria AA, Mitu AI, Chowdhury MSM, Shammi J, Khan MA (2021) Evaluation of the combined application of *Purpureocillium lilacinum* PLSAU-1 and *Glomus* sp. against *Meloidogyne incognita*: implications for arsenic phytotoxicity on eggplant. Eur J Plant Pathol 159:139–152

Kolawole GO, Haastrup TM, Olabiyi TI (2018) Can arbuscular mycorrhiza fungi and NPK fertilizer suppress nematodes and improve tuber yield of yam (*Dioscorea rotundata* 'cv' ewuru)? Eurasian J Soil Sci 7:181–186

Li H-Y, Yang G-D, Shu H-R, Yang Y-T, Ye B-X, Nishida I, Zheng C-C (2006) Colonization by the arbuscular mycorrhizal fungus *Glomus versiforme* induces a defense response against the root-knot nematode *Meloidogyne incognita* in the grapevine (*Vitis amurensis* Rupr.), which includes transcriptional activation of the class III chitinase gene VCH3. Plant Cell Physiol 47:154–163

Malviya D, Singh P, Singh UB, Paul S, Bisen PK, Rai JP, Verma RL, Fiyaz RA, Kumar A, Kumari P, Dei S, Ahmed MR, Bagyaraj DJ, Singh HV (2023) Arbuscular mycorrhizal fungi-mediated activation of plant defense responses in direct seeded rice (*Oryza sativa* L.) against root-knot nematode *Meloidogyne graminicola*. Front Microbiol 14:1104490

Messa VR, Costa ACT, Kuhn OJ, Stroze CT (2020) Nematophagous and endomycorrhizal fungi in the control of *Meloidogyne incognita* in soybean. Rizhosphere 15:100222

Morandi D (1996) Occurrence of phytoalexins and phenolic compounds in endomycorrhizal interactions, and their potential role in biological control. Plant Soil 185:241–251

Odeyemi IS, Afolami SO, Adekoyejo AB (2013) Integration of *Glomus mosseae* with *Chromolaena odorata* powder for suppression of *Meloidogyne incognita* on maize (Zea mays L.). Arch Phytopathol Plant Protect 46:1589–1597

Perry RN, Wright DJ (1998) The Physiology and Biochemistry of free-living and plant-parasitic nematodes. CABI Publishing, London

Pozo MJ, Azcón-Aguilar C (2007) Unraveling mycorrhiza-induced resistance. Curr Opin Plant Biol 10:393–398

Püschel D, Bitterlich M, Rydlová J, Jansa J (2021) Drought accentuates the role of mycorrhiza in phosphorus uptake. Soil Biol Biochem 157:108243

Rodriguez-Heredia M, Djian-Caporalino C, Ponchet M, Lapeyre L, Canaguier R, Fazari A, Marteu N, Industri B, Offroychave M (2020) Protective effects of mycorrhizal association in tomato and pepper against *Meloidogyne incognita* infection, and mycorrhizal networks for early mycorrhization of low mycotrophic plants. Phytopathol Mediterr 59(2):377–384

Sá CSB, Campos MAS (2020) Arbuscular mycorrhizal fungi decrease Meloidogyne enterolobii infection of Guava seedlings. J Helminthol 94:1–5

Saad AFSA, Massoud MA, Ibrahim HS, Khalil MS (2012) Activity of Nemathorin, natural product and bioproducts against root-knot nematodes on tomato. Arch Phytopathol Plant Protect 45: 955–962

Sasanelli N, Anton A, Takacs T, D'Addabbo TD, Biro I, Malov X (2009) Influence of arbuscular mycorrhizal fungi on the nematicidal properties of leaf extracts of *Thymus vulgaris* L. Helminthologia 46:230–240

Sery DJM, Kouadjo ZGC, Voko BRR, Zeze A (2016) Selecting native arbuscular mycorrhizal fungi to promote cassava growth and increase yield under field conditions. Front Microbiol 7:1–13

Sharma IP, Sharma AK (2017a) Physiological and biochemical changes in tomato cultivar PT-3 with dual inoculation of micorrhiza and PGPR against root-knot nematode. Symbiosis 71:175–182

Sharma IP, Sharma AK (2017b) Co-inoculation of tomato with an arbuscular mycorrhizal fungus improves plant immunity and reduces root-knot nematode infection. Rhizosphere 4:25–28

Siddiqui ZA, Mahmood I (1998) Effect of plant growth promoting bacterium, an AM fungus and soil types on the morphometrics and reproduction of *Meloidogyne javanica* on tomato. Appl Soil Ecol 8:77–84

Silva MA, Cavalcante UMT, Silva FSB, Soares SAG, Maia LC (2004) Crescimento de mudas de maracujazeiro-doce (*Passiflora alata* Curtis) associadas a fungos micorrízicos arbusculares (Glomeromycota). Acta Bot Bras 18:981–985

Silva BA, Cruz RMS, Miamote A, Alberton O, Silva C, Dias-Arieira CR (2021) Interaction between mycorrhizal fungi and 'Meloidogyne javanica' on the growth and essential oil composition of basil ('Ocimum basilicum'). Aust J Crop Sci 15:416–421

Silva MTR, Silva BA, Alberton O, Schwengber RP, Dias-Arireira CR (2022) Rhizophagus clarus controls *Meloidogyne* javanica and enhances the activity of defense-related enzymes in tomato. Hortic Bras 40:162–167

Sohrabi F, Sheikholeslami M, Heydari R, Rezaee S, Sharifi R (2020) Investigating the effect of *Glomus mosseae*, *Bacillus subtilis* and *Trichoderma harzianum* on plant growth and controlling *Meloidogyne javanica* in tomato. Indian Phytopathol 73:293–300. https://doi.org/10.1007/s42360-020-00227-w

Tedersoo L, Sánchez-Ramírez S, Koljalg U, Bahram M, Doring M, Schigel D, May T, Ryberg M, Abarenkov K (2018) High-level classification of the Fungi and a tool for evolutionary ecological analyses. Fungal Divers 90:135–159

Tiwari S, Pandey S, Chauhan PS, Pandey R (2017) Biocontrol agents in co-inoculation manages root knot nematode [*Meloidogyne icognita* (Kofoid and White) Chitwood] and enhances essential oil content in *Ocimum brasilicum* L. Ind Crop Prod 97:292–301

Udo IA, Uko AE, Obok EE, Ubi JO, Umoetok SBA (2022) Management of Meloidogyne incognita and salinity on sweet pepper (Capsicum annuum L.) with different arbuscular mycorrhizal fungus species. J Appl Biol Biotechnol 10(4):66–72

Vallejos-Torres G, Espinoza E, Marín-Díaz J, Solis R, Arévalo LA (2020) The role of arbuscular mycorrhizal fungi against root-knot nematode infections in coffee plants. J Soil Sci Plant Nutr 21:364–373. https://doi.org/10.1007/s42729-020-00366-z

Vos C, Geerinckx K, Mkandawire R, Panis B, De Waele D, Elsen A (2012a) Arbuscular mycorrhizal fungi affect both penetration and further life stage development of root-knot nematodes in tomato. Mycorrhiza 22:157–163

Vos C, Claerhout S, Mkandawire R, Panis B, De Waele D, Elsen A (2012b) Arbuscular mycorrhizal fungi reduce root-knot nematode penetration through altered root exudation of their host. Plant Soil 354:335–345

Vos C, Schouteden N, van Tuinen D, Chatagnier O, Elsen A, de Waele D, Panis B, Gianinazzi-Pearson V (2013) Mycorrhiza-induced resistance against the root-Knot nematode *Meloidogyne* incógnita involves priming of defense gene responses in tomato. Soil Biol Biochem 60:45–54

Wuyts N, Swennen R, De Waele D (2006) Effects of plant phenylpropanoid pathway products and selected terpenoids and alkaloids on the behaviour of the plant-parasitic nematodes Radopholus similis, Pratylenchus penetrans and *Meloidogyne* incognita. Nematology 8:89–101

Chapter 11
Glomalin and Carbon Sequestration in Terrestrial Ecosystems

Nasser Aliasgharzad and Elham Malekzadeh

Abstract The fungi in *Glomeromycota* are mutualistic symbionts of plant roots and produce a special glycoprotein called "glomalin" on their spores and mycelium cell wall. The glomalin enters adjacent soil after cell wall death and decomposition. It contains 3–5% nitrogen and 36–59% carbon with considerable amounts of iron (0.8–8%). Glomalin is considered a recalcitrant source of carbon. The half-life of glycoprotein is approximately 50 years, so it has a relatively long persistence in soil. Therefore, it could contribute to the sequestration of carbon in land-based ecosystems. The rate of carbon flow from the plant to the underground parts and then to the fungal symbionts affects the amount of glomalin synthesis by the fungi. The impact of different environmental factors such as nutrient availability, tillage, atmospheric CO_2 level, drought, salinity, and heavy metal toxicity stresses on carbon allocation to the fungi and its consequence on the amount of glomalin production are addressed here. Also, the contribution of glomalin in carbon sequestration in soils is discussed.

Keywords Arbuscular mycorrhizal fungi · Climate change · Glomalin · Soil carbon pool

11.1 Introduction

Arbuscular mycorrhizal (AM) fungi belonging to the phylum *Glomeromycota* establish endomycorrhizal symbiosis with roots of numerous plant species (Smith and Read 2008). Some plant families such as Chenopodiaceae, Brassicaceae, and Amaranthaceae are considered non-mycorrhizal plants (Brundrett 2009). AM fungi (AMF) are obligate root symbionts with a lack of host plant specificity and benefit

N. Aliasgharzad (✉)
Department of Soil Science, Faculty of Agriculture, University of Tabriz, Tabriz, Iran
e-mail: n-aliasghar@tabrizu.ac.ir

E. Malekzadeh
Department of Soil Science, Gorgan University of Agricultural Sciences and Natural Resources, Gorgan, Iran

G. J. Ahammed, R. Hajiboland (eds.), *Arbuscular Mycorrhizal Fungi and Higher Plants*, https://doi.org/10.1007/978-981-99-8220-2_11

from plant photosynthates as carbon and energy sources. In turn, they supply plants with mineral nutrients prevailingly phosphorus. In this type of endomycorrhiza, intraradical hyphae penetrate the root cortex and produce vesicles and arbuscules as specialized organs inside the root cells, while extraradical hyphae extend into the soil and produce spores. The main role of extraradical hyphae is to absorb nutrients from the soil and deliver them to the root cells (Smith and Read 1997). These fungi are the most widespread group occurring from frozen to tropic, saline to nonsaline, acidic to alkaline, and dry to wet soils. The higher abundance and diversity of them are found in broad-leaved forests and grasslands compared to cultivated ecosystems (Smith and Read 1997). Since AMF communities mostly reside in the topsoil, tillage reduces AMF. Therefore, deep tillage may result in the dilution of AMF propagules in large soil volumes, and this may decrease root colonization. Moreover, plowing greatly damages the AMF hyphae when carried out in the fall, which leads to the detachment of the hyphae from the host plant, which diminishes fungal survival. Compared to conventional tillage, zero or minimum tillage conserves mycorrhizal network in soil and increases mycelia viability (Kabir 2005). In this way, the crop sown the next season can be attached to the network faster to be colonized with AMF.

Soil depth and land use influence the AMF abundance in soil. Egboka et al. (2022) studied the spore densities of AMF in Ibadan and Ikwuano, Nigeria, at 0–45 cm soil depths with 15 cm intervals that had land use types of fallow and cassava and pineapple cultivation. In both areas, the fallow land use had a significantly higher spore density compared with cassava and pineapple cultivation. Spore abundance significantly declined across the soil depths in Ibadan, and the average number of spores per 100 g of soil at 0–15, 15–30, and 30–45 cm depths were 54 ± 6, 45 ± 3, and 39 ± 5, respectively. However, the number of spores in 100 g of soil in Ikwuano ranged from 64 to 67 in the studied depths, which had no significant difference. Indeed, they found a positive correlation ($r = 0.91$) between spore density and soil exchangeable K^+ but a negative correlation ($r = -0.83$) with total N (Egboka et al. 2022). Other researchers have reported that AMF communities in deeper soil layers have different spore density and species diversity compared with those from the top layers (Muleta et al. 2008; Oehl et al. 2005; Yang et al. 2010). In contrast, a study carried out in a site contaminated with heavy metals showed that spore number and root colonization percent increased as soil depth increased to 60 cm (Gucwa-Przepióra et al. 2013).

Ecological studies have revealed that intensified land uses lead to decreased AMF populations and their species diversity (Oehl et al. 2003). The species richness of AMF is usually lower in arable lands with annuals than in forest and grasslands with perennial plants (Snoeck et al. 2010). However, Ndoye et al. (2012) pointed out that changing land use from grassland to Acasia cultivation enhanced AMF diversity and spore abundance.

It has been established that different plant species respond differently to AMF species; therefore, the diversity and abundance of indigenous AMF will decrease as a result of continued monoculture (Sanders et al. 1995). Aliasgharzad et al. (2001) studied the abundance of AMF spores and their root length colonization (RLC) in the rhizosphere of four glycophytes (onion, alfalfa, wheat, and barley), along with *Salicornia* sp. and *Salsola* sp. (two halophyte species) in the Tabriz Plain in the northwest of Iran. The area has saline soils, and the mean electrical conductivity (EC) of 7.3, 12.3, 12.1, 21.1, and 92.0 dS m^{-1} were recorded in rhizosphere soils of onion, alfalfa, wheat, barley, and halophytes, respectively. The spores count in the rhizosphere significantly varied depending on the plant species, and the mean values of 144.8, 129.8, 119.5, and 115.3 spores per 10 g dry soil were found in onion, alfalfa, wheat, and barley, respectively. In comparison, the rhizosphere soils with *Salsola* sp. and *Salicornia* sp. plants showed relatively lower spore density (~100 spores per 10 g dry soil). The percent of RLC was also influenced by plant species and soil salinity, and the mean values of 32.8, 30.8, 11.2, and 4.5% were reported for onion, alfalfa, wheat, and barley roots, respectively. It is noteworthy that the roots of these halophytes were not colonized by AMF, because both species belong to the Chenopodiceae which are known as non-mycorhizal plants (Brundrett 2009).

11.2 Glomalin: Definition and Structure

The biochemical structure of glomalin is unknown and is usually defined by its extraction method (Wright et al. 1996; Rillig 2004). Glomalin is found on the hyphal and spore cellular wall of AM fungi (Wright and Upadhyaya 1996; Driver et al. 2005; Aliasgharzad et al. 2016), which was discovered by Wright et al. (1996) on the hyphae of AM fungi using indirect immunofluorescence method (Wright 2000; Nichols and Wright 2004). Evidence suggests that AM fungi, and not plant roots, are responsible for glomalin production (Rillig and Steinberg 2002; Matos et al. 2022). Two mechanisms are suggested for the release of glomalin to the soil: (1) secretion from the hyphae and arbuscules of the fungi to colonized roots, organic matter, soil particles, and the surrounding environment of external mycelium and (2) its release after decomposition of AMF structures (Wright et al. 1996; Wright and Upadhyaya 1996; Wright 2000). Despite an unknown biochemical structure, glomalin is a glycoprotein consists of various inorganic elements such as phosphorus, iron, magnesium, calcium, and potassium (Ullah et al. 2019) and different substance groups (Fig. 11.1). Glomalin contains 0.8–8% iron (Ullah et al. 2019), which is essential for its accumulation or function. Iron is also the reason behind the yellow to reddish color of glomalin extract (Wright and Upadhyaya 1998; Rillig et al. 2001). It has been reported that in soils with low content of iron, the accumulation of glycoprotein was lower, and an isolate of AMF called *Gigaspora gigantea* (MA453A) was unable to grow under iron limitation. This is confirmed by field reports, as high soil pH resulted in decreased glomalin concentration and aggregate stability due to the low availability of iron (Wright and Upadhyaya 1998). It has also been reported that the presence of iron in the glomalin structure increases its thermal stability and antimicrobial properties (Prasad et al. 2018).

The dynamics of glomalin accumulation on extracellular hyphae, auxiliary cells, spores, or internal structures (intracellular hyphae, arbuscules, and vesicles) are still

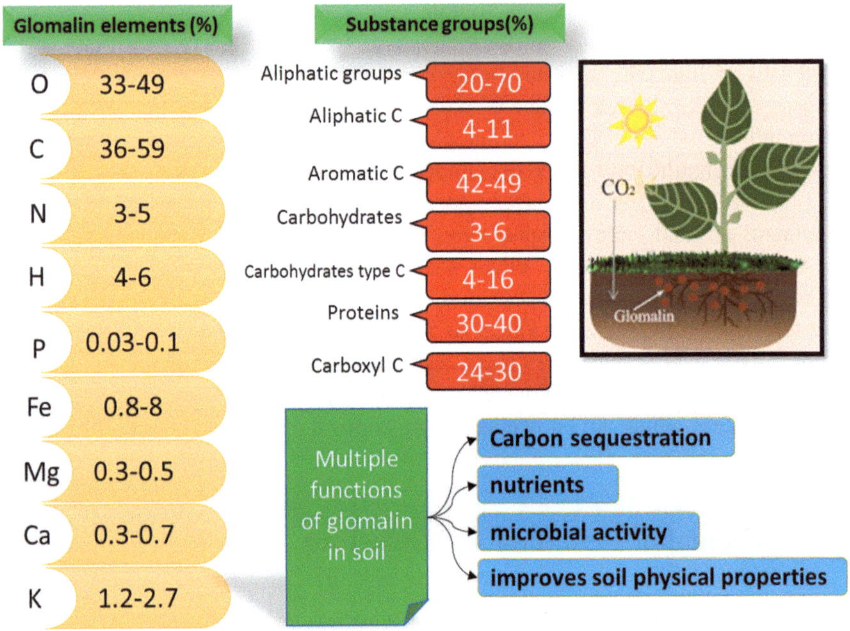

Fig. 11.1 Elemental composition, substance groups, and multiple functions of glomalin in soil. (Modified and redrawn from Ullah et al. 2019; Singh et al. 2022; Agnihotri et al. 2022)

lacking information. Glomalin is stable, is not solved in water, and can resist thermal decomposition; however, exposure to high temperature (121 °C) and alkaline buffers (citrate, borate, or pyrophosphate at pH 7 to 9) facilitate its decomposition (Wright et al. 1996; Wright and Upadhyaya 1996). The glue-like nature of glomalin enables this compound to bind to the surface of soil particles and cell membranes, so it probably forms a hydrophobic coating on the AM fungi hyphae to decrease the loss of soluble substances (Wright and Upadhyaya 1998, 1999). AM fungi-produced glomalin can resist chemical (acid) and trypsin hydrolysis (Wright et al. 1996). Based on the lectin-binding affinity and HPCE, [1] glomalin is a glycoprotein with a major asparagine-linked (N-linked) chain of carbohydrates (Wright et al. 1998). Denaturation and deglycosylation of glomalin by HPCE were unsuccessful, since before the decomposition, the protein that remains completely intact is deposited in the solution and separated by centrifugation. As a result, some components of glomalin are probably resistant to denaturation and deglycosylation (Nichols 2003). Thermal treatment denatures heat-nonstable proteins extracted along with glomalin (Wright et al. 1996). These denatured proteins as well as some other small proteins can be removed in the course of primary purification through acid precipitation, redissolution in alkaline solution, and dialysis processes. Since glomalin can

[1] High-performance capillary electrophoresis.

resist decomposition, it can be a part of organic matter found in labile and resistant sources in soil after 10 years at minimum (Rillig et al. 2001; Steinberg and Rillig 2003). Molecular stability of glomalin is due to its chemical properties including hydrophobicity and the presence of iron bonds. Hydrophobicity and iron bonds of glomalin make it resistant to micribial degradation. (Agnihotri et al. 2022).

Nichols (2003) performed methods such as removal of iron, measurement of iron content, purification by phenyl-HPLC [2] column, and separation of amino acids and carbohydrates for analyzing the components of glomalin. In the phenyl column, hydrophobic groups including amino acids of tryptophan, leucine, phenylalanine, isoleucine, valine, alanine, and tyrosine) and aliphatics are brought to the protein surface by increasing the surface tension in the buffer solution with ammonium sulfate salt. The phenyl column is highly hydrophobic, and proteins adhere to the column rather than each other. The surface tension decreases as a result of the reduced concentration of salts in the buffer solution, and the protein is separated from the column. Purification of the hydrophobic glycoproteins or oligosaccharides in solutions can be done by lectin affinity chromatography. According to the results of the analysis, iron and organic matter may be attached to the soil glomalin and/or during its extraction processes. This makes it difficult to analyze the structure of glomalin, as the glomalin molecule becomes more stable and complex as a result of the binding of organic matters and iron (Nichols 2003).

11.3 Methods for Assaying Glomalin

The extraction of glomalin from hyphae and soil is undertaken using sodium citrate buffer, followed by autoclaving for 30–60 min or more (Wright et al. 1996; Wright and Upadhyaya 1996). Different extraction methods are used for different fractions of glomalin including total glomalin (TG) and easily extractable glomalin (EEG). The extraction of EEG is carried out by autoclaving soil/hyphae in 20 mM sodium citrate buffer with a pH of 7 for 30 min, while TG is extracted with the help of 50 mM sodium citrate buffer (pH = 8) for 60 min. Total glomalin is extracted in 1 h; however, it may take more time (Wright and Upadhyaya 1998). In addition to the difference in the extract concentration and the time needed for TG extraction, the process continues until the color of the extract turns amber or colorless which is obtained after three or five, seven (Wright and Upadhyaya 1998), or nine cycles (Rillig et al. 2003). Immediately after autoclaving, the extracted sample is separated from the soil using a centrifuge to make sure that the soil particles are separated from glomalin. Due to the proteinous nature of glomalin, the extract is kept at 4 °C (Wright et al. 1996).

[2] High-performance liquid chromatography (HPLC).

Bradford assay is a general method for protein measurement (Bradford 1976). The basis for Bradford's test is the binding of protein to Coomassie Brilliant Blue G-250, which changes the color of the reaction from red to blue after binding to the protein (Bradford 1976; Wright et al. 1996). The intensity of this color change is read at 595 nm which pertains to the concentration of protein in the glomalin extract. The protein content in the extract is calculated as total or easily extractable Bradford-reactive soil protein (T-BRSP or EE-BRSP) (Rillig 2004). Some researchers believed that total glomalin represents the glomalin produced by fungi that has accumulated in the soil over many years, while easily extractable glomalin is newly produced in the soil (Siami et al. 2022; Agnihotri et al. 2022; Wright and Upadhyaya 1998). However, Staunton et al. (2020) rejected this hypothesis and emphasized that the difference in components is due to the increasing of strong stabilization of glomalin-related soil proteins (GRSP) on mineral surfaces by increasing contact period with soil.

Indirect ELISA by monoclonal antibody MAb32B11 which is prepared against *Glomus intraradices* spores is another method to determine the amount of glomalin (Wright et al. 1996). In this method, the anti-glomalin antibody is added to the glomalin extract which binds to the antigenic site (the specific site for the binding of antibody) of the glomalin. Subsequently, the secondary antibody which is the anti-mouse IgM antibody, conjugated by biotin, is added and binds to the antigenic site of the primary MAb32B11 antibody. A solution that contains protein (e.g., ExtAvidin) or enzyme (e.g., peroxidase) and the colored substrate is then added. Protein molecules are attached to biotinylated-secondary antibody, and the enzyme reacts with the substrate, and a blue-green color is produced. The intensity of color change is read at 405 or 410 nm and calculated by an equation of standard curve prepared by glomalin obtained from pot culture or soil samples with 100% immunoreactivity in the concentration range of 0.005–0.04 μg (Wright et al. 1996; Nichols and Wright 2004; Rillig 2004).

Total glomalin measured by ELISA is defined as total-immunoreactive soil protein (T-IRSP), and its easily extractable component is defined as EE-immunoreactive soil protein (EE-IRSP) (Wright and Upadhyaya 1998; Rillig 2004). Rillig (2004) described and used the terms EE-IRSP, BRSP, EE-BRSP, and IRSP instead of the term "glomalin" to describe glomalin extracted from soil. Since other non-glomalin proteins are simultaneously extracted in a hot alkaline buffer (Zbiral et al. 2017), Rillig (2004) recommended the term "GRSP" for this soil extract. Also, similar terms are assigned to various glomalin components extracted from different sources, i.e., from mycorrhizal roots and hyphae. For example, the term "Bradford-reactive root protein" is replaced with "root glomalin" which is related to the glomalin in the root extract estimated by Bradford assay (Rosier et al. 2008). The values for glomalin resulting from the ELISA assay are usually compared with those of Bradford to calculate the percent of immunoreactive protein. This percent is obtained by the division of the ELISA data by the Bradford data ×100. A higher immunoreactivity percentage indicates greater reactivity of extracted glomalin fraction with antibody (Wright et al. 1996; Nichols and Wright 2004; Rillig 2004).

Although Bradford method is not specific for glomalin detection because of the presence of residual humic substances, tannic acids, plant proteins, and heat shock proteins in the glomalin extract (Zbiral et al. 2017; Whiffen et al. 2007; Rosier et al. 2006), data obtained by this method are positively correlated with those obtained by ELISA assay (Wright and Upadhyaya 1996, 1998, 1999; Harner et al. 2004). Many researchers use only the Bradford assay for estimation of the glomalin content (Aliasgharzad et al. 2016; Cisse et al. 2020; Staunton et al. 2020), because it is more rapid and simpler and requires less technical work compared to the ELISA. Cisse et al. (2020) recommended that to avoid the interferences of phenolic compounds in the Bradford assay, soil extract is diluted, and pH-dependent color absorbance is subtracted before quantifying the protein by the Bradford reagent. Rosier et al. (2006) reported that proteins yielded by other organisms can be considered Bradford- or immunoreactive root proteins. These findings indicate that BRSP is not solely related to AM fungi and requires further research (Whiffen et al. 2007).

Near-infrared reflectance spectroscopy (NIRS) is a rapid, cheap, nondestructive, and environmentally friendly technique to determine the basic characteristics of some soil compounds, their chemical and biological properties, and quality indicators (Zbiral et al. 2017). It has been observed that GRSP can be determined with sufficient precision and accuracy simultaneously by measuring nitrogen and oxidizable carbon in agricultural, pasture, and forest soils (Zbiral et al. 2017). Pohanka and Vlcek (2018) used a biosensor based on a piezoelectric "quartz crystal microbalance" (QCM) and a glomalin-specific monoclonal antibody to measure glomalin in an extract. QCM is a type of sensor that uses the "piezoelectric effect" of a thin quartz crystal between two electrodes. A mass change following the attachment of an analyte (the target molecule, here glomalin) to the QCM surface produces a vibrational frequency that is converted into an electrical signal. The piezoelectric biosensor is a simple device that can be used in unequipped laboratories or in the field experiments. In the biosensor method, in contrast to the spectrophotometer (Bradford method), the sample extracted by autoclave can be used directly without using any reagent (Pohanka and Vlcek 2018).

11.4 The Impact of Environmental Factors on Glomalin Production

There are evidences indicating that 40–50% of the assimilated carbon by plants through photosynthesis is transferred to the AM fungi (Harris and Paul 1987), although other work estimates it around 10–20% (Jakobsen et al. 2002). Environmental conditions such as soil nutrient content, atmospheric CO_2 level, salinity, drought, heavy metals, etc. could influence plant photosynthesis, thereby carbon allocation to the fungal partner. The rate of glomalin production by the fungi is a function of photosynthate amounts delivered from plant roots to the fungi. The impact of important environmental factors on glomalin production is addressed here.

11.4.1 Soil Nutrients

Among nutrients, soil phosphorus (P) and nitrogen (N) affect glomalin production via direct or indirect mechanisms. It has been widely accepted that higher levels of the available P in soil inhibit mycorrhizal establishment in roots, thereby reducing fungal biomass inside the root and adjacent soil. Therefore, glomalin production by the fungi is reduced indirectly by increasing available P in the soil. Considering the chemical structure of glomalin (glycoprotein), sufficient availability of N to the mycorrhizal system will directly encourage glomalin production. In the soils with low nutrient availability, the AMF play an important role in delivering nutrients to the plant which in turn receive more carbon from the plant. The extent of carbon allocation to the AMF is a very complex phenomenon and depends on environmental conditions and the plant and fungal species (Egboka et al. 2022). Olsson et al. (2010) reported that the dynamics of carbon during the mycorrhizal symbiosis pertains to carbon costs and phosphorus benefits. In a greenhouse study, they used $^{13}CO_2$ to trace carbon flow from the clover root to the fungus. Only a little amount of plant carbon was allocated to the fungus under the conditions that plant mycorrhizal benefit was decreased by P addition. Balík et al. (2020) conducted a long-term experiment to examine the impact of different fertilizers on the soil glomalin content. The content of EEG, TG, and G_{NIRS} (glomalin determined using the near-infrared reflectance spectroscopy) were determined in topsoil. They reported that the mineral and organic fertilizers and their combination led to a significant increase in EEG, TG, and G_{NIRS} compared to the fertilizer-free control. Moreover, there was a relatively stronger correlation between EEG, TG, and G_{NIRS} with soil humic acid than with soil organic matter. In a greenhouse study with corn plants inoculated with *Rhizophagus clarus* or *Rhizophagus intraradices*, Aliasgharzad et al. (2016) reported that by increasing nitrogen levels added to the soil, from 0 to 100 and 200 mg/kg (as urea), EEG increased by 75% and 112%, respectively. This increase for TG was 59% and 76%, respectively, indicating the pronounced effect of N fertilizer on EEG than TG. However, the addition of P showed different trends. Addition of phosphorus at rates of 20 and 40 mg/kg resulted in 27% increase and 6% decrease in EEG and a 24% increase and 13% decrease in TG, respectively, compared with the P-free conditions. The efficiency of *R. clarus* in glomalin production was higher than *R. intraradices*. Moreover, increasing the P level up to 40 mg/kg inhibited both root mycorrhizal colonization and glomalin production (Aliasgharzad et al. 2016).

11.4.2 Atmospheric CO_2 Level

Arbuscular mycorrhizal symbiosis often exhibits higher proliferation, under elevated CO_2 (eCO_2) due to enhanced rate of plant photosynthesis. However, it depends also on the photosynthetic physiology of the plant (i.e., C_3, C_4) and the availability of nutrients in the soil, in a way that C_4 is more influenced than C_3 plants in terms of carbon allocation to the root and mycorrhizal colonization under eCO_2 (Frew et al. 2021). In a pot study, soybean plants were inoculated with AMF at 350 (ambient) and 550 ppm (elevated) CO_2 levels. Elevated CO_2 led to an increase in the AMF spore density in soil and higher root colonization (Adeyemi et al. 2020). It seems that increased carbon assimilation by plants under eCO_2 would result in more carbon flow to the underground part of plants, thereby increased carbon allocation to AMF. As mentioned before, glomalin is produced by AMF and precipitated on spore and hyphal cell wall. Hence, more glomalin will enter the soil under higher concentrations of atmospheric CO_2. Zhang et al. (2015) examined the responses of easily extractable and total glomalin-related soil protein (EE-GRSP and T-GRPS) to eCO_2 (approx. 700 μmol mol^{-1} CO_2) and/or nitrogen fertilization in a subtropical forest. Increased CO_2 levels led to a significant increase in T-GRSP by 35%. However, elevated CO_2 reduced EE-GRSP by 5% in the top 10 cm of the soil layer. The concentration of EE-GRSP and T-GRSP went up as a result of eCO_2 and nitrogen fertilization treatment. They concluded that the greater accumulation ratios of T-GRSP (22.6 ± 13.6%) in comparison with SOC (15.9 ± 9.4%) under eCO_2 might be related to the rapid glomalin deposition in the soil (Zhang et al. 2015).

11.4.3 Salinity and Drought

Previous works have shown that the production of glomalin in AMF can be a protective reaction in fungal organs against stressors. Indeed, it has been well established that glomalin acts as a heat shock protein (hsp60) for protecting fungi against adverse environmental factors (Gadkar and Rillig 2006). Therefore, it is reasonable that AMF produce more glomalin under stressful conditions such as salinity or drought. Hammer and Rillig (2011) conducted an in vitro experiment in which the fungus *Glomus intraradices* was subjected to different levels of NaCl and glycerol, resembling the salinity and osmotic stresses, respectively. The application of NaCl caused a strong increase in the production of glomalin, whereas the addition of glycerol led to no response. The lack of response in glycerol treatment might be related to its role as a soil carbon source rather than as an osmolyte. Polyethylene glycol (PEG) is usually used for this purpose which is not a carbon source for microorganisms and acts as an effective osmolyte in hydroponic plant culture and microbial culture media (Eliane et al. 2019).

In contrast, under field conditions, (Zhang et al. 2017b) stated that soil salinity had a negative correlation with GRSP concentration. Also, they analyzed the impact of soil parameters on T-GRSP and EE-GRSP using regression models and concluded that soil salinity accounts for 46% and 25% variation in T-GRSP and EE-GRSP, respectively. Purified T-GRSP of the soil with high salinity had higher nitrogen (13.13%) and lower carbon content (43.41%) compared to that of low-salinity soil. Total glomalin extracted from an extremely saline soil showed a higher binding capacity to Ca^{2+} and PO_4^{3-} which enhance its stability in soil by 29.8% and 14.1%, respectively. The sodic-saline soils which suffer from low aggregate stability would benefit from glomalin to construct stable aggregates. They concluded that, despite the glomalin synthesis reduction in saline conditions, the positive alteration in its chemical nature and behavior could compensate its effect on soil aggregation (Zhang et al. 2017a, b).

Other investigations revealed that mild salinity promotes glomalin production, but it declined at higher salinity perhaps due to severe inhibition of AMF at high salt concentrations. In a study conducted by Ahmadi et al. (2014) on corn plants inoculated with AMF, EE-GRSP soil concentrations increased under controlled salt-stress conditions. They pointed out that, by increasing salinity level up to 8 dS/m, both root colonization percentage and EE-GRSP were significantly decreased; however, the extent of glomalin produced per unit length of colonized root was increased.

11.4.4 Heavy Metals

Glomalin-encoding genes could be overexpressed upon exposure to stressors such as heavy metals (HM), because (as mentioned above) glomalin is considered a heat shock protein, and increased levels of HM may improve its production (Gadkar and Rillig 2006). As a glycoprotein, glomalin could sequestrate HM to protect fungi from their toxic effects. This phenomenon can be considered a biostabilization mechanism in which mycorrhizal plants can survive and tolerate HM in contaminated soils (Malekzadeh et al. 2016a). In an experiment with clover plants grown in the sand and colonized by *Rhizophagus irregularis*, toxic Pb concentrations (0, 150, 300, and 450 mM as $Pb(NO_3)_2$) were applied to the pots. Immunoreactive (IR) and Bradford reactive (BR) glomalin showed no correlation with hyphal biomass regardless of the Pb level. However, the content of both glomalin increased significantly with increasing Pb application level to the pots (Fig. 11.2). There was also a positive correlation between root colonization percentage and glomalin content. The maximum IR glomalin (10.04 μg/mg hyphae) was observed at 450 mM Pb, which was significantly higher compared to the lower Pb levels. Indeed, the total Pb sequestrated by root glomalin (228.91 mg Pb/mg glomalin) rose as the Pb level increased up to 300 mM (Malekzadeh et al. 2016b). In contrast, Qiu et al. (2022) reported that the soil glomalin was negatively correlated with the bioavailability of Cd, Sb, Cu, As, Pb, Zn, and Ni ($r = -0.60$, $p < 0.001$) under field conditions. The contribution of soil glomalin to the sequestration of these heavy metals was higher than that of root glomalin, albeit the amount of sequestration was plant species-specific.

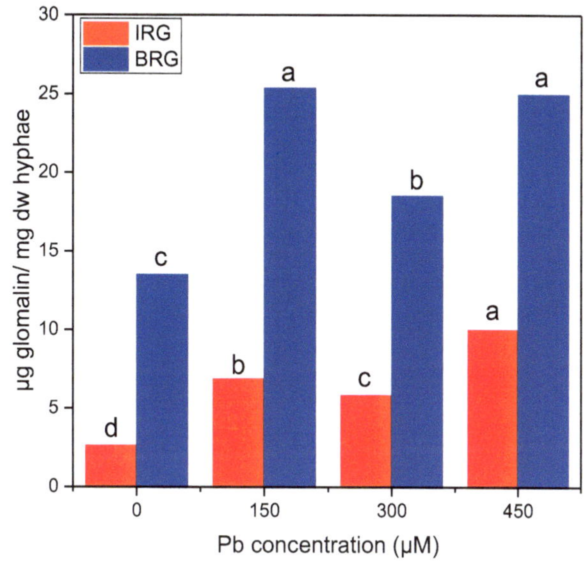

Fig. 11.2 Immunoreactive (IRG) and Bradford reactive (BRG) glomalin contents at different levels of Pb (μM). (Extracted and redrawn from Malekzadeh et al. 2016b)

Cadmium as a highly toxic element for plants and other organisms enters soils via P fertilizers and industrial activities and adversely affects mycorrhizal symbiosis. In a pot culture experiment, Malekzadeh et al. (2016a) inoculated clover plants with arbuscular mycorrhizal fungus (*Rhizophagus irregularis*) and treated with 0, 15, 30, and 45 μM of Cd. In a separate in vitro experiment, they also applied 0, 0.001, 0.01, and 0.1 mM of Cd to plates containing the T-DNA transformed carrot root. The IR and BR glomalin contents in both experiments increased with increasing Cd concentration. The total Cd sequestrated by glomalin was significantly increased by increasing the Cd concentration in both experiments (Fig. 11.3). According to these data, enhanced production of glomalin by AM fungus under Cd stress serves as a mechanism to protect both the host and fungus from Cd toxicity.

11.5 Glomalin and Carbon Sequestration

Carbon is an essential element on earth, which supports all dimensions of life by transferring between living organisms and the environment. Increasing CO_2 emission due to the burning of fossil fuels, deforestation, land use change, degradation and erosion of soil, and mismanagement of agricultural lands, along with other greenhouse gases, has caused a rise in the average temperature of the earth which is a threat to plant diversity and affects aquatic and terrestrial ecosystems (Mishra and Kizhakkepurakkal 2014; IPCC 2007).

It is necessary to identify the practical technologies for sequestration of CO_2 for a long time such as their sequestration in terrestrial ecosystems (Lal 2020). Since the terrestrial storage is part of the active carbon cycle, it can potentially store

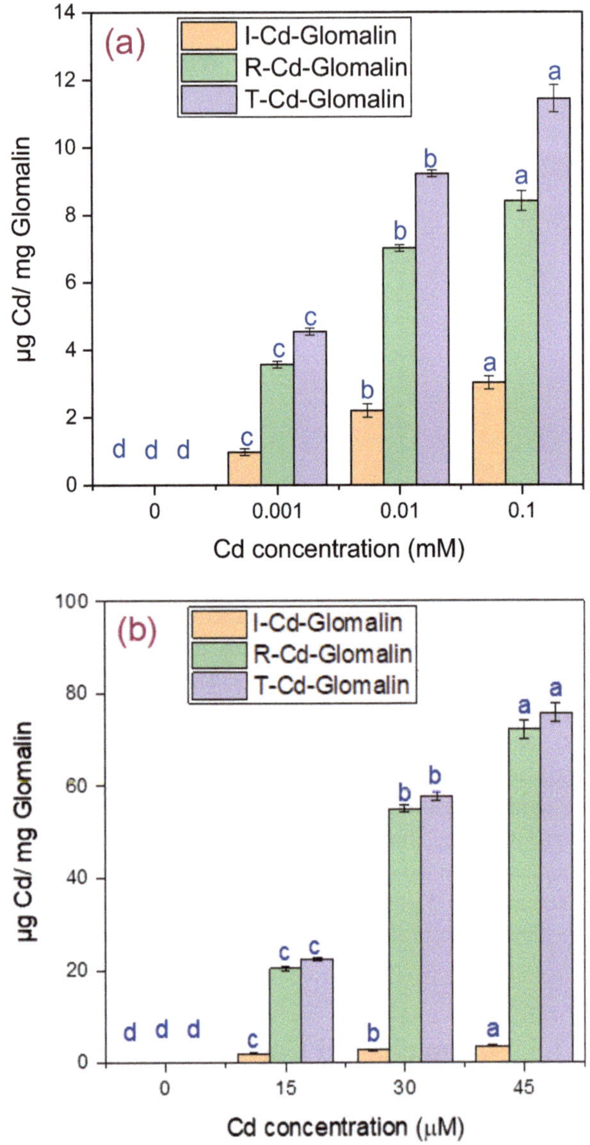

Fig. 11.3 Irreversible (I-Cd-glomalin), reversible (R-Cd-glomalin), and total (T-Cd-glomalin) Cd sequestrated by glomalin in the hyphal compartment, in response to increased levels of cadmium during (**a**) in vitro and (**b**) pot cultures. (Extracted and redrawn from Malekzadeh et al. 2016a)

atmospheric carbon at high concentrations. Carbon sequestration is defined as the capture of atmospheric CO_2 by plants and other reserves on the planet (Lal 2004). Soil organic carbon storage is a "salient component" of terrestrial ecosystems that serves as the main regulator of the biosphere-atmosphere carbon flow. A proposed way to mitigate the concentration of CO_2 in the atmosphere is to enhance the storage of carbon in soils at a global scale. The storage of carbon in soils is two to three times more than the carbon stored in the atmosphere and plant biomass, so it is possible to manage the flow of carbon from plant to soil (Scharlemann et al. 2014). The management of carbon sequestration in the form of soil organic carbon primarily depends on plant yield and the distribution of photosynthetic products (Zhu and Miller 2003) and then the role of microbial communities in releasing carbon to the atmosphere by respiration or its flow into the soil through decomposition, mineralization, and immobilization processes (Parihar et al. 2020). Some researchers suggested that AM fungi could sequester more than 70% of carbon in the soil, but this has been less considered in global climate change models (Averill et al. 2014). A portion of the plant's photosynthetic products is allocated to the roots which is even higher in the symbiosis by AM fungi (Nautiyal et al. 2019). Harris and Paul (1987) reported that carbon allocated to AM fungi to be 40–50% of photosynthates transferred to the root, while Treseder and Allen (2000) estimated it to be more than 85%. Mycorrhizal hyphae help redistribute carbon from shoots to roots by consuming photosynthetic products and producing extracellular polymers and amino acids to store organic carbon in the soil (Wang et al. 2016; Wilson et al. 2009; Antoninka et al. 2011). Allocation of carbon through increasing net primary production (NPP) in aerial and root parts is an important factor in carbon sequestration, and about 47% of NPP is allocated to AM fungi (Treseder 2016). The biomass of extracellular hyphae in soil is approximately 0.5–0.03 mg/g, which includes about 20–30% of the microbial biomass and 15% of the organic carbon reserve of soils (Leake et al. 2004). According to data showing 50% carbon content of the extracellular hyphae at a depth of 30 cm, the amount of soil organic carbon produced by AM fungi has been estimated to be 54–900 kg/ha (Treseder and Cross 2006). Also, the intracellular structures of AM fungi constitute approximately 0.5 pg of organic carbon in topsoil consisting of 4% of the global microbial carbon sink (Treseder and Cross 2006). Bago et al. (2000) reported that AM fungi globally regulate the flow of approximately 5 billion tons of carbon per year to soil. This process constitutes a significant portion of the carbon cycle. Therefore, AM fungi help to increase the soil organic carbon reserve by adding large amounts of organic residues in the form of extra- and intracellular mycelium, glomalin, and other hyphal secretions.

The AM fungi communities under the tillage layer have often been neglected which can provide a good potential for organic carbon storage in subsoil. More than 50% of the total fungal biomass is located under a depth of 30 cm (Higo et al. 2013), and the soil depth to which the mycorrhizal roots could be developed may reach up to 8 m (de Araujo Pereira et al. 2018). Also, laboratory observations show that by elevating CO_2 concentration, mycorrhizal colonization has increased at a depth of 14–45 cm (Rillig and Field 2003). AM fungi ensure the health of host plant in situations where the topsoil is dry or depleted from nutrients by increasing the plant availability to the nutrients and water in subsoil (Kautz et al. 2013). Therefore, the plant may benefit more from the subsoil per carbon unit spent on the fungal symbiont compared to that from the topsoil (Sosa-Hernández et al. 2019). AM fungi are expected to expand their hyphal networks in subsoil due to the suboptimal conditions for roots, which lead to more colonization of soil micropores, thereby more carbon sequestration in the subsoil layers.

The survival time of organic carbon in the soil increases by increasing soil depth (over 3 m) and reaches more than 10,000 years (Sosa-Hernández et al. 2019). The main reasons for this phenomenon are: (1) the subsoil usually contains smaller amount of energy sources and nutrients which limits microbial activity and reduces decomposition rate of soil organic matter; (2) the subsoil has a high compaction and bulk density, lower porosity, and air which reduces the habitat and abundance of microorganisms; (3) the amount of clay increases in the subsoil which form an organic-mineral complex with organic matter and helps to stabilize organic carbon; and (4) in the subsoil, a higher proportion of organic matter is placed in microaggregates and leads to its decomposition more slowly compared to the topsoil (Torres-Sallan et al. 2017; Weil and Brady 2016; Sosa-Hernández et al. 2019). As fungal hyphae penetrate deeper into soil, fungal secretions are transported to greater distances from the root system which can be an important strategy for storage of soil organic carbon. Clemmensen et al. (2013) reported that over 70% of subsoil carbon is derived from roots. Also, the organic matter of the subsoil is mostly of microbial origin, because the microbe-derived organic matter is better combined with the soil minerals compared to the organic matter of plant origin (Rumpel and Kögel-Knabner 2011). Since the clay minerals and sesquioxides content is higher in the subsoil, higher potential for carbon sequestration is observed in this soil layer compared to the topsoil (Sosa-Hernández et al. 2019).

The relationship between glomalin, soil organic carbon, and stability of aggregates and its role in keeping the soil fertile has been frequently reported (Zhu and Miller 2003; Nautiyal et al. 2019; Emran et al. 2012). Glomalin physically prevents the degradation rate of soil organic carbon by forming permanent aggregates (Zhao et al. 2018). Soil aggregates can preserve over 5% of soil carbon and nitrogen from degradation caused by water and wind erosion (Emran et al. 2012). Stable aggregates effectively protect organic matter by increasing the water and air permeability of the soil, improving soil consistency and preventing erosion to a large extent (Ortas et al. 2013). According to Ferrero Holtz et al. (2016), as the total organic carbon decreases, the contribution of glomalin to the total carbon increases. This indicates an increase in "resistant carbon forms" with a decrease in total soil organic carbon. Increased content of resistant carbon in soil leads to longer carbon retention duration in the soil, so glomalin not only directly contributes to carbon sequestration as a carbon source but also does it indirectly due to its resistant nature (Zhang et al. 2017a). The annual losses of soil organic carbon globally are 1.3 and 1.0 billion tons by water and wind erosions, respectively (Lal 2020). Therefore, any factor that affects mycorrhizal development and glomalin content can also affect soil aggregate stability and the soil organic matter value (Nautiyal et al. 2019). According to several studies, soil aggregate stability is improved through the establishment of hyphae in aggregate and the production of extracellular polymers (Rillig and Mummey 2006). Also, by increasing the soil depth, because the subsoil is less disturbed and the hyphae network is not destroyed, it can lead to a long survival time of protected organic matter inside soil aggregates. Therefore, the stability of soil aggregate by mycorrhizal hyphae and secretions including glomalin can play a critical role in protecting organic matter and its deposition in the soil on a global scale (Sosa-Hernández et al. 2019).

11.6 Conclusions

Global warming, resulting from the rapid increase of CO_2 concentration in the atmosphere, is threatening food production in agricultural ecosystems, hence human life at all. Soils are main reservoirs for organic carbon which could be lost by deforestation and agricultural practices. The rate of carbon flow through the atmosphere to the soil is a function of the photosynthesis rate in plants. Most plants allocate a considerable portion of assimilated carbon to the fungal partner in roots which led to enhanced production of glomalin by the fungi. Nutrient availability, especially nitrogen, can also stimulate glomalin production, while higher phosphorus reduces it via inhibition of root mycorrhizal colonization. Elevated CO_2 levels accelerate plant photosynthesis rate which in turn promotes glomalin production as well. Glomalin—which is homologous with heat shock protein (hsp60)—can be overexpressed under stressful conditions such as salinity, drought, or heavy metal contaminations, although this promotion is most pronounced under moderate stress conditions. Considering all the above factors affecting glomalin production, it is necessary to manage the natural and agricultural ecosystems in a manner that encourage glomalin production, as a relatively stable carbon pool in soils. Nowadays, many of agricultural soils worldwide are faced with drought and salinity, and AM fungi not only could support plant growth under these stressful conditions but are also stimulated to produce more glomalin. The efforts should be focused on farmers' awareness and acceptance of the use of AM fungi for plant production. It's obvious that the accumulation of recalcitrant-C sources in soil would lead to a decline in global greenhouse gas emissions which is in line with the international goal to limit climate change.

References

Adeyemi NO, Atayese MO, Dare M, Olubode A (2020) Effects of elevated carbon dioxide on arbuscular mycorrhizal fungi activities and soil microbial properties in soybean (*Glycine max* L. Merrill) rhizosphere. Acta Fytotech Zootech 23(3):109–116. https://doi.org/10.15414/afz.2020.23.03.109-116

Agnihotri R, Sharma MP, Prakash A, Ramesh A, Bhattacharjya S, Patra AK, Manna MC, Kurganova I, Kuzyakov Y (2022) Glycoproteins of arbuscular mycorrhiza for soil carbon sequestration: review of mechanisms and controls. Sci Total Environ 806:150571. https://doi.org/10.1016/j.scitotenv.2021.150571

Ahmadi GS, Aliasgharzad N, Tavasoli A (2014) Effects of NaCl salinity levels on the glomalin produced by Glomerales in symbiosis with corn plant. J Water Soil 28:92–100. (In Persian with English abstract)

Aliasgharzad N, Saleh Rastin N, Towfighi H, Alizadeh A (2001) Occurrence of arbuscular mycorrhizal fungi in saline soils of the Tabriz plain of Iran in relation to some physical and chemical properties of soil. Mycorrhiza 11(3):119–122. https://doi.org/10.1007/s005720100113

Aliasgharzad N, Afshari Z, Najafi N (2016) Carbon sequestration by glomeral fungi in soil is influenced by phosphorus and nitrogen fertilization. Int J Adv Sci Eng Inf Technol 6(1):1–6. https://doi.org/10.18517/ijaseit.6.1.588

Antoninka A, Reich PB, Johnson NC (2011) Seven years of carbon dioxide enrichment, nitrogen fertilization and plant diversity influence arbuscular mycorrhizal fungi in a grassland ecosystem. New Phytol 192:200–214. https://doi.org/10.1111/j.1469-8137.2011.03776.x

Averill C, Turner BL, Finzi AC (2014) Mycorrhiza-mediated competition between plants and decomposers drives soil carbon storage. Nature 505(7484):543–545. https://doi.org/10.1038/nature12901

Bago B, Pfeffer PE, Shachar-Hill Y (2000) Carbon metabolism and transport in arbuscular mycorrhizas. Plant Physiol 124:949–958. https://doi.org/10.1104/pp.124.3.949

Balík J, Sedlář O, Kulhánek M, Černý J, Smatanová M, Suran P (2020) Effect of organic fertilizers on glomalin content and soil organic matter quality. Plant Soil Environ 66(11):590–597. https://doi.org/10.17221/385/2020-PSE

Bradford MM (1976) A rapid sensitive method for the quantification of microgram quantities of protein utilizing the principle of protein-dye binding. Anal Biochem 72:248–254. https://doi.org/10.1006/abio.1976.9999

Brundrett MC (2009) Mycorrhizal associations and other means of nutrition of vascular plants: understanding the global diversity of host plants by resolving conflicting information and developing reliable means of diagnosis. Plant Soil 320:37–77. https://doi.org/10.1007/s11104-008-9877-9

Cisse G, Essi M, Nicolas M, Staunton S (2020) Bradford quantification of glomalin-related soil protein in coloured extracts of forest soils. Geoderma 372:114394. https://doi.org/10.1016/j.geoderma.2020.114394.hal-02865912

Clemmensen KE, Bahr A, Ovaskainen O, Dahlberg A, Ekblad A, Wallander H, Stenlid J, Finlay RD, Wardle DA, Lindahl BD (2013) Roots and associated fungi drive long-term carbon sequestration in boreal forest. Science 339:1615–1618. https://doi.org/10.1126/science.1231923

de Araujo Pereira AP, Santana MC, Bonfim JA, de Lourdes Mescolotti D, Cardoso EJBN (2018) Digging deeper to study the distribution of mycorrhizal arbuscular fungi along the soil profile in pure and mixed *Eucalyptus grandis* and *Acacia mangium* plantations. Appl Soil Ecol 128:1–11. https://doi.org/10.1016/j.apsoil.2018.03.015

Driver JD, Holben WE, Rillig MC (2005) Characterization of glomalin as a hyphal wall component of arbuscular mycorrhizal fungi. Soil Biol Biochem 37:101–106. https://doi.org/10.1016/j.soilbio.2004.06.011

Egboka NT, Fagbola O, Nkwopara UN, Okoli NH, Afangide AI, Nwosu TV (2022) Density of arbuscular mycorrhizal fungi and nutrient status of soils in selected land use types and soil depths. Sarhad J Agric 38(2):633–647. https://doi.org/10.17582/journal.sja/2022/38.2.633.647

Eliane M, Modeste K, André S, Edmond K, Mongomaké K (2019) Effect of water stress induced by polyethylene glycol 6000 on somatic embryogenesis in Cocoa (*Theobroma cacao* L.). Agric Sci 10:1240–1254. https://doi.org/10.4236/as.2019.109092

Emran M, Gispert M, Pardini G (2012) Patterns of soil organic carbon, glomalin and structural stability in abandoned Mediterranean terraced lands. Eur J Soil Sci 63:637–649. https://doi.org/10.1111/j.1365-2389.2012.01493.x

Ferrero Holtz EW, Gonzalez MG, Giuffre L, Ciarlo E (2016) Glomalins and their relationship with soil carbon. Int J Appl Sci Technol 6(2):1–5

Frew A, Price JN, Oja J, Vasar M, Opik M (2021) Impacts of elevated atmospheric CO_2 on arbuscular mycorrhizal fungi and their role in moderating plant allometric partitioning. Mycorrhiza 31:423–430. https://doi.org/10.1007/s00572-021-01025-6

Gadkar V, Rillig MC (2006) The arbuscular mycorrhizal fungal protein glomalin is a putative homolog of heat shock protein 60. FEMS Microbiol Lett 263(1):93–101. https://doi.org/10.1111/j.1574-6968.2006.00412.x. PMID: 16958856

Gucwa-Przepióra E, Błaszkowski J, Kurtyka R, Małkowski Ł, Małkowski E (2013) Arbuscular mycorrhiza of *Deschampsia cespitosa* (Poaceae) at different soil depths in highly metal-contaminated site in southern Poland. Acta Soc Bot Pol 82:251–258. https://doi.org/10.5586/asbp.2013.033

Hammer EC, Rillig MC (2011) The influence of different stresses on glomalin levels in an arbuscular mycorrhizal fungus-Salinity increases glomalin content. PLoS One 6(12):e28426. https://doi.org/10.1371/journal.pone.0028426

Harner MJ, Ramsey PW, Rillig MC (2004) Protein accumulation and distribution in floodplain soils and river foam. Ecol Lett 7:829–836. https://doi.org/10.1111/j.1461-0248.2004

Harris KK, Paul EA (1987) Carbon requirements of vesicular arbuscular mycorrhizae. In: Safir GR (ed) Ecophysiology of VA mycorrhizal plants. CRC Press, Boca Raton, pp 93–105

Higo M, Isobe K, Yamaguchi M, Drijber RA, Jeske ES, Ishii R (2013) Diversity and vertical distribution of indigenous arbuscular mycorrhizal fungi under two soybean rotational systems. Biol Fertil Soils 49:1085–1096. https://doi.org/10.1007/s00374-013-0807-5

Intergovernmental Panel on Climate Change (IPCC) (2007) In: Metz B, Davidson OR, Bosch PR, Dave R, Meyer LA (eds) Climate change 2007: working group III: mitigation of climate change. Cambridge University Press, Cambridge

Jakobsen I, Smith SE, Smith FA (2002) Function and diversity of arbuscular mycorrhizae in carbon and mineral nutrition. In: van der Heijden MGA, Sanders IR (eds) Mycorrhizal ecology. Ecological studies. Springer, Heidelberg, pp 75–92

Kabir Z (2005) Tillage or no-tillage: impact on mycorrhizae. Can J Plant Sci 85:23–29. https://doi.org/10.4141/P03-160

Kautz T, Amelung W, Ewert F, Gaiser T, Horn R, Jahn R et al (2013) Nutrient acquisition from arable subsoils in temperate climates: a review. Soil Biol Biochem 57:1003–1022. https://doi.org/10.1016/j.soilbio.2012.09.014

Lal R (2004) Soil carbon sequestration impacts on global climate change and food security. Science 304(5677):1623–1627. https://doi.org/10.1126/science.1097396

Lal R (2020) Soil erosion and gaseous emissions. Appl Sci 10:2784. https://doi.org/10.3390/app10082784

Leake J, Johnson D, Donnelly D, Muckle G, Boddy L, Read D (2004) Networks of power and influence: the role of mycorrhizal mycelium in controlling plant communities and agroecosystem functioning. Can J Bot 82:1016–1045. https://doi.org/10.1139/b04-060

Malekzadeh E, Aliasgharzad N, Majidi J, Aghebati-Maleki L, Abdolalizadeh J (2016a) Cd-induced production of glomalin by arbuscular mycorrhizal fungus (*Rhizophagus irregularis*) as estimated by monoclonal antibody assay. Environ Sci Pollut Res 23:20711–20718. https://doi.org/10.1007/s11356-016-7283-z

Malekzadeh E, Aliasgharzad N, Majidi J, Abdolalizadeh J, Aghebati-Maleki L (2016b) Contribution of glomalin to Pb sequestration by arbuscular mycorrhizal fungus in a sand culture system with clover plant. Eur J Soil Biol 74:45–51. https://doi.org/10.1016/j.ejsobi.2016.03.003

Matos PS, da Silva CF, Damian JM, Cerri CEP, Pereira MG, Zonta E (2022) Beneficial services of glomalin and arbuscular mycorrhizal fungi in degraded soils in Brazil. Sci Agric 79(5): e20210064. https://doi.org/10.1590/1678-992X-2021-0064

Mishra S, Kizhakkepurakkal SAR (2014) Arbuscular mycorrhizal fungi (AM fungi): a potential tool for augmenting carbon dioxide sequestration. In: Fulekar MH, Kale RK (eds) Recent trends in life science. IK International Publishing House Pvt. Ltd, New Delhi, pp 59–75

Muleta D, Assefa F, Nemomissa S, Granhall U (2008) Distribution of arbuscular mycorrhizal fungi in soils of small holder agroforestry monocultural coffee systems in southwestern Ethiopia. Biol Fertil Soils 44:653–659. https://doi.org/10.1007/s00374-007-0261-3

Nautiyal P, Rajput R, Pandey D, Arunachalam K, Arunachalam A (2019) Role of glomalin in soil carbon storage and its variation across land uses in temperate Himalayan regime. Biocatal Agric Biotechnol 21:101311. https://doi.org/10.1016/j.bcab.2019.101311

Ndoye F, Kane A, Ngonkeu EL, Bakhoum N, Sanon A, Diouf D, Ourèye SM, Baudoin E, Noba K, Prin Y (2012) Changes in land use system and environmental factors affect arbuscular mycorrhizal fungal density and diversity, and enzyme activities in rhizospheric soils of *Acacia senegal* (L.) Wild. ISRN Ecol 563191:13. https://doi.org/10.5402/2012/563191

Nichols KA (2003) Characterization of glomalin a glycoprotein produced by arbuscular mycorrhizal fungi. PhD Dissertation, University of Maryland, College Park, Maryland, p 285

Nichols KA, Wright SF (2004) Contributions of fungi to soil organic matter in agroecosystems. In: Magdoff F, Weil RR (eds) Soil organic matter in sustainable agriculture. CRC Press, Boca Raton, pp 179–198

Oehl F, Sieverding E, Ineichen K, Mäder P, Boller T, Wiemken A (2003) Impact of land use intensity on the species diversity of arbuscular mycorrhizal fungi in agroecosystems of central Europe. Appl Environ Microbiol 69:2816–2824. https://doi.org/10.1128/AEM.69.5.2816

Oehl F, Sieverding E, Ineichen K, Ris EA, Boller T, Wiemken A (2005) Community structure of arbuscular mycorrhizal fungi at different soil depths in extensively and intensively managed agroecosystems. New Phytol 165:273–283. https://doi.org/10.1111/j.1469-8137.2004.01235.x

Olsson PA, Rahm J, Aliasgharzad N (2010) Carbon dynamics in mycorrhizal symbiosis is linked to carbon costs and phosphorus benefits. FEMS Microbiol Ecol 72(1):125–131. https://doi.org/10.1111/j.1574-6941.2009.00833.x

Ortas I, Akpinar C, Lal R (2013) Long-term impacts of organic and inorganic fertilizers on carbon sequestration in aggregates of an entisol in Mediterranean Turkey. Soil Sci 178:12–23. https://doi.org/10.1097/SS.0b013e3182838017

Parihar M, Rakshit A, Meena VS, Gupta VK, Rana K, Choudhary M, Tiwari G, Mishra PK, Pattenayak A, Bisht JK, Jatav SS, Khati P, Jatav HS (2020) The potential of arbuscular mycorrhizal fungi in C cycling: a review. Arch Microbiol 202:1581–1596. https://doi.org/10.1007/s00203-020-01915-x

Pohanka M, Vlcek V (2018) Assay of glomalin using a quartz crystal microbalance biosensor. Electroanalysis 30:1–7. https://doi.org/10.1002/elan.201700772

Prasad M, Chaudhary M, Srinivasan R, Mahawer SK (2018) Glomalin: a miracle protein for soil sustainability. Indian Farm 5(9):1092–1100

Qiu L, Lin H, Song B, Kong T, Sun W, Sun X, Zhang Y, Li B (2022) Glomalin-related soil protein (GRSP) in metal sequestration at Pb/Zn-contaminated sites. J Soils Sediments 22:577–593. https://doi.org/10.1007/s11368-021-03092-w

Rillig MC (2004) Arbuscular mycorrhizae, glomalin, and soil aggregation. Can J Soil Sci 84:355–363. https://doi.org/10.4141/S04-003

Rillig MC, Field CB (2003) Arbuscular mycorrhizae respond to plants exposed to elevated atmospheric CO2 as a function of soil depth. Plant Soil 254:383–391. https://doi.org/10.1023/A:1025539100767

Rillig MC, Mummey DL (2006) Mycorrhizas and soil structure. New Phytol 171:41–53

Rillig MC, Steinberg PD (2002) Glomalin production by an arbuscular mycorrhizal fungus: a mechanism of habitat modification? Soil Biol Biochem 34:1371–1374. https://doi.org/10.1016/S0038-0717(02)00060-3

Rillig MC, Wright SF, Nichols KA, Schmidt WF, Torn MS (2001) Large contribution of arbuscular mycorrhizal fungi to soil carbon pools in tropical forest soils. Plant Soil 233:167–177. https://doi.org/10.1023/A:1010364221169

Rillig MC, Ramsey PW, Morris S, Paul EA (2003) Glomalin, an arbuscularmycorrhizal fungal soil protein, responds to land-use change. Plant Soil 253:293–299. https://doi.org/10.1023/A:1024807820579

Rosier CL, Hoye AT, Rillig MC (2006) Glomalin-related soil protein: assessment of current detection and qualification tools. Soil Biol Biochem 38:2205–2211. https://doi.org/10.1016/j.soilbio.2006.01.021

Rosier CL, Piotrowski SJ, Hoye AT, Rillig MC (2008) Intraradical protein and glomalin as a tool for quantifying arbuscular mycorrhizal root colonization. Pedobiologia 52:41–50. https://doi.org/10.1016/j.pedobi.2008.02.002

Rumpel C, Kögel-Knabner I (2011) Deep soil organic matter—a key but poorly understood component of terrestrial C cycle. Plant Soil 338:143–158. https://doi.org/10.1007/s11104-010-0391-5

Sanders IR, Alt M, Groppe K, Boller T, Wiemken A (1995) Identification of ribosomal DNA polymorphisms among and within spores of the Glomales: application to studies on the genetic diversity of arbuscular mycorrhizal fungal communities. New Phytol 130:419–427

Scharlemann JP, Tanner EV, Hiederer R, Kapos V (2014) Global soil carbon: understanding and managing the largest terrestrial carbon pool. Carbon Manag 5:81–91. https://doi.org/10.4155/cmt.13.77

Siami A, Aliasgharzad N, Maleki LA, Najafi N, Shahbazi F, Biswas A (2022) Recalcitrant C source mapping utilizing solely terrain-related attributes and data mining techniques. Agronomy 12:1653. https://doi.org/10.3390/agronomy12071653

Singh AK, Zhu X, Chen C, Wu J, Yang B, Zakari S, Jiang XJ, Singh N, Liu W (2022) The role of glomalin in mitigation of multiple soil degradation problems. Crit Rev Environ Sci Technol 52(9):1604–1638. https://doi.org/10.1080/10643389.2020.1862561

Smith SE, Read DJ (1997) Mycorrhizal symbiosis, 2nd edn. Academic, San Diego

Smith SE, Read DJ (2008) Mycorrhizal symbiosis. Academic, San Diego

Snoeck D, Abolo D, Jagoret P (2010) Temporal changes in VAM fungi in the cocoa agroforestry systems of Central Cameroon. Agrofor Syst 78:323–328. https://doi.org/10.1007/s10457-009-9254-6

Sosa-Hernández MA, Leifheit EF, Ingraffia R, Rillig MC (2019) Subsoil arbuscular mycorrhizal fungi for sustainability and climate-smart agriculture: a solution right under our feet? Front Microbiol 10:744. https://doi.org/10.3389/fmicb.2019.00744

Staunton S, Saby NPA, Arrouays D, Quiquampoix H (2020) Can soil properties and land use explain glomalin-related soil protein (GRSP) accumulation? A nationwide survey in France. Catena 193:104620. https://doi.org/10.1016/j.catena.2020.104620

Steinberg PD, Rillig MC (2003) Differential decomposition of arbuscular mycorrhizal fungal hyphae and glomalin. Soil Biol Biochem 35:191–194. https://doi.org/10.1016/S0038-0717(02)00249-3

Torres-Sallan G, Schulte RPO, Lanigan GJ, Byrne KA, Reidy B, Simó I, Six J, Creamer RE (2017) Clay illuviation provides a long-term sink for C sequestration in subsoils. Sci Rep 7:45635. https://doi.org/10.1038/srep45635

Treseder KK (2016) Model behavior of arbuscular mycorrhizal fungi: predicting soil carbon dynamics under climate change. Botany 94:417–423. https://doi.org/10.1139/cjb-2015-0245

Treseder KK, Allen MF (2000) Mycorrhizal fungi have a potential role in soil carbon storage under elevated CO2 and nitrogen deposition. New Phytol 147:189–200. https://doi.org/10.1046/j.1469-8137.2000.00690.x

Treseder KK, Cross A (2006) Global distributions of arbuscular mycorrhizal fungi. Ecosystems 9:305–316. https://doi.org/10.1007/s10021-005-0110-x

Ullah S, Muhammad B, Amin R, Abbas H, Muneer MA (2019) Sensitivity of arbuscular mycorrhizal fungi in old-growth forest: direct effect on growth and soil carbon storage. Appl Ecol Environ Res 17(6):13749–13758. https://doi.org/10.15666/aeer/1706_1374913758

Wang ZG, Bi YL, Jiang B, Zhakypbek Y, Peng SP, Liu WW, Liu H (2016) Arbuscular mycorrhizal fungi enhance soil carbon sequestration in the coalfields, northwest China. Sci Rep 6:34336. https://doi.org/10.1038/srep34336

Weil RR, Brady NC (2016) Soil aeration and temperature. In: The nature and properties of soils. Pearson, Columbus, pp 284–325

Whiffen LK, Midgley DJ, McGee PA (2007) Polyphenolic compounds interfere with quantification of protein in soil extracts using the Bradford method. Soil Biol Biochem 39:691–694. https://doi.org/10.1016/j.soilbio.2006.08.012

Wilson GW, Rice CW, Rillig MC, Springer A, Hartnett DC (2009) Soil aggregation and carbon sequestration are tightly correlated with the abundance of arbuscular mycorrhizal fungi: results from long-term field experiments. Ecol Lett 12:452–461. https://doi.org/10.1111/j.1461-0248.2009.01303.x

Wright SF (2000) A fluorescent antibody assay for hyphae and glomalin from arbuscular mycorrhizal fungi. Plant Soil 226:171–177. https://doi.org/10.1023/A:1026428300172

Wright SF, Upadhyaya A (1996) Extraction of an abundant and unusual protein from soil and comparison with hyphal protein from arbuscular mycorrhizal fungi. Soil Sci 161:575–586. https://doi.org/10.1097/00010694-199609000-00003

Wright SF, Upadhyaya A (1998) A survey of soils for aggregate stability and glomalin, a glycoprotein produced by hyphae of arbuscular mycorrhizal fungi. Plant Soil 198:97–107. https://doi.org/10.1023/A:1004347701584

Wright SF, Upadhyaya A (1999) Quantification of arbuscular mycorrhizal fungi activity by the glomalin concentration on hyphal traps. Mycorrhiza 8:283–285. https://doi.org/10.1007/s005720050247

Wright SF, Franke-Snyder M, Morton JB, Upadhyaya A (1996) Time-course study and partial characterization of a protein on hyphae of arbuscular mycorrhizal fungi during active colonization of roots. Plant Soil 181:193–203. https://doi.org/10.1007/BF00012053

Wright SF, Upadhyaya A, Buyer JS (1998) Comparison of N-linked oligosaccharides of glomalin from arbuscular mycorrhizal fungi and soils by capillary electrophoresis. Soil Biol Biochem 30: 1853–1857

Yang FY, Li GZ, Zhang DE, Christie P, Li XL, Gai JP (2010) Geographical and plant genotype effects on the formation of arbuscular mycorrhiza in *Avena sativa* and *Avena nuda* at different soil depths. Biol Fertil Soils 46:435–443. https://doi.org/10.1007/s00374-010-0450-3

Zbiral J, Cizmar D, Maly S, Obdrzalkova E (2017) Determination of glomalin in agriculture and forest soils by near-infrared spectroscopy. Plant Soil Environ 63(5):226–230. https://doi.org/10.17221/181/2017-PSE

Zhang J, Tang X, He X, Liu J (2015) Glomalin-related soil protein responses to elevated CO_2 and nitrogen addition in a subtropical forest: potential consequences for soil carbon accumulation. Soil Biol Biochem 83:142–149. https://doi.org/10.1016/j.soilbio.2015.01.023

Zhang J, Tang X, Zhong S, Yin G, Gao Y, He X (2017a) Recalcitrant carbon components in glomalin-related soil protein facilitate soil organic carbon preservation in tropical forests. Sci Rep 7:2391. https://doi.org/10.1038/s41598-017-02486-6

Zhang Z, Wang Q, Wang H, Nie S, Liang Z (2017b) Effects of soil salinity on the content, composition, and ion binding capacity of glomalin-related soil protein (GRSP). Sci Total Environ 581:657–665. https://doi.org/10.1016/j.scitotenv.2016.12.176

Zhao FZ, Fan XD, Ren CJ, Zhang L, Han XH, Yang GH, Wang J, Doughty R (2018) Changes of the organic carbon content and stability of soil aggregates affected by soil bacterial community after afforestation. Catena 171:622–631. https://doi.org/10.1016/j.catena.2018.08.006

Zhu YG, Miller RM (2003) Carbon cycling by arbuscular mycorrhizal fungi in soil-plant systems. Trends Plant Sci 8(9):407–709. https://doi.org/10.1016/s1360-1385(03)00184-5

Chapter 12
Arbuscular Mycorrhizal Fungi in Organic Versus Conventional Farming

Sara Fareed Mohamed Wahdan, Aya G. A. Asran, Mayar Abdellatef, Mohamed A. M. Atia, and Li Ji

Abstract Arbuscular mycorrhizal fungi (AMF) establish symbiotic relationships with plant roots, enhancing nutrient uptake and promoting plant growth. This chapter discusses various factors that influence AM fungal populations under different farming systems. These factors include soil management routines such as tillage intensity, pesticide usage, fertilizer application, and crop rotation strategies. The aim is to compare the effects of organic farming practices that prioritizes sustainable approaches with those employed in conventional agriculture. The chapter looks at how organic farming methods differ from conventional farming in terms of AM fungal communities, diversity, and functionality. It describes practical strategies for improving mycorrhizal associations through inoculation techniques or by implementing specific agronomic practices that promote a favorable environment for these beneficial microorganisms. Overall, this chapter illuminates the role of AMF in influencing soil health and plant productivity across various agricultural approaches. It emphasizes the importance of taking these symbiotic relationships into account when designing sustainable farming systems that maximize yields while minimizing environmental impacts.

Keywords AMF · Monoculture · Crop rotation · Soil management · Beneficial microorganism · Symbiosis

---

S. F. M. Wahdan (✉) · A. G. A. Asran · M. Abdellatef
Department of Botany and Microbiology, Faculty of Science, Suez Canal University, Ismailia, Egypt
e-mail: sarah_wahdan@science.suez.edu.eg

M. A. M. Atia
Genome Mapping Department, Agricultural Genetic Engineering Research Institute (AGERI), Agricultural Research Center (ARC), Giza, Egypt

L. Ji
School of Forestry, Central South University of Forestry and Technology, Changsha, People's Republic of China

G. J. Ahammed, R. Hajiboland (eds.), *Arbuscular Mycorrhizal Fungi and Higher Plants*, https://doi.org/10.1007/978-981-99-8220-2_12

12.1 Introduction

Frank (1885) first used the term "mykorhiza" (mycorrhiza) and noted that nutrients and water were transported to plant roots by endophytic hyphae through mutualistic symbiotic relationships (Frank 1885). Arbuscular mycorrhizal fungi (AMF), as one of the utmost earliest obligate symbioses, are considered the most important soil organisms for agroecosystem sustainability (Martin et al. 2017).

Mycorrhizal symbiosis benefits plants in several ways, including improving their access to immobile nutrients (Basu et al. 2018; Hodge and Fitter 2010; Javot et al. 2007), tolerance to abiotic stress (Aroca et al. 2007; Chourasiya et al. 2021), and protection against pathogens (Morris et al. 2019). Meanwhile, they can play a key role in soil aggregation (Rillig and Mummey 2006), nutrient cycling (Lanfranco et al. 2018), and soil stability (Morris et al. 2019) and have prodigious potential for improving agricultural sustainability (Basu et al. 2018; Bender et al. 2016; Lehmann et al. 2020). As a result of their implications to plant performance and soil health, they play a crucial role in sustaining ecosystem productivity (Castillo et al. 2016; Ma et al. 2021).

More importantly, extensive hyphae networks could be produced by AMF, thereby enhancing water and nutrient uptake by the roots (Smith and Smith 2011). Mycorrhizae can increase the effective root absorption area of the host plants by producing mycelium and forming a mycelial network. Mycorrhizae can also enhance the environmental adaptability and stress resistance of host plants to cope with constraint from an adverse environment. Additionally, AMF are more important in terms of their potential to improve plant growth and health in future climate change scenarios (Wahdan et al. 2021; Johnson et al. 2013). The significance of AMF in improving plant nutrition (especially phosphorus (P)) and improving plant stress tolerance to both biotic and abiotic stress in natural ecosystems or low-input organic agriculture has been widely recognized (Jeffries et al. 2003).

12.2 AMF and Agroecosystem

12.2.1 Plant Growth and Productivity

Although the "high investment and high production" mode of conventional agriculture in the past has significantly increased the yield of crops, with the development of modern agriculture, the "low investment and high production" mode of intensive agriculture is becoming a key strategy of sustainable agricultural development (Shen et al. 2013). To meet the increasing demands of global food production, conserve environmental quality, and respond to more hostile environmental changes, agriculture management is more complex than ever, and simultaneously accomplishing high nutrient utilization efficiency and increasing crop yields have appeared key challenges (Cassman et al. 2003).

AMF provide multiple advantages to plant hosts because they are in direct contact with crop roots (Zhang et al. 2017). AMF are not only important for improving plant resistance to stress and sustaining biogeochemical cycling and ecological functions (Wu et al. 2018) but also for increasing crop yields (Cavagnaro et al. 2015; Rillig et al. 2016). Mounting evidences have suggested that mycorrhizal inoculation can promote the accumulation of grains and aboveground biomass (Ren et al. 2019; Zhao et al. 2015a, b), which being considered to be an effective agronomic measure (Pellegrino and Bedini 2014). The potential mechanism of increased crop yield after inoculation with AMF has received more attention (Bowles et al. 2016; Ren et al. 2019; Zhao et al. 2015). Several studies proved that AMF colonization promotes crop yields and can be a critical factor in making agroecosystems more sustainable (Bender et al. 2016; Thirkell et al. 2017). Recently, much discussion has been devoted to whether AM symbiosis is appropriate for agricultural production (Ryan and Graham 2018; Rillig et al. 2019). AMF symbionts can enhance the water absorption capacity of roots by producing hyphal structures (Zhao et al. 2015a, b) and increase host plant nutrient availability, directly contributing to increased crop yields (Hestrin et al. 2019). Besides, inoculation with AMF can boost the root biomass of host plants and thus produce more root exudates, which may facilitate microbial-driven nitrogen (N) mineralization, and further increase nutrient availability (Mo et al. 2020).

Despite numerous data on the benefits of inoculation of the host plants by AMF, the role of AMF on host plant yield is still under debate. It is known that the respiration rate is increased upon mycorrhization of plants that consumes part of the photosynthetic products (Mortimer et al. 2008). Thus, the enhanced carbon (C) assimilation observed in the mycorrhizal plants is most likely offset by higher respiratory consumption, and the crop yield did not change significantly (Xavier and Germida 2003). Some other studies have revealed that root colonization with AMF reduces crop productivity, crop biomass production, and yield (Berruti et al. 2016; Ryan and Graham 2018; Jayne and Quigley 2014). Divergent influences of AMF on crop productivity might be relevant to experimental conditions, AMF species identity, climate, soil moisture, and soil fertility (Gosling et al. 2016; Jayne and Quigley 2014; Roger et al. 2013; Thirkell et al. 2017). Therefore, further studies are required to elucidate how inoculation with AMF differs in plant production and physiological properties under various conditions.

12.2.2 *Quality of Crops*

AMF can enhance plant uptake of nutrients and promote plant health (Gonzalez-Dugo 2010). After colonization of plant roots with fungi, AMF can significantly promote the acquisition of soil mineral nutrients, especially P. The ^{14}C tracking experiment found that the translocation of carbohydrates from plant to AM symbiotic structures (from root cells to AM mycelium) can promote the uptake of P by the fungi and its further transport to the plant (Bücking and Shachar-Hill 2005).

Additionally, inoculation of date palms with *Albahypha drummondii*, *Pervetustus simplex*, *Septoglomus xanthium*, *Claroideoglomus etunicatum*, *Rhizoglomus irregular*, and *Funneliformis mosseae* resulted in a higher shoot length and stem diameter as compared with control plant (El Hilali et al. 2022). Inoculation of tomato (*Solanum lycopersicum* L.) with *Rhizophagus intraradices*, *R. irregularis*, *Funneliformis mosseae, and Glomus iranicus* resulted in a higher growth index and P content in AMF-treated plants as compared with the control group (Shafiei et al. 2022). Furthermore, the growth rate and consumption index of *Tuta absoluta* larvae feeding AMF tomato leaves were much lower than that on plants without AMF (Shafiei et al. 2022).

Moreover, AMF promote plant water uptake; increase water use efficiency, especially under drought stress conditions; and enhance plant drought resistance (Birhane et al. 2012). In wheat plants, AMF inoculation induces an increase in plant biomass accumulation under long-term growth conditions under drought stress (Al-Karaki et al. 2004).

12.2.3 AMF as Biocontrol Agent

AMF play a vital role in boosting the plant resistance to above- and/or belowground pests and diseases (Whipps et al. 2008). AMF-plant symbiosis has been revealed to profoundly alter plant primary and secondary metabolism, including the salicylic acid (SA) and jasmonate (JA) signaling pathways that are critical to plant defenses (Cameron et al. 2013). Plants that have been colonized by AMF prior to pest or pathogen attacks may be systemically primed through defense compound reallocation (Jung et al. 2012), thus allowing rapid and increased expression of defense genes compared with non-AMF plants (Song et al. 2015). AMF can enhance the ability of host plants against fungal, bacterial, viral, and nematode diseases. AMF can significantly inhibit the dispersal of root-knot nematodes and decrease gall numbers and the infection rate and the damage caused by nematodes (Lax et al. 2011). *Glomus mosseae* increased the resistance of different tomato varieties to *Fusarium*, but the extent of resistance was not consistent (Steinkellner et al. 2012). Cucumber, bent grass (*Agrostis stolonifera*), and tomato were inoculated with *Phoma* sp. GS8-2 and *Glomus mosseae*, respectively. *Glomus mosseae* reduced *Rhizoctonia*-caused cucumber root rot and *Fusarium oxysporum*-caused tomato root rot but aggravated bent grass brown spot (Saldajeno et al. 2012). Cosme et al. (2011) explored the effect of inoculation of *Glomus intraradices* on the oviposition of *Lissorhoptrus oryzophilus* and found that this insect prefers to lay eggs on mycorrhizal plants. The possible explanation is that AMF colonization improves the root N and P contents; thus, the mycorrhizal plants with better growth were preferred to lay eggs; however, the harm of adults did not increase.

From the agricultural point of view, however, trade-offs that reduce yields may occur between AMF, crops, and pests. In addition to benefiting crop nutrition, AMF colonization also enhances the attraction, quantity characteristics, and quality measures of plants against herbivores (Hartley and Gange 2009) and therefore improves capabilities of herbivores (Kempel et al. 2010). Moreover, phloem-feeding insects frequently outperform non-AM controls on AMF-colonized plants (Hartley and Gange 2009). AMF management strategies should consider these trade-offs and balance them accordingly.

12.2.4 Contribution of AMF to Soil Health

Arbuscular mycorrhiza can directly affect the cycle of inorganic elements such as N, P, K, Ca, and Mg through chemical, physical, and biological activities and play an important role in the biogeochemical cycling of soil nutrients. The impact of arbuscular mycorrhiza on inorganic N is mainly through the hyphae absorbing N from the substrate and transferring it to the host plant and improving the rate of N fixation in N-fixing host plants by alleviating various stresses. The impact of AMF on promoting plant growth is closely associated with the improvement of plant P nutrition by mycorrhizal colonization. In soils with insufficient P supply, the inoculation of AMF can greatly improve the absorption and use of P by the symbiotic partner plants and alleviate the phenomenon of P deficiency in the rhizospheric zone (Recorbet et al. 2013).

AMF improve soil health through external hyphae too, thereby sustaining the constancy of the soil food web and increasing the stability of the soil structure (Finlay 2008), which provides numerous benefits to the host plants, such as defense against pathogens, enhanced salinity tolerance, decreased pH and heavy metals, and biofortification with trace elements in the crops (Ryan and Graham 2018). The huge mycelial network formed by the symbiosis of AMF and crops can extend to deeper and wider soils beyond the rhizosphere trophic zone of plants, thus improving the utilization of elements by plants (Lehmann et al. 2014). Besides facilitating host plant uptake of nutrients and water, AMF expand nutrient entrapment through their mycelial network, increasing soil aggregation through mycelium and glomalin, thereby improving soil nutrient storage and retention (Rillig and Mummey 2006) as well as reducing soil erosion and the risk of nutrient leaching.

12.2.5 Abiotic Stress Alleviation

Abiotic stress (e.g., drought, high temperature, salinity, heavy metal stress, and unfavorable soil pH) leads to degradation of soil and poses a serious threat to agricultural production. Therefore, abiotic stress is considered to be one of the main reasons for crop yield reduction worldwide (Singh et al. 2011). AMF can enhance the tolerance of plants to abiotic stress. The extracellular hyphae of AMF can promote the uptake and consumption of nutrients and water by plants, and it is one of the critical factors in facilitating plants to avoid drought damage (Tyagi et al.

2017). Increased drought resistance and better crop performance can be due to the abundance of antioxidative enzymes (superoxide dismutase, catalase, peroxidase) and soluble sugars symbiotically produced by AMF (Huang et al. 2011). AMF also improve plant drought tolerance by improving soil structural stability. It has been reported that AMF hyphae have a rich filamentous hyphae network, which can enhance soil structure and also change soil structure by producing glycoproteins to form aggregates (Singh et al. 2013). Therefore, AMF symbiosis can enhance the drought tolerance of plants by enhancing the stability of soil structure, thereby increasing soil water-holding capacity (Ruiz-Lozano 2003).

The occurrence of AMF greatly enhanced the tolerance of plants to extreme temperature. Under the condition of high temperature stress, root colonization with AMF can enhance the activity of antioxidant enzymes and increase the content of soluble protein and proline in maize (*Zea mays* L.). Plants inoculated with AMF increased the activity of SOD, POD, and CAT by 50%, 40%, and 21%, respectively, indicating that AMF can resist high temperature stress damage by improving the antioxidant capacity of plants, thereby promoting plant growth (Mathur et al. 2018). In addition, low temperature or high temperature induces excessive ROS production under stress, causing oxidative stress response to damage plants. Usually, AMF can effectively promote the production of SOD, POD, CAT, and GR after colonizing host plants and relieve the damage caused by reactive oxygen species. Specifically, they reduce peroxidation of membrane lipids as well as membrane permeability and increase osmotic adjustment substance accumulation in response to temperature stress (Ahanger and Agarwal 2017; Hajiboland et al. 2019).

Under heavy metal stress, AMF change the growth pattern of plants by exploiting and exploring unstressed parts of the environment, alleviating the toxicity of heavy metals through various pathways (Gonzalez-Guerrero et al. 2008). The roots of AMF-inoculated plants can greatly contain heavy metals, which can inhibit the transport of heavy metals from the root system to the aboveground tolerance to adapt to heavy metal stress (Słomka et al. 2011; Zhang et al. 2010). However, mycelia have limited uptake potential of heavy metals, and in highly polluted places, the possibility of mycelium avoiding the hazards of toxic metals is very limited. Therefore, mycorrhizae have developed different strategies to resist heavy metal stress: AMF can secrete glomycin (Ferrol et al. 2009); glomycin participates in heavy metal inactivation by chelating heavy metals in soil. Audet and Charest (2007) revealed that AMF shift from "enhanced uptake" at soil with low concentrations of heavy metals to "metal binding" at soil with high concentrations of heavy metals. The remediation effect of AMF on heavy metals is also reflected in the ability to induce host plants to produce antioxidant enzymes, upregulate the expression of related peroxidase genes, and reduce the detrimental effects of ROS on plants under heavy metal stress (Rozpądek et al. 2014).

Salt stress can negatively affect plant photosynthetic capacity, enzyme activity, protein synthesis, and mineral nutrition and produce osmotic stress and ion stress on plants, thereby affecting the physiological and biochemical balance of plants (Hashem et al. 2016; Iqbal et al. 2015). Salt stress can promote ROS production, cause plant oxidative damage, and interfere with the normal metabolism of plants

(Akyol et al. 2020). AMF can improve the antioxidant capacity of host, reduce H_2O_2 content, inhibit lipid peroxidation, and enhance the salt stress of plants by inducing plants to produce and accumulate osmotic adjustment substances such as glycine betaine, proline, and soluble sugar tolerance (Akyol et al. 2020; Porcel et al. 2015; Zhu 2003). Hidri et al. have found that AMF can improve nutrient (P) acquisition; reduce sodium absorption; enhance water absorption; and improve photosynthesis, chlorophyll content, and availability of antioxidant molecules under saline conditions. Hajiboland et al. (2010) reported that improved salt stress tolerance in tomato plants inoculated with AMF was associated with increased uptakc of P, potassium (K), and calcium (Ca) and reduced sodium (Na) toxicity.

12.3 Response of Arbuscular Mycorrhizal Fungi to Agricultural Management Practice

AMF develop symbiosis with most plant species and can assist the host plant in many ways, including better nutrient uptake, drought and salt tolerance, and disease resistance (Smith and Read 2008). These fungi are crucial to the preservation of the soil and the sustenance of plants. The development and survival of fungal propagules, including spores, hyphae, and colonized roots, are necessary for AM fungus to persist in various ecosystems. In the presence of host plants and no soil disturbance, hyphae are considered to be the primary source of inoculum, but spores are thought to be a relatively resistant fungal structure that may be viewed as "long-term" propagules when no viable host plants are present. For agricultural crops, there are management techniques that are widely acknowledged. Numerous agricultural management practices, such as soil tillage, fertilization, and plant protection techniques, have deleterious impact on mycorrhizal symbioses in agroecosystems (Säle et al. 2015; Jansa et al. 2002; Gryndler et al. 2006) (Fig. 12.1). These traditional management techniques can decrease soil biodiversity, especially AMF (Tsiafouli et al. 2015). The technique of organic agriculture, in contrast, can enhance soil quality in terms of boosting the diversity and biomass of soil microbiome (Mäder et al. 2002; Verbruggen et al. 2010) (Fig. 12.1). This section focuses on the discussion of the effects of agricultural management practices on AMF.

12.3.1 Agrochemical Management Practice

Agrochemicals are chemical formulations used in industrial agriculture. It refers to chemical fertilizers and biocides, which are substances used for control of hazardous organisms. A biocide can be (1) pesticides including insecticides, herbicides, fungicides, and nematicides; (2) antimicrobials including antibacterials, antifungals, antiviral, antiprotozoals, and antiparasites; and (3) synthetic fertilizers. In addition to

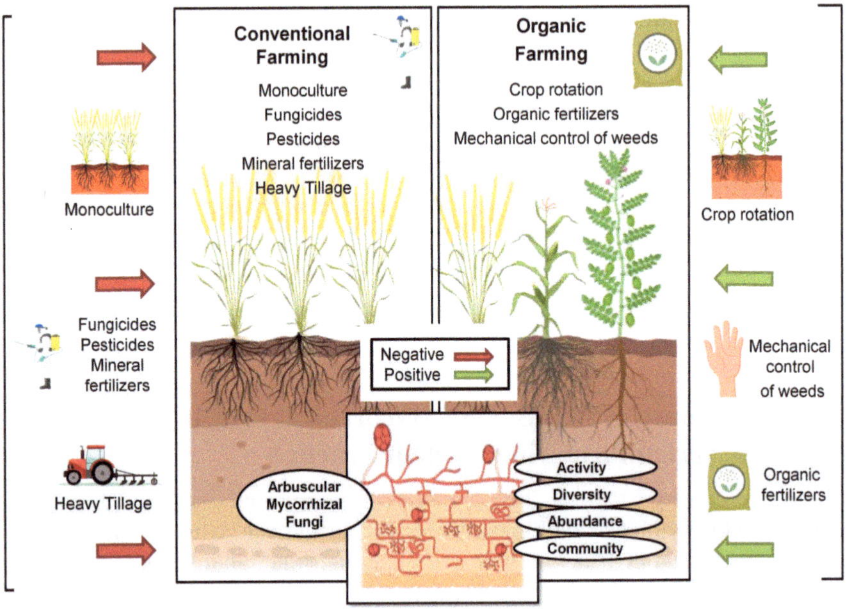

Fig. 12.1 Impact of different agricultural management practices on arbuscular mycorrhizal fungi

potentially contaminating the soil, the application of agrochemicals may have detrimental effects on AMF community (Kurle and Pfleger 1994; Abd-Alla et al. 2000; Lekberg and Koide 2005; Pasaribu et al. 2011; Mathimaran et al. 2007) leading to a decline in AMF spore diversity and abundance (Oehl et al. 2004).

12.3.1.1 Pesticides

Insecticides, fungicides, nematicides, and herbicides are the four major categories of pesticides. Pesticides that are applied in agricultural systems could harm both human and environment when used in an unrestrained manner. The quantity of pesticides used and, consequently, their presence in food depend on the farming system which is generally higher in conventional integrated pest management (IPM) than that in organic systems (Atkinson 2009). Therefore, agricultural production has been referred to as "the ghost of conventional agriculture past" (Riedo et al. 2021). Consequently, the application of beneficial soil organisms like AMF contrasts with previous and present conventional agricultural procedures because of the deleterious effects of excessive levels of pesticides on soil microbial communities (Dodd 2000; Montesinos 2003; Jacobsen and Hjelmsø 2014). Such unfavorable outcomes are caused not only by the direct harm of pesticides to soil life, but it also alters the soil ecosystem by favoring weeds and some specific crops through the emergence of pesticide resistance (Baek et al. 2021). When active compounds (pesticides) are administered as a soil drench, seed treatment, and foliar spray or when these substances exist in runoff from leaf application or drift to the soil, they affect the viability and functionality of AMF structures like spores and hyphae in soil. Additionally, variations in the physiology of the host plant may have indirect impacts on AMF symbiosis.

12.3.1.1.1 Fungicides

It is not surprising that fungicides might have a deleterious impact on AMF. There is a varied sensitivity of AMF to fungicides that explains why negative, neutral, and positive impacts of fungicides have been recorded. Some fungicides target specific families of fungi (e.g., Ascomycota, to which the majority of fungal diseases belong). However, some fungicides, including thiazoles (benomyl and carbendazim), have a substantial adverse impact on Glomeromycota, the phylum of fungi to which AMF belongs. Thus, benomyl causes a reduction in AMF activity in the field (Allison et al. 2007; Chiocchio et al. 2000; Hartnett and Wilson 2002; O'Connor et al. 2009). Additionally, benomyl has a negative effect on nutrient mineralization and decomposition by reducing the diversity and abundance of saprotrophic soil fungus. An alternative fungicide, Topsin M (also known as topsin), was described for use in ecological study including the reduction of AMF (Wilson and Williamson 2008). The dosage used, as well as the fungicide's interactions with the soil matrix (such as adsorption on organic matter or solubility in soil solution), influences how well fungicides work.

12.3.1.1.2 Herbicides

Because the majority of current efforts are focused on aboveground control, herbicide impact on soil biota has got little attention (Kremer 2014). The three herbicides that are most frequently used are paraquat, oxyfluorfen, and glyphosate. The herbicide oxyfluorfen was found to have a negative or indifferent effect on AMF. Recently, genetically modified crops (such as cotton, maize, soybean, and canola) have been released onto the market and include a gene that renders such plants resistant to glyphosate; its use is probably going to grow. Consequently, glyphosate usage as a weed control method was increased particularly in no-till or conservation agriculture (Watrud et al. 2011; Baek et al. 2021). Due to its alleged quick decomposition in soil and minimal toxicity to vertebrates, glyphosate has been viewed as a reasonably advantageous foliar-acting herbicide (Duke and Powles 2008; Weidenhamer and Callaway 2010). Studies examining the effects of glyphosate on AMF reveal a suppression of AM fungal spore germination as well as germ tube growth and a reduction of mycorrhizal count in soil (Zaller et al. 2014) but only at concentrations higher than those advised for usage in the field. When used at acceptable concentrations, glyphosate has no effect on mycorrhiza (Baumgartner et al. 2010; Pasaribu et al. 2011). Although sublethal doses of herbicides may not cause the plant to die, they can diminish photosynthetic rates to the point where the symbiosis is compromised.

12.3.1.1.3 Insecticides and Nematicides

The effects of fungicides and herbicides on AMF are far greater than the effects of insecticides and nematicides. Insecticides and nematicides, according to earlier evaluations, had either no adverse effects or moderately positive effects on AMF (Trappe et al. 1984; Hamel and Strullu 2006). However, differential impact of systemic and contact insecticides was documented, just as they were in the case of fungicides (Sarr et al. 2013; Deliopoulos et al. 2008). For example, phoxim, an insecticide/acaricide, was reported to prevent AMF colonization in carrot (*Daucus carota*); however, it was not the case in green onion (*Allium fistulosum*) (Wang et al. 2011a, b). Aldicarb, a nematicide, had no impact on mycorrhizal colonization of potato (*Solanum tuberosum*) (Goulson 2013). Currently, neonicotinoid insecticide class is one of the most popularly used globally due to their broad spectrum of action on many pests of various crops. Numerous civilizations use neonicotinoids extensively. They have drawn a lot of attention because of their extremely detrimental effects on insect diversity, which also have major repercussions for ecosystem services like pollination (Hladik et al. 2018; Malfatti et al. 2023). Neoicotinoid effects on AMF have just recently been the subject of studies. For instance, two neonicotinoids, imidacloprid as well as thiamethoxam, were evaluated on the spore germination of two AMF species, *Rhizophagus clarus* and *Glomus albida*. The results showed that both neonicotinoids dramatically decreased spore germination of AMF in a dose-dependent way (Malfatti et al. 2023).

12.3.1.2 Fertilizers

Due to the critical symbiotic interaction between hyphae of mycorrhizal fungi and host plants, AMF might be sensitive to alternations in soil nutrients (Hack et al. 2019). Numerous earlier researches have demonstrated that modifying the soil microenvironment with the addition of mineral fertilizer can have (positive, negative, or inconsequential) impacts on the proliferation of AMF. A field study revealed that while P application increased AMF abundance, N application predominantly changed the species composition of AMF (Chen et al. 2014). Nevertheless, according to Xiao et al. (2019), the application of N had an impact on AMF abundance, while the addition of P had an impact on AMF diversity, and the augmentation of N and P had no discernible impact on the AMF community composition. Besides, high soil nutrient levels, such as N and P, can promote AMF sporulation, so adding organic fertilizer helps the soil flora thrive (Qin et al. 2015). Furthermore, soil pH and K have a significant impact on the AMF community structure (Qin et al. 2015).

12.3.2 Soil Tillage

Rotary soil tillage is regarded as a traditional farming method. Rotary tillage is important because it breaks up and softens the soil while also equally distributing organic materials and nutrients. No-till farming, on the other hand, is a method of crop production in which the soil is not tilled. Crop growth and yield have been observed to be lowered (Huynh et al. 2019), unchanged (Büchi et al. 2017), or increased (da Silva et al. 2022) when compared to rotary tillage cultivation. As previously mentioned, tillage is a factor that influences soil hardness and the alterations in root growth which can either improve or decrease yield (Ren et al. 2018). Additionally, it has been proposed that variations in the species of soil microbes may have an impact on crop yield, which is another element that may change crop growth and yield during plowing (Ma et al. 2020). AMF produce exterior hyphae that provide the host with P that could not be taken up through plant roots (Smith and Read 2008). Intensive soil tillage may have a deleterious impact on AMF biomass and community composition. The disruption of the AMF mycelial network has been proven to be the primary reason why traditional farming generally diminishes the quantity of AMF when compared to no-tillage systems (Gu et al. 2020; Jansa et al. 2006; Castelli et al. 2014). Additionally, substantial research has been done to show how tillage techniques and AMF colonization can change AMF communities in soil and plant roots (Tatewaki et al. 2021; Liu et al. 2022a, b; Lu et al. 2018; Gu et al. 2020; Tatewaki et al. 2021).

12.3.3 Crop Rotation

Crop rotation is a popular agronomic soil management technique that aims to manage nutrient requirements, maintain healthy soil, reduce pest and disease pressure, and reduce reliance on agricultural chemicals in order to produce high yields that are sustainable (Angus et al. 2015). Crop rotation slows the growth of weeds and plant diseases (Krupinsky et al. 2002; Liebman and Dyck 1993). It also hastens the buildup of soil organic C and soil N (Havlin et al. 1990; Van Eerd et al. 2014). Additionally, a system with a greater variety of rotating crops is linked to a greater diversity and richness of soil microbial life (Venter et al. 2016). The benefits of crop rotation systems are influenced by the plant species chosen and the order in which they are planted (Yang and Kay 2001; West and Post 2002). Despite its significance, it can also impact soil microbial communities, particularly AMF, which are imperative for plant growth. The existence of nonhost crops in the rotation has been proven to have an impact on root colonization and spore abundance (Arihara and Karasawa 2000). Extremely little AMF variety has reportedly been found in monoculture (Burrows and Pfleger 2002; Oehl et al. 2003). Crop rotations appear to be a simple and low-cost farming technique associated with increased AMF sporulation in many upland farming systems. AMF sporulation increased in a maize-common bean crop rotation system compared to a maize monoculture, and AMF sporulation correlates with the crop previously planted on the site (Oruru and Njeru 2016). Similarly, crop rotation with mycorrhizal plants and AMF inoculation both improve maize growth performance in limed, acid sulfate soil (Higo et al. 2010). Crop rotation also aids plant nutrient absorption. Improved N uptake by maize plants in a maize-common bean crop rotation system is one example (Oruru and Njeru 2016). Similarly, Ratnayake et al. (2017) revealed that rotations of crops can boost nutrient availability in soil and stocks of C compared with monocultures.

12.4 Comparing Arbuscular Mycorrhizal Fungi in Organic and Conventional Farming

12.4.1 Benefits and Limitations of Organic and Conventional Farming

Organic and conventional farming both have advantages and disadvantages. Although conventional farming is frequently associated with higher yields and lower costs, it is harmful to the environment, reduces biodiversity, and poses potential health risks. In addition, traditional farming controls synthetic inputs (such as pesticides, chemical fertilizers, and amendments) consistently, disregarding the naturally occurring spatial-longitudinal heterogeneity of soil and crop conditions within fields. As a result, organic farming is emerging as a viable option for productive and sustainable agriculture although it is characterized by high cost and relatively unstable yield (Jaacks et al. 2022). Organic farming relies on the use of organic fertilizers, biofertilizers, and organominerals, which helps to avoid the negative environmental effects and toxicities associated with chemicalized farming. In addition, organic agriculture increases the agroecosystem's resistance to the negative effects of climate change by enhancing the sequestration and conservation of C in soil systems (Gamage et al. 2023).

12.4.2 Influence of Organic and Conventional Farming on Arbuscular Mycorrhizal Diversity and Community Composition

Agricultural management strategies and differences in land-use types can have a significant impact on AMF populations (Martinez and Johnson 2010; Zhao et al. 2015a, b). Understanding the effects of organic and conventional agricultural practices on the diversity and community composition of mycorrhizal symbiosis is essential for agricultural production and sustainable soil utilization. Several studies have been undertaken in agricultural contexts for comparing the effect of organic and

Table 12.1 Recent studies comparing the effect of organic and conventional management approaches on arbuscular mycorrhizal fungi

Studied plants	Major findings	Applied techniques	References
Maize root	The colonization intensity, operational taxonomic unit, and taxonomic diversity were all greater on conventional farms compared with that on organic farms	Sequencing of V4–V5 hyper-variable regions of the fungal 18S rRNA gene	Chen et al. (2022)
Wheat	Conventional agriculture had a substantial impact on the AMF population but led to a decline in AMF diversity	Sequencing of 18S rRNA gene amplicons	Wahdan et al. (2021)
Cereal crops	Organic farming had a positive effect on the AMF diversity, which influenced plant productivity	Sequencing of the LSU region of ribosomal DNA	Manoharan et al. (2017)
Apple	AMF abundance was substantially greater in organically managed than in conventionally managed orchards. Also, organic orchards harbored significantly different AMF communities	454 pyrosequencing of SSU rRNA gene amplicons	van Geel et al. (2015)
Maize and potato	A decline was observed in the diversity of mycorrhizal taxa in conventionally maintained fields	T-RFLP of 25S rRNA gene fragments	Verbruggen et al. (2012)
Maize and potato	The average number of AMF taxa was greater in areas treated organically. In addition, AMF abundance significantly increased after transition to organic agriculture	Terminal restriction fragment-length polymorphism (T-RFLP) of PCR-amplified large subunit rRNA gene fragments	Verbruggen et al. (2010)
Onion	Farming systems did not influence AMF diversity	Sequencing of the partial 18S-ITS1–5.8S-ITS2 rDNA region	Galván et al. (2009)
Potatoes, winter wheat, and beetroots	Mycorrhizal spore abundance and species diversity were considerably greater in organic systems than that in conventional ones	Morphological identification	Oehl et al. (2004)

conventional management approaches on AMF diversity (Table 12.1). Almost all studies indicated that organic farming is capable of sustaining greater diversity of AMF than conventional farming and showed the necessity of an improved diversity for long-term crop productivity (Manoharan et al. 2017).

12.5 Digging Deeper: Factors to Consider When Applying Arbuscular Mycorrhizal Fungi in Organic Farming

The environmentally benign, innovative, and sustainable method of organic farming increases agricultural output and the quality of life for many farmers. The role of AMF in helping soil fertility, sustainability, and crop production improvement is one of the most researched and well-known topics in agronomy. When incorporating AMF into organic farming practices, several factors need to be taken into consideration.

Firstly, arbuscular mycorrhizal fungal communities can be influenced by the identity of the host plant. Furthermore, plant traits, such as root exudates and morphological characteristics, play a critical role in determining the specificity of AM symbiosis (Ramana et al. 2023). Hence, various studies have shown that the presence of suitable AMF species is required in a given soil system. For instance, Gao et al. (2020) reported a high specificity between roots of cotton plant and the mycorrhizal fungus, *Rhizophagus irregularis* CD1. They also reported an increase in growth and yield, boll number per plant, as well as the maturity of the fiber cotton when inoculated with *Rhizophagus irregularis* CD1 as compared with non-inoculated plant. After mycorrhizal inoculation, both the expression of particular P transporter family genes and the P concentration in cotton biomass were dramatically increased.

Secondly, the presence of other soil microorganisms should also be considered. Some microorganisms, such as certain bacteria and fungi, may cooperate with AMF to facilitate the host plant growth (Feng et al. 2023). Therefore, it is essential to maintain a balanced microbial community to support the symbiotic relationship between AMF and plants. For instance, a recent study revealed that some isolates of AMF (e.g., *Gigaspora* sp.) and bacteria (e.g., *Sphingomonas*) assisted their host plants thrive in the native soil (Feng et al. 2023). The combined use of *Bacillus subtilis* and AM fungi resulted in superior yield-related indexes and biofortification in wheat grains (Yadav et al. 2020).

Thirdly, combining AMF with suitable sustainable agronomic practices can maximize the expected benefits from plant-arbuscular mycorrhizal symbiosis. For instance, the use of biofertilizers to agricultural soils has a positive impact on crop productivity and soil fertility. Combining the exogenous AMF *Rhizoglomus irregulare* (DAOM) with phospho-compost resulted in considerably increased shoot and root dry weights as well as yield in lettuce plants compared to control plants (Anli et al. 2022). In addition, soil characteristics, particularly soil total organic C, N, P, and GRSP content, were considerably enhanced in soils treated with biofertilizers and *Rhizoglomus irregulare* (DAOM) after 2 months of field testing compared to the control (Anli et al. 2022).

Finally, regular monitoring and evaluation of AMF colonization and effectiveness are essential. This can be accomplished by sampling and analysing roots, and assessing plant growth and nutrient acquisition. To maximize the benefits provided by AMF, management practices may need to be adjusted based on the results.

12.6 Arbuscular Mycorrhizal Fungi and Climate-Smart Organic Agriculture

Climate change is one of the most significant concerns influencing the way of life of people worldwide. Climate change and some agricultural management practices, such as tilling and excessive chemical use, have all led to land degradation and crop output decline. AMF are capable of mitigating future environmental disturbances. Field investigations and greenhouse tests under simulated climate warming indicate that temperature increases may have a positive effect on AM symbiosis (Hu et al. 2015). Furthermore, it has been suggested that the stimulation of AMF by elevated atmospheric carbon dioxide (CO_2) is a key mechanism for promoting soil C sequestration by boosting C inputs to soil and by shielding organic C from degradation via aggregation (Cheng et al. 2012). In a unique field experiment that simulates future climate, Wahdan et al. (2021) found that organic agricultural practices significantly increased total richness of AMF under future climate conditions compared to all other treatments, indicating that organic farming not only mitigates the effects of climate change but also increases richness of AMF under future climate conditions (Wahdan et al. 2021).

12.7 Future Directions in Sustainable Forestry: AMF Response to Management Practices in Forests

AMF play a major role in the establishment and succession of plant communities, driving the restoration process of degraded land (Manaut et al. 2015; Karthikeyan and Krishnakumar 2012). AMF is one of the pivotal factors determining species diversity, community structure, and productivity in terrestrial ecosystems, and they regulate the resource allocation among adjacent plants through the underground mycelial network, thereby affecting the coexistence and competition among plants (Smith et al. 2010). Zobel and Öpik (2014) stated that the inherent relationships between AMF and the plant community can be explained by "driver" and "passenger" hypotheses, which illustrated the driving effect of AMF on the plant. Generally, AMF will govern the plant community structure of a certain successional stage, and the specific plant community of this successional stage will affect the colonization of the AMF communities, and the varied AMF communities will further affect the plant community in the next successional stage (Renker et al. 2004). In addition, plant-soil feedback is also a manifestation of AMF-driven plant community succession. Positive feedback can promote community development in the early stage, while negative feedback can facilitate plant species replacement and drive community succession (Kikvidze et al. 2010).

The influence of AMF on plant fitness can be attributed to improved seedling establishment, plant growth, a higher tolerance to belowground plant pathogens, better water acquisition, and improved soil structure and stability in forest ecosystems (Newsham et al. 1995). Through the removal of nutrients, disturbance of soil or forest floor, and altered microclimates, intensive forest management practices can negatively affect AMF biodiversity (Sharmah and Jha 2014). It has been suggested that intensive forest clearcutting may affect the distribution and composition of AMF communities (Moora et al. 2014; Mummey et al. 2010), favoring some AMF taxa and disfavoring others (Gottshall et al. 2017). Intensive management practices can significantly decrease AMF biomass in soil and glomalin-related soil protein (GRSP) content, of which understory removal, mineral fertilization, and soil tillage are major practices with profound negative effects on AMF (Qin et al. 2017). AMF can produce glomalin, which plays the role of "super glue" in the formation of stable soil structure, which can significantly increase the water percolating capacity and stability of soil (Burrows 2014).

In addition, the abundance of AMF taxa may be correlated with the level of management, and AMF diversity varies depending on management techniques as seen by the high AMF biodiversity in non-managed fields and low diversity in intensively managed fields (Sharmah and Jha 2014). In Moso bamboo forests, AMF abundance and community composition are altered by long-term intensive management, leading to a decrease in soil aggregation (Qin et al. 2017). Moreover, when AMF communities are strongly altered, Moso bamboo forests have been intensively managed for long-term experience soil erosion and organic C decomposition. Because of intensive management practices, AMF communities differ from those of old stands, but the ecosystem can still provide a "symbiotic service" needed to restore old-growth understory plant communities (Uibopuu et al. 2009).

In abandoned slash-and-burn forests, which have been disturbed for 2 or 3 years, AMF communities were similar to those in mature forests (de León et al. 2018). Studies indicate that AMF communities and forest regeneration may be threatened by clearcutting tropical rain forests (Stürmer and Siqueira 2011), whereas slash-and-burn management may have relatively minor effects (Rillig 2004; Stürmer and Siqueira 2011). According to Aguilar-Fernández et al. (2009), slash-and-burn maintained similar levels of species richness in forest and pasture plots. AMF communities that develop after forest conversion to pasture seem to be defined more by the persistent effects derived from conversion and management which can be manifested by variations in composition of plant species, soil quality, and microclimates than by the slash-and-burn process (Sharmah and Jha 2014).

The composition of AMF communities was significantly affected by land use, with communities in the clearcut region being the most distinct (de León et al. 2018). Under various land-use regimes, changes in AMF communities may be caused by both biotic and abiotic interactions. Host plants assist AMF establishment via different types of propagules that arrive at a location of interaction (Fonseca et al. 2017), although abiotic factors filtering environment are more significant components in community development than AMF competition (Bouffaud et al. 2016). According to Oehl et al. (2010), soil type and land-use intensity are the primary factors determining the composition and richness of AMF, and different agricultural practices including monoculture and narrow cereal mono-crop rotations can significantly reduce the biodiversity and richness of AMF in comparison to grasslands.

In terms of plant community structures and production, the biological and functional variety of AMF is crucial to forest ecosystems (Jiang et al. 2017). AMF populations in differently managed forests had distinct effects on the development of various forest plant species; in general, the old growth inoculum had a greater beneficial impact than the inoculum from the young forest. The response of plant communities to the severity of forest management may be mediated by mycorrhizal interactions since land use influences the composition of AM fungus populations (Uibopuu et al. 2012). Thinning is a crucial silvicultural strategy that has an impact on a variety of forest features, including overstory, understory, soil properties (Trentini et al. 2017; Zhou et al. 2016), and microbial communities (Dang et al. 2018). In comparison with the late successional species, the early successional species considerably increased the rate of AMF colonization (Zangaro et al. 2003). The findings of Lu et al. suggested that the expansion of Chinese fir may have reached a turning point in the 32-year period as the number of shared OTUs continuously climbed from 9-year, 17-year, and 23-year to 32-year while decreasing in 45-year (Lu et al. 2019). The colonization, biomass, and diversity of AMF considerably increased with the emergence of *Cunninghamia lanceolata*, and the biomass contribution of AMF to glomalin-related soil protein (GRSP) showed a strong association with soil organic C (Liu et al. 2022a, b). Zhou et al. found that the topsoil C storage was significantly enhanced by in-forest planting of *Sarcandra glabra*, while the diversity and composition of the related AMF community were not significantly affected (Zhou et al. 2022). In-forest planting of *Sarcandra glabra* was beneficial for the forest and ecologically safe for the *C. lanceolata* AMF (Zhou et al. 2022). Wang et al. found that the rate of AMF colonization was substantially and negatively linked with soil P, suggesting that low P soil would be a better choice for using AMF in afforestation (Wang et al. 2019).

12.8 Conclusions

AMF play a vital role in ecosystem function, soil structure maintenance through C sequestration, and water retention. As a result, whether farmers use organic or conventional methods has an impact not only on agricultural productivity but also on overall environmental sustainability. This chapter investigated the impact of organic vs. conventional farming practices on AMF using existing research and empirical evidence.

For starters, organic farming has been found to be more conducive to promoting AMF diversity and abundance due to its emphasis on sustainable and eco-friendly practices. The absence of synthetic fertilizers and pesticides promotes a more balanced and natural ecosystem, which promotes the sporulation and activity of AMF. As a result, organic farming systems have the potential to significantly improve soil fertility, nutrient cycling, and plant health. Conventional farming, on the other hand, has been shown to have negative effects on AMF populations because of its heavy reliance on agrochemical inputs. Synthetic pesticides and fertilizers can disrupt the symbiotic association between AMF and plant host, resulting in a reduction in AMF diversity and activity. This, in turn, can have deleterious impacts on soil health and nutrient availability.

More research is needed in the future to gain a better understanding of the mechanisms that govern interactions between farming practices and AMF communities. This knowledge will be extremely useful in directing agricultural policies and practices toward more environmentally sound and sustainable approaches. Efforts should also be made to educate farmers, policymakers, and consumers about the significance of AMF and the potential benefits of organic farming in promoting their growth.

References

Abd-Alla MH, Omar SA, Karanxha S (2000) The impact of pesticides on arbuscular mycorrhizal and nitrogen-fixing symbioses in legumes. Appl Soil Ecol 14:191–200. https://doi.org/10.1016/S0929-1393(00)00056-1

Aguilar-Fernández M, Jaramillo VJ, Varela-Fregoso L et al (2009) Short-term consequences of slash-and-burn practices on the arbuscular mycorrhizal fungi of a tropical dry forest. Mycorrhiza 19:179–186. https://doi.org/10.1007/s00572-009-0229-2

Ahanger MA, Agarwal R (2017) Potassium up-regulates antioxidant metabolism and alleviates growth inhibition under water and osmotic stress in wheat (*Triticum aestivum* L). Protoplasma 254:1471–1486. https://doi.org/10.1007/s00709-016-1037-0

Akyol TY, Yilmaz O, Uzilday B et al (2020) Plant response to salinity: an analysis of ROS formation, signaling, and antioxidant defense. Turk J Bot 44:1–13

Al-Karaki G, McMichael B, Zak J (2004) Field response of wheat to arbuscular mycorrhizal fungi and drought stress. Mycorrhiza 14:263–269. https://doi.org/10.1007/s00572-003-0265-2

Allison VJ, Rajaniemi TK, Goldberg DE, Zak DR (2007) Quantifying direct and indirect effects of fungicide on an old-field plant community: an experimental null-community approach. Plant Ecol 190:53–69. https://doi.org/10.1007/s11258-006-9190-8

Angus JF, Kirkegaard JA, Hunt JR et al (2015) Break crops and rotations for wheat. Crop Pasture Sci 66(6):523–552

Anli M, Boutasknit A, Ait-El-Mokhtar M et al (2022) Improving lettuce yield and quality of an agricultural soil using a combination of arbuscular mycorrhizal fungus and phosphate-green wastes compost. Gesunde Pflanzen 74:205–217. https://doi.org/10.1007/s10343-021-00603-0

Arihara J, Karasawa T (2000) Effect of previous crops on arbuscular mycorrhizal formation and growth of succeeding maize. Soil Sci Plant Nutr 46(1):43–51. https://doi.org/10.1080/00380768.2000.10408761

Aroca R, Porcel R, Ruiz-Lozano JM (2007) How does arbuscular mycorrhizal symbiosis regulate root hydraulic properties and plasma membrane aquaporins in *Phaseolus vulgaris* under drought, cold or salinity stresses? New Phytol 173:808–816. https://doi.org/10.1111/j.1469-8137.2006.01961.x

Atkinson D (2009) Soil microbial resources and agricultural policies. In: Azcón-Aguilar C, Barea JM, Gianinazzi S, Gianinazzi-Pearson V (eds) Mycorrhizas-functional processes and ecological impact. Springer, Berlin/Heidelberg, pp 1–16

Audet P, Charest C (2007) Dynamics of arbuscular mycorrhizal symbiosis in heavy metal phytoremediation: meta-analytical and conceptual perspectives. Environ Pollut 147:609–614. https://doi.org/10.1016/j.envpol.2006.10.006

Baek Y, Bobadilla LK, Giacomini DA et al (2021) Evolution of glyphosate-resistant weeds. Rev Environ Contam Toxicol 255:93–128. https://doi.org/10.1007/398_2020_55

Basu S, Rabara RC, Negi S (2018) AMF: the future prospect for sustainable agriculture. Physiol Mol Plant Pathol 102:36–45. https://doi.org/10.1016/j.pmpp.2017.11.007
Baumgartner K, Fujiyoshi P, Smith R et al (2010) Weed flora and dormant-season cover crops have no effects on arbuscular mycorrhizae of grapevine. Weed Res 50(5):456–466. https://doi.org/10.1111/j.1365-3180.2010.00793.x
Bender SF, Wagg C, van der Heijden MG (2016) An underground revolution: biodiversity and soil ecological engineering for agricultural sustainability. Trends Ecol Evol 31:440–452. https://doi.org/10.1016/j.tree.2016.02.016
Berruti A, Lumini E, Balestrini R et al (2016) Arbuscular mycorrhizal fungi as natural biofertilizers: let's benefit from past successes. Front Microbiol 6:1559. https://doi.org/10.3389/fmicb.2015.01559
Birhane E, Sterck FJ, Fetene M et al (2012) Arbuscular mycorrhizal fungi enhance photosynthesis, water use efficiency, and growth of frankincense seedlings under pulsed water availability conditions. Oecologia 169:895–904. https://doi.org/10.1007/s00442-012-2258-3
Bouffaud M-L, Creamer RE, Stone D et al (2016) Indicator species and co-occurrence in communities of arbuscular mycorrhizal fungi at the European scale. Soil Biol Biochem 103:464–470. https://doi.org/10.1016/j.soilbio.2016.09.022
Bowles TM, Barrios-Masias FH, Carlisle EA et al (2016) Effects of arbuscular mycorrhizae on tomato yield, nutrient uptake, water relations, and soil carbon dynamics under deficit irrigation in field conditions. Sci Total Environ 566:1223–1234. https://doi.org/10.1016/j.scitotenv.2016.05.178
Büchi L, Wendling M, Amossé C et al (2017) Long and short term changes in crop yield and soil properties induced by the reduction of soil tillage in a long term experiment in Switzerland. Soil Tillage Res 174:120–129. https://doi.org/10.1016/j.still.2017.07.002
Bücking H, Shachar-Hill Y (2005) Phosphate uptake, transport and transfer by the arbuscular mycorrhizal fungus *Glomus intraradices* is stimulated by increased carbohydrate availability. New Phytol 165:899–912. https://doi.org/10.1111/j.1469-8137.2004.01274.x
Burrows R (2014) Glomalin production and infectivity of arbuscular-mycorrhizal fungi in response to grassland plant diversity. Am J Plant Sci 5(1):103–111. https://doi.org/10.4236/ajps.2014.51013
Burrows RL, Pfleger FL (2002) Arbuscular mycorrhizal fungi respond to increasing plant diversity. Can J Bot 80(2):120–130. https://doi.org/10.1139/b01-138
Cameron DD, Neal AL, van Wees SC et al (2013) Mycorrhiza-induced resistance: more than the sum of its parts? Trends Plant Sci 18:539–545. https://doi.org/10.1016/j.tplants.2013.06.004
Cassman KG, Dobermann A, Walters DT et al (2003) Meeting cereal demand while protecting natural resources and improving environmental quality. Annu Rev Environ Resour 28:315–358. https://doi.org/10.1146/annurev.energy.28.040202.122858
Castelli M, Urcoviche RC, Gimenes RMT et al (2014) Arbuscular mycorrhizal fungi diversity in maize under different soil managements and seed treatment with fungicide. J Food Agric Environ 12:486–491
Castillo C, Borie F, Oehl F et al (2016) Arbuscular mycorrhizal fungi biodiversity: prospecting in Southern-Central zone of Chile. A review. J Soil Sci Plant Nutr 16:400–422
Cavagnaro TR, Bender SF, Asghari HR et al (2015) The role of arbuscular mycorrhizas in reducing soil nutrient loss. Trends Plant Sci 20:283–290. https://doi.org/10.1016/j.tplants.2015.03.004
Chen YL, Zhang X, Ye JS et al (2014) Six-year fertilization modifies the biodiversity of arbuscular mycorrhizal fungi in a temperate steppe in Inner Mongolia. Soil Biol Biochem 69:371–381. https://doi.org/10.1016/j.soilbio.2013.11.020
Chen J, Li J, Yang Y, Wang Y, Zhang Y, Wang P (2022) Effects of conventional and organic agriculture on soil arbuscular mycorrhizal fungal community in low-quality farmland. Front Microbiol 13:914627. https://doi.org/10.3389/fmicb.2022.914627
Cheng L, Booker FL, Tu C et al (2012) Arbuscular mycorrhizal fungi increase organic carbon decomposition under elevated CO_2. Science 337:1084–1087. https://doi.org/10.1126/science.1224304

Chiocchio V, Venedikian N, Martinez AE et al (2000) Effect of the fungicide benomyl on spore germination and hyphal length of the arbuscular mycorrhizal fungus *Glomus mosseae*. Int Microbiol 3:173–175

Chourasiya D, Gupta MM, Sahni S et al (2021) Unraveling the AM fungal community for understanding its ecosystem resilience to changed climate in agroecosystems. Symbiosis 84(3):295–310. https://doi.org/10.1007/s13199-021-00761-9

Cosme M, Stout MJ, Wurst S (2011) Effect of arbuscular mycorrhizal fungi (*Glomus intraradices*) on the oviposition of rice water weevil (*Lissorhoptrus oryzophilus*). Mycorrhiza 21:651–658. https://doi.org/10.1007/s00572-011-0399-6

da Silva GF, Calonego JC, Luperini BCO et al (2022) Soil—plant relationships in soybean cultivated under conventional tillage and long-term no-tillage. Agronomy 11(19):2657. https://doi.org/10.3390/plants11192657

Dang P, Gao Y, Liu J et al (2018) Effects of thinning intensity on understory vegetation and soil microbial communities of a mature Chinese pine plantation in the Loess Plateau. Sci Total Environ 630:171–180. https://doi.org/10.1016/j.scitotenv.2018.02.197

de León DG, Neuenkamp L, Moora M et al (2018) Arbuscular mycorrhizal fungal communities in tropical rain forest are resilient to slash-and-burn agriculture. J Trop Ecol 34:186–199. https://doi.org/10.1017/S0266467418000184

Deliopoulos T, Haydock PPJ, Jones PW (2008) Interaction between arbuscular mycorrhizal fungi and the nematicide aldicarb on hatch and development of the potato cyst nematode, *Globodera pallida*, and yield of potatoes. Nematology 10:783–799. https://doi.org/10.1163/156854108786161427

Dodd JC (2000) The role of arbuscular mycorrhizal fungi in agro- and natural ecosystems. Outlook Agric 29(1):63–70. https://doi.org/10.5367/000000000101293059

Duke SO, Powles SB (2008) Glyphosate: a once-in-a century herbicide. Pest Manag Sci 64:319–325. https://doi.org/10.1002/ps.1518

El Hilali R, Symanczik S, El Kinany S et al (2022) Cultivation, identification, and application of arbuscular mycorrhizal fungi associated with date palm plants in Drâa-Tafilalet oasis. Rhizosphere 22:100521. https://doi.org/10.1016/j.rhisph.2022.100521

Feng Z, Liu X, Qin Y et al (2023) Cooperation of arbuscular mycorrhizal fungi and bacteria to facilitate the host plant growth dependent on soil pH. Front Microbiol 14:1116943. https://doi.org/10.3389/fmicb.2023.1116943

Ferrol N, González-Guerrero M, Valderas A et al (2009) Survival strategies of arbuscular mycorrhizal fungi in Cu-polluted environments. Phytochem Rev 8:551–559

Finlay RD (2008) Ecological aspects of mycorrhizal symbiosis: with special emphasis on the functional diversity of interactions involving the extraradical mycelium. J Exp Bot 59:1115–1126. https://doi.org/10.1093/jxb/ern059

Fonseca MB, Dias T, Carolino MM et al (2017) Belowground microbes mitigate plant-plant competition. Plant Sci 262:175–181. https://doi.org/10.1016/j.plantsci.2017.06.006

Frank B (1885) Über die auf Wurzelsymbiose beruhende Ernährung gewisser Bäume durch unterirdische Pilze. https://doi.org/10.1111/j.1438-8677.1885.tb04240.x

Galván GA, Parádi I, Burger K et al (2009) Molecular diversity of arbuscular mycorrhizal fungi in onion roots from organic and conventional farming systems in the Netherlands. Mycorrhiza 19: 317–328. https://doi.org/10.1007/s00572-009-0237-2

Gamage A, Gangahagedara R, Gamage J, Jayasinghe N, Kodikara N, Suraweera P, Merah O (2023) Role of organic farming for achieving sustainability in agriculture. Farming System 1 (1): 100005, https://doi.org/10.1016/j.farsys.2023.100005

Gao X, Guo H, Zhang Q et al (2020) Arbuscular mycorrhizal fungi (AMF) enhanced the growth, yield, fiber quality and phosphorus regulation in upland cotton (*Gossypium hirsutum* L.). Sci Rep 10(1):2084. https://doi.org/10.1038/s41598-020-59180-3

Gonzalez-Dugo V (2010) The influence of arbuscular mycorrhizal colonization on soil–root hydraulic conductance in *Agrostis stolonifera* L. under two water regimes. Mycorrhiza 20: 365–373. https://doi.org/10.1007/s00572-009-0294-6

Gonzalez-Guerrero M, Melville LH, Ferrol N et al (2008) Ultrastructural localization of heavy metals in the extraradical mycelium and spores of the arbuscular mycorrhizal fungus *Glomus intraradices*. Can J Microbiol 54:103–110. https://doi.org/10.1139/W07-119

Gosling P, Jones J, Bending GD (2016) Evidence for functional redundancy in arbuscular mycorrhizal fungi and implications for agroecosystem management. Mycorrhiza 26:77–83. https://doi.org/10.1007/s00572-015-0651-6

Gottshall CB, Cooper M, Emery SM (2017) Activity, diversity and function of arbuscular mycorrhizae vary with changes in agricultural management intensity. Agric Ecosyst Environ 241:142–149. https://doi.org/10.1016/j.agee.2017.03.011

Goulson D (2013) An overview of the environmental risks posed by neonicotinoid insecticides. J Appl Ecol 50:977–987. https://doi.org/10.1111/1365-2664.12111

Gryndler M, Larsen J, Hršelová H et al (2006) Organic and mineral fertilization, respectively, increase and decrease the development of external mycelium of arbuscular mycorrhizal fungi in a long-term field experiment. Mycorrhiza 16:159–166. https://doi.org/10.1007/s00572-005-0027-4

Gu S, Wu S, Guan Y et al (2020) Arbuscular mycorrhizal fungal community was affected by tillage practices rather than residue management in black soil of northeast China. Soil Tillage Res 198:104552. https://doi.org/10.1016/j.still.2019.104552

Hack CM, Porta M, Schäufele R et al (2019) Arbuscular mycorrhiza mediated effects on growth, mineral nutrition and biological nitrogen fixation of *Melilotus alba* Med. in a subtropical grassland soil. Appl Soil Ecol 134:38–44. https://doi.org/10.1016/j.apsoil.2018.10.008

Hajiboland R, Aliasgharzadeh N, Laiegh SF et al (2010) Colonization with arbuscular mycorrhizal fungi improves salinity tolerance of tomato (*Solanum lycopersicum* L.) plants. Plant Soil 331:313–327. https://doi.org/10.1007/s11104-009-0255-z

Hajiboland R, Joudmand A, Aliasgharzad N, Tolra R, & Poschenrieder C (2019). Arbuscular mycorrhizal fungi alleviate low-temperature stress and increase freezing resistance as a substitute for acclimation treatment in barley. Crop & Pasture Science, 70:218–233. https://doi.org/10.1071/cp18385

Hamel C, Strullu D (2006) Arbuscular mycorrhizal fungi in field crop production: potential and new direction. Can J Plant Sci 86:941–950. https://doi.org/10.4141/P05-099

Hartley SE, Gange AC (2009) Impacts of plant symbiotic fungi on insect herbivores: mutualism in a multitrophic context. Annu Rev Entomol 54:323–342. https://doi.org/10.1146/annurev.ento.54.110807.090614

Hartnett DC, Wilson GWT (2002) The role of mycorrhizas in plant community structure and dynamics: lessons from grasslands. In: Smith SE, Smith FA (eds) Diversity and integration in mycorrhizas. Developments in plant and soil sciences, vol 94. Springer, Dordrecht. https://doi.org/10.1007/978-94-017-1284-2_31

Hashem A, Abd-Allah EF, Alqarawi AA et al (2016) The interaction between arbuscular mycorrhizal fungi and endophytic bacteria enhances plant growth of *Acacia gerrardii* under salt stress. Front Microbiol 7:1089. https://doi.org/10.3389/fmicb.2016.01089

Havlin JL, Kissel DE, Maddux LD et al (1990) Crop rotation and tillage effects on soil organic carbon and nitrogen. Soil Sci Soc Am J 54(2):448–452. https://doi.org/10.2136/sssaj1990.03615995005400020026x

Hestrin R, Hammer EC, Mueller CW et al (2019) Synergies between mycorrhizal fungi and soil microbial communities increase plant nitrogen acquisition. Commun Biol 2:233. https://doi.org/10.1038/s42003-019-0481-8

Higo M, Isobe K, Kang DJ et al (2010) Inoculation with arbuscular mycorrhizal fungi or crop rotation with mycorrhizal plants improves the growth of maize in limed acid sulfate soil. Plant Prod Sci 13(1):74–79. https://doi.org/10.1626/pps.13.74

Hladik ML, Main AR, Goulson D (2018) Environmental risks and challenges associated with neonicotinoid insecticides. Environ Sci Technol 52:3329–3335. https://doi.org/10.1021/acs.est.7b06388

Hodge A, Fitter AH (2010) Substantial nitrogen acquisition by arbuscular mycorrhizal fungi from organic material has implications for N cycling. Proc Natl Acad Sci 107:13754–13759. https://doi.org/10.1073/pnas.1005874107

Hu Y, Wu S, Sun Y et al (2015) Arbuscular mycorrhizal symbiosis can mitigate the negative effects of night warming on physiological traits of *Medicago truncatula* L. Mycorrhiza 25:131–142. https://doi.org/10.1007/s00572-014-0595-2

Huang Z, Zou Z, He C et al (2011) Physiological and photosynthetic responses of melon (*Cucumis melo* L.) seedlings to three *Glomus* species under water deficit. Plant Soil 339:391–399. https://doi.org/10.1007/s11104-010-0591-z

Huynh HT, Hufnagel J, Wurbs A et al (2019) Influences of soil tillage, irrigation and crop rotation on maize biomass yield in a 9-year field study in Müncheberg, Germany. Field Crops Res 241: 107565. https://doi.org/10.1016/j.fcr.2019.107565

Iqbal N, Umar S, Khan NA (2015) Nitrogen availability regulates proline and ethylene production and alleviates salinity stress in mustard (*Brassica juncea*). J Plant Physiol 178:84–91. https://doi.org/10.1016/j.jplph.2015.02.006

Jaacks LM, Serupally R, Dabholkar S et al (2022) Articles Impact of large-scale, the government legislated and funded organic farming training on pesticide use in Andhra Pradesh, India: a cross-sectional study. Lancet Planet Health 6:e310–e319. https://doi.org/10.1016/S2542-5196(22)00062-6

Jacobsen CS, Hjelmsø MH (2014) Agricultural soils, pesticides and microbial diversity. Curr Opin Biotechnol 27:15–20. https://doi.org/10.1016/j.copbio.2013.09.003

Jansa J, Mozafar A, Anken T et al (2002) Diversity and structure of AMF communities as affected by tillage in a temperate soil. Mycorrhiza 12:225–234. https://doi.org/10.1007/s00572-002-0163-z

Jansa J, Wiemken A, Frossard E (2006) The effects of agricultural practices on arbuscular mycorrhizal fungi. Geol Soc Lond, Spec Publ 266(1):89–115. https://doi.org/10.1144/GSL.SP.2006.266.01.08

Javot H, Penmetsa RV, Terzaghi N et al (2007) A *Medicago truncatula* phosphate transporter indispensable for the arbuscular mycorrhizal symbiosis. Proc Natl Acad Sci 104:1720–1725. https://doi.org/10.1073/pnas.0608136104

Jayne B, Quigley M (2014) Influence of arbuscular mycorrhiza on growth and reproductive response of plants under water deficit: a meta-analysis. Mycorrhiza 24:109–119. https://doi.org/10.1007/s00572-013-0515-x

Jeffries P, Gianinazzi S, Perotto S et al (2003) The contribution of arbuscular mycorrhizal fungi in sustainable maintenance of plant health and soil fertility. Biol Fertil Soils 37:1–16. https://doi.org/10.1007/s00374-002-0546-5

Jiang J, Moore JA, Priyadarshi A et al (2017) Plant-mycorrhizal interactions mediate plant community coexistence by altering resource demand. Ecology 98:187–197. https://doi.org/10.1002/ecy.1630

Johnson NC, Angelard C, Sanders IR et al (2013) Predicting community and ecosystem outcomes of mycorrhizal responses to global change. Ecol Lett 16:140–153. https://doi.org/10.1111/ele.12085

Jung SC, Martinez-Medina A, Lopez-Raez JA et al (2012) Mycorrhiza-induced resistance and priming of plant defenses. J Chem Ecol 38:651–664. https://doi.org/10.1007/s10886-012-0134-6

Karthikeyan A, Krishnakumar N (2012) Reforestation of bauxite mine spoils with *Eucalyptus tereticornis* Sm. seedlings inoculated with arbuscular mycorrhizal fungi. Ann For Res 55: 207–216. https://doi.org/10.15287/afr.2012.61

Kempel A, Schmidt AK, Brandl R et al (2010) Support from the underground: induced plant resistance depends on arbuscular mycorrhizal fungi. Funct Ecol 24:293–300. https://doi.org/10.1111/j.1365-2435.2009.01647.x

Kikvidze Z, Armas C, Fukuda K et al (2010) The role of arbuscular mycorrhizae in primary succession: differences and similarities across habitats. Web Ecol 10:50–57. https://doi.org/10.5194/we-10-50-2010

Kremer RJ (2014) Environmental implications of herbicide resistance: soil biology and ecology. Weed Sci 62:415–426. JSTOR. http://www.jstor.org/stable/43700669

Krupinsky JM, Bailey KL, McMullen MP et al (2002) Managing plant disease risk in diversified cropping systems. Agron J 94(2):198–209. https://doi.org/10.2134/agronj2002.1980

Kurle JE, Pfleger FL (1994) The effects of cultural practices and pesticides on VAM fungi. Mycorrhizae and plant health. APS Press, Minnesota, pp 101–131

Lanfranco L, Fiorilli V, Gutjahr C (2018) Partner communication and role of nutrients in the arbuscular mycorrhizal symbiosis. New Phytol 220:1031–1046. https://doi.org/10.1111/nph.15230

Lax P, Becerra AG, Soteras F et al (2011) Effect of the arbuscular mycorrhizal fungus *Glomus intraradices* on the false root-knot nematode *Nacobbus aberrans* in tomato plants. Biol Fertil Soils 47:591–597. https://doi.org/10.1007/s00374-010-0514-4

Lehmann A, Veresoglou SD, Leifheit EF et al (2014) Arbuscular mycorrhizal influence on zinc nutrition in crop plants–a meta-analysis. Soil Biol Biochem 69:123–131. https://doi.org/10.1016/j.soilbio.2013.11.001

Lehmann J, Bossio DA, Kögel-Knabner I et al (2020) The concept and future prospects of soil health. Nat Rev Earth Environ 1:544–553. https://doi.org/10.1038/s43017-020-0080-8

Lekberg Y, Koide R (2005) Arbuscular mycorrhizal fungi, rhizobia, available soil P and nodulation of groundnut (*Arachis hypogaea*) in Zimbabwe. Agric Ecosyst Environ 110(3–4):143–148. https://doi.org/10.1016/j.agee.2005.03.011

Liebman M, Dyck E (1993) Crop rotation and intercropping strategies for weed management. Ecol Appl 3(1):92–122. https://doi.org/10.2307/1941795

Liu W, Ma K, Wang X et al (2022a) Effects of no-tillage and biologically-based organic fertilizer on soil arbuscular mycorrhizal fungal communities in winter wheat field. Appl Soil Ecol 178: 104564. https://doi.org/10.1016/j.apsoil.2022.104564

Liu Z, Han Y, Lai F et al (2022b) Contribution of arbuscular mycorrhizal fungal communities to soil carbon accumulation during the development of *Cunninghamia lanceolata* plantations. Forests 13:2099. https://doi.org/10.3390/f13122099

Lu X, Lu X, Liao Y (2018) Effect of tillage treatment on the diversity of soil arbuscular mycorrhizal fungal and soil aggregate-associated carbon content. Front Microbiol 9:2986. https://doi.org/10.3389/fmicb.2018.02986

Lu N, Xu X, Wang P et al (2019) Succession in arbuscular mycorrhizal fungi can be attributed to a chronosequence of *Cunninghamia lanceolata*. Sci Rep 9:18057. https://doi.org/10.1038/s41598-019-54452-z

Ma G, Kang J, Wang J et al (2020) Bacterial community structure and predicted function in wheat soil from the North China plain are closely linked with soil and plant characteristics after seven years of irrigation and nitrogen application. Front Microbiol 11:506. https://doi.org/10.3389/fmicb.2020.00506

Ma Y, Zhang H, Wang D et al (2021) Differential responses of arbuscular mycorrhizal fungal communities to long-term fertilization in the wheat rhizosphere and root endosphere. Appl Environ Microbiol 87:e00349–e00321. https://doi.org/10.1128/aem.00349-21

Mäder P, Fliessbach A, Dubois D et al (2002) Soil fertility and biodiversity in organic farming. Science 296(5573):1694–1697. https://doi.org/10.1126/science.1071148

Malfatti ALR, Filho LCJO, Carniel LSC et al (2023) Risk assessment tests of neonicotinoids on spore germination of arbuscular mycorrhizal fungi *Gigaspora albida* and *Rhizophagus clarus*. J Soils Sedim 23:1295–1303. https://doi.org/10.1007/s11368-022-03419-1

Manaut N, Sanguin H, Ouahmane L et al (2015) Potentialities of ecological engineering strategy based on native arbuscular mycorrhizal community for improving afforestation programs with carob trees in degraded environments. Ecol Eng 79:113–119. https://doi.org/10.1016/j.ecoleng.2015.03.007

Manoharan L, Rosenstock NP, Williams A et al (2017) Agricultural management practices influence AMF diversity and community composition with cascading effects on plant productivity. Appl Soil Ecol 115:53–59. https://doi.org/10.1016/j.apsoil.2017.03.012
Martin FM, Uroz S, Barker DG (2017) Ancestral alliances: plant mutualistic symbioses with fungi and bacteria. Science 356:eaad4501. https://doi.org/10.1126/science.aad4501
Martinez TN, Johnson NC (2010) Agricultural management influences propagule densities and functioning of arbuscular mycorrhizas in low-and high-input agroecosystems in arid environments. Appl Soil Ecol 46:300–306. https://doi.org/10.1016/j.apsoil.2010.07.001
Mathimaran N, Ruh R, Jama B et al (2007) Impact of agricultural management on arbuscular mycorrhizal fungal communities in Kenyan ferralsol. Agric Ecosyst Environ 119(1–2):22–32. https://doi.org/10.1016/j.agee.2006.06.004
Mathur S, Sharma MP, Jajoo A (2018) Improved photosynthetic efficacy of maize (*Zea mays*) plants with arbuscular mycorrhizal fungi (AMF) under high temperature stress. J Photochem Photobiol B 180:149–154. https://doi.org/10.1016/j.jphotobiol.2018.02.002
Mo F, Han J, Wen XX et al (2020) Quantifying regional effects of plastic mulch on soil nitrogen pools, cycles, and fluxes in rain-fed agroecosystems of the Loess Plateau. Land Degrad Dev 31: 1675–1687. https://doi.org/10.1002/ldr.3548
Montesinos E (2003) Plant-associated microorganisms: a view from the scope of microbiology. Int Microbiol 6:221–223. https://doi.org/10.1007/s10123-003-0141-0
Moora M, Davison J, Öpik M et al (2014) Anthropogenic land use shapes the composition and phylogenetic structure of soil arbuscular mycorrhizal fungal communities. FEMS Microbiol Ecol 90:609–621. https://doi.org/10.1111/1574-6941.12420
Morris EK, Morris D, Vogt S et al (2019) Visualizing the dynamics of soil aggregation as affected by arbuscular mycorrhizal fungi. ISME J 13:1639–1646. https://doi.org/10.1038/s41396-019-0369-0
Mortimer P, Pérez-Fernández M, Valentine A (2008) The role of arbuscular mycorrhizal colonization in the carbon and nutrient economy of the tripartite symbiosis with nodulated *Phaseolus vulgaris*. Soil Biol Biochem 40:1019–1027. https://doi.org/10.1016/j.soilbio.2007.11.014
Mummey DL, Clarke JT, Cole CA et al (2010) Spatial analysis reveals differences in soil microbial community interactions between adjacent coniferous forest and clear cut ecosystems. Soil Biol Biochem 42:1138–1147. https://doi.org/10.1016/j.soilbio.2010.03.020
Newsham K, Fitter A, Watkinson A (1995) Multi-functionality and biodiversity in arbuscular mycorrhizas. Trends Ecol Evol 10(10):407–411. https://doi.org/10.1016/S0169-5347(00)89157-0
O'Connor P, Manjarrez M, Smith SE (2009) The fate and efficacy of benomyl applied to field soils to suppress activity of arbuscular mycorrhizal fungi. Can J Microbiol 55(7):901–904. https://doi.org/10.1139/w09-035
Oehl F, Sieverding E, Ineichen K et al (2003) Impact of land use intensity on the species diversity of arbuscular mycorrhizal fungi in agroecosystems of Central Europe. Appl Environ Microbiol 69(5):2816–2824. https://doi.org/10.1128/aem.69.5.2816-2824.2003
Oehl F, Sieverding E, Mäder P et al (2004) Impact of long-term conventional and organic farming on the diversity of arbuscular mycorrhizal fungi. Oecologia 138:574–583. https://doi.org/10.1007/s00442-003-1458-2
Oehl F, Laczko E, Bogenrieder A et al (2010) Soil type and land use intensity determine the composition of arbuscular mycorrhizal fungal communities. Soil Biol Biochem 42:724–738. https://doi.org/10.1016/j.soilbio.2010.01.006
Oruru MB, Njeru EM (2016) Upscaling arbuscular mycorrhizal symbiosis and related agroecosystems services in smallholder farming systems. Biomed Res Int 2016:4376240. https://doi.org/10.1155/2016/4376240
Pasaribu A, Mohamad RB, Awang Y et al (2011) Growth and development of symbiotic Arbuscular mycorrhizal fungi, *Glomus mossea* (Nicol. and Gerd.), in alachlor and glyphosate treated soils. Afr J Biotechnol 10(55):11520–11526. https://doi.org/10.5897/AJB11.1200

Pellegrino E, Bedini S (2014) Enhancing ecosystem services in sustainable agriculture: biofertilization and biofortification of chickpea (*Cicer arietinum* L.) by arbuscular mycorrhizal fungi. Soil Biol Biochem 68:429–439. https://doi.org/10.1016/j.soilbio.2013.09.030

Porcel R, Redondo-Gómez S, Mateos-Naranjo E et al (2015) Arbuscular mycorrhizal symbiosis ameliorates the optimum quantum yield of photosystem II and reduces non-photochemical quenching in rice plants subjected to salt stress. J Plant Physiol 185:75–83. https://doi.org/10.1016/j.jplph.2015.07.006

Qin H, Lu K, Strong PJ et al (2015) Long-term fertilizer application effects on the soil, root arbuscular mycorrhizal fungi and community composition in rotation agriculture. Appl Soil Ecol 89:35–43. https://doi.org/10.1016/j.apsoil.2015.01.008

Qin H, Chen J, Wu Q et al (2017) Intensive management decreases soil aggregation and changes the abundance and community compositions of arbuscular mycorrhizal fungi in Moso bamboo (*Phyllostachys pubescens*) forests. For Ecol Manag 400:246–255

Ramana JV, Tylianakis JM, Ridgway HJ et al (2023) Root diameter, host specificity and arbuscular mycorrhizal fungal community composition among native and exotic plant species. New Phytol 239:301–310. https://doi.org/10.1111/nph.18911

Ratnayake RR, Perera B, Rajapaksha R et al (2017) Soil carbon sequestration and nutrient status of tropical rice based cropping systems: Rice-Rice, Rice-Soya, Rice-Onion and Rice-Tobacco in Sri Lanka. Catena 150:17–23. https://doi.org/10.1016/j.catena.2016.11.006

Recorbet G, Abdallah C, Renaut J et al (2013) Protein actors sustaining arbuscular mycorrhizal symbiosis: underground artists break the silence. New Phytol 199:26–40. https://doi.org/10.1111/nph.12287

Ren B, Li X, Dong S, Liu P et al (2018) Soil physical properties and maize root growth under different tillage systems in the North China Plain. Crop J 6:669–676. https://doi.org/10.1016/j.cj.2018.05.009

Ren A-T, Zhu Y, Chen Y-L et al (2019) Arbuscular mycorrhizal fungus alters root-sourced signal (abscisic acid) for better drought acclimation in *Zea mays* L. seedlings. Environ Exp Bot 167:103824. https://doi.org/10.1016/j.envexpbot.2019.103824

Renker C, Zobel M, Öpik M et al (2004) Structure, dynamics, and restoration of plant communities: do Arbuscular Mycorrhizae matter? In: Temperton VK, Hobbs RJ, Nuttle T et al (eds) Assembly rules and restoration ecology: bridging the gap between theory and practice. Island Press, pp 189–229

Riedo J, Wettstein FE, Rösch A et al (2021) Widespread occurrence of pesticides in organically managed agricultural soils—the ghost of a conventional agriculture past? Environ Sci Technol 55:2919–2928. https://doi.org/10.1021/acs.est.0c06405

Rillig MC (2004) Arbuscular mycorrhizae and terrestrial ecosystem processes. Ecol Lett 7:740–754. https://doi.org/10.1111/j.1461-0248.2004.00620.x

Rillig MC, Mummey DL (2006) Mycorrhizas and soil structure. New Phytol 171:41–53. https://doi.org/10.1111/j.1469-8137.2006.01750.x

Rillig MC, Sosa-Hernández MA, Roy J et al (2016) Towards an integrated mycorrhizal technology: harnessing mycorrhiza for sustainable intensification in agriculture. Front Plant Sci 7:1625. https://doi.org/10.3389/fpls.2016.01625

Rillig MC, Aguilar-Trigueros CA, Camenzind T et al (2019) Why farmers should manage the arbuscular mycorrhizal symbiosis. New Phytol 222:1171–1175. https://doi.org/10.1111/nph.15602

Roger A, Colard A, Angelard C et al (2013) Relatedness among arbuscular mycorrhizal fungi drives plant growth and intraspecific fungal coexistence. ISME J 7:2137–2146. https://doi.org/10.1038/ismej.2013.112

Rozpądek P, Wężowicz K, Stojakowska A et al (2014) Mycorrhizal fungi modulate phytochemical production and antioxidant activity of *Cichorium intybus* L. (Asteraceae) under metal toxicity. Chemosphere 112:217–224. https://doi.org/10.1016/j.chemosphere.2014.04.023

Ruiz-Lozano JM (2003) Arbuscular mycorrhizal symbiosis and alleviation of osmotic stress. New perspectives for molecular studies. Mycorrhiza 13(6):309–317. https://doi.org/10.1007/s00572-003-0237-6

Ryan MH, Graham JH (2018) Little evidence that farmers should consider abundance or diversity of arbuscular mycorrhizal fungi when managing crops. New Phytol 220:1092–1107. https://doi.org/10.1111/nph.15308

Saldajeno MGB, Ito M, Hyakumachi M (2012) Interaction between the plant growth-promoting fungus *Phoma sp.* GS8-2 and the arbuscular mycorrhizal fungus *Glomus mosseae*: impact on biocontrol of soil-borne diseases, microbial population, and plant growth. Australas Plant Pathol 41:271–281. https://doi.org/10.1007/s13313-011-0101-7

Säle V, Aguilera P, Laczko E et al (2015) Impact of conservation tillage and organic farming on the diversity of arbuscular mycorrhizal fungi. Soil Biol Biochem 84:38–52. https://doi.org/10.1016/j.soilbio.2015.02.005

Sarr B, Ndiaye F, Ndiaye M et al (2013) Effet de deux types d'insecticides sur la mycorhization arbusculaire et le développement de deux variétés de pomme de terre (*Solanum tuberosum*). Int J Biol Chem Sci 7:1902–1909. https://doi.org/10.4314/ijbcs.v7i5.10

Shafiei F, Shahidi-Noghabi S, Sedaghati E (2022) The impact of arbuscular mycorrhizal fungi on tomato plant resistance against *Tuta absoluta* (Meyrick) in greenhouse conditions. J Asia Pac Entomol 25(3):101971. https://doi.org/10.1016/j.aspen.2022.101971

Sharmah D, Jha D (2014) Diversity of arbuscular mycorrhizal fungi in disturbed and undisturbed forests of Karbi Anglong Hills of Assam, India. Agric Res 3:229–238. https://doi.org/10.1007/s40003-014-0110-1

Shen J, Li C, Mi G, Li L et al (2013) Maximizing root/rhizosphere efficiency to improve crop productivity and nutrient use efficiency in intensive agriculture of China. J Exp Bot 64:1181–1192. https://doi.org/10.1093/jxb/ers342

Singh LP, Gill SS, Tuteja N (2011) Unraveling the role of fungal symbionts in plant abiotic stress tolerance. Plant Signal Behav 6:175–191. https://doi.org/10.4161/psb.6.2.14146

Singh PK, Singh M, Tripathi BN (2013) Glomalin: an arbuscular mycorrhizal fungal soil protein. Protoplasma 250:663–669. https://doi.org/10.1007/s00709-012-0453-z

Słomka A, Kuta E, Szarek-Łukaszewska G et al (2011) Violets of the section Melanium, their colonization by arbuscular mycorrhizal fungi and their occurrence on heavy metal heaps. J Plant Physiol 168:1191–1199. https://doi.org/10.1016/j.jplph.2011.01.033

Smith SE, Read DJ (2008) Arbuscular mycorrhizaes. In: Smith SE, Read DJ (eds) Mycorrhizal symbiosis, 3rd edn. Academic, London, pp 13–187

Smith SE, Smith FA (2011) Roles of arbuscular mycorrhizas in plant nutrition and growth: new paradigms from cellular to ecosystem scales. Annu Rev Plant Biol 62:227–250. https://doi.org/10.1146/annurev-arplant-042110-103846

Smith SE, Facelli E, Pope S et al (2010) Plant performance in stressful environments: interpreting new and established knowledge of the roles of arbuscular mycorrhizas. Plant Soil 326:3–20. https://doi.org/10.1007/s11104-009-9981-5

Song Y, Chen D, Lu K et al (2015) Enhanced tomato disease resistance primed by arbuscular mycorrhizal fungus. Front Plant Sci 6:786. https://doi.org/10.3389/fpls.2015.00786

Steinkellner S, Hage-Ahmed K, García-Garrido JM et al (2012) A comparison of wild-type, old and modern tomato cultivars in the interaction with the arbuscular mycorrhizal fungus *Glomus mosseae* and the tomato pathogen *Fusarium oxysporum* f. sp. *lycopersici*. Mycorrhiza 22:189–194. https://doi.org/10.1007/s00572-011-0393-z

Stürmer SL, Siqueira JO (2011) Species richness and spore abundance of arbuscular mycorrhizal fungi across distinct land uses in Western Brazilian Amazon. Mycorrhiza 21:255–267. https://doi.org/10.1007/s00572-010-0330-6

Tatewaki Y, Higo M, Isobe K (2021) Community structure of arbuscular mycorrhizal fungi in the roots of maize under different of preceding winter cropping and tillage systems in south Kanto region of Japan. Soil Microorg 75:23–31. (In Japanese)

Thirkell TJ, Charters MD, Elliott AJ et al (2017) Are mycorrhizal fungi our sustainable saviours? Considerations for achieving food security. J Ecol 105:921–929. https://doi.org/10.1111/1365-2745.12788

Trappe JM, Molina R, Castellano M (1984) Reactions of mycorrhizal fungi and mycorrhiza formation to pesticides. Annu Rev Phytopathol 22:331–359. https://doi.org/10.1146/annurev.py.22.090184.001555

Trentini CP, Campanello PI, Villagra M et al (2017) Thinning of loblolly pine plantations in subtropical Argentina: impact on microclimate and understory vegetation. For Ecol Manag 384:236–247. https://doi.org/10.1016/j.foreco.2016.10.040

Tsiafouli MA, Thébault E, Sgardelis SP et al (2015) Intensive agriculture reduces soil biodiversity across Europe. Glob Chang Biol 21(2):973–985. https://doi.org/10.1111/gcb.12752

Tyagi J, Sultan E, Mishra A et al (2017) The impact of AMF symbiosis in alleviating drought tolerance in field crops. In: Varma A, Prasad R, Tuteja N (eds) Mycorrhiza—nutrient uptake, biocontrol, ecorestoration. Springer, Cham. https://doi.org/10.1007/978-3-319-68867-1_11

Uibopuu A, Moora M, Saks U et al (2009) Differential effect of arbuscular mycorrhizal fungal communities from ecosystems along management gradient on the growth of forest understorey plant species. Soil Biol Biochem 41:2141–2146. https://doi.org/10.1016/j.soilbio.2009.07.026

Uibopuu A, Moora M, Öpik M et al (2012) Temperate forest understorey species performance is altered by local arbuscular mycorrhizal fungal communities from stands of different successional stages. Plant Soil 356:331–339. https://doi.org/10.1007/s11104-011-1116-0

Van Eerd LL, Congreves KA, Hayes A et al (2014) Long-term tillage and crop rotation effects on soil quality, organic carbon, and total nitrogen. Can J Soil Sci 94(3):303–315. https://doi.org/10.4141/cjss2013-093

van Geel M, Ceustermans A, van Hemelrijck W et al (2015) Decrease in diversity and changes in community composition of arbuscular mycorrhizal fungi in roots of apple trees with increasing orchard management intensity across a regional scale. Mol Ecol 24:941–952. https://doi.org/10.1111/mec.13079

Venter ZS, Jacobs K, Hawkins et al (2016) The impact of crop rotation on soil microbial diversity: a meta-analysis. Pedobiologia 59(4):215–223. https://doi.org/10.1016/j.pedobi.2016.04.001

Verbruggen E, Van Der Heijden MGA, Weedon JT, Kowalchuk GA, Röling WFM (2012), Community assembly, species richness and nestedness of arbuscular mycorrhizal fungi in agricultural soils. Molecular Ecology, 21: 2341–2353. https://doi.org/10.1111/j.1365-294X.2012.05534.x

Verbruggen E, Röling WFM, Gamper HA et al (2010) Positive effects of organic farming on below-ground mutualists: large-scale comparison of mycorrhizal fungal communities in agricultural soils. New Phytol 186:968–979. https://doi.org/10.1111/j.1469-8137.2010.03230.x

Wahdan SFM, Reitz T, Heintz-Buschart A et al (2021) Organic agricultural practice enhances arbuscular mycorrhizal symbiosis in correspondence to soil warming and altered precipitation patterns. Environ Microbiol 23:6163–6176. https://doi.org/10.1111/1462-2920.15492

Wang FY, Shi ZY, Tong RJ et al (2011a) Dynamics of phoxim residues in green onion and soil as influenced by arbuscular mycorrhizal fungi. J Hazard Mater 185:112–116. https://doi.org/10.1016/j.jhazmat.2010.09.004

Wang FY, Tong RJ, Shi ZY et al (2011b) Inoculations with arbuscular mycorrhizal fungi increase vegetable yields and decrease phoxim concentrations in carrot and green onion and their soils. PLoS One 6:16949. https://doi.org/10.1371/journal.pone.0016949

Wang J, Wang GG, Zhang B et al (2019) Arbuscular mycorrhizal fungi associated with tree species in a planted forest of Eastern China. Forests 10(5):424. https://doi.org/10.3390/f10050424

Watrud LS, King G, Londo JP et al (2011) Changes in constructed Brassica communities treated with glyphosate drift. Ecol Appl 21:525–538. https://doi.org/10.1890/09-2366.1

Weidenhamer JD, Callaway RM (2010) Direct and indirect effects of invasive plants on soil chemistry and ecosystem function. J Chem Ecol 36(1):59–69. https://doi.org/10.1007/s10886-009-9735-0

West TO, Post WM (2002) Soil organic carbon sequestration rates by tillage and crop rotation: a global data analysis. Soil Sci Soc Am J 66(6):1930–1946. https://doi.org/10.2136/sssaj2002.1930

Whipps J, Hand P, Pink D, Bending GD (2008) Phyllosphere microbiology with special reference to diversity and plant genotype. J Appl Microbiol 105:1744–1755. https://doi.org/10.1111/j.1365-2672.2008.03906.x

Wilson GWT, Williamson MM (2008) Topsin-M: the new benomyl for mycorrhizal-suppression experiments. Mycologia 100:548–554. https://doi.org/10.3852/08-024r

Wu S, Hu Y, Zhang X, Sun Y et al (2018) Chromium detoxification in arbuscular mycorrhizal symbiosis mediated by sulfur uptake and metabolism. Environ Exp Bot 147:43–52. https://doi.org/10.1016/j.envexpbot.2017.11.010

Xavier LJ, Germida JJ (2003) Selective interactions between arbuscular mycorrhizal fungi and *Rhizobium leguminosarum* bv. *viceae* enhance pea yield and nutrition. Biol Fertil Soils 37:261–267. https://doi.org/10.1007/s00374-003-0605-6

Xiao D, Che R, Liu X et al (2019) Arbuscular mycorrhizal fungi abundance was sensitive to nitrogen addition but diversity was sensitive to phosphorus addition in karst ecosystems. Biol Fertil Soils 55:457–469. https://doi.org/10.1007/s00374-019-01362-x

Yadav R, Ror P, Rathore P et al (2020) Bacteria from native soil in combination with arbuscular mycorrhizal fungi augment wheat yield and biofortification. Plant Physiol Biochem 150:222–233. https://doi.org/10.1016/j.plaphy.2020.02.039

Yang XM, Kay BD (2001) Rotation and tillage effects on soil organic carbon sequestration in a typic Hapludalf in Southern Ontario. Soil Tillage Res 59(3–4):107–114. https://doi.org/10.1016/S0167-1987(01)00162-3

Zaller JG, Heigl F, Ruess L, Grabmaier A (2014) Glyphosate herbicide affects belowground interactions between earthworms and symbiotic mycorrhizal fungi in a model ecosystem. Sci Rep 4:5634. https://doi.org/10.1038/srep05634

Zangaro W, Nisizaki S, Domingos J, Nakano E (2003) Mycorrhizal response and successional status in 80 woody species from south Brazil. J Trop Ecol 19:315–324. https://doi.org/10.1017/S0266467403003341

Zhang H-H, Tang M, Chen H, Zheng C-L, Niu Z-C (2010) Effect of inoculation with AM fungi on lead uptake, translocation and stress alleviation of *Zea mays* L. seedlings planting in soil with increasing lead concentrations. Eur J Soil Biol 46:306–311. https://doi.org/10.1016/j.ejsobi.2010.05.006

Zhang R, Vivanco JM, Shen Q (2017) The unseen rhizosphere root–soil–microbe interactions for crop production. Curr Opin Microbiol 37:8–14. https://doi.org/10.1016/j.mib.2017.03.008

Zhao C, Fu S, Mathew RP, Lawrence KS, Feng Y (2015a) Soil microbial community structure and activity in a 100-year-old fertilization and crop rotation experiment. J Plant Ecol 8:623–632. https://doi.org/10.1093/jpe/rtv007

Zhao R, Guo W, Bi N, Guo J, Wang L, Zhao J, Zhang J (2015b) Arbuscular mycorrhizal fungi affect the growth, nutrient uptake and water status of maize (*Zea mays* L.) grown in two types of coal mine spoils under drought stress. Appl Soil Ecol 88:41–49. https://doi.org/10.1016/j.apsoil.2014.11.016

Zhou L, Cai L, He Z, Wang R, Wu P, Ma X (2016) Thinning increases understory diversity and biomass, and improves soil properties without decreasing growth of Chinese fir in southern China. Environ Sci Pollut Res 23:24135–24150. https://doi.org/10.1007/s11356-016-7624-y

Zhou H, Ouyang T, Liu L, Xia S, Jia Q (2022) In-Forest planting of high-value herb *Sarcandra glabra* enhances soil carbon storage without affecting the diversity of the arbuscular mycorrhiza fungal community and composition of *Cunninghamia lanceolata*. Microorganisms 10(9):1844. https://doi.org/10.3390/microorganisms10091844

Zhu J-K (2003) Regulation of ion homeostasis under salt stress. Curr Opin Plant Biol 6:441–445. https://doi.org/10.1016/s1369-5266(03)00085-2

Zobel M, Öpik M (2014) Plant and arbuscular mycorrhizal fungal (AMF) communities–which drives which? J Veg Sci 25:1133–1140. https://doi.org/10.1111/jvs.12191

Chapter 13
Arbuscular Mycorrhizal Fungi Under Intercrop, Regenerative, and Conventional Agriculture Systems

Melissa Robdrup, Michelle Hubbard, Linda Yuya Gorim, and Monika A. Gorzelak

Abstract Arbuscular mycorrhizal fungi (AMF) increase in diversity and abundance in agricultural systems that emphasize soil health practices, including regenerative agriculture and intercropping. Regenerative agriculture in principle includes any practice that increases biodiversity and living roots and integrates livestock while reducing tillage, bare soil, and agrichemical inputs. Intercropping increases biodiversity in an annual system and reduces disease prevalence and weeds while improving soil conditions and yielding more than the equivalent monocrop. These principles and practices simultaneously support AMF proliferation in soils and in turn AMF provide multiple benefits to crops. AMF colonize roots, trading photosynthates for nutrients acquired beyond the reach of the plant root system. While colonizing roots, they trigger innate plant immunity and confer resistance to some insect, fungal, and bacterial pests. Colonized plants hold more water and thus are more resistant to drought. In soils with ample AMF propagules, multiple plants are likely to become connected to their neighbors by a common mycorrhizal network (CMN). Plants connected by a CMN are likely to share beneficial microbes, resistance to disease, and resources. A better understanding of crop root traits and AMF is important to building a wholistic picture of ecological interactions that can be leveraged to maintain agricultural production in intercropped, regenerative, and conventional systems.

M. Robdrup · M. A. Gorzelak (✉)
Agriculture and Agri-Food Canada (AAFC), Lethbridge Current Research and Development Centre, Lethbridge, AB, Canada
e-mail: monika.gorzelak@agr.gc.ca

M. Hubbard
Agriculture and Agri-Food Canada (AAFC), Swift Current Research and Development Centre, Swift Current, SK, Canada

L. Y. Gorim
Department of Agricultural, Food and Nutritional Sciences. 410 Agriculture/Forestry Centre, University of Alberta, Edmonton, AB, Canada

G. J. Ahammed, R. Hajiboland (eds.), *Arbuscular Mycorrhizal Fungi and Higher Plants*, https://doi.org/10.1007/978-981-99-8220-2_13

Keywords Arbuscular mycorrhizal fungi (AMF) · Common mycorrhizal networks (CMN) · Intercropping · Mycorrhizas · Soil health · Regenerative agriculture · Roots

13.1 Introduction

Arbuscular mycorrhizal fungi (AMF) are plant root symbionts with the potential to confer multiple benefits to crops. Roots are colonized by AMF from soil-dwelling spores, a process that can take weeks. A colonized plant benefits from increased access to nutrients as the fungal hyphal network extends beyond its own root system to acquire and transport nutrients and water to their host plant. In turn, the plant host supports AMF growth by furnishing photosynthates (Smith and Read 2008). Other benefits of this symbiosis include plant resistance to disease and abiotic stress (Chandrasekaran and Paramasivan 2022; Yu et al. 2022). Conventional cropping systems produce high yields, in part by circumnavigating this natural process (Parrott 2010), combating pathogenic fungi through fungicide application (Jin et al. 2013), delivering readily available nutrients directly to roots through fertilization (Gryndler et al. 2006; Lin et al. 2012; Spagnoletti et al. 2018), and breaking up soil and fungal hyphal networks through tillage (Bowles et al. 2017). Rising input costs and increased recognition of environmental and climate costs associated with these practices, alternative agriculture practices, including those referred to as regenerative, are being adopted by some farmers. Regenerative practices place emphasis on roots, biodiversity, and soil health (LaCanne and Lundgren 2018). One such practice, intercropping (growing two or more crops with at least some overlap in both space and time), incorporates biodiversity in systems, such as annual grain production, which are more typically kept in a monoculture rotation (Vandermeer et al. 1998). Examining AMF in agriculture systems necessitates viewing agriculture as firmly entrenched within an ecosystem, as the proliferation of these obligate biotrophs is entirely dependent on the host plant and the host's response to the environment. The benefits of AMF-crop symbiosis may be best realized by working with ecological processes. Here, we define and discuss intercropping and regenerative agriculture from the point of view of supporting AMF growth and in contrast to conventional practices. We examine the importance of roots and breeding for root characteristics which ultimately support AMF. We look at the potential of AMF to support sustainable cropping by leveraging this evolutionarily persistent and well-established symbiosis over costly synthetic inputs. Ultimately, more work needs to be done to delineate effective ways of using AMF to support agricultural production; given current understanding, practices such as those espoused in the regenerative agriculture movement, including intercropping, can help support a diverse AMF community which will confer benefits to crops.

Annual row cropping systems include many practices that have been shown to reduce AMF. Forming a symbiosis with AMF requires investment of resources, including energy, from the plant. Energy is also needed to maintain the plant-AMF relationship, as the sites of nutrient exchange, the arbuscules, are cycled every few days once the symbiosis is established. Plants furnished with sufficient or overabundant mineral nutrients, as is frequently the case with applications of fertilizer in row cropping, are less likely to invest the resources in an AMF partnership. Indeed, research has shown fertilization reduces colonization by AMF in annual row cropping systems (Miller et al. 1995; Gryndler et al. 2006; Lin et al. 2012; Spagnoletti et al. 2018). Crop energy savings through reduced AMF colonization, given proper mineral nutrition, can result in high yields, provided that environmental conditions are ideal. Given a changing climate with increasing fluctuations and more extreme weather events, counting on an ideal environment is increasingly risky. Resilience to drought is increased in plants colonized by AMF as the hyphal network stores and transports up to an additional 20% water to plant hosts (Ruth et al. 2011). AMF can also prime plant defenses against root and foliar pathogens, increasing resistance to diseases (Spagnoletti et al. 2018). The application of fungicides, a common practice in row cropping to reduce fungal pathogens, simultaneously reduces beneficial microbes such as AMF (Jin et al. 2013; Buysens et al. 2015; de Novais et al. 2019). Consistent applications of both fungicides and fertilizer will eventually deplete the soil of AMF propagules, reducing its capacity to provide inoculum. While commercial inoculums are available, they rarely consist of more than a few species and can be nonviable (Salomon et al. 2022). Commercial inoculum is also costly for larger-scale operations (de Santana et al. 2014). Furthermore, tilling can disrupt hyphal networks formed in the soil that would take weeks to regrow (Säle et al. 2015; Bowles et al. 2017; Rosner et al. 2018). Tillage is not necessary for row cropping, but disturbance is necessary to seed and harvest. Given the increasing costs of inputs, a recognition of changes to climate, and the importance of supporting resilient agroecosystems, approaches that push the boundaries of conventional agriculture and attempt to incorporate the ecosystem principles into agriculture are being considered on a wider scale. Regenerative practices are intended to support the innate capacity of soil symbionts to provide for crop nutrition, disease resilience, and drought tolerance, as well as some surprising emergent phenomenon as described herein could potentially be leveraged for agricultural benefits given more research.

13.2 Regenerative Agriculture

Regenerative agriculture is an alternative means of producing food that, its advocates claim, may have lower – or even net positive – environmental and/or social impacts (Newton et al. 2020). The term regenerative agriculture was first coined by Rodale (1983) to highlight how industrialized agriculture was severely reducing its natural resource base, and that without rebuilding that base, "sustainable agriculture" and "conservation agriculture" were insufficient for supporting the food and natural resource needs of a growing human population (Fenster et al. 2021). Several definitions of regenerative agriculture exist, which are tied to processes, outcomes, or both (Newton et al. 2020). This agriculture system has caught the agricultural community's attention, especially small-to-medium-scale growers looking to reduce input costs in their production systems. Advocates for this system believe that it is key to mitigating some of the observed climatic challenges and sequestering more carbon, thereby making cropping systems more resilient. However, other commentators are more cautious regarding the potential contribution of regenerative agriculture to sustainability objectives (Ranganathan et al. 2020).

Regenerative farming systems are designed by producers to increase soil quality and biodiversity in farmland while producing nourishing farm products profitably based on agreed principles (LaCanne and Lundgren 2018). The five unifying principles include (1) abandoning tillage (no-till), (2) eliminating bare soil, (3) fostering plant diversity on the farm, (4) integrating livestock with cropping operations, and (5) reducing or eliminating the use of synthetic agrichemicals. The application of these principles also happen to supports the growth and proliferation of AMF (Bowles et al. 2020). Reducing agrichemicals (Miller et al. 1995; Gryndler et al. 2006; Lin et al. 2012; Spagnoletti et al. 2018), tillage (Bowles et al. 2017), and bare soil while increasing plant diversity and combining cropping and livestock production have all been demonstrated to increase AMF diversity and abundance (de Souza et al. 2019; Bowles et al. 2020; Sarto et al. 2020; Pires et al. 2021). Common characteristics of regenerative farms in North America include natural or organic inputs and a focus on local/regional solutions (reviewed by Newton et al. (2020)). Regenerative farms are diverse and complementary in their enterprises and adaptive in their management choices, ensuring that a farm is resilient and profitable in the face of adversity (Fenster et al. 2021). However, some misconceptions exist about what constitutes a regenerative farm. The five principles are key to the development of a regenerative matrix score that can be used to set apart regenerative from conventional farms to avoid confusion as this movement develops (Fenster et al. 2021).

13.3 Intercropping

Intercropping is defined as growing two or more crops together in space and time (Vandermeer et al. 1998). This practice can take many different forms but results in an increase in living roots, reduced spatiotemporal bare soil, and increased plant diversity. The companion crops can be planted and harvested at the same time or planted and harvested sequentially. This latter format is sometimes referred to as relay cropping. The overlap between the crops makes relay cropping distinct from crop rotation, in which a single crop is taken from seeding to harvest prior to a subsequent crop being planted. While crop rotation facilitates the diversification of agricultural systems in time, intercropping increases diversity further. Indeed, intercropping can involve the cocultivation of three or more crops. The published scientific literature suggests that the beneficial impact of intercropping often outweighs any negatives (Huss et al. 2022).

13.3.1 Successes

Many benefits of intercropping have been demonstrated experimentally. One such benefit is overyielding or the production of higher yields per unit area through the use of intercropping than would be achieved by cultivating each companion crop as a monocrop (Pelzer et al. 2012; Barker and Dennett 2013; Xiao et al. 2019). Overyielding leads to land equivalency ratios (LERs) greater than 1.0. LERs are calculated by dividing the yield of each crop in the intercrop by the yield of that crop where it to be grown as a monocrop. Intercropping can also reduce production risk by enhancing yield stability (Raseduzzaman and Jensen 2017). Economic advantages can also be realized by intercropping (Pelzer et al. 2012). Reduced need for agricultural inputs, especially nitrogen fertilizer in legume-non-legume intercrops, such as pea-wheat (Pelzer et al. 2012) and faba bean-wheat (Xiao et al. 2019), can contribute to this advantage. Intercropping can also improve crop phosphorous nutrition (Liao et al. 2020; Messaoudi et al. 2020; Liu et al. 2022).

Foliar crop diseases, such as mycosphaerella blight of pea (Fernández-Aparicio et al. 2010; Schoeny et al. 2010), ascochyta blight of chickpea (Zhou et al. 2023), and chocolate spot of faba beans (Fernández-Aparicio et al. 2011; Guo et al. 2020), have been reduced through intercropping. Lower levels of foliar disease likely diminish the need for fungicide application, thereby providing further economic advantages. Intercropping has had mixed success at managing soilborne disease. Fusarium wilt of faba bean and chickpea have been successfully reduced by intercropping faba bean and wheat (Guo et al. 2020; Yang et al. 2022; Zhang et al. 2023) or chickpea with mustard (Jamil and Ashraf 2021).

Soil properties can be impacted by intercropping. For example, Cong et al. (2015) found that intercropping maize with wheat or with faba bean produced soil organic matter, litter, and roots that decomposed more rapidly, as compared to monocrop maize. Consistently, Kapayou et al. (2023) observed more rapid soil respiration in a three-way (maize-bean-squash) intercropping system, relative to monocrops. However, intercropping can also result in more carbon being stored in the soil and reduced greenhouse gas release (Yang et al. 2023).

Intercropping can reduce lodging, especially in crop varieties that are prone to this issue. For example, Podgórska-Lesiak and Sobkowicz (2013) found that barley could provide mechanical support to leafy pea varieties. Reduced lodging can, in turn, diminish the risk of foliar disease and facilitate harvest.

Weed management can be advanced through intercropping. In a meta-analysis, Verret et al. (2017) found that intercropping with legumes can reduce weed pressure. Many pulse crops tend to be uncompetitive relative to non-legume grain crops, and as such, pulse-cereal intercropping can provide weed management advantages over monocrops of pulses. For example, Corre-Hellou et al. (2011) found that pea-barley was better able to outcompete weeds than a pea monocrop. Multiple crops growing together also have the potential to more completely utilize resources, such as solar radiation, water, and nutrients, than a monocrop, thus leaving fewer resources available to weeds (Liebman and Dyck 1993). Other potential mechanisms of weed control in intercropping include allelopathy where chemicals are released from plant roots and associated AMF fungal networks to inhibit or reduce the growth of another plant (Weston and Duke 2003; Barto et al. 2011).

Management of insect pests can be facilitated by intercropping (Huss et al. 2022) in systems including corn-soybean (Li et al. 2022), onion-barley (Uesugi et al. 2023), and grass-peppermint (Gowton et al. 2021). Populations of beneficial insects, including predator species, can increase in intercrops, relative to monocrops (Dingha et al. 2021; Li et al. 2022; Pierre et al. 2022; Uesugi et al. 2023), at least partially explaining the reduction in undesirable insects. Other potential mechanisms include the production of volatile compounds that deter insects (Khan et al. 2008; Gowton et al. 2021), hiding or camouflaging crops (visually and/or olfactorily) of interest to insect pests, dilution, and physical barriers to insects travelling into the crop and slowing spread within the crop (reviewed by Huss et al. (2022)).

13.3.2 Failures

Intercropping is not a panacea. It has failed to reduce root rot in field pea (unpublished data), despite a sound theoretical basis that intercropping with brassica species may lead to reduced pathogen load. Brassica species contain compounds known as glucosinolates (reviewed by Fahey et al. (2001)) that can break down to isothiocyanates, which are in turn capable of inhibiting the growth and infection processes of *Aphanomyces euteiches* (Smolinska et al. 1997; Dandurand et al. 2000; Hossain et al. 2014), an important pathogen contributing the root rot in pea (Chatterton et al. 2019). Intercropping has also failed to reduce root rot in bean (Ocimati et al. 2017), potentially because the root rot pathogens were also able to infect the companion crops intercropped with the bean. Lack of herbicide weed control options can be a challenge for intercropping, especially if the mixture includes both broadleaf (dicot) and monocrot crops. The availabilities of fungicides registered for all crops can also be a limitation. While intercropping can lead to yield gains, it can also result in yield losses. For example LERs below 1.0 have been documented in pea-oat intercropping (Neugschwandtner and Kaul 2014).

13.3.3 Intercropping Future Research Needs

The mechanisms by which foliar disease is reduced by intercropping are not clearly understood. However, there are many potential mechanisms, including dilution of susceptible hosts, barrier to spore dispersal provided by the companion crop, changes in canopy microclimate, especially relative humidity, or triggering of plant defenses (Fernández-Aparicio et al. 2010, 2011; Schoeny et al. 2010; Guo et al. 2020). For the barrier and/or dilution mechanisms to occur, the companion crop(s) providing the barrier or dilution should be nonhosts to the disease(s) in question. This is indeed the case for the foliar diseases reduced by intercropping in the published literature.

Despite the many benefits that intercropping has the potential to provide, farmers face a variety of obstacles to implementation. One hindrance is the complexity and unknowns associated with management, such as the need to invest more time and/or money in labor or equipment (Huss et al. 2022). Lack of knowledge—both on the farmers part and in the farming and research community—can also reduce the likelihood of implementation. Likely due to the small acres devoted to intercropping in regions where broad-acre, mechanized agriculture is practiced, both equipment and crop insurance options that are well suited to intercropping are lacking.

To increase intercropping uptake and success, additional research, extension, policy development, and investment are needed. On the research front, the performance of intercropping could be improved by the selection or breeding of crop varieties that are well suited to intercropping (Moore et al. 2022, 2023). This is highlighted by the findings of Tavoletti and Merletti (2022) and Pankou et al. (2022) on the performance of various varieties of durum wheat and faba bean or wheat and pea in intercropping. Consistently, Moutier et al. (2022) found that both wheat and pea or faba bean variety had a significant impact on performance in an intercrop. Haug et al. (2021) suggest that incomplete factorial experiments could facilitate selection of varieties best suited to intercropping.

Extension, both within the research community from research to agronomist and producer, is needed to build on past and current successes and overcome hurdles. The number of scientific publications on intercropping has increased in recent years, permitting researchers to learn from and build on each other's work. Farmer-to-farmer and farmer-to-researcher information sharing is also vital. If farmers are more aware of research results, they are better able to make informed management decisions. When researchers are aware of what farmers are currently doing and what challenges they face, they can propose and conduct relevant research.

As more scientific data points to the potential sustainability benefits of intercropping, the arguments in favor of developing policies or incentives that promote intercropping become stronger. In addition, if more farmers adopt intercropping, the demand for equipment that can readily be set to seed or harvest two or more crops simultaneously is likely to grow. If equipment is developed to meet this demand, the barrier to entry for farmers just considering intercropping is lowered. Additional scientific and commercial data will also facilitate the development of practical programs to ensure intercrops. This will in turn reduce a risk associated with intercropping, potentially encouraging more intercropping.

13.4 The Role of Roots in Cropping Systems

One approach to tackle cropping system challenges such as root rot diseases and lodging is to incorporate crop root traits into breeding programs (Lynch 1995). In some parts of the world, breeding selection incorporates targeted root traits. For example, in the 1970s in the Canadian prairies, high-yield durum and spring wheat cultivars were selected under drought stress (Hurd 1964, 1968, 1974). In Australia, where rainfed wheat production systems are reliant on stored soil water for maximizing yields, breeding selection incorporates field assessments involving root traits (Rich et al. 2016). The role plant roots play in rangeland on the Canadian prairies has been widely studied (Kembel and Cahill Jr. 2011; Ljubotina and Cahill Jr. 2019; Otfinowski and Coffey 2020). Conversely, most crop improvement strategies such as conventional breeding, cultural practices, and assessments on genotype by environment by management (G *x* E *x* M) interactions are based on data collected solely from aboveground (McGrail et al. 2020). There is a need for crop breeding programs to engage and select not only locally adapted crop cultivars but crop cultivars with root traits for increased nutrient acquisition, especially nitrogen uptake, and increased ability to sequester more carbon to tackle climate change.

Roots play a vital role in connecting plants to soils and in the transformation and circulation of elements and organic compounds across the ecosystem; they are the first organs in the plant to perceive and respond to different environmental factors including water and nutrient resources that underpin agricultural productivity (Freschet et al. 2018; Tracy et al. 2020). For example, the grain yield of Czech winter wheat varieties in dry years is generally positively correlated with root system size (Heřmanská et al. 2014). Roots anchor plants and prevent toxic elements from reaching the plants' reproductive organs (Bailey et al. 2002; Choi and Cho 2019). Shoots may drive water uptake, but root architecture determines plant access to water and often sets limits on shoot function (Liu et al. 2010). For example, a significant positive correlation was found between cover crop root biomass, P and K uptake in eastern Austria (Bodner et al. 2010). Plant roots host soil microorganisms (de Dorlodot et al. 2007) and can form associative or symbiotic relationships with beneficial microbes in the rhizosphere (Smith and De Smet 2012; Canarini et al. 2019; Imran et al. 2021). Plant roots act as a medium of transfer of atmospheric carbon (carbon compounds) into the soil to aid in long-term carbon sequestration, a measure to tackle global climate change (Pausch and Kuzyakov 2018), hence the concept of "live roots" in cropping systems (Blesh and Martin 2018), which is one of the regenerative agriculture principles.

The plastic nature of plant roots allows them to adapt to different soil environments, thereby affecting crop morphology, physiology, and growth (Bassirirad 2006; Wu et al. 2020). For example, in natural plant communities, species-specific plasticity has been observed in roots developing into nutrient-rich patches with significant root responses to high N (NO_3^-) microsites (Bassirirad 2006). The latter has implications for cropping systems usually involving crops and weeds competing for nutrients during crop establishment. Hence, a vigorous root system with accompanied inherent plasticity is recognized as a critical trait that enhances both water and nutrient uptake, especially under abiotic stresses resulting in improved crop performance (Wu et al. 2016; Desgroux et al. 2017). Root distribution pattern with depth is another trait that plays an important role in the estimation of soil root carbon input and its balance in agroecosystems (Fan et al. 2016). Root hair length and density enhance root contact with soil particles and also increase root interactions with soil microorganisms (Carminati et al. 2017). Rongsawat et al. (2021) found correlations between root hairs and increased wheat biomass production and yield. Root branching density favors soil exploration, which facilittes uptake of mobile soil resources including nitrate and water (Pedersen et al. 2009; Eissenstat et al. 2015). Freschet et al. (2021) in their review exhaustively described several root traits with respect to plant functions such as tolerance to lodging, drought, initiation and establishment of mycorrhizal symbioses, ecosystem phosphorus cycling, and others. However, the selection and inclusion of root traits in any investigation is research question and goal dependent.

Plant root systems are classified based on their sites of origin into four classes: tap, lateral, basal, and shoot-borne roots (Zobel and Waisel 2010). The root systems of cereals (wheat and Italian ryegrass) and pulses (soybean) have been described in detail by Zobel and Waisel (2010) and that of canola by Wu et al. (2016). It is generally believed that while both monocots and dicot seedlings have a defined tap root during early life, the defined tap root is mostly only retained by dicot plants (Waidmann et al. 2020). Functional differences between root classes are likely as lateral and tap roots support different microbial populations presumably through differences in exudates with implications for nutrient uptake (Zobel and Waisel 2010).

The topsoil where most roots are located also harbors spores of disease-causing organisms such as root rots and clubroot. Therefore, identifying cultivars that mobilize roots at depths where the spores of disease-causing organisms are absent and additional moisture can be accessed with or without the aid of mycorrhiza is a strategy that can be exploited by breeding programs to develop new-generation cultivars with unique adaptations. More research is required to better understand how roots' vertical distribution patterns and their exudates interact with diazotrophs in the presence or absence of nutrients, their contribution to carbon sequestration, and how these complex interactions relate to root longevity.

13.4.1 Phenotyping Methodologies Employed in Root Studies

Root trait as part of breeding programs is seen as a key component to develop resilient cropping systems (Martens et al. 2015; Paez-Garcia et al. 2015). However, available root methodologies are labor- and resource-intensive; and results tend to be variable due to environment, crop cultivar, and root plasticity (Sharma and Carena 2016; Wu et al. 2020). Greenhouse root phenotyping methodologies were developed to reduce inherent environmental variability and labor costs. An example is the WinRHIZO™ software, an image analysis system designed for washed root measurements, although other less labor-intensive and nondestructive approaches have also been used. Demerits of indoor-based phenotyping techniques are that roots have different spatial arrangements than do field-grown plants due to container size constraints and/or artificial media used (Araus and Cairns 2014). It has been suggested that robust plant root phenotyping studies should include a combination of field, greenhouse, or laboratory approaches in order to elucidate the developments of root architecture (Paez-Garcia et al. 2015). The use of an LCR meter that measures inductance (L), electrical capacitance (C), and resistance (R) is a promising strategy that bypasses the labor-intensive phase of root washing and processing, enabling root functionality measurements (Heřmanská et al. 2014; Wu and Ma 2016; Wu et al. 2020).

Across the globe, attention to plant root systems has gained momentum with minirhizotrons and 3D root phenotyping platform development which facilitates noninvasive assessments (Fahlgren et al. 2015; Kuijken et al. 2015; Jeudy et al. 2016). Imaging coupled with root system analyzers (e.g., Ez-Rhizo and WinRHIZO™, smart rot and X-ray computed tomography) has increasingly been used (Martins et al. 2020). The development of high-throughput machine learning and computer vision has enabled root trait extraction and measurement useful in breeding programs targeting more vigorous and resilient crop varieties (Falk et al. 2020). In Canada, technologies employed have been limited to shovelomics, soil columns, rhizotrons, and WinRHIZO™ except in a few cases—LCR meter (Wu et al. 2016) and CI-600 In Situ Root Imager (Bourgault et al. 2022). A few studies have looked at harnessing root traits to mitigate drought stress (Ashe et al. 2017; Gorim and Vandenberg 2017); root traits, root lodging parameters, and canola yield (Wu and Ma 2016, Wu et al. 2020); water and nutrient uptake (Liu et al. 2010, 2011; Gan et al. 2011; Cutforth et al. 2013); and effects of low and high N levels versus rhizobia inoculation on root traits in wild and cultivated lentil (Vargas Palacio 2021) and relate root mass to photosynthetic efficiency in durum wheat varieties with different pedigrees (Ashe et al. 2017). In the United States and Australia, identifying and incorporating root ideotypes rather than individual root traits into breeding programs is a recommended approach that can also be pursued in prairie breeding programs (Lynch 2013; Rao et al. 2021). The latter together with new technologies and phenotyping platforms provide an opportunity to phenotype a large number of germplasm, contributing to the development of adaptable cultivars. Future research should focus on the interaction of root traits and root exudates of crop cultivars, identifying root microbiomes, including AMF, associated with desirable outcomes such as carbon sequestration.

13.5 Mycorrhizas as Key Root Symbionts

The most common root symbiont on earth is the arbuscular mycorrhiza from the subphylum Glomeromycotina (within the phylum Mucoromycota). Fossil evidence (Remy et al. 1994; Redecker et al. 2000) and molecular clock evidence (Heckman et al. 2001) point to arbuscular mycorrhizas as participating in the colonization of land by plants 400 or up to 600 million years ago (mya), respectively. Phylogenetic analyses considering the contemporaneous radiations of both plants and fungi show a fungal-mediated colonization of land by embryophytes coinciding with Glomeromycotina diversification ca 484 mya (Lutzoni et al. 2018). Since this initial diversification, it appears that Glomeromycotina has a speciation rate of one order of magnitude lower than other eukaryotes, despite being ubiquitous (Perez-Lamarque et al. 2022). Glomeromycotina are globally dispersed, with 34% of species occurring on all seven continents, despite being associated with plants which display significant endemism (Davison et al. 2015; Barbosa et al. 2017). The observation that AMF are widely dispersed (Davison et al. 2015) is difficult to reconcile without a clear mechanism of dispersal (Egan et al. 2014, Correia et al. 2019, Chaudhary et al. 2020). AMF have yet to be observed forming reproductive structures and had been assumed asexual until a recent discovery of meiosis genes and mating-type loci (Ropars et al. 2016). Along with other early diverging fungi, AMF form coenocytic hyphae which form no cross-walls effectively allowing for the mixing of thousands of nuclei throughout the fungal body (Kokkoris et al. 2020). To understand the distribution of diversity between nuclei, single nucleus sequencing was employed to discover that while dikaryons, a mixture of two nuclear types suggesting a reproductive state (Chen et al. 2018) do exist, most known AMF have identical haploid nuclei. These multiple lines of evidence point to sexual or parasexual reproduction in AMF (Yildirir et al. 2020) which is likely necessary to sustain the genetic variability to successfully colonize 70–90% of all land plants.

AMF are obligate symbionts, acquiring all their carbon from photosynthetic plants in exchange for soil-derived nutrients. The AMF life cycle begins with spore germination in soil in response to root hormones followed by establishment of root cortical cell colonization, arbuscule formation, and intra- and extraradical hyphal growth and completing the cycle with sporulation (Gutjahr and Parniske 2013). Inducing spore germination involves strigolactones exuded from roots, which are upregulated under phosphate starvation. These signaling molecules also induce hyphal branching in AMF. Germination triggers AMF to release various chitooligosaccharides that signal plants to increase lateral root formation, trigger transcription of genes that will allow hyphal penetration, and increase starch accumulation in roots. A hyphopodium is formed on a surface root cell which is required for a penetration peg that allows the hyphae to enter the cell. The plant is actively involved in this process creating a prepenetration apparatus which is required for initial penetration by hyphae and arbuscule formation. Perhaps, the most dramatic feature of this symbiosis is the coordinated movement of plant cell organelles to create space for and to synthesize the periarbscular membrane which surrounds the arbuscule and ultimately mediates transport of carbon and nutrient compounds. Arbuscule formation is well defined in terms of gene expression changes at each stage (Gutjahr and Parniske 2013). Arbuscules are ephemeral with an arbuscule lifespan of 1–3 days (Gutjahr and Parniske 2017). The quick turnover likely renders the symbiosis efficient and may be the mechanism for preventing cheating in the symbiotic exchange of nutrients (Floss et al. 2017; Gutjahr and Parniske 2017).

On the landscape scale, the picture of AMF community assembly is not clear (Powell and Bennett 2016). Counterintuitively, studies often find higher AMF diversity at the small local scale than when comparing sites far apart (Kivlin et al. 2011; Rincon et al. 2021). In a similar environment, when AMF are experiencing the same abiotic pressures while colonizing the same host, we would expect that selection would favor fewer species and a common composition resulting in lower beta diversity. Yet, this is rarely the case (except see (Davison et al. 2016). While

AMF assembly rules are not yet clear, they are important to optimizing mycorrhizas in cropping systems, especially more plant diverse cropping approaches such as intercropping. In intercropping, the presence of roots from two or more crops may provide a more heterogeneous belowground environment, potentially facilitating the establishment of a healthier or more diverse AMF network.

13.5.1 Common Mycorrhizal Networks

Common mycorrhizal networks (CMNs) are an emergent phenomenon of arbuscular mycorrhizal colonization of plant roots and may be a factor contributing to the effects observed such as increased plant resistance to disease, overyielding, and breeding/selection of varieties that are well suited to intercropping systems. CMNs are created when the hyphae of one AM species connects the roots of two or more plants (Simard and Durall 2004; Babikova et al. 2013a). In doing so, they are reported to assist in seedling establishment by colonizing new roots, thereby regulating the plant community composition and diversity (Simard and Durall 2004, Babikova et al. 2013a). In tomato plants connected to a CMN, defense responses to both fungal and bacterial pathogens are faster and stronger than in control plants not connected to a CMN (Song et al. 2015b; Fujita et al. 2022). When challenged with spider mites and aphids, the presence of a CMN was thought to be a factor in transferring resistance signals to uninfected neighbors to prime their defenses (Song et al. 2010, 2013, 2014, 2015a). There may be an AMF action in the transmission or stimulation of volatile organic chemicals (VOCs), as well. The presence of pea aphids resulted in excretion of VOCs that were transferred to neighboring plants that attracted parasitoids, a natural enemy of aphids (Babikova et al. 2013a,b, 2014b). Since aphids and other phloem-sucking pests remove carbon from the plant, it is hypothesized that the AMF will secure their carbon supply by ensuring defense signals are transferred to connected plants (Babikova et al. 2014b).

In addition to transferring beneficial chemicals, AMF have also been found to transfer allelochemicals, which can negatively impact deleterious microbes, pathogens, and competitor plants. The roots of some plants synthesize and exude these competition-reducing secondary metabolites into the rhizosphere where they are taken up by AMF hyphae and transported to neighboring plants (Barto et al. 2011). Caution is advised in planting species that are known to be inhibitory within a close distance since a CMN would suppress the growth of the subordinate plants. However, this could be advantageous as weed control and possibly be leveraged in intercropping, if more were known about this phenomenon. Another interesting use of CMNs that has been proposed is planting sentinels in a cash crop. The sentinel plant will be the first to be infected or attacked, so it is important to plant the species based on insects and pathogens most likely encountered by the crop of interest. The sentinel will transfer defense signals via the CMN to the crop plants and prime them for attack. Secretion of VOCs throughout the crop could also suffice (Heil et al. 2000).

In either monoculture or intercropping, plants do not grow in isolation; their roots and AMF likely interact, possibly forming CMNs if agricultural practices allow. The emergent phenomenon of CMNs could be responsible for increased disease resistance and overyielding in intercropping (Barto et al. 2012; Simard et al. 2012; Johnson and Gilbert 2015; Qiao et al. 2020; Figueiredo et al. 2021). AMF also play a role in kin recognition by transferring volatile organic chemicals (VOCs) through their CMN and influencing the chemical composition of VOC emitted by plants (Schausberger et al. 2012; Babikova et al. 2014b; Zhang et al. 2022). The potential benefits to crops of AMF colonization including improved nutrient acquisition, increased resilience to abiotic stressors, and improved soil structure should be considered irrespective of CMN formation. Most studies of AMF impact on crop performance do not consider CMN formation owing to the difficulty of demonstrating its presence in field studies. However, the unexpected impacts of intercropping could be at least partially explained by CMNs.

Hyphae extending into the soil provide microhabitats that support a myriad of bacteria and are referred to as extraradical hyphae. Extraradical hyphae interface with the soil matrix through a film of water present on the hyphal surface called the hyphosphere (Andrade et al. 1997). The hyphosphere itself is divided into the endo-hyphosphere and the exo-hyphosphere: endo pertains to the inside of the hyphae and exo pertains to the outside of the hyphae (Faghihinia et al. 2022). The endo-hyphosphere contains AMF endosymbiotic microbes (Toljander et al. 2006; Agnolucci et al. 2015; Wang et al. 2016; Jiang et al. 2021), while the exo-hyphosphere is further divided into the hyphoplane or surface of the hyphae and ecto-hyphosphere, being the surrounding soil (Faghihinia et al. 2022).

A microhabitat is created by endo-hyphosphere exudates that function to attract and feed specific microbes including those from the Talaromyces, Burkholderia, Pseudomonadota, Actinomycetota, Gemmatimonadota, and Bacteroidota bacterial phyla (Drigo et al. 2010; Doilom et al. 2020; Faghihinia et al. 2022). Key players in this system are phosphate-solubilizing bacteria which facilitate phosphate mineralization for transport to the host plant while contributing to the growth of both the plant and the AMF and are supported by carbon from hyphae (Jiang et al. 2021). As the hyphae grow, move, and senesce throughout the soil, they disperse microbes. Some microbes are attached to the hyphoplane, while others use it as an intra-soil transportation system, moving along the film (Jiang et al. 2021). A mycorrhizal hyphal network thus supports a micro-ecosystem with the potential to transfer nutrients, share beneficial bacteria, and confer benefits to connected crop plants in a CMN.

13.5.2 AMF in Agriculture Systems

AMF impact primary macronutrients including carbon, nitrogen, phosphorus, and potassium; they influence abiotic stressors including drought, soil salinity, and heavy metals; they influence soil structure; and they influence plant response to disease pressure through mycorrhizal-induced resistance (MIR). We consider these interactions in-turn below.

13.5.3 *Primary Macronutrients*

13.5.3.1 Carbon

Carbon is one of the most important macronutrients on the planet. As plants photosynthesize, 20% of the carbohydrates generated are transferred to AMF extraradical hyphae (Bago et al. 2000). Excess carbon is stored as triacylglycerol (TAG) within the vesicles of intraradical hyphae (Bago et al. 2000). One of the most important functions of AMF in relation to soil health is the accumulation and storage of rhizodeposits as various forms of organic carbon (Cheng et al. 2012; Averill et al. 2014). Being a mutualistic keystone relationship, that of AMF and the root requires balance between carbon stabilization and mineralization (Bonfante and Anca 2009; Rillig et al. 2015; Gavito et al. 2019; Jeewani et al. 2021).

The amount of carbon transferred from the plant to the AMF is usually of concern to producers. However, evidence shows that when the extraradical mycelium is excised from the plant root, even when the plant is fully nourished and watered, photosynthetic rates decrease for up to 2 h (Gavito et al. 2019). Thus, the allocation of carbon to AMF by the plant is compensated for by an increase in photosynthesis (Schweiger et al. 2014; Gavito et al. 2019). It is possible that in seedlings that are establishing, the sink strength may be stronger than their photosynthetic capabilities. Overall, however, it is to the benefit of the plant at any and all growth stages to be in symbiosis with AMF for defense and microbial recruitment (Gavito et al. 2019). Specific proteins are part of the plant's regulatory mechanisms to prevent over-extrusion of carbon by the fungus (Salmeron-Santiago et al. 2021).

Specific proteins are part of the plants' regulatory mechanisms to prevent over-extrusion of carbon by the fungus (Salmeron-Santiago et al. 2021). Carbon allocation by the plant is used as a bargaining chip when AMF are connected to more than one plant or species in the CMN. Preferential nutrient provision may be given to the plant that provides the most carbon to AMF, known as high-quality hosts. In contrast, low-quality hosts provide less carbon (Bucking et al. 2016). Preferential allocation still provides carbon to the low-quality host, most likely as a carbon insurance policy, so that if one host dies, the fungus is still connected via the CMN to other plants (Kiers et al. 2011; Bucking et al. 2016). This strategy also ensures that the high-quality host encounters competition and provides its due share of carbon to the AMF in exchange for nutrients (Bucking et al. 2016). However, when the only available host is of low quality meaning that low amounts of carbon are provided to the plant, the AMF will favor this plant and provide more nutrients to benefit the host (Bucking et al. 2016).

13.5.4 Nitrogen

AMF can supply up to 25% of the plant's nitrogen requirements reducing the need for fertilizer (Liu et al. 2021). Legumes fix approximately 20% of the required nitrogen within an ecosystem by way of their symbiotic relationship with N-fixing rhizobacterium (Adomako et al. 2022). A CMN in legume-non-legume intercrops may further facilitate increased nitrogen use efficiency by making some of the N fixed by the legume available to the non-legume. An investigation into the effects of phosphorus addition to the soil in the presence of AMF and its ability to acquire nitrogen revealed that added phosphorus decreases the beneficial action of AMF. In nitrogen-limited systems, either commensalism or parasitism can occur between the plants and AMF (Liu et al. 2021). Therefore, it is important to add nitrogen to crops that may be limited to ensure that carbon stores are not being depleted by AMF.

13.5.5 Phosphorus

Phosphorus is an important element in the proper growth and development of plants and is often the factor that is most likely to limit plant growth. As a part of all cell membranes and the production of ATP, the cells' respiratory currency, phosphorus must be bioavailable in the soil for plants to access and use it (Liu et al. 2021). Within the soil microbiome, the most important microbe for phosphorus provisioning are AMF (Adomako et al. 2022). Since AMF extend the root system up to 2 feet beyond the rhizosphere nutrient depletion zone, AMF can supply up to 80% of the plant's phosphorus requirements. AMF influence interspecific interactions in a legume system as they catalyze the generation of phosphate ions using alkaline phosphatase making P more available (Liu et al. 2021). This could be particularly important when legumes and non-legumes are intercropped.

AMF will utilize nitrogen for itself before supplying this to the plant; however, phosphorus is generated in surplus and is easy for the fungi to transfer (Liu et al. 2021). When soil P is limiting, the AMF symbiosis becomes even more necessary. Even though it may seem like plants are investing a lot of carbon (4–20%) into AMF, these fungi are more efficient in their use of carbon when considering soil exploration and extraction of nutrients compared to plant roots, saving the plant energy on root production (Adomako et al. 2022). The addition of too much phosphorus to the soil will suppress the actions of AMF (Liu et al. 2021), in addition to losses from leaching and surface water runoff.

P solubilizing bacteria (PSB) utilize AMF as a conduit, thereby increasing the amount of phosphorus available to be absorbed by the AMF, in addition to stimulating hyphal growth increasing fungal fitness, and suppressing soilborne diseases (Jiang et al. 2021; Adomako et al. 2022). In their quest for acquiring organic phosphate, PSB make up approximately 40% of the bacterium in the soil (Adomako et al. 2022).

In phosphorus-limited systems, mutualism between plant and AMF is the dominant relationship. Since legumes make a tripartite symbiosis with AMF and nodule-forming rhizobium species, more phosphorus is required because the nodules are a strong nitrogen sink. Legume species rely more heavily on AMF for the supply of phosphorus while they are fixing their own nitrogen, improving their competitive edge. Without the addition of phosphorus, the nutrient limitation for the legume becomes exacerbated as the plant has a higher root-shoot biomass (Liu et al. 2021).

13.5.6 *Potassium*

Potassium (K) is essential to plant growth and chemical processes. Mycorrhizas help plants access and uptake K and through the CMN provision the allocation of this nutrient to the plant that requires it most in an intercropped system. This allows more roots to be colonized with a higher percentage of arbuscules in the intercropped system in addition to better supporting seedling growth. The CMN can regulate K exchange in agriculture by supporting the growth of taller crops over shorter ones using asymmetric K provisioning. CMNs are advantageous in intercropping in agriculture due to the higher percentage of root colonization versus a monocropped system (Gao et al. 2021).

13.5.7 *Abiotic Stressors*

13.5.7.1 Drought Stress

Drought stress is one of the major limiting factors for plant growth around the world (Chandrasekaran and Paramasivan 2022). In the plant, drought stress causes the production of reactive oxygen species (ROS) that accumulate intracellularly, damaging cells, preventing photosynthesis, and ultimately resulting in apoptosis (Bahadur et al. 2019; Chandrasekaran and Paramasivan 2022). Besides irrigation, mycorrhizas can mitigate some effects of drought stress (up to 17%) (Chandrasekaran and Paramasivan 2022) by providing water along its hyphosphere to the roots, closing stomata to improve water use efficiency (WUE) to prevent loss due to transpiration, increasing antioxidant production, maintaining pathogen and pest defenses (Orine et al. 2022), and increasing nutrient supply to the plant (Bahadur et al. 2019, Chandrasekaran and Paramasivan 2022).

AMF species present in native soils will be better adapted to local environments and, during times of drought, are better able to not only survive themselves but also provide drought stress tolerance to its associated plants by inducing genes within the drought pathway. Roots colonized by AMF are more plastic and resist water loss. In times of drought, plant roots exude strigolactones as a signal to AMF that they require hydration (Bahadur et al. 2019). By maintaining proper nutritional status, the plant is better able to cope with drought by sustaining higher turgor pressure that supports continued photosynthesis (Chandrasekaran and Paramasivan 2022) that is typically impeded under drought stress. Since AMF extend the rhizosphere depletion zone, once the soil moisture has been exploited, AMF are able to access and return soil moisture from distances beyond the rhizosphere (Bahadur et al. 2019). With prolonged drought, however, even AMF and their symbiosis will be negatively affected, with a pronounced decline in the number of intraradical spores and arbuscules.

In drought situations, common during times of climate change and normal weather phenomena, AMF are thought to alter stomatal regulation and the hydraulics of the root system, improving WUE. Approximately 35% of water obtained by the plant is from AMF that extend their hyphal network to obtain water when it is scarce (Adomako et al. 2022). Interactions of drought, AMF, and intercropping remain to be explored. Intercropping is generally thought to be better suited to wetter environments. However, intercropping can still provide benefits under drought conditions (Pourali et al. 2023).

13.5.8 Salinity

Over one billion hectares (ha) or 20% of crop lands are considered to be saline due to the overaccumulation of sodium chloride (NaCl), nickel (Ni), iron (Fe), and others, with losses increasing yearly (Porcel et al. 2011; Evelin et al. 2019; Ma et al. 2019). Salinity occurs through both natural and anthropogenic processes, including successional planting of shallow-rooted plants, high rates of evaporation, weathering of rocks, vicinity to bodies of salt water, poor irrigation sources, reduced soil porosity, and poor soil drainage. While plants have adapted mechanisms to deal with high salinity, such as maintaining turgor pressure, increasing WUE to maintain photosynthesis, and inducing phytohormones to detoxify oxidation stress, AMF further support the plant's ability to survive in saline soils (Porcel et al. 2011; Evelin et al. 2019).

AMF hyphae provide the stressed plant with phosphorus and can access low concentrations of P in the soil while upregulating genetic expression of plant nitrogen transporters to increase its intracellular concentration (Porcel et al. 2011, Evelin et al. 2019). AMF also maintain osmoregulation through increased nutrient supply, allowing the plants to increase turgor pressure, preventing cellular dehydration that could kill the plant (Evelin et al. 2019). Photosynthesis can be negatively affected by a reduction in leaf size, whereas plants with mycorrhizas can maintain higher photosynthetic rates due to an increase in chlorophyll in the leaves because of improved micronutrient uptake (Evelin et al. 2019). Therefore, mycorrhizas are a powerful tool in the support of crops in saline soils by supporting the adaptive functions that already exist within the plant.

13.5.9 Heavy Metals

High concentrations of heavy metals in soil lead to plant stress. The presence of AMF in soils contaminated with heavy metals ameliorates the negative effects of heavy metals on crops (Janeeshma and Puthur 2020). This has been attributed in part to the ability of glomalin-like substances to bind toxic metals, including copper, lead, zinc, arsenic, and cadmium (Janeeshma and Puthur 2020; Holátko et al. 2021). Bioremediation of soils using AMF (mycoremediation) is the most effective method of supporting plant growth in these soils. Two mycoremediation methods are phytoextraction (uptake of heavy metals from soil) and phytostabilization (heavy metal accumulation in roots that do not travel to the shoots; AM preferred method). Biomagnification can occur during phytostabilization and should be used with caution in agriculture where the roots are consumed by humans. Plants with specific root systems are designed for phytostabilization and are typically hyperaccumulators of metals. The benefit of AMF in this system is the enhancement of metal tolerance by these hyperaccumulators. In phytoremediation, it is possible to re-extract the metals from the plant tissue since the roots will transport metals to the shoots. To maximize this effect, AMF colonization increases metal uptake and transports it to the shoots in hyperaccumulators, to remove the highest amount of metal from the soil (Janeeshma and Puthur 2020).

Independent of plants, AMF can withstand metal toxicity by using glomalin-related soil proteins (GRSP) to complex with the metals using chelation, immobilizing and inactivating them and reducing their damaging effects (Janeeshma and Puthur 2020, Holátko et al. 2021). Within fungal structures such as mycelium, vesicles, and spores, depending on the species of AMF and type of metal, AMF store heavy metals in association with uranium, which prevents metal transportation across the periarbuscular membrane to the plant. Outside in extraradical mycelium, AMF are able to bind metals to their hyphae using adsorption that immobilizes heavy metals, removing them from the soil within 30 min of contact. The metals are then crystalized by AMF taking up 10x more metal than hyperaccumulator plant roots (Janeeshma and Puthur 2020). Ultimately, AMF help remove, immobilize, and prevent transportation of heavy metals from contaminated soils, allowing for unimpeded plant growth and bioremediation.

13.5.10 Soil Aggregates and Structure

Soil aggregates are structures formed when soil particles bind together (Rillig et al. 2015). Hydrostable soil aggregates are formed by sticky substances like glomalin-related soil protein (GRSP) that self-polymerize and are secreted by AMF and other fungi (Barea et al. 2002; Rillig et al. 2015; Holátko et al. 2021). GRSPs are hydrophobic glycoproteins that attract soil particles, binding them together creating aggregates (Barea et al. 2002). The entanglement of mycelium also contributes to aggregate formation, improving soil structure and stability (Rillig et al. 2015). By creating pores in the soil matrix in the presence of aggregates, water infiltration improves to reduce standing water during heavy precipitation events, maintaining aeration and improving plant health (Barea et al. 2002). Improved soil structure allows for better root growth and prevention of lodging and compaction and thereby an increased ability to store carbon (Rillig et al. 2015).

13.5.11 Mycorrhizal-Induced Resistance (MIR)

Mycorrhizal fungi induce the defense responses of plants by priming their resistance against fungal, bacterial, and viral pathogens and several pests (Pozo et al. 2002; Barber et al. 2013; Song et al. 2013, 2015a; Babikova et al. 2014a; Schouteden et al. 2015; Tao et al. 2016; Hill et al. 2018; Goicoechea 2020; Li et al. 2021; Fujita et al. 2022; Yu et al. 2022). By decreasing the response time and increasing strength of responses, AMF act somewhat like a vaccine, allowing the plant to suppress and/or fight the pathogen/pest better than non-colonized plants. AMF species *Rhizophagus irregularis* and *Funneliformis mosseae* exert especially strong MIR (Pozo et al. 2002; Song et al. 2010, 2013, 2014, 2015a; Fujita et al. 2022). MIR triggers increased signaling within both the salicylic acid (SA) and jasmonic acid (JA) pathways involved in stress and defense responses (Song et al. 2015a; Wang et al. 2020).

Tomato plants colonized at low levels by the AMF species *Gigaspora margarita* were more resistant to both fungal (*Botrytis cinerea*) and bacterial (*Pseudomonas syringae* pv. *tomato* DC3000) pathogens, mounting a faster, stronger defense response versus non-colonized plants (Fujita et al. 2022). When defending against a chewing caterpillar (*Helicoverpa armigera*), being colonized by AMF, *G. mosseae* depressed larval development while increasing defense-related genes that overexpressed JA at the four-leaf stage (Song et al. 2013). Pathogenic fungal infections in tomato plants are reduced when colonized by AMF, especially *G. mosseae*. Both localized (within the root) and systemic damage were reduced due to plant reinforcement of the cell wall with callose (Pozo et al. 2002).

AMF have been used as a biological control for parasitic nematodes, and the efficacy of this depends on the combination of AMF species, crop, and environment. AMF compete for nutrients and space within the soil and roots while providing the plant with enhanced defenses as a result of increased nutrition, altered VOC profiles and root exudates, and priming induced systemic resistance (ISR) (Schouteden et al. 2015). In altering root exudates, the plant can attract beneficial soil microbes that can also contribute to the destruction or suppression of pathogens and pests (Carvalhais et al. 2015; Plett et al. 2021).

A drawback for some crops is that mycorrhizas can alter VOC profiles that attract rather than repel aphids possibly due to an increased foliar phosphorus content (Schausberger et al. 2012; Babikova et al. 2014a) and the suppression of key sesquiterpenes that repel aphids, specifically (*E*)-β-farnesene and (E)-caryophyllene.

However, if aphid infestation occurred prior to AMF colonization, the aphids themselves produced VOCs that repelled other aphids, and the plant produced a neutral VOC profile that was neither repellant or attractive to aphids. In addition, aphids decreased the level of AMF colonization, hypothesized to be the result of a reduction in carbon allocation to AMF (Babikova et al. 2014a).

13.6 AMF Impact on Ecosystems

Interspecific competition is mediated by the spatial heterogeneity of the nutrient distribution within the soil. Soil microbes are known to balance this heterogeneity using their mineralization and transport abilities. The interaction between plants, microbes, and heterogeneous soil nutrients is key to ecosystem function and stability, supported by AMF ability to transport nutrient-solubilizing microbes around the soil (Whiteside et al. 2019; Adomako et al. 2022). The distribution of nutrients within the soil impacts interspecific interactions that are then mediated by AMF (Adomako et al. 2022). There is also evidence that nutrients are acquired from decomposed organic matter by AMF and its associated microbiome (Liu et al. 2021) that are moved from nutrient-rich to nutrient-poor environments using AMF as a trader (Whiteside et al. 2019; Gao et al. 2021).

Soil microbes are an important nutrient mobilization cohort that compete with plants for soil nutrients (Adomako et al. 2022). Plants can utilize these nutrients by rhizophagy wherein the bacteria enter the root tips, reactive oxygen species (ROS) cause the expulsion of nutrients from the bacteria into the root, and the bacteria are ejected out of the root hairs to rebuild their cell membrane and acquire more nutrients as the cycle continues (White et al. 2018). The actions of microbes affect interspecific competition since AMF move these bacteria around in the soil matrix via the hyphosphere (Adomako et al. 2022; Faghihinia et al. 2022). AMF stimulate the microbial "priming effect" by feeding other microbes that have the capacity to break down organic matter to obtain nutrients (Liu et al. 2021). AMF rely on the enzymatic action of saprophytic fungi to break down bound nutrients and make them bioavailable to the plant (Adomako et al. 2022); however, AMF will repress saprotrophs as they create hyphae in the direction of organic matter, thereby outcompeting them for nutrients (Liu et al. 2021).

13.6.1 Summary of Benefits to Crops with AMF in Agriculture Systems

AMF provide soil-derived nutrients, including nitrogen and phosphorus, along with micronutrients and water, to the plant in exchange for photosynthetic carbon (Parniske 2008). This is accomplished as the AMF increase the root surface area for nutrient absorption beyond what the roots are capable of alone (Adomako et al. 2022). In addition, as AMF influence the soil system by moving, growing, and multiplying within the soil, their hyphosphere transports bacteria, nematodes, archaea, and protists that may act as bioprotectants against root pathogens and pests (Faghihinia et al. 2022; Zhang et al. 2022). Transport of mineralized nutrients also occurs within and along the fungal hyphae, benefitting plant biomass production and, ultimately, photosynthetic capability, leading to increased plant health and yields. AMF increase plant water uptake and water retention during drought (Adomako et al. 2022). The increased provision of nutrients has also been correlated with the increase in defense response mechanisms for mycorrhizal plants, known as mycorrhizal-induced resistance. The presence of the AMF in the roots encourages strengthening of cellular walls systemically, while a stronger, faster defense response occurs when the plant is attacked by a pathogen or pest (Pozo and Azcon-Aguilar 2007; Jung et al. 2012; Cameron et al. 2013). In addition, the plant is better supported when encountering abiotic stressors, including drought, salinity, and heavy metals (Bahadur et al. 2019; Janeeshma and Puthur 2020). AMF further benefit agricultural soils by stabilizing nutrients within the soil to prevent leaching of fertilizers (Adomako et al. 2022) into waterways, which has a poor societal image and negative environmental impact. Moreover, by utilizing AMF in agricultural soils, less fertilizer will be required since AMF can supply the crops with the majority of their N and P requirements.

The use and nurturing of the AMF-plant relationship is beneficial to agricultural production economically since nutrients are supplemented through this symbiosis. By exploring the smaller pores and longer reaches beyond the nutrient depletion zone of the rhizosphere, AMF can access nutrients the roots cannot. Paired with their ability to move microbes around within the soil, AMF as an economical tool should be considered by producers. This financial saving extends to increased yields as the plant invests more energy into aboveground biomass production, increasing photosynthesis, and less into root growth and soil exploration (Adomako et al. 2022).

13.7 Conclusion

AMF have a significant impact on plant proliferation, especially in natural soils and natural systems. Regenerative agriculture and the practice of intercropping both support AMF diversity and abundance in soils more so than conventional agricultural practices. Regenerative management actively pursues improvement in production by leveraging ecological principles, soil beneficial microbiomes, and wholistic nutrient cycling by reducing tillage, eliminating bare soil, fostering plant diversity, integrating livestock into cropping, and reducing agrichemicals. Intercropping increases aboveground diversity, often resulting in overyielding, disease reduction, lower input requirements, and improved soil conditions. Engaging in regenerative or intercropping practices will encourage a diverse and abundant AMF community that increases the chances of beneficial interactions with crops. Crop breeding for intercropping and regenerative approaches should focus on root traits and beneficial AMF interactions to facilitate building farming practices that attempt to ameliorate the environmental impacts of agriculture while maintaining sufficient productivity and food production.

References

Adomako MO, Roiloa S, Yu FH (2022) Potential roles of soil microorganisms in regulating the effect of soil nutrient heterogeneity on plant performance. Microorganisms 10:2399

Agnolucci M, Battini F, Cristani C, Giovannetti M (2015) Diverse bacterial communities are recruited on spores of different arbuscular mycorrhizal fungal isolates. Biol Fertil Soils 51: 379–389

Andrade G, Mihara K, Linderman R, Bethlenfalvay G (1997) Bacteria from rhizosphere and hyphosphere soils of different arbsuscular-mycorrhizal fungi. Plant Soil 192:71–79

Araus JL, Cairns JE (2014) Field high-throughput phenotyping: the new crop breeding frontier. Trends Plant Sci 19:52–61

Ashe P, Shaterian H, Akhov L, Kulkarni M, Selvaraj G (2017) Contrasting root and photosynthesis traits in a large-acreage Canadian durum variety and its distant parent of Algerian origin for assembling drought/heat tolerance attributes. Front Chem 5:121

Averill C, Turner BL, Finzi AC (2014) Mycorrhiza-mediated competition between plants and decomposers drives soil carbon storage. Nature 505:543–545

Babikova Z, Gilbert L, Bruce TJ, Birkett M, Caulfield JC, Woodcock C, Pickett JA, Johnson D (2013a) Underground signals carried through common mycelial networks warn neighbouring plants of aphid attack. Ecol Lett 16:835–843

Babikova Z, Gilbert L, Bruce T, Dewhirst SY, Pickett JA, Johnson D, Field K (2014a) Arbuscular mycorrhizal fungi and aphids interact by changing host plant quality and volatile emission. Funct Ecol 28:375–385

Babikova Z, Johnson D, Bruce T, Pickett JA, Gilbert L (2013b) How rapid is aphid-induced signal transfer between plants via common mycelial networks? Commun Integr Biol 6:e25904

Babikova Z, Johnson D, Bruce T, Pickett J, Gilbert L (2014b) Underground allies: how and why do mycelial networks help plants defend themselves?: what are the fitness, regulatory, and practical implications of defence-related signaling between plants via common mycelial networks? BioEssays 36:21–26

Bago B, Pfeffer PE, Shachar-Hill Y (2000) Carbon metabolism and transport in arbuscular mycorrhizas. Plant Physiol 124:949–957

Bahadur A, Batool A, Nasir F, Jiang S, Mingsen Q, Zhang Q, Pan J, Liu Y, Feng H (2019) Mechanistic insights into arbuscular mycorrhizal fungi-mediated drought stress tolerance in plants. Int J Mol Sci 20:4199

Bailey PHJ, Currey JD, Fitter AH (2002) The role of root system architecture and root hairs in promoting ancharage against uprooting forces in *Alium cepa* and root mutants of *Arabadopsis thaliana*. J Exp Bot 53:333–340

Barber NA, Kiers ET, Theis N, Hazzard RV, Adler LS (2013) Linking agricultural practices, mycorrhizal fungi, and traits mediating plant-insect interactions. Ecol Appl 23:1519–1530

Barbosa MV, Pereira EA, Cury JC, Carneiro MAC (2017) Occurrence of arbuscular mycorrhizal fungi on King George Island, South Shetland Islands, Antarctica. An Acad Bras Cienc 89:1737–1743

Barea J-M, Azcon R, Azcon-Aguilar C (2002) Mycorhizosphere interactions to improve plant fitness and soil quality. Antonie Van Leeuwenhoek 81:343–351

Barker S, Dennett MD (2013) Effect of density, cultivar and irrigation on spring sown monocrops and intercrops of wheat (Triticum aestivum L.) and faba beans (Vicia faba L.). Eur J Agron 51: 108–116

Barto EK, Hilker M, Muller F, Mohney BK, Weidenhamer JD, Rillig MC (2011) The fungal fast lane: common mycorrhizal networks extend bioactive zones of allelochemicals in soils. PLoS One 6:e27195

Barto EK, Weidenhamer JD, Cipollini D, Rillig MC (2012) Fungal superhighways: do common mycorrhizal networks enhance below ground communication? Trends Plant Sci 17:633–637

Bassirirad H (2006) Root system characteristics and control of nitrogen uptake. J Crop Improv 15: 25–51

Blesh J, Martin A (2018) Functional traits in cover crop mixtures: biological nitrogen fixation and multifunctionality. J Appl Ecol 55:38–48

Bodner G, Himmelbauer M, Loiskandl W, Kaul HP (2010) Improved evaluation of cover crop species by growth and root factors. Agron Sustain Dev 30:455–464

Bonfante P, Anca IA (2009) Plants, mycorrhizal fungi, and bacteria: a network of interactions. Annu Rev Microbiol 63:363–383

Bourgault M, Lamb P, McPhee K, McGee RJ, Vandenberg A, Warkentin T (2022) Genotypic variability in root length in pea (*Pisum sativum* L.) and lentil (*Lens culinaris* Medik.) cultivars in a semi-arid environment based on mini-rhizotron image capture. Plant Phenome J 5:e20037

Bowles TM, Jackson LE, Loeher M, Cavagnaro TR, Nuñez M (2017) Ecological intensification and arbuscular mycorrhizas: a meta-analysis of tillage and cover crop effects. J Appl Ecol 54:1785–1793

Bowles TM, Mooshammer M, Socolar Y, Calderón F, Cavigelli MA, Culman SW, Deen W, Drury CF, Garcia AGY, Gaudin ACM, Harkcom WS, Lehman RM, Osborne SL, Robertson GP, Salerno J, Schmer MR, Strock J, Grandy AS (2020) Long-term evidence shows that crop-rotation diversification increases agricultural resilience to adverse growing conditions in North America. One Earth 2:284–293

Bucking H, Mensah JA, Fellbaum CR (2016) Common mycorrhizal networks and their effect on the bargaining power of the fungal partner in the arbuscular mycorrhizal symbiosis. Commun Integr Biol 9:e1107684

Buysens C, Dupre de Boulois H, Declerck S (2015) Do fungicides used to control Rhizoctonia solani impact the non-target arbuscular mycorrhizal fungus Rhizophagus irregularis? Mycorrhiza 25:277–288

Cameron DD, Neal AL, van Wees SC, Ton J (2013) Mycorrhiza-induced resistance: more than the sum of its parts? Trends Plant Sci 18:539–545

Canarini A, Kaiser C, Merchant A, Richter A, Wanek W (2019) Root exudation of primary metabolites: mechanisms and their roles in plant responses to environmental stimuli. Front Plant Sci 10:157

Carminati A, Benard P, Ahmed MA, Zarebanadkouki M (2017) Liquid bridges at the root-soil interface. Plant Soil 417:1–15

Carvalhais LC, Dennis PG, Badri DV, Kidd BN, Vivanco JM, Schenk PM (2015) Linking Jasmonic acid signaling, root exudates, and rhizosphere microbiomes. Mol Plant-Microbe Interact 28: 1049–1058

Chandrasekaran M, Paramasivan M (2022) Arbuscular mycorrhizal fungi and antioxidant enzymes in ameliorating drought stress: a meta-analysis. Plant Soil 480:295–303

Chatterton S, Harding MW, Bowness R, McLaren DL, Banniza S, Gossen BD (2019) Importance and causal agents of root rot on field pea and lentil on the Canadian prairies, 2014–2017. Can J Plant Pathol 41:98–114

Chaudhary VB, Nolimal S, Sosa-Hernandez MA, Egan C, Kastens J (2020) Trait-based aerial dispersal of arbuscular mycorrhizal fungi. New Phytol 228:238

Chen EC, Mathieu S, Hoffrichter A, Sedzielewska-Toro K, Peart M, Pelin A, Ndikumana S, Ropars J, Dreissig S, Fuchs J, Brachmann A, Corradi N (2018) Single nucleus sequencing reveals evidence of inter-nucleus recombination in arbuscular mycorrhizal fungi. elife 7:e39813

Cheng L, Booker FL, Tu C, Burkey KO, Zhou L, Shew DH, Rufty TW, Hu S (2012) Arbuscular mycorrhizal fungi increase organic carbon decomposition under elevated CO_2. Science 337: 1084–1087

Choi HS, Cho HT (2019) Root hairs enhance Arabidopsis seedling survival upon soil disruption. Sci Rep 9:11181

Cong WF, Hoffland E, Li L, Janssen BH, van der Werf W (2015) Intercropping affects the rate of decomposition of soil organic matter and root litter. Plant Soil 391:399–411

Corre-Hellou G, Dibet A, Hauggaard-Nielsen H, Crozat Y, Gooding M, Ambus P, Dahlmann C, von Fragstein P, Pristeri A, Monti M, Jensen ES (2011) The competitive ability of pea-barley intercrops against weeds and the interactions with crop productivity and soil N availability. Field Crop Res 122:264–272

Correia M, Heleno R, da Silva LP, Costa JM, Rodriguez-Echeverria S (2019) First evidence for the joint dispersal of mycorrhizal fungi and plant diaspores by birds. New Phytol 222:1054–1060

Cutforth HW, Angadi SV, McConkey BG, Miller PR, Ulrich D, Gulden R, Volkmar KM, Entz MH, Brandt SA (2013) Comparing rooting characteristics and soil water withdrawal patterns of wheat with alternative oilseed and pulse crops grown in the semiarid Canadian prairie. Can J Soil Sci 93:147–160

Dandurand LM, Mosher RD, Knudsen GR (2000) Combined effects of Brassica napus seed meal and Trichoderma harzianum on two soilborne plant pathogens. Can J Microbiol 46:1051–1057

Davison J, Moora M, Jairus T, Vasar M, Öpik M, Zobel M (2016) Hierarchical assembly rules in arbuscular mycorrhizal (AM) fungal communities. Soil Biol Biochem 97:63–70

Davison J, Moora M, Opik M, Adholeya A, Ainsaar L, Ba A, Burla S, Diehiou A, Hiiesalu I, Jairus T, Johnson NK, Koorem K, Kochar M, Ndiaye C, Partel M, Reier U, Saks U, Singh R, Vasar M, Zobel M (2015) Global assessment of arbuscular mycorrhizal fungus diversity reveals very low endemism. Science 349:970–973

de Dorlodot S, Forster B, Pages L, Price A, Tuberosa R, Draye X (2007) Root system architecture: opportunities and constraints for genetic improvement of crops. Trends Plant Sci 12:474–481

de Novais CB, Avio L, Giovannetti M, de Faria SM, Siqueira JO, Sbrana C (2019) Interconnectedness, length and viability of arbuscular mycorrhizal mycelium as affected by selected herbicides and fungicides. Appl Soil Ecol 143:144–152

de Santana AS, Cavalcante UMT, de Sa Barreto Sampaio EV, Maia LC (2014) Production, storage and costs of inoculum of arbuscular mycorrhizal fungi (AMF). Rev Bras Bot 37:159–165

de Souza A, Moraes JM, Zanchi CS, Pires GC, Moretti CF, Barbosa MV, Silva AO, Pacheco LP, Carbone Carneiro MA, Oliveira RL, Kemmelmeier K, Souza ED (2019) Arbuscular mycorrhizal fungi in integrated crop livestock systems with intercropping in the pasture phase in the Cerrado. Rhizosphere 11:100165

Desgroux A, Baudais VN, Aubert V, Le Roy G, de Larambergue H, Miteul H, Aubert G, Boutet G, Duc G, Baranger A, Burstin J, Manzanares-Dauleux M, Pilet-Nayel ML, Bourion V (2017) Comparative genome-wide-association mapping identifies common loci controlling root system architecture and resistance to Aphanomyces euteiches in pea. Front Plant Sci 8:2195

Dingha BN, Omaliko PC, Amoah BA, Jackai LE, Shrestha D (2021) Evaluation of cowpea (Vigna unguiculata) in an intercropping system as pollinator enhancer for increased crop yield. Sustainability 13:9612

Doilom M, Guo JW, Phookamsak R, Mortimer PE, Karunarathna SC, Dong W, Liao CF, Yan K, Pem D, Suwannarach N, Promputtha I, Lumyong S, Xu JC (2020) Screening of phosphate-solubilizing fungi from air and soil in Yunnan, China: four novel species in aspergillus, Gongronella, Penicillium, and Talaromyces. Front Microbiol 11:585215

Drigo B, Pijl AS, Duyts H, Kielak AM, Gamper HA, Houtekamer MJ, Boschker HT, Bodelier PL, Whiteley AS, van Veen JA, Kowalchuk GA (2010) Shifting carbon flow from roots into associated microbial communities in response to elevated atmospheric CO_2. Proc Natl Acad Sci U S A 107:10938–10942

Egan C, Li D-W, Klironomos J (2014) Detection of arbuscular mycorrhizal fungal spores in the air across different biomes and ecoregions. Fungal Ecol 12:26–31

Eissenstat DM, Kucharski JM, Zadworny M, Adams TS, Koide RT (2015) Linking root traits to nutrient foraging in arbuscular mycorrhizal trees in a temperate forest. New Phytol 208:114–124
Evelin H, Devi TS, Gupta S, Kapoor R (2019) Mitigation of salinity stress in plants by arbuscular mycorrhizal Symbiosis: current understanding and new challenges. Front Plant Sci 10:470
Faghihinia M, Jansa J, Halverson LJ, Staddon PL (2022) Hyphosphere microbiome of arbuscular mycorrhizal fungi: a realm of unknowns. Biol Fertil Soils 59:17–34
Fahey JW, Zalcmann AT, Talalay P (2001) The chemical diversity and distribution of glucosinolates and isothiocyanates among plants. Phytochemistry 56:5–51
Fahlgren N, Gehan MA, Baxter I (2015) Lights, camera, action: high-throughput plant phenotyping is ready for a close-up. Curr Opin Plant Biol 24:93–99
Falk KG, Jubery TZ, Mirnezami SV, Parmley KA, Sarkar S, Singh A, Ganapathysubramanian B, Singh AK (2020) Computer vision and machine learning enabled soybean root phenotyping pipeline. Plant Methods 16:5
Fan J, McConkey B, Wang H, Janzen H (2016) Root distribution by depth for temperate agricultural crops. Field Crop Res 189:68–74
Fenster TLD, LaCanne CE, Pecenka JR, Schmid RB, Bredeson MM, Busenitz KM, Michels AM, Welch KD, Lundgren JG (2021) Defining and validating regenerative farm systems using a composite of ranked agricultural practices. F1000Res 10:115
Fernández-Aparicio M, Amri M, Kharrat M, Rubiales D (2010) Intercropping reduces Mycosphaerella pinodes severity and delays upward progress on the pea plant. Crop Prot 29: 744–750
Fernández-Aparicio M, Shtaya MJY, Emeran AA, Allagui MB, Kharrat M, Rubiales D (2011) Effects of crop mixtures on chocolate spot development on faba bean grown in mediterranean climates. Crop Prot 30:1015–1023
Figueiredo AF, Boy J, Guggenberger G (2021) Common mycorrhizae network: a review of the theories and mechanisms behind underground interactions. Front Fungal Biol 2:48
Floss DS, Gomez SK, Park HJ, MacLean AM, Muller LM, Bhattarai KK, Levesque-Tremblay V, Maldonado-Mendoza IE, Harrison MJ (2017) A transcriptional program for Arbuscule degeneration during AM Symbiosis is regulated by MYB1. Curr Biol 27:1206–1212
Freschet GT, Roumet C, Comas LH, Weemstra M, Bengough AG, Rewald B, Bardgett RD, De Deyn GB, Johnson D, Klimesova J, Lukac M, McCormack ML, Meier IC, Pages L, Poorter H, Prieto I, Wurzburger N, Zadworny M, Bagniewska-Zadworna A, Blancaflor EB, Brunner I, Gessler A, Hobbie SE, Iversen CM, Mommer L, Picon-Cochard C, Postma JA, Rose L, Ryser P, Scherer-Lorenzen M, Soudzilovskaia NA, Sun T, Valverde-Barrantes OJ, Weigelt A, York LM, Stokes A (2021) Root traits as drivers of plant and ecosystem functioning: current understanding, pitfalls and future research needs. New Phytol 232:1123–1158
Freschet GT, Violle C, Bourget MY, Scherer-Lorenzen M, Fort F (2018) Allocation, morphology, physiology, architecture: the multiple facets of plant above- and below-ground responses to resource stress. New Phytol 219:1338–1352
Fujita M, Kusajima M, Fukagawa M, Okumura Y, Nakajima M, Akiyama K, Asami T, Yoneyama K, Kato H, Nakashita H (2022) Response of tomatoes primed by mycorrhizal colonization to virulent and avirulent bacterial pathogens. Sci Rep 12:4686
Gan Y, Liu L, Wang X, Fird G (2011) Vertical distribution profiles and temporal growth patterns of roots in selected oilseeds, pulses and spring wheat. Crop Pasture Sci 62:457–466
Gao D, Pan X, Khashi M, Rahman U, Zhou X, Wu F (2021) Common mycorrhizal networks benefit to the asymmetric interspecific facilitation via K exchange in an agricultural intercropping system. Biol Fertil Soils 57:959–971
Gavito ME, Jakobsen I, Mikkelsen TN, Mora F (2019) Direct evidence for modulation of photosynthesis by an arbuscular mycorrhiza-induced carbon sink strength. New Phytol 223:896–907
Goicoechea N (2020) Mycorrhizal fungi as bioprotectors of crops against Verticillium wilt-A hypothetical scenario under changing environmental conditions. Plants (Basel) 9:1468

Gorim LY, Vandenberg A (2017) Root traits, nodulation and root distribution in soil for five wild lentil species and Lens culinaris (Medik.) grown under well-watered conditions. Front Plant Sci 8:1632

Gowton CM, Cabra-Arias C, Carrillo J (2021) Intercropping with peppermint increases ground dwelling insect and pollinator abundance and decreases Drosophila suzukii in fruit. Front Sustain Food Syst 5:700842

Gryndler M, Larsen J, Hrselova H, Rezacova V, Gryndlerova H, Kubat J (2006) Organic and mineral fertilization, respectively, increase and decrease the development of external mycelium of arbuscular mycorrhizal fungi in a long-term field experiment. Mycorrhiza 16:159–166

Guo Z, Dong Y, Dong K, Zhu J, Ma L (2020) Effects of nitrogen management and intercropping on faba bean chocolate spot disease development. Crop Prot 127:104972

Gutjahr C, Parniske M (2013) Cell and developmental biology of arbuscular mycorrhiza symbiosis. Annu Rev Cell Dev Biol 29:593–617

Gutjahr C, Parniske M (2017) Cell biology: control of partner lifetime in a plant-fungus relationship. Curr Biol 27:R420–R423

Haug B, Messmer MM, Enjalbert J, Goldringer I, Forst E, Flutre T, Mary-Huard T, Hohmann P (2021) Advances in breeding for mixed cropping—incomplete factorials and the producer/associate concept. Front Plant Sci 11:620400

Heckman DS, Geiser DM, Eidell BR, Stauffer RL, Kardos NL, Hedges SB (2001) Molecular evidence for the early colonization of land by fungi and plants. Science 293:1129–1133

Heil M, Hilpert A, Kaiser W, Linsenmair KE (2000) Reduced growth and seed set following chemical induction of pathogen defence: does systemic acquired resistance (SAR) incur allocation costs? J Ecol 88:645–654

Heřmanská A, Středa T, Chloupek O (2014) Improved wheat grain yield by a new method of root selection. Agron Sustain Dev 35:195–202

Hill EM, Robinson LA, Abdul-Sada A, Vanbergen AJ, Hodge A, Hartley SE (2018) Arbuscular mycorrhizal fungi and plant chemical defence: effects of colonisation on aboveground and belowground metabolomes. J Chem Ecol 44:198–208

Holátko J, Brtnický M, Kučerík J, Kotianová M, Elbl J, Kintl A, Kynický J, Benada O, Datta R, Jansa J (2021) Glomalin—truths, myths, and the future of this elusive soil glycoprotein. Soil Biol Biochem 153:108116

Hossain S, Bergkvist G, Berglund K, Glinwood R, Kabouw P, Mårtensson A, Persson P (2014) Concentration- and time-dependent effects of isothiocyanates produced from brassicaceae shoot tissues on the pea root rot pathogen aphanomyces euteiches. J Agric Food Chem 62:4584–4591

Hurd EA (1964) Root study of three wheat varieties and their resistance to drought and damage by soil cracking. Can J Plant Sci 44:240–248

Hurd EA (1968) Growth of roots of seven varieties of spring wheat at high and low moisture levels. Agron J 60:201–205

Hurd EA (1974) 1974 phenotype and drought tolerance in wheat. Agric Meteorol 14:39–55

Huss CP, Holmes KD, Blubaugh CK (2022) Benefits and risks of intercropping for crop resilience and Pest management. J Econ Entomol 115:1350–1362

Imran A, Hakim S, Tariq M, Nawaz MS, Laraib I, Gulzar U, Hanif MK, Siddique MJ, Hayat M, Fraz A, Ahmad M (2021) Diazotrophs for lowering nitrogen pollution crises: looking deep into the roots. Front Microbiol 12:637815

Jamil A, Ashraf S (2021) Impacts of agronomic practices in the management of fusarium wilt of chickpea. Australas Plant Pathol 50:441–450

Janeeshma E, Puthur JT (2020) Direct and indirect influence of arbuscular mycorrhizae on enhancing metal tolerance of plants. Arch Microbiol 202:1–16

Jeewani PH, Luo Y, Yu G, Fu Y, He X, Van Zwieten L, Liang C, Kumar A, He Y, Kuzyakov Y, Qin H, Guggenberger G, Xu J (2021) Arbuscular mycorrhizal fungi and goethite promote carbon sequestration via hyphal-aggregate mineral interactions. Soil Biol Biochem 162:108417

Jeudy C, Adrian M, Baussard C, Bernard C, Bernaud E, Bourion V, Busset H, Cabrera-Bosquet L, Cointault F, Han S, Lamboeuf M, Moreau D, Pivato B, Prudent M, Trouvelot S, Truong HN,

Vernoud V, Voisin AS, Wipf D, Salon C (2016) RhizoTubes as a new tool for high throughput imaging of plant root development and architecture: test, comparison with pot grown plants and validation. Plant Methods 12:31

Jiang F, Zhang L, Zhou J, George TS, Feng G (2021) Arbuscular mycorrhizal fungi enhance mineralisation of organic phosphorus by carrying bacteria along their extraradical hyphae. New Phytol 230:304–315

Jin H, Germida JJ, Walley FL (2013) Suppressive effects of seed-applied fungicides on arbuscular mycorrhizal fungi (AMF) differ with fungicide mode of action and AMF species. Appl Soil Ecol 72:22–30

Johnson D, Gilbert L (2015) Interplant signalling through hyphal networks. New Phytol 205:1448–1453

Jung SC, Martinez-Medina A, Lopez-Raez JA, Pozo MJ (2012) Mycorrhiza-induced resistance and priming of plant defenses. J Chem Ecol 38:651–664

Kapayou DG, Herrighty EM, Hill CG, Camacho VC, Nair A, Winham DM, McDaniel MD (2023) Reuniting the three sisters: collaborative science with native growers to improve soil and community health. Agric Hum Values 40:65–82

Kembel SW, Cahill JF Jr (2011) Independent evolution of leaf and root traits within and among temperate grassland plant communities. PLoS One 6:e19992

Khan ZR, Midega CAO, Amudavi DM, Hassanali A, Pickett JA (2008) On-farm evaluation of the 'push-pull' technology for the control of stemborers and striga weed on maize in western Kenya. Field Crop Res 106:224–233

Kiers ET, Duhamel M, Beesetty Y, Mensah JA, Franken O, Verbruggen E, Fellbaum CR, Kowalchuk GA, Hart MM, Bago A, Palmer TM, West SA, Vandenkoornhuyse P, Jansa J, Buchking H (2011) Reciprocal rewards stabilize cooperation in the mycorrhizal symbiosis. Science 333:880–882

Kivlin SN, Hawkes CV, Treseder KK (2011) Global diversity and distribution of arbuscular mycorrhizal fungi. Soil Biol Biochem 43:2294–2303

Kokkoris V, Stefani F, Dalpe Y, Dettman J, Corradi N (2020) Nuclear dynamics in the arbuscular mycorrhizal fungi. Trends Plant Sci 25:765

Kuijken RC, van Eeuwijk FA, Marcelis LF, Bouwmeester HJ (2015) Root phenotyping: from component trait in the lab to breeding. J Exp Bot 66:5389–5401

LaCanne CE, Lundgren JG (2018) Regenerative agriculture: merging farming and natural resource conservation profitably. PeerJ 6:e4428

Li H, Bolscher T, Winnick M, Tfaily MM, Cardon ZG, Keiluweit M (2021) Simple plant and microbial exudates destabilize mineral-associated organic matter via multiple pathways. Environ Sci Technol 55:3389–3398

Li L, Chen F, Xing G (2022) Effects of fertilizer level and intercropping planting pattern with corn on the yield-related traits and insect Community of Soybean. Agronomy 12:3080

Liao D, Zhang C, Li H, Lambers H, Zhang F (2020) Changes in soil phosphorus fractions following sole cropped and intercropped maize and faba bean grown on calcareous soil. Plant Soil 448: 587–601

Liebman M, Dyck E (1993) Crop rotation and intercropping strategies for weed management. Ecol Appl 3:92–122

Lin X, Feng Y, Zhang H, Chen R, Wang J, Zhang J, Chu H (2012) Long-term balanced fertilization decreases arbuscular mycorrhizal fungal diversity in an arable soil in North China revealed by 454 pyrosequencing. Environ Sci Technol 46:5764–5771

Liu L, Gan Y, Bueckert R, Van Rees K (2011) Rooting systems of oilseed and pulse crops I: temporal growth patterns across the plant developmental periods. Field Crop Res 122:256–263

Liu L, Gan Y, Bueckert R, Van Rees K, Warkentin T (2010) Fine root distributions in oilseed and pulse crops. Crop Sci 50:222–226

Liu J, Li Y, Han C, Yang D, Yang J, Cade-Menun BJ, Chen Y, Sui P (2022) Maize-soybean intercropping facilitates chemical and microbial transformations of phosphorus fractions in a calcareous soil. Front Microbiol 13:1028969

Liu H, Wu Y, Xu H, Ai Z, Zhang J, Liu G, Xue S (2021) Mechanistic understanding of interspecific interaction between a C4 grass and a C3 legume via arbuscular mycorrhizal fungi, as influenced by soil phosphorus availability using a (13) C and (15) N dual-labelled organic patch. Plant J 108:183–196

Ljubotina MK, Cahill JF Jr (2019) Effects of neighbour location and nutrient distributions on root foraging behaviour of the common sunflower. Proc Biol Sci 286:20190955

Lutzoni F, Nowak MD, Alfaro ME, Reeb V, Miadlikowska J, Krug M, Arnold AE, Lewis LA, Swofford DL, Hibbett D, Hilu K, James TY, Quandt D, Magallon S (2018) Contemporaneous radiations of fungi and plants linked to symbiosis. Nat Commun 9:5451

Lynch J (1995) Root architecture and plant productivity. Plant Physiol 109:7–13

Lynch JP (2013) Steep, cheap and deep: an ideotype to optimize water and N acquisition by maize root systems. Ann Bot 112:347–357

Ma Y, Rajkumar M, Oliveira RS, Zhang C, Freitas H (2019) Potential of plant beneficial bacteria and arbuscular mycorrhizal fungi in phytoremediation of metal-contaminated saline soils. J Hazard Mater 379:120813

Martens JRT, Entz MH, Wonneck MD (2015) Review: redesigning Canadian prairie cropping systems for profitability, sustainability, and resilience. Can J Plant Sci 95:1049–1072

Martins SM, Brito GG, Gonçalves WDC, Tripode BMD, Lartaud M, Duarte JB, Morello CDL, Giband M (2020) PhenoRoots: an inexpensive non-invasive phenotyping system to assess the variability of the root system architecture. Sci Agric 77:1

McGrail R, Van Sanford D, McNear D (2020) Trait-based root phenotyping as a necessary tool for crop selection and improvement. Agronomy 10:1328

Messaoudi H, Gérard F, Dokukin P, Djamai H, Rebouh NY, Latati M (2020) Effects of intercropping on field-scale phosphorus acquisition processes in a calcareous soil. Plant Soil 449:331–341

Miller MH, McGonigle TP, Addy HD (1995) Functional ecology of vesicular arbuscular mycorrhizas as influenced by phosphate fertilization and tillage in an agricultural ecosystem. Crit Rev Biotechnol 15:241–255

Moore VM, Peters T, Schlautman B, Brummer EC (2023) Toward plant breeding for multicrop systems. Proc Natl Acad Sci U S A 120:120

Moore VM, Schlautman B, Fei SZ, Roberts LM, Wolfe M, Ryan MR, Wells S, Lorenz AJ (2022) Plant breeding for intercropping in temperate field crop systems: a review. Front Plant Sci 13: 843065

Moutier N, Baranger A, Fall S, Hanocq E, Marget P, Floriot M, Gauffreteau A (2022) Mixing ability of intercropped wheat varieties: stability across environments and tester legume species. Front Plant Sci 13:1495

Neugschwandtner RW, Kaul HP (2014) Sowing ratio and N fertilization affect yield and yield components of oat and pea in intercrops. Field Crop Res 155:159–163

Newton P, Civita N, Frankel-Goldwater L, Bartel K, Johns C (2020) What is regenerative agriculture? A review of scholar and practitioner definitions based on processes and outcomes. Front Sustain Food syst 4:194

Ocimati W, Tusiime G, Opio F, Ugen MA, Buruchara R (2017) Sorghum (Sorghum bicolor) as a bean intercrop or rotation crop contributes to the survival of bean root rot pathogens and perpetuation of bean root rots. Plant Pathol 66:1480–1486

Orine D, Defossez E, Vergara F, Uthe H, van Dam NM, Rasmann S (2022) Arbuscular mycorrhizal fungi prevent the negative effect of drought and modulate the growth-defence trade-off in tomato plants. J Sustain Agric Environ 1:177–190

Otfinowski R, Coffey V (2020) Can root traits predict communities of soil nematodes in restored northern prairies? Plant Soil 453:459–471

Paez-Garcia A, Motes CM, Scheible WR, Chen R, Blancaflor EB, Monteros MJ (2015) Root traits and phenotyping strategies for plant improvement. Plants (Basel) 4:334–355

Pankou C, Lithourgidis A, Menexes G, Dordas C (2022) Importance of selection of cultivars in wheat–pea intercropping systems for high productivity. Agronomy 12:2367

Parniske M (2008) Arbuscular mycorrhiza: the mother of plant root endosymbioses. Nat Rev Microbiol 6:763–775

Parrott L (2010) Measuring ecological complexity. Ecol Indic 10:1069–1076

Pausch J, Kuzyakov Y (2018) Carbon input by roots into the soil: quantification of rhizodeposition from root to ecosystem scale. Glob Chang Biol 24:1–12

Pedersen A, Zhang K, Thorup-Kristensen K, Jensen LS (2009) Modelling diverse root density dynamics and deep nitrogen uptake—a simple approach. Plant Soil 326:493–510

Pelzer E, Bazot M, Makowski D, Corre-Hellou G, Naudin C, Al Rifaï M, Baranger E, Bedoussac L, Biarnès V, Boucheny P, Carrouée B, Dorvillez D, Foissy D, Gaillard B, Guichard L, Mansard MC, Omon B, Prieur L, Yvergniaux M, Justes E, Jeuffroy MH (2012) Pea-wheat intercrops in low-input conditions combine high economic performances and low environmental impacts. Eur J Agron 40:39–53

Perez-Lamarque B, Opik M, Maliet O, Afonso Silva AC, Selosse MA, Martos F, Morlon H (2022) Analysing diversification dynamics using barcoding data: the case of an obligate mycorrhizal symbiont. Mol Ecol 31:3496–3512

Pierre JF, Latournerie-Moreno L, Garruña R, Jacobsen KL, Laboski CAM, Us-Santamaría R, Ruiz-Sánchez E (2022) Effect of maize–legume intercropping on maize physio-agronomic parameters and beneficial insect abundance. Sustainability (Switzerland) 14:12385

Pires GC, Eloá de Lima M, Zanchi CS, Moretti de Freitas C, Andrade de Souza JM, Andrea de Camargo T, Pacheco LP, Wruck FJ, Carbone Carneiro MA, Kemmelmeier K, de Moraes A, Damacena de Souza E (2021) Arbuscular mycorrhizal fungi in the rhizosphere of soybean in integrated crop livestock systems with intercropping in the pasture phase. Rhizosphere 17: 100270

Plett JM, Solomon J, Snijders F, Marlow-Conway J, Plett KL, Bithell SL (2021) Order of microbial succession affects rhizobia-mediated biocontrol efforts against Phytophthora root rot. Microbiol Res 242:126628

Podgórska-Lesiak M, Sobkowicz P (2013) Prevention of pea lodging by intercropping barley with peas at different nitrogen fertilization levels. Field Crop Res 149:95–104

Porcel R, Aroca R, Ruiz-Lozano JM (2011) Salinity stress alleviation using arbuscular mycorrhizal fungi. A review. Agron Sustain Dev 32:181–200

Pourali S, Aghayari F, Ardakani MR, Paknejad F, Golzardi F (2023) Benefits from Intercropped Forage Sorghum–red clover under drought stress conditions. Gesunde Pflanzen 75:1769

Powell JR, Bennett AE (2016) Unpredictable assembly of arbuscular mycorrhizal fungal communities. Pedobiologia 59:11–15

Pozo MJ, Azcon-Aguilar C (2007) Unraveling mycorrhiza-induced resistance. Curr Opin Plant Biol 10:393–398

Pozo MJ, Cordier C, Dumas-Gaudot E, Gianinazzi S, Barea JM, Azcon-Aguilar C (2002) Localized versus systemic effect of arbuscular mycorrhizal fungi on defence responses to Phytophthora infection in tomato. J Exp Bot 53:525–534

Qiao X, Guo X, Li A (2020) Common mycorrhizal networks contribute to overyielding in faba bean/coix intercropping systems. Agron J 112:2598–2607

Ranganathan J, Waite R, Searchinger T (2020) Regenerative agriculture: good for soil health but limited potential to mitigate climate change. World Resources Institute, Washington, DC

Rao S, Armstrong R, Silva-Perez V, Tefera AT, Rosewarne GM (2021) Pulse root Ideotype for water stress in temperate cropping system. Plants (Basel) 10:692

Raseduzzaman M, Jensen ES (2017) Does intercropping enhance yield stability in arable crop production? A meta-analysis. Eur J Agron 91:25–33

Redecker D, Kodner R, Graham LE (2000) Glomalean fungi from the Ordovician. Science 289: 1920–1921

Remy W, Tayor TN, Hass H, Kerp H (1994) Four hundred-million-year old vesicular arbuscular mycorrhizae. Proc Natl Acad Sci U S A 91:11841–11843

Rich SM, Wasson AP, Richards RA, Katore T, Prashar R, Chowdhary R, Saxena DC, Mamrutha HM, Zwart A, Misra SC, Sai Prasad SV, Chatrath R, Christopher J, Watt M (2016) Wheats developed for high yield on stored soil moisture have deep vigorous root systems. Funct Plant Biol 43:173–188

Rillig MC, Aguilar-Trigueros CA, Bergmann J, Verbruggen E, Veresoglou SD, Lehmann A (2015) Plant root and mycorrhizal fungal traits for understanding soil aggregation. New Phytol 205: 1385–1388

Rincon C, Droh G, Villard L, Masclaux FG, N'Guetta A, Zeze A, Sanders IR (2021) Hierarchical spatial sampling reveals factors influencing arbuscular mycorrhizal fungus diversity in Cote d'Ivoire cocoa plantations. Mycorrhiza 31:289–300

Rodale R (1983) Breaking new ground: the search for sustainable agriculture. Futurist 17:15–20

Rongsawat T, Peltier JB, Boyer JC, Very AA, Sentenac H (2021) Looking for root hairs to overcome poor soils. Trends Plant Sci 26:83–94

Ropars J, Toro KS, Noel J, Pelin A, Charron P, Farinelli L, Marton T, Kruger M, Fuchs J, Brachmann A, Corradi N (2016) Evidence for the sexual origin of heterokaryosis in arbuscular mycorrhizal fungi. Nat Microbiol 1:16033

Rosner K, Bodner G, Hage-Ahmed K, Steinkellner S (2018) Long-term soil tillage and cover cropping affected arbuscular mycorrhizal fungi, nutrient concentrations, and yield in sunflower. Agron J 110:2664–2672

Ruth B, Khalvati M, Schmidhalter U (2011) Quantification of mycorrhizal water uptake via high-resolution on-line water content sensors. Plant Soil 342:459–468

Säle V, Aguilera P, Laczko E, Mäder P, Berner A, Zihlmann U, van der Heijden MGA, Oehl F (2015) Impact of conservation tillage and organic farming on the diversity of arbuscular mycorrhizal fungi. Soil Biol Biochem 84:38–52

Salmeron-Santiago IA, Martinez-Trujillo M, Valdez-Alarcon JJ, Pedraza-Santos ME, Santoyo G, Pozo MJ, Chavez-Barcenas AT (2021) An updated review on the modulation of carbon partitioning and allocation in arbuscular mycorrhizal plants. Microorganisms 10:75

Salomon MJ, Demarmels R, Watts-Williams SJ, McLaughlin MJ, Kafle A, Ketelsen C, Soupir A, Bücking H, Cavagnaro TR, van der Heijden MGA (2022) Global evaluation of commercial arbuscular mycorrhizal inoculants under greenhouse and field conditions. Appl Soil Ecol 169: 104225

Sarto MVM, Borges WLB, Sarto JRW, Pires CAB, Rice CW, Rosolem CA (2020) Soil microbial community and activity in a tropical integrated crop-livestock system. Appl Soil Ecol 145: 103350

Schausberger P, Peneder S, Jurschik S, Hoffman D (2012) Mycorrhiza changes plant volatiles to attract spider mite enemies. Funct Ecol 26:441–449

Schoeny A, Jumel S, Rouault F, Lemarchand E, Tivoli B (2010) Effect and underlying mechanisms of pea-cereal intercropping on the epidemic development of ascochyta blight. Eur J Plant Pathol 126:317–331

Schouteden N, De Waele D, Panis B, Vos CM (2015) Arbuscular mycorrhizal fungi for the biocontrol of plant-parasitic nematodes: a review of the mechanisms involved. Front Microbiol 6:1280

Schweiger R, Baier MC, Muller C (2014) Arbuscular mycorrhiza-induced shifts in foliar metabolism and photosynthesis mirror the developmental stage of the symbiosis and are only partly driven by improved phosphate uptake. Mol Plant-Microbe Interact 27:1403–1412

Sharma S, Carena MJ (2016) BRACE: A method for high throughput maize phenotyping of root traits for short-season drought tolerance. Crop Sci 56:2996–3004

Simard SW, Beiler KJ, Bingham MA, Deslippe JR, Philip LJ, Teste FP (2012) Mycorrhizal networks: mechanisms, ecology and modelling. Fungal Biol Rev 26:39–60

Simard SW, Durall DM (2004) Mycorrhizal networks: a review of their extent, function, and importance. Can J Bot 82:1140–1165

Smith S, De Smet I (2012) Root system architecture: insights from Arabidopsis and cereal crops. Philos Trans R Soc Lond Ser B Biol Sci 367:1441–1452

Smith SE, Read DJ (2008) Mycorrhizal symbiosis, 3rd edn. Academic, Cambridge, MA

Smolinska U, Morra MJ, Knudsen GR, Brown PD (1997) Toxicity of glucosinolate degradation products from Brassica napus seed meal toward Aphanomyces euteiches f. sp. pisi. Phytopathology 87:77–82

Song Y, Chen D, Lu K, Sun Z, Zeng R (2015a) Enhanced tomato disease resistance primed by arbuscular mycorrhizal fungus. Front Plant Sci 6:786

Song YY, Simard SW, Carroll A, Mohn WW, Zeng RS (2015b) Defoliation of interior Douglas-fir elicits carbon transfer and stress signalling to ponderosa pine neighbors through ectomycorrhizal networks. Sci Rep 5:8495

Song YY, Ye M, Li C, He X, Zhu-Salzman K, Wang RL, Su YJ, Luo SM, Zeng RS (2014) Hijacking common mycorrhizal networks for herbivore-induced defence signal transfer between tomato plants. Sci Rep 4:3915

Song YY, Ye M, Li CY, Wang RL, Wei XC, Luo SM, Zeng RS (2013) Priming of anti-herbivore defense in tomato by arbuscular mycorrhizal fungus and involvement of the jasmonate pathway. J Chem Ecol 39:1036–1044

Song YY, Zeng RS, Xu JF, Li J, Shen X, Yihdego WG (2010) Interplant communication of tomato plants through underground common mycorrhizal networks. PLoS One 5:e13324

Spagnoletti FN, Leiva M, Chiocchio V, Lavado RS (2018) Phosphorus fertilization reduces the severity of charcoal rot (Macrophomina phaseolina) and the arbuscular mycorrhizal protection in soybean. J Plant Nutr Soil Sci 181:855–860

Tao L, Ahmad A, Roode JC, Hunter MD, Heijden M (2016) Arbuscular mycorrhizal fungi affect plant tolerance and chemical defences to herbivory through different mechanisms. J Ecol 104: 561–571

Tavoletti S, Merletti A (2022) A comprehensive approach to evaluate durum Wheat–Faba bean mixed crop performance. Front Plant Sci 13:733116

Toljander JF, Artursson V, Paul LR, Jansson JK, Finlay RD (2006) Attachment of different soil bacteria to arbuscular mycorrhizal fungal extraradical hyphae is determined by hyphal vitality and fungal species. FEMS Microbiol Lett 254:34–40

Tracy SR, Nagel KA, Postma JA, Fassbender H, Wasson A, Watt M (2020) Crop improvement from phenotyping roots: highlights reveal expanding opportunities. Trends Plant Sci 25:105–118

Uesugi R, Konishi-Furihata R, Tabuchi K, Yoshimura H, Shimoda T (2023) Predacious natural enemies associated with suppression of onion Thrips, Thrips tabaci (Thysanoptera: Thripidae), in intercropped onion-barley agroecosystems. Environ Entomol 52:183–196

Vandermeer J, Van Noordwijk M, Anderson J, Ong C, Perfecto I (1998) Global change and multi-species agroecosystems: concepts and issues. Agric Ecosyst Environ 67:1–22

Vargas Palacio AG (2021) Exploring the genetic resources of *lens* and *rhizobium* to improve the biological nitrogen fixation (BNF) ability in the lentil crop. University of Saskatchewan, Saskatoon

Verret V, Gardarin A, Pelzer E, Médiène S, Makowski D, Valantin-Morison M (2017) Can legume companion plants control weeds without decreasing crop yield? A meta-analysis. Field Crop Res 204:158–168

Waidmann S, Sarkel E, Kleine-Vehn J (2020) Same same, but different: growth responses of primary and lateral roots. J Exp Bot 71:2397–2411

Wang F, Shi N, Jiang R, Zhang F, Feng G (2016) In situ stable isotope probing of phosphate-solubilizing bacteria in the hyphosphere. J Exp Bot 67:1689–1701

Wang J, Song L, Gong X, Xu J, Li M (2020) Functions of Jasmonic acid in plant regulation and response to abiotic stress. Int J Mol Sci 21:1446

Weston LA, Duke SO (2003) Weed and crop allelopathy. Crit Rev Plant Sci 22:367–389

White JF, Kingsley KL, Verma SK, Kowalski KP (2018) Rhizophagy cycle: an oxidative process in plants for nutrient extraction from symbiotic microbes. Microorganisms 6:95

Whiteside MD, Werner GDA, Caldas VEA, Van't Padje A, Dupin SE, Elbers B, Bakker M, Wyatt GAK, Klein M, Hink MA, Postma M, Vaitla B, Noe R, Shimizu TS, West SA, Kiers ET (2019)

Mycorrhizal fungi respond to resource inequality by moving phosphorus from rich to poor patches across networks. Curr Biol 29:2043–2050
Wu W, Duncan RW, Ma B-L (2016) Quantification of canola root morphological traits under heat and drought stresses with electrical measurements. Plant Soil 415:229–244
Wu W, Ma BL (2016) A new method for assessing plant lodging and the impact of management options on lodging in canola crop production. Sci Rep 6:31890
Wu W, Shah F, Duncan RW, Ma BL (2020) Grain yield, root growth habit and lodging of eight oilseed rape genotypes in response to a short period of heat stress during flowering. Agric For Meteorol 287:107954
Xiao J, Dong Y, Yin X, Ren J, Tang L, Zheng Y (2019) Wheat growth is stimulated by interspecific competition after Faba bean attains its maximum growth rate. Crop Sci 59:293–306
Yang W, Li Y, Zhao Q, Guo Y, Dong Y (2022) Intercropping alleviated the phytotoxic effects of cinnamic acid on the root cell wall structural resistance of faba bean and reduced the occurrence of fusarium wilt. Physiol Plant 174:e13827
Yang L, Luo Y, Lu B, Zhou G, Chang D, Gao S, Zhang J, Che Z, Cao W (2023) Long-term maize and pea intercropping improved subsoil carbon storage while reduced greenhouse gas emissions. Agric Ecosyst Environ 349:108444
Yildirir G, Malar CM, Kokkoris V, Corradi N (2020) Parasexual and sexual reproduction in arbuscular mycorrhizal fungi: room for both. Trends Microbiol 28:517–519
Yu L, Zhang W, Geng Y, Liu K, Shao X (2022) Cooperation with arbuscular mycorrhizal fungi increases plant nutrient uptake and improves defenses against insects. Front Ecol Evol 10: 833389
Zhang X, Yan J, Khashi M, Rahman U, Wu F (2022) The impact of root exudates, volatile organic compounds, and common mycorrhizal networks on root system architecture in root-root interactions. J Plant Interact 17:685–694
Zhang Z, Yang W, Li Y, Zhao Q, Dong Y (2023) Wheat–faba bean intercropping can control fusarium wilt in faba bean under F. Commune and ferulic acid stress as revealed by histopathological analysis. Physiol Mol Plant Pathol 124:101965
Zhou Y, Chen C, Franck WL, Khan Q, Franck S, Crutcher FK, McVay K, McPhee K (2023) Intercropping chickpea–flax for yield and disease management. Agron J 115:726–743
Zobel RW, Waisel Y (2010) A plant root system architectural taxonomy: a framework for root nomenclature. Plant Biosyst 144:507–512

Chapter 14
Applications of Arbuscular Mycorrhizal Fungi for Sustainable Agricultural Systems

B. M. Herath, K. W. Y. R. Kalamulla, T. A. N. Mayadunna, M. A. E. G. Perera, I. K. Jayamanna, and P. N. Yapa

Abstract Arbuscular mycorrhizal fungi (AMF) represent a critical component of terrestrial ecosystems, forming symbiotic relationships with the majority of plant species. This mutualistic association has profound implications for sustainable agriculture, as it enhances nutrient uptake, improves plant health, and contributes to soil structure and fertility. This chapter provides an overview of the diverse applications of AMF in sustainable agricultural systems, highlighting their role in enhancing crop productivity, reducing environmental impacts, and promoting long-term agricultural sustainability. By harnessing the potential of AMF, farmers can build more sustainable agricultural systems that address the challenges of food security, environmental degradation, and climate change.

Keywords Plant growth · Nutrient uptake · Soil health · Nutrient cycling · Environmental sustainability

14.1 Introduction

The pursuit of sustainable agricultural practices has become increasingly important in the face of numerous challenges, including population growth, climate change, and environmental degradation. One approach gaining attraction in the agricultural community is the utilization of arbuscular mycorrhizal fungi (AMF) to enhance plant growth and improve the overall sustainability of farming systems (Riaz et al. 2021).

AMF are a type of fungi that form a mutualistic symbiotic relationship with the roots of most land plants. They belong to the phylum Glomeromycota and are one of the most widespread and ecologically important groups of mycorrhizal fungi

B. M. Herath · K. W. Y. R. Kalamulla · T. A. N. Mayadunna · P. N. Yapa (✉)
Department of Biological Sciences, Faculty of Applied Sciences, Rajarata University of Sri Lanka, Mihintale, Sri Lanka
e-mail: neelamanie@as.rjt.ac.lk

M. A. E. G. Perera · I. K. Jayamanna
Department of Bioprocess Technology, Faculty of Technology, Rajarata University of Sri Lanka, Mihintale, Sri Lanka

G. J. Ahammed, R. Hajiboland (eds.), *Arbuscular Mycorrhizal Fungi and Higher Plants*, https://doi.org/10.1007/978-981-99-8220-2_14

(Agnihotri et al. 2022). AMF have been found to associate with around 80% of all known plant species, including agricultural crops, grasses, trees, and shrubs. Through this symbiosis, AMF extend their hyphae into the soil, increasing the root surface area and facilitating the uptake of nutrients, particularly phosphorus, from the soil (Yadav et al. 2013). In return, the plant provides the AMF with carbohydrates, fostering a mutually beneficial relationship. This nutrient exchange between the plant and the fungus results in improved plant growth, increased resistance to environmental stresses, and enhanced nutrient utilization efficiency (Aryal et al. 2003).

Secondary roles of AMF include reduction of root invasion by microbial soil-borne plant pathogens (Newsham et al. 1995); reduction in plant uptake of phytotoxic heavy metals (Herath et al. 2023), improved host plant water balance in periods of ample water and drought (Auge 2001), and soil particle aggregation through the cohesive action of a glomalin water-stable glycoprotein (Rillig and Mummey 2006).

The application of AMF in sustainable agricultural systems offers several advantages (Fig. 14.1). Firstly, AMF can significantly reduce the need for synthetic fertilizers by enhancing nutrient uptake and utilization. This leads to reduced fertilizer runoff and associated environmental pollution, addressing concerns related to water quality and ecosystem health. By relying less on chemical inputs, farmers can minimize their ecological footprint and contribute to the preservation of natural resources (Fard et al. 2020).

Moreover, AMF have been shown to improve soil health and structure. The hyphal network formed by AMF enhances soil aggregation, promoting better water infiltration and retention, reducing erosion, and enhancing overall soil fertility. This improvement in soil quality contributes to long-term sustainability and resilience in agricultural systems (Fall et al. 2022).

In addition to nutrient acquisition and soil health benefits, AMF have demonstrated their ability to enhance plant tolerance to various abiotic stresses, including drought, salinity, and heavy metal toxicity (Herath et al. 2021). These environmental stresses are expected to become more prevalent under changing climatic conditions, posing significant challenges to crop production (Blanchet et al. 2016; Harikumar 2015). AMF symbiosis can help plants better cope with these stresses by improving water and nutrient availability, protecting against oxidative damage, and regulating plant hormone levels. This resilience contributes to the sustainability and productivity of agricultural systems in the face of climate change (Nazari et al. 2020).

The application of AMF in sustainable agriculture can take various forms, such as inoculating soil with AMF propagules, incorporating AMF into hydroponic or greenhouse systems, or utilizing AMF-inoculated substrates for seedling production. These techniques allow for the establishment of AMF associations early in the plant's life cycle, maximizing the benefits throughout its growth (Bergstrand 2022; Chai et al. 2019).

This chapter examines the key concepts of the utilization of AMF in sustainable agricultural systems. It offers a promising approach to address the challenges facing modern farming practices. By enhancing nutrient uptake, improving soil health, and

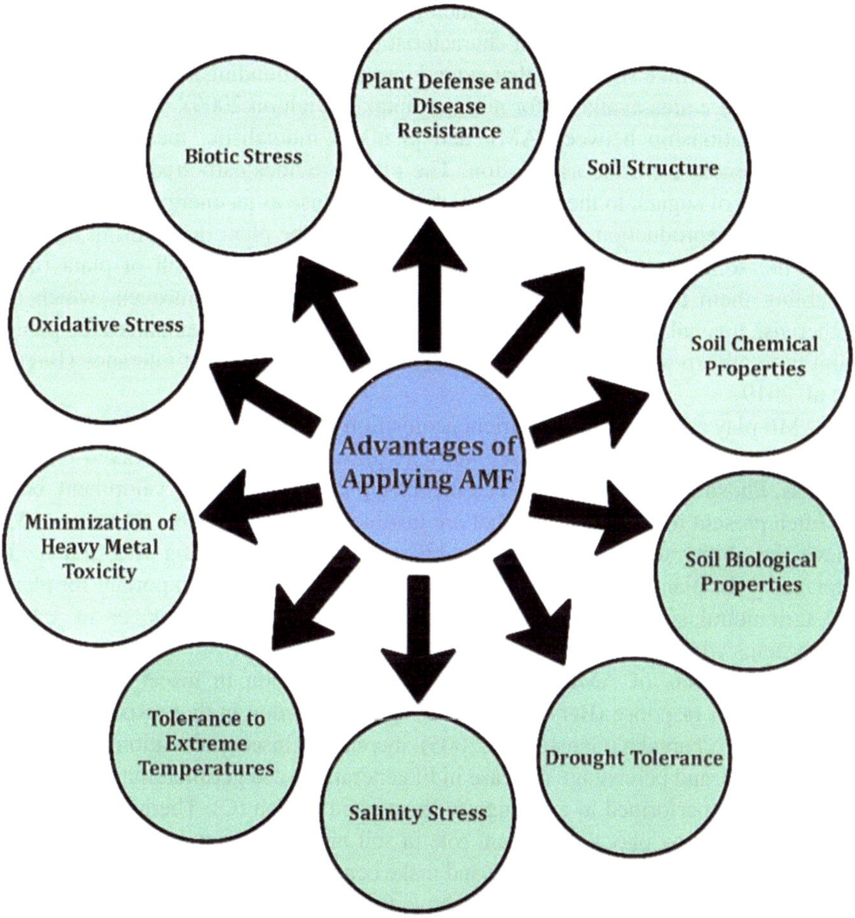

Fig. 14.1 An illustrated depiction highlighting the benefits of applying AMF on plants. Arrowheads indicate improvement of soil properties or tolerance to stress.

increasing plant resilience to environmental stresses, AMF applications contribute to the development of more sustainable and resilient farming systems. As the demand for sustainable agriculture continues to grow, integrating AMF into farming practices represents a valuable strategy to enhance productivity, protect the environment, and ensure food security for future generations.

14.2 Effect of AMF on Plant-Soil System

AMF are characterized by several distinctive features that set them apart from other types of fungi. One of their key characteristics is the formation of specialized structures called "arbuscules" within the root cells of the host plant. Arbuscules are branched, treelike structures that facilitate nutrient exchange between the fungus and the plant. Another important characteristic of AMF is the formation of hyphae, which are threadlike structures that extend into the surrounding soil, greatly enhancing the surface area available for nutrient uptake (Dighton 2009).

The relationship between AMF and plants is mutualistic, meaning that both partners benefit from the association. The plant provides carbohydrates, primarily in the form of sugars, to the AMF, which the fungi use as an energy source for their growth and reproduction. In return, the AMF assist the plant in acquiring nutrients from the soil. The hyphae of AMF extend far beyond the reach of plant roots, enabling them to access nutrients, such as phosphorus and nitrogen, which are otherwise unavailable to the plant. The mycelium of AMF also enhances the plant's ability to absorb water from the soil, thereby improving drought tolerance (Begum et al. 2019).

AMF play a crucial role in nutrient acquisition for plants, particularly in nutrient-deficient soils. One of the primary nutrients in which AMF are involved is phosphorus. Phosphorus is an essential element for plant growth and development, but it is often present in soils in forms that are insoluble and inaccessible to plants. AMF have the ability to solubilize and mobilize phosphorus, making it available for uptake by the plant. This phosphorus acquisition is particularly important for plants in agricultural systems where fertilizers are not readily available or in natural ecosystems where nutrient cycling is limited (Qi et al. 2022).

Further effects of AMF reported include a reduction in insect herbivory by induced plant response (Bennett et al. 2009) and variation in that response relative to nitrogen (N) uptake (Gange et al. 2005), increase in insect pollination (Gange and Smith 2005), and percentage increase in F_1 generation seed germination. All of these functions are performed in exchange for host plant carbon (C). There is evidence to suggest AMF may play a significant role in soil nitrogen (N) and carbon (C) cycles (Leigh et al. 2008; Jones et al. 2009) and make considerable contributions to terrestrial ecosystem C sinks. In addition to the above functions, AM fungi can also influence, perhaps even organize and structure, plant community patterns and soil microbiota community populations (Rillig and Mummey 2006; Toljander et al. 2007).

AMF also contribute to the acquisition of other nutrients, such as nitrogen, potassium, and micronutrients like zinc and copper. Through their extensive hyphal network, AMF can explore larger soil volumes and scavenge nutrients from a broader range of locations. They can break down organic matter and release nutrients bound in organic compounds, making them accessible to plants. Moreover, AMF can enhance the efficiency of nutrient uptake by improving the root system's surface area and nutrient-absorbing capacity (Neumann and George 2010). In addition to nutrient acquisition, AMF can also confer other benefits to plants. They can enhance plant growth, improve tolerance to environmental stresses such as drought and salinity, and contribute to disease resistance by activating the plant's defense mechanisms (Table 14.1) (Willis et al. 2013).

Table 14.1 Crop responses to inoculation with AMF under different stress conditions

Crop	Stress conditions	Fungal species	Observed responses	References
Zea mays L. (corn)	Drought	*Rhizophagus intraradices*, strain BGCBJ09	Enhanced plant dry weight, absorption of phosphorus (P), nitrogen (N), potassium (K), and magnesium (mg) in the aboveground portion, and improved water utilization efficiency	Zhao et al. (2015)
	Heavy metal stress	*Glomus isolates*	Enhanced dry weight and essential element content (K, P, and Mg), and clear disparities in cellular distribution of heavy metals and essential elements	Kaldorf et al. (1999)
	Heat stress	*Funneliformis* (glomus) species	Heterogeneity in the regulation of photosystem II	Mathur and Jajoo (2020)
	Heat stress	*Rhizophagus intraradices*, *Funneliformis mosseae*, *F. geosporum*	Enhanced foliage elongation, heightened plant stature, amplified leaf count, elevated chlorophyll a level, increased photosynthetic efficiency, enhanced stomatal conductance, and accelerated transpiration rate	Mathur et al. (2016)
Triticum aestivum L. (wheat)	Drought	*Glomus mosseae*, *Glomus fasciculatum*, *Gigaspora decipiens*	Augmented plant growth and overall chlorophyll pigment content	Pal and Pandey (2016)
	Drought	*Glomus mosseae*	Elevated osmotic potential, chlorophyll content, and fluorescence, along with increased activities of antioxidant enzymes, ascorbic acid, enzymes related to nitrogen (N) and phosphorus (P) metabolism, and higher levels of nitrogen (N), phosphorus (P), and potassium (K) contents	Rani (2016)
	Salinity	*G. etunicatum*, *F. mosseae*, *R. irregularis*	Enhanced plant growth, improved nutrient absorption and crop yield, and decreased levels of Na^+ and Cl^-	Daei et al. (2009)

(continued)

Table 14.1 (continued)

Crop	Stress conditions	Fungal species	Observed responses	References
	Heat stress	*R. fasciculatus*, *F. mosseae*	The biomass, nitrogen (N), potassium (K), and water contents have shown an elevation, whereas there has been a decrease in the levels of antioxidant compounds such as glutathione, ascorbate, and H_2O_2	Marulanda et al. (2007)
	Heat stress	*Rhizophagus irregularis*, *Funneliformis mosseae*, *Funneliformis geosporum*, *Claroideoglomus claroideum*	Enhanced root functionality: Boosted grain quantity, nutrient allocation, and nutrient composition	Cabral et al. (2016)
Oryza sativa L. (Rice)	Salinity	*Claroideoglomus etunicatum*	Enhanced quantum yield of PSII photochemistry, net photosynthetic rate, and stomatal conductance	Porcel et al. (2015)
Solanum lycopersicum (tomato)	Heat stress	*Rhizophagus irregularis*	Improved photosynthetic efficiency, enhanced root water flow capacity, and increased aquaporin levels and phosphorylation status	Calvo-Polanco et al. (2016)
	Salinity	*Rhizophagus irregularis*	Improved shoot fresh weight, leaf area, leaf number, root fresh weight, and concentrations of growth hormones	Khalloufi et al. (2017)
	Salinity	*Glomus intraradices*	Enhanced dry matter accumulation, increased ion absorption, improved growth characteristics, and elevated chlorophyll levels	Hajiboland et al. (2010)
	Heavy metal stress	*F. mosseae* (syn. *Glomus mosseae*), *R. intraradices* (syn. *Glomus intraradices*), *C. etunicatum* (syn. *Glomus etunicatum*)	AMF alleviated oxidative stress by decreasing the production of malonaldehyde and hydrogen peroxide. Additionally, it enhanced the plant's defense system, offering	Hashem et al. (2016)

(continued)

Table 14.1 (continued)

Crop	Stress conditions	Fungal species	Observed responses	References
			effective protection against Cd stress	
Poncirus trifoliata (trifoliate orange)	Drought	*Funneliformis mosseae*, *Paraglomus occultum*	Enhanced extension of hyphae, accelerated absorption rate of hyphal water, and elevated leaf water potential	Zhang et al. (2018)
Hordeum vulgare L. (barley)	Heat stress	*G. versiforme*, *R. irregularis*	Enhancing survival rates and easing low-temperature stress	Hajiboland et al. (2019)
Cucumis sativus L. (cucumber)	Salinity	*Glomus etunicatum*, *Glomus intraradices*, *Glomus mosseae*	Enhanced biomass, synthesis of photosynthetic pigments, and augmented antioxidant enzyme activity	Hashem et al. (2018)
Capsicum annuum L. (sweet and chili pepper)	Salinity	*R. irregularis*	The leaf area, mineral content, proline, sugars, and cell membrane integrity were enhanced, while the shoot content of Na was decreased	Beltrano et al. (2013)

14.3 AMF Application Techniques in Agriculture

14.3.1 Production of Biofertilizers

Agricultural management practices significantly influence the dynamics of AMF in agricultural fields. The following subsection explores the effects of tillage practices, pesticide and fertilizer applications, as well as the potential benefits of organic farming practices on AMF symbiosis (Bonfante and Genre 2010; Cardoso Filho et al. 2017; Gao et al. 2019).

Pesticide and fertilizer applications can have profound effects on AMF abundance and diversity. Some pesticides, particularly fungicides and nematicides, can directly or indirectly affect AMF colonization and hyphal growth. The indiscriminate use of pesticides can disrupt the delicate balance of soil microbial communities, including AMF, leading to decreased colonization and functional diversity. Similarly, excessive or imbalanced fertilizer applications, particularly high phosphorus levels, can reduce the reliance on AMF symbiosis and limit their colonization. Implementing integrated pest management strategies and adopting judicious fertilizer practices that consider AMF interactions can help mitigate negative impacts on AMF in agricultural systems (Akiyama and Hayashi 2006; Audet and Charest 2007).

Organic farming practices have shown potential benefits for enhancing AMF symbiosis and promoting sustainable agriculture. Organic farming systems, characterized by reduced chemical inputs, rely on organic amendments, crop rotation, and cover cropping to enhance soil fertility and microbial diversity (Ortas and Rafique 2017). These practices foster a conducive environment for AMF colonization and mycorrhizal functioning. Organic farms often exhibit higher levels of AMF diversity, improved nutrient cycling, and increased plant resilience to environmental stresses. Incorporating organic farming principles can contribute to the preservation and promotion of AMF symbiosis in agricultural systems (Sharma et al. 2017; Shtark et al. 2010).

By considering and implementing appropriate agricultural management practices, farmers can optimize AMF interactions, enhance nutrient cycling, and promote sustainable agriculture. Understanding the effects of these practices on AMF abundance, diversity, and functional capabilities is crucial for maximizing the benefits of AMF symbiosis in agricultural fields (Chen et al. 2018; Ercolin and Reinhardt 2011).

Plants require essential nutrients to carry out various physiological functions and maintain optimal growth and development. These nutrients can be broadly categorized into two groups: macronutrients and micronutrients. Macronutrients, including nitrogen (N), phosphorus (P), potassium (K), calcium (Ca), magnesium (Mg), and sulfur (S), are required in relatively large quantities. Micronutrients, such as iron (Fe), manganese (Mn), zinc (Zn), copper (Cu), boron (B), molybdenum (Mo), and chlorine (Cl), are needed in trace amounts. These nutrients play crucial roles in processes such as photosynthesis, respiration, and synthesis of proteins, nucleic acids, and other essential molecules (Sparks and Benfey 2017).

Plants have evolved various strategies to acquire these nutrients from the sur-rounding environment. These strategies can be categorized into two main types:

- Direct uptake: Some nutrients, such as carbon (C) and oxygen (O), are obtained directly from the atmosphere through stomata in leaves. Additionally, hydrogen (H) is acquired through water absorption by the roots (Auge 2001).
- Indirect uptake: Most nutrients are acquired indirectly through the roots. The root system plays a vital role in nutrient acquisition by employing several strategies (Kameoka et al. 2019).

As an indirect nutrition uptake strategy, many plants form mutualistic symbiotic relationships with fungi called mycorrhizae. These fungi form a network of hyphae that extend into the soil, increasing the root surface area for nutrient absorption. In return, the fungi receive carbohydrates from the plant (Sparks and Benfey 2017).

Production of soil-based biofertilizers concerning environmental conditions is the natural way to enrich the particular agricultural land. Depending on the farmer's requirement, the customized AMF-based biofertilizers can be produced by mixing one or more AMF species with other beneficial microorganisms such as plant growth-promoting rhizobacteria (PGPR) and phosphate-solubilizing bacteria (PSB) (Karima and Samia 2020). As a conventional method of biofertilizer production, the selected AMF species are first multiplied using selective host plants on pot culture under greenhouse conditions called the trap culture method, and then, AMF root inoculums and spore-baring soil are introduced to the plant with a suitable carrier (Kalamulla et al. 2022). Kalamulla et al. (2022) revealed that the application of indigenous AMF genera Glomus, Claroideoglomus, and Acaulospora isolated from rice rhizosphere together with *Azospirillum* sp. increased the biometric and yield parameters in rice plants similar to the application of compost and inorganic fertilizer to the rice cultivation. Other than using mixed AMF inoculums, scientists used one selective AMF sp. to produce specific biofertilizers using single spore technique, and further multiplication was done using trap culture methods with selective host plants (Karima and Samia 2020). The conventional method of spore separation using microscope is time-consuming and skill demanding. Scientists developed a microfluidic device for manual separation of spores from the centrifuged suspension of mixed spores via manual temporary flow diversion mechanism (Srisom et al. 2020).

Greenhouse conditions do not always provide a contaminant-free environment for inoculum preparation of AMF. Then, researchers developed the in vitro monoxenic cultivation system with synthetic medium under strictly controlled conditions together with the tissue culture techniques to obtain contamination-free, pure sterile AMF propagules (Karima and Samia 2020). In vitro, production of AMF biofertilizers consists of the higher propagule number in a small amount of culture medium. Two types of in vitro systems have been developed to produce sterilized AMF propagules:

- AMF grown on transformed plant root (by Ri-TDA of *Agrobacterium rhizogenes*).
- AMF can be grown on autotrophic plants which are grown such that the aerial part of the plant grows outside the Petri dish (Bécard and Fortin 1988).

St-Arnaud et al. (1996) created a two-compartment in vitro system in the absence of host roots to enhance the spore production of *Glomus intraradices* in a modified minimal medium. In the experimental setup, the proximal compartment was made up with adding sucrose and mycorrhizal Ri T-DNA transformed *Daucus carota* roots, and the distal compartment was maintained without adding sucrose and roots. The distal compartment was permitted to grow only endosymbionts because of lacking sugar in the medium. 6–8 weeks after subculturing, nearly 15,000 viable spores per plate were obtained (St-Arnaud et al. 1996). The efficiency and continuation of this technique can be increased by the replacement of the gel in the medium and resupply of sucrose to the proximal compartment every 2 months (Douds 2002). This technique is more adapted to the mass production of industrial AMF inoculums in biofertilizer production. With the suitable formulation technique, the produced inoculums (mycorrhizae containing root fragments, spores, and mycelium) (Karima and Samia 2020) then can be introduced to the target field by considering the time of inoculum viability (Sahu and Brahmaprakash 2016).

14.3.2 Seed Coating with AMF

Application of formulated AMF-based biofertilizer to the larger area of open agricultural field increased the cost per plant, because of nontargeted spreading of the introduced biofertilizer. For the effective crop performance, seed quality is a major aspect of sustainable agriculture. Seed coating with AMF inoculum is an effective delivery method while increasing the efficiency of action and reducing the quantity of inoculum needed (Oliveira et al. 2016). Seed coating is one of the ancient agricultural methods used to increase the agricultural productivity of the crop. But usage of different chemical fertilizers inside the coat and chemical-based seed coating materials increased the environmental disturbance. Seed coating with the beneficial soil microbes with the biodegradable coating material opens the path to sustainable agricultural practices without reduction of the crop yield (Sohali et al. 2022). Accurately selecting coating material while enhancing the AMF spore viability is not an easy task. According to the seed coating technique, AMF can directly infect the newly germinated seed within a short period of time after the germination (Nurrobifahmi et al. 2021).

To improve the agricultural production of cowpea (*Vigna unguiculata* L. Walp) that were grown in the semiarid lands, the cowpea seeds were coating with the AMF, *Rhizophagus irregularis* and *Pseudomonas libanensis* using silicon dioxide and starch as a coating material according to the pan coating methods (Ma et al. 2019). The application of tapioca starch and polyvinyl alcohol (PVA) as a seed coating material enhances the AMF spore viability. PVA provides a hydrophilic layer to the AMF spore and prevents dehydration under higher soil temperatures (Nurrobifahmi et al. 2021). Other than the improvement of the plant productivity and yield component via increasing the nutrient supplement, seed coating with the AMF confers seed resistance against the pathogen. Application of the spores of *Rhizophagus fasciculatus* and *Rhizophagus aggregatus* increased the seed resistance to insects while extending the germinability period up to 4 months in maize and sorghum (Sene et al. 2021).

14.3.3 Seed Bio-Priming with AMF

Inoculation of the seeds with the beneficial microorganism's combination of seed hydration refers to seed bio-priming. It improves crop productivity and yield parameters via increasing seed viability, vigor indices, plant nutrition, and protection against pathogens (Singh et al. 2010). Other than using of plant growth-promoting bacteria, AMF can also use as a microbial inoculum in seed bio-priming. In this priming technique, seeds are soaked in a solution of mycorrhizae with water and allow the seeds to soak for 8–12 h. In this process, the bio-stimulants are adhered to the seed and finally established at planting. Seed bio-priming with silicon nanoparticles and mycorrhizal fungi increased the salinity tolerance in wheat plants by increasing plant physiological parameters (leaf water potential, soluble proteins, and sugar) and antioxidant activity (Ahmadi-Nouraldinvand et al. 2023). Yadav et al. (2018) tested a mixture of beneficial microbes, including *Glomus intraradices* for bio-priming of corn (*Zea mays* L.) seeds to test the plant productivity and yield attributes. Combined use of *Trichoderma viride* and *Glomus intraradices* along with 75% recommended NPK dose provided a more effective combination for baby corn production. As used in the seed bio-priming technique, mycorrhizae can be directly added to the hydroponic systems. Addition of a higher dosage of the chemical to the plant in a hydroponic system causes death of the plant. But the addition of the mycorrhizae does not harm the plant in any direction (Yadav et al. 2018).

14.3.4 Soil Drenching

AMF application through soil drenching involves several steps to ensure successful colonization and establishment of mycorrhizal associations (Berruti et al. 2016). Firstly, an AMF inoculant is prepared before soil drenching. This inoculant can contain AMF propagules like spores or mycelium or liquid AMF cultures (Schaefer et al. 2021). The inoculant can be obtained from commercial sources or propagated in a laboratory or nursery. It is important to follow the manufacturer's instructions or established protocols to prepare the inoculant and ensure its viability and effectiveness (Schaefer et al. 2021).

During soil drenching, the AMF inoculant is directly applied to the soil, surrounding the plant roots. Typically, the inoculant is mixed with water or a suitable carrier solution to facilitate even distribution. The drenching solution is poured onto the soil surface or applied through irrigation systems, allowing it to percolate down into the root zone. This contact between the AMF inoculant and the roots facilitates colonization and the formation of mycorrhizal associations (Oehl et al. 2003). The timing and frequency of soil drenching depend on factors such as plant species, growth stage, and environmental conditions. It is often recommended to apply the AMF inoculant during the early stages of plant growth or transplanting to maximize colonization. Subsequent drenching applications may be repeated periodically throughout the growing season to maintain and enhance the mycorrhizal associations (Smith and Read 2010; Akpinar et al. 2019).

To ensure the effectiveness of soil drenching, several factors should be considered. The soil should be adequately moist before and after drenching to facilitate the movement of the inoculant and the establishment of mycorrhizal associations (Alrajhei et al. 2022). The appropriate dosage of the AMF inoculant should be determined based on the plant species, soil conditions, and the specific inoculant used. Proper distribution of the drenching solution across the soil is important to ensure consistent colonization of the roots. Monitoring and evaluation of the effectiveness of the AMF application can be done by assessing plant growth, nutrient uptake, and other relevant parameters (Azevedo Correa 2019).

Soil drenching is a convenient and effective method for introducing AMF inoculants to the root zone, promoting mycorrhizal associations, and enhancing nutrient uptake in agricultural systems (Barbosa et al. 2019). It is crucial to follow appropriate application methods, timing, and monitoring to ensure the desired benefits of mycorrhizal symbiosis. It is also advisable to consult the manufacturer's instructions or agricultural experts for specific AMF inoculant products for best results (Bedini et al. 2013).

14.3.5 Rhizobox or Rhizotron Systems

AMF application techniques in rhizobox or rhizotron systems are used to study the interactions between plant roots and AMF under controlled conditions. These systems allow researchers to observe and manipulate the colonization and development of mycorrhizal associations (Meena et al. 2018). Before setting up the rhizobox or rhizotron system, the soil medium used in the compartments can be pre-inoculated with AMF inoculants, which contain AMF propagules such as spores or mycelium. These inoculants are mixed into the soil medium to ensure the presence of AMF for colonization during the experiment. This technique allows the controlled introduction of AMF into the system (Gianinazzi and Schüepp 1994).

During planting in the rhizobox or rhizotron system, the roots of seedlings or transplants can be immersed or dipped in an AMF inoculant solution. Coating the roots with the inoculant ensures direct contact between the AMF and roots, facilitating rapid colonization and establishment of mycorrhizal symbiosis. This method enables the study of early stages of AMF-root interactions in a controlled environment (Wattenburger et al. 2020). Another approach is to topically apply AMF inoculants to the soil surface of the system. The inoculant is spread or sprinkled around the planted roots, and subsequent watering or irrigation helps it move into the soil, allowing AMF to colonize the roots. This localized application provides insights into the effects of mycorrhizal colonization on root growth and nutrient uptake (Fig. 14.2 and 14.3) (Verbruggen and Toby Kiers 2010).

In certain cases, AMF inoculants can be injected directly into specific areas of interest within the rhizobox compartments. This targeted injection allows researchers to study the localized effects of mycorrhizal colonization on root development and nutrient acquisition (Fig. 14.2 and 14.3). It proves particularly useful when investigating the spatial distribution of AMF colonization within the system (Juntahum et al. 2022). These AMF application techniques in rhizobox or rhizotron systems enable researchers to control and manipulate the presence of AMF, facilitating the study of mycorrhizal associations and their impact on plant growth and nutrient uptake (Fig. 14.2 and 14.3) (Kadam et al. 2020). By observing and analyzing root development, nutrient acquisition, and other parameters, researchers can gain insights into the dynamics of the plant-mycorrhizal symbiosis in a controlled experimental setup (Ma et al. 2022).

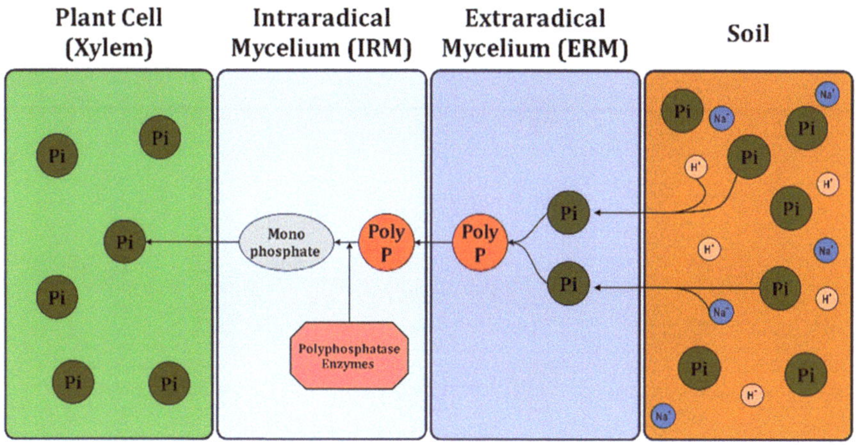

Fig. 14.2 Transport of phosphate and transfer mechanisms in mycorrhizal interactions for host cell nutrient uptake. During mycorrhizal interactions, the transportome responsible for phosphate (Pi) uptake and transfer undergoes a series of processes. Initially, inorganic phosphate is taken up from the soil solution through the plasma membrane. To energize this process, the fungal organism utilizes an H+ and Na + symport mechanism. Following uptake, cytoplasmic Pi is accumulated in the vacuoles as polyphosphates (Poly P). These Poly P molecules are then transferred through hyphae using motile vacuoles toward the intraradical mycelium (IRM). Within the intraradical mycelium, it is likely that polyphosphatase enzymes degrade Poly P into monophosphates. These monophosphates are subsequently transferred to the plant cells, where they are utilized to regenerate Pi, completing the transfer process

It is important to note that the specific AMF inoculant, application method, and timing should be chosen based on the objectives of the study, the plant species being investigated, and the experimental design. Consulting with AMF inoculant manufacturers or experts in the field can provide guidance on selecting the appropriate techniques for rhizobox or rhizotron systems (Zhang et al. 2019; Zhu et al. 2016).

14.3.6 Greenhouse and Hydroponic Systems

AMF application techniques can be adapted for greenhouse and hydroponic systems to enhance plant growth and nutrient uptake (Singh et al. 2010).

14.3.6.1 Greenhouse Systems

In greenhouse systems that utilize soil-based substrates, AMF inoculants can be mixed into the substrate during potting or tray filling (Bergstrand 2022). The inoculants may contain AMF propagules, such as spores or mycelium, which establish mycorrhizal associations with the plant roots as they grow. The inoculation

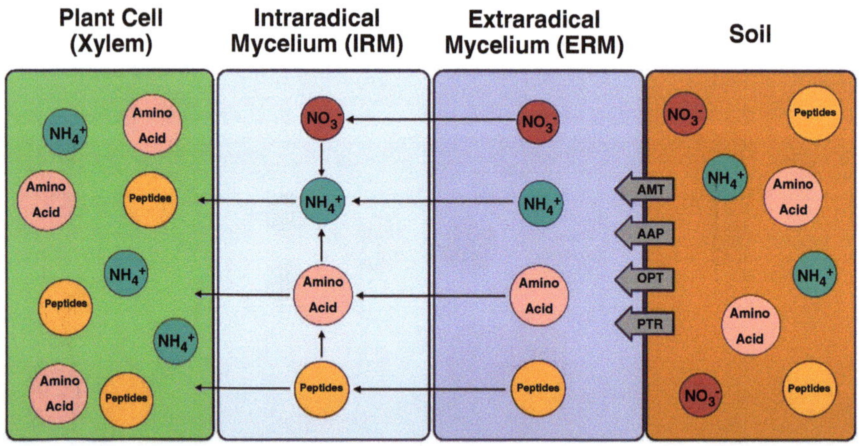

Fig. 14.3 The existing understanding of nitrogen transfer mechanisms in mycorrhizal interactions reveals the presence of five distinct compartments involved in the transfer of nitrogen compounds, namely, ammonium, nitrate, amino acids, and peptides. These compartments include the soil, extraradical mycelium (ERM), intraradical mycelium (IRM), and the plant cell (Xylem). Specific transporters facilitate the movement of nitrogen compounds within these compartments, such as NRT (nitrate transporter) for nitrate, AMT (ammonium transporter) for ammonium, AAP (amino acid transporter) for amino acids, OPT (oligopeptide transporter) for peptides, and PTR (peptide transporter) for peptide transfer

ensures that the AMF are present in the substrate, facilitating early colonization of plant roots and improved nutrient uptake (Porcel et al. 2015).

Prior to transplanting seedlings or young plants into the greenhouse, their roots can be dipped into an AMF inoculant solution. The roots are immersed in the solution for a brief period, allowing the inoculant to adhere to the roots (Ashaolu and Ashaolu 2020). This ensures direct contact between the AMF and the roots, promoting rapid colonization and establishment of mycorrhizal symbiosis in the greenhouse environment (Ruiz-Lozano and Aroca 2010). AMF inoculants can also be applied topically to the substrate surface in greenhouse systems. This involves spreading or sprinkling the inoculant onto the substrate around the base of the plants (Chai et al. 2019). Subsequent irrigation or misting helps the inoculant move down into the substrate and come into contact with the plant roots. This method provides an additional source of AMF inoculum for the roots to establish mycorrhizal associations (Sahodaran and Ray 2018).

14.3.6.2 Hydroponic Systems

In hydroponic systems, where plants are grown in nutrient-rich water without soil or substrate, AMF inoculants can be applied as a dip for seedlings before they are transferred to the hydroponic system (Aryal et al. 2003). The roots of the seedlings are immersed in an AMF inoculant solution, allowing for direct contact between the

AMF and the roots. This promotes early colonization and establishment of mycorrhizal symbiosis in the hydroponic system, enhancing nutrient uptake efficiency (Chandrasekaran 2022). AMF inoculants can also be added directly to the nutrient solution in hydroponic systems. The inoculants can contain AMF propagules, such as spores or mycelium, or even liquid AMF cultures (Silveira et al. 2021). The inoculant is mixed into the nutrient solution, ensuring that the AMF are present and available for colonization by the plant roots. This method allows for continuous exposure of the roots to the AMF, facilitating the establishment of mycorrhizal associations and improving nutrient uptake in the hydroponic system (Ebbisa 2022).

These AMF application techniques in greenhouse and hydroponic systems aimed to establish mycorrhizal associations between the plant roots and the beneficial fungi, even in the absence of traditional soil. The mycorrhizal symbiosis enhances nutrient acquisition, particularly phosphorus uptake, and can improve plant growth, vigor, and resilience in these controlled environments also (Berruti et al. 2016). It is important to note that the choice of AMF inoculant, application method, and timing should be based on the specific requirements of the plant species, the growth medium used, and the overall system conditions. Consulting with AMF inoculant manufacturers or agricultural experts can provide guidance on selecting the appropriate techniques for greenhouse and hydroponic systems (Jung and Martinez-Medina 2020; Ebbisa 2022).

14.3.7 Crop, Cultivar, Cover Crop, and Rotation

Selecting crops or cultivars that have a strong affinity for mycorrhizal associations and possess efficient root architectures for accessing adequate phosphorus (P) is an important consideration. Also, forming active symbioses with AMF is a critical consideration (Berruti et al. 2016). It is important to note that certain crop families, like Brassicaceae or Amaranthaceae, do not form mycorrhizal associations and may even release compounds that inhibit fungal growth in the soil. In such cases, it is beneficial to avoid crop rotation with these nonmycorrhizal families and consider planting mycorrhizal cover crops in between these crops before the next cropping cycle (Karasawa and Takebe 2012).

Furthermore, the domestication process of some crops may have diminished their ability to respond positively to AMF, especially when soil P levels are high. An interesting finding by Jung et al. (2012) revealed that both wild and domesticated species of 27 crops responded similarly to AMF under low Pi (phosphorus) conditions. However, when exposed to high P conditions, the growth response of 14 pairs of wild varieties to AMF was not significantly different, whereas it notably reduced growth in domesticated species. Consequently, it is evident that domesticated crops tend to benefit from mycorrhizal associations mainly when soil P concentrations are low.

While AM fungi do not exhibit strict host specificity, they do exhibit preferences that can vary based on geographical distribution and land use (Bainard et al. 2014;

Torrecillas et al. 2013). Interestingly, even though different host plants may have contrasting characteristics, the AM fungal communities associated with them may respond similarly to soil phosphorous gradients. Therefore, crop rotation can positively influence AMF function, especially when introducing highly mycotrophic crops from the Fabaceae or Poaceae families. However, it is important to note that AMF species composition may vary depending on the plant species used, and it may take some time for the fungal community to adapt and be replaced (Higo et al. 2010, 2015). Nevertheless, the subsequent mycotrophic crop is likely to benefit from this transition, leading to reduced requirements for phosphorus application compared to leaving the land fallow without any vegetation (Jemo et al. 2014).

14.4 Conclusions

The application of AMF in sustainable agricultural systems holds great promise for addressing the challenges of modern farming practices. AMF offer multiple benefits, including enhanced nutrient uptake, improved soil health, and increased plant resilience to environmental stresses. By establishing symbiotic relationships with plant roots, AMF enhance nutrient acquisition efficiency, reducing the reliance on synthetic fertilizers and minimizing environmental pollution. The hyphal network formed by AMF improves soil structure, water retention, and nutrient cycling, contributing to long-term sustainability. Additionally, AMF help plants withstand abiotic stresses, such as drought and salinity, which are expected to intensify with climate change. The various AMF application techniques, such as soil inoculation, seedling treatments, and integration into hydroponic or greenhouse systems, provide flexibility in incorporating these beneficial fungi into diverse agricultural practices. By harnessing the potential of AMF, farmers and researchers can practice sustainable way of farming that promote productivity, conserve resources, protect the environment, and ensure food security for future generations. The integration of AMF into agricultural practices represents a valuable strategy in the transition toward more sustainable and resilient farming systems.

References

Agnihotri R, Sharma MP, Prakash A, Ramesh A, Bhattacharjya S, Patra AK, Manna MC, Kurganova I, Kuzyakov Y (2022) Glycoproteins of arbuscular mycorrhiza for soil carbon sequestration: review of mechanisms and controls. Sci Total Environ 806:150571. https://doi.org/10.1016/j.scitotenv.2021.150571

Ahmadi-Nouraldinvand F, Sharifi RS, Siadat SA, Khalilzadeh R (2023) Reduction of salinity stress in wheat through seed bio-priming with mycorrhiza and growth-promoting bacteria and its effect on physiological traits and plant antioxidant activity with silicon nanoparticles application. Silicon 16:1

Akiyama K, Hayashi H (2006) Strigolactones: chemical signals for fungal symbionts and parasitic weeds in plant roots. Ann Bot (Lond) 97:925–931

Akpinar C, Demirbas A, Ortas I (2019) The effect of different compost compositions on arbuscular mycorrhizal colonization and nutrients concentration of leek (*Allium porrum* L.) plant. Commun Soil Sci Plant Anal 50:2309–2320. https://doi.org/10.1080/00103624.2019.1659299

Alrajhei K, Saleh I, Abu-Dieyeh MH (2022) Biodiversity of arbuscular mycorrhizal fungi in plant roots and rhizosphere soil from different arid land environment of Qatar. Plant Direct 6:e369. https://doi.org/10.1002/pld3.369

Aryal UK, Xu HL, Fujita M (2003) Rhizobia and am fungal inoculation improve bean plants' growth and nutrient uptake under organic fertilisation. J Sustain Agric 21:27–39. https://doi.org/10.1300/J064v21n03_04

Ashaolu TJ, Ashaolu JO (2020) Perspectives on the trends, challenges, and benefits of green, innovative and organic (GSO) foods. Int J Gastron Food Sci 22:100273. https://doi.org/10.1016/j.ijgfs.2020.100273

Audet P, Charest C (2007) Dynamics of arbuscular mycorrhizal symbiosis in heavy metal phytoremediation: meta-analytical and conceptual perspectives. Environ Pollut 147:609–614

Auge RM (2001) Water relations, drought and vesicular-arbuscular mycorrhizal symbiosis. Mycorrhiza 11:3–42

Azevedo Correa EJ (2019) Arbuscular mycorrhizal fungi and farm management practices. World J Agric Soil Sci 2:1–10. https://doi.org/10.33552/WJASS.2019.02.000534

Bainard LD, Bainard JD, Hamel C, Gan Y (2014) Spatial and temporal structuring of arbuscular mycorrhizal communities is differentially influenced by abiotic factors and host crop in a semi-arid prairie agroecosystem. FEMS Microbiol Ecol 88:333–344. https://doi.org/10.1111/1574-6941.12300

Barbosa MV, Pedroso DDF, Curi N, Carneiro MAC et al (2019) Do different arbuscular mycorrhizal fungi affect the formation and stability of soil aggregates? Cienc Agrotecnologia 43:1–9. https://doi.org/10.1590/1413-7054201943003519

Bécard G, Fortin JA (1988) Early events of vesicular–arbuscular mycorrhiza formation on Ri T-DNA transformed roots. New Phytol 108(2):211–218

Bedini S, Avio L, Sbrana C, Turrini A, Migliorini P, Vazzana C, Giovannetti M (2013) Mycorrhizal activity and diversity in a long-term organic Mediterranean agroecosystem. Biol Fertil Soils 49: 781–790. https://doi.org/10.1007/s00374-012-0770-6

Begum N, Qin C, Ahanger MA, Raza S, Khan MI, Ashraf M, Ahmed N, Zhang L (2019) Role of arbuscular mycorrhizal fungi in plant growth regulation: implications in abiotic stress tolerance. Front Plant Sci 10:1068. https://doi.org/10.3389/fpls.2019.01068

Beltrano J, Ruscitti M, Arango M, Ronco M (2013) Effects of arbuscular mycorrhiza inoculation on plant growth, biological and physiological parameters and mineral nutrition in pepper grown under deferent salinity and P levels. J Soil Sci Plant Nutr 13:123–141

Bennett AE, Bever JD, Deane Bowers M (2009) Arbuscular mycorrhizal fungal species suppress inducible plant responses and alter defensive strategies following herbivory. Oecologia 160: 771–779

Bergstrand KJ (2022) Organic fertilizers in greenhouse production systems—a review. Sci Horticult 295:110855. https://doi.org/10.1016/j.scienta.2021.110855

Berruti A, Lumini E, Balestrini R, Bainciotto V (2016) Arbuscular mycorrhizal fungi as natural biofertilizers: let's benefit from past successes. Front Microbiol 6:1559. https://doi.org/10.3389/fmicb.2015.01559

Blanchet G, Gavazov K, Bragazza L, Sinaj S (2016) Responses of soil properties and crop yields to different inorganic and organic amendments in a Swiss conventional farming system. Agric Ecosyst Environ 230:116–126. https://doi.org/10.1016/j.agee.2016.05.032

Bonfante P, Genre A (2010) Mechanisms underlying beneficial plant-fungus interactions in mycorrhizal symbiosis. Nat Commun 1:48. https://doi.org/10.1038/ncomms1046

Cabral C, Ravnskov S, Tringovska I, Wollenweber B (2016) Arbuscular mycorrhizal fungi modify nutrient allocation and composition in wheat (*Triticum aestivum* L.) subjected to heat-stress. Plant Soil 408(1–2):385–399. https://doi.org/10.1007/s11104-016-2942-x

Calvo-Polanco M, Sánchez-Romera B, Aroca R, Asins MJ, Declerck S, Dodd IC, Martinez-Andujar C, Albacete A, Ruiz-Lozano JM (2016) Exploring the use of recombinant inbred lines in combination with beneficial microbial inoculants (AM fungus and PGPR) to improve drought stress tolerance in tomato. Environ Exp Bot 131:47–57. https://doi.org/10.1016/j.envexpbot.2016.06.015

Cardoso Filho JA, Sobrinho RR, Pascholati SF (2017) Arbuscular mycorrhizal symbiosis and its role in plant nutrition in sustainable agriculture. In: Meena VS et al (eds) Agriculturally important microbes for sustainable agriculture. Springer, Singapore. https://doi.org/10.1007/978-981-10-5343-6_5

Chai R, Ye X, Ma C, Wang Q, Tu R, Zhang L, Gao H (2019) Greenhouse gas emissions from synthetic nitrogen manufacture and fertilisation for main upland crops in China. Carbon Balance Manag 14:20. https://doi.org/10.1186/s13021-019-0133-9

Chandrasekaran M (2022) Arbuscular mycorrhizal fungi mediated enhanced biomass, root morphological traits and nutrient uptake under drought stress: a meta-analysis. J Fungi 8:660. https://doi.org/10.3390/jof8070660

Chen M, Arato M, Borghi L, Nouri E, Reinhardt D (2018) Beneficial services of arbuscular mycorrhizal fungi—from ecology to application. Front Plant Sci 9:1–14. https://doi.org/10.3389/fpls.2018.01270

Daei G, Ardekani MR, Rejali F, Teimuri S, Miransari M (2009) Alleviation of salinity stress on wheat yield, yield components, and nutrient uptake using arbuscular mycorrhizal fungi under field conditions. J Plant Physiol 166:617–625

Dighton J (2009) Mycorrhizae. In: Encyclopedia of microbiology, pp 153–162. https://doi.org/10.1016/b978-012373944-5.00327-8

Douds DD (2002) Increased spore production by *Glomus intraradices* in the split-plate monoxenic culture system by repeated harvest, gel replacement, and resupply of glucose to the mycorrhiza. Mycorrhiza 12:163–167

Ebbisa A (2022) Arbuscular mycorrhizal fungi (AMF) in optimizing nutrient bioavailability and reducing agrochemicals for maintaining sustainable agroecosystems. In: de Sousa R (ed) Mycorrhiza. New insights. IntechOpen, London. https://doi.org/10.5772/intechopen.106995

Ercolin F, Reinhardt D (2011) Successful joint ventures of plants: arbuscular mycorrhiza and beyond. Trends Plant Sci 16(7):356–362. https://doi.org/10.1016/j.tplants.2011.03.006

Fall AF, Nakabonge G, Ssekandi J, Founounce-Mboup H, Apori SO, Ndiaye A, Badji A, Ngom K (2022) Roles of arbuscular mycorrhizal fungi on soil fertility: contribution in the improvement of physical, chemical, and biological properties of the soil. Front Fungal Biol 3:723892. https://doi.org/10.3389/ffunb.2022.723892

Fard SE, Yarnia M, Farahvash F, Behrouzyar EK, Rashidi V (2020) Arbuscular mycorrhizas and phosphorus fertilizer affect photosynthetic capacity and antioxidant enzyme activity in peppermint under different water conditions. Acta Agrobot 73:1–13. https://doi.org/10.5586/aa.7345

Gange AC, Brown VK, Aplin DM (2005) Ecological specificity of arbuscular mycorrhizae: evidence from foliar- and seed-feeding insects. Ecology 86:603–611

Gange A, Smith AK (2005) Arbuscular mycorrhizal fungi influence visitation rates of pollinating insects. Ecol Entomol 30:600–606

Gao W, Wang P, Wu Q (2019) Functions and application of glomalin-related soil proteins: a review. Sains Malays 48(1):111–119. https://doi.org/10.17576/jsm-2019-4801-13

Gianinazzi S, Schüepp H (eds) (1994) Impact of arbuscular mycorrhizas on substainable agriculture and natural ecosystems. Springer, Cham

Hajiboland R, Joudmand A, Aliasgharzadeh N, Laiegh SF, Poschenrieder C (2010) Colonization with arbuscular mycorrhizal fungi improves salinity tolerance of tomato *Solanum lycopersicum* L. plants. Plant Soil 331:313–327. https://doi.org/10.1007/s11104-009-0255-z

Hajiboland R, Joudmand A, Aliasgharzadeh N, Laiegh SF, Poschenrieder C (2019) Arbuscular mycorrhizal fungi alleviate low-temperature stress and increase freezing resistance as a substitute for acclimation treatment in barley. Crop Pasture Sci 70:218–233

Harikumar VS (2015) Arbuscular mycorrhizal associations in sesame under low-input cropping systems. Arch Agron Soil Sci 61:347–359. https://doi.org/10.1080/03650340.2014.932906

Hashem A, Allah EA, Alqarawi AA, Al Huqail AA, Egamberdieva D, Wirth S (2016) Alleviation of cadmium stress in *Solanum lycopersicum* L. by arbuscular mycorrhizal fungi via induction of acquired systemic tolerance. Saudi J Biol Sci 23:272–281

Hashem A, Alqarawi AA, Radhakrishnan R, Al-Arjani AF, Aldehaish HA, Egamberdieva D, Abd Allah EF (2018) Arbuscular mycorrhizal fungi regulate the oxidative system, hormones and ionic equilibrium to trigger salt stress tolerance in *Cucumis sativus* L. Saudi J Biol Sci 25(6): 1102. https://doi.org/10.1016/j.sjbs.2018.03.009

Herath BM, Bamunuarachchige C, Stephenson SL, Elgorban AM, Asad S, Kumla J, Suwannarach N, Karunarathna SC, Yapa PN (2023) Soil heavy metal absorption potential of *Azolla pinnata* and *Lemna gibba* with arbuscular mycorrhizal fungi in Rice (*Oryza sativa* L.) farming. Sustainability 15(5):4320. https://doi.org/10.3390/su15054320

Herath BMMD, Madushan KWA, Lakmali JPD (2021) Arbuscular mycorrhizal fungi as a potential tool for bioremediation of heavy metals in contaminated soil. J Adv Res Rev 10(3):217–228. https://doi.org/10.30574/wjarr.2021.10.3.0255

Higo M, Isobe K, Kang D, Ujie K, Drijber RA (2010) Inoculation with arbuscular mycorrhizal fungi or crop rotation with mycorrhizal plants improves the growth of maize in limed acid sulfate soil. Plant Prod Sci 13:74–79. https://doi.org/10.1626/pps.13.74

Higo M, Isobe K, Kondo T, Yamaguchi M, Takeyama S, Drijber RA, Torigoe Y (2015) Temporal variation of the molecular diversity of arbuscular mycorrhizal communities in three different winter cover crop rotational systems. Biol Fertil Soils 51:21–32. https://doi.org/10.1007/s00374-014-0945-4

Jemo M, Souleymanou A, Frossard E, Jansa J (2014) Cropping enhances mycorrhizal benefits to maize in a tropical soil. Soil Biol Biochem 79:117–124. https://doi.org/10.1016/j.soilbio.2014.09.014

Jones DL, Nguyen C, Finlay RD (2009) Carbon flow in the rhizosphere: carbon trading at the soil-root interface. Plant Soil 321:5–33

Jung SC, Martinez-Medina A (2020) Mycorrhiza-induced resistance and priming of plant defenses. Front Plant Sci 11:581956. https://doi.org/10.3389/fpls.2020.581956

Jung SC, Martinez-Medina A, Lopez-Raez JA, Pozo MJ (2012) Mycorrhiza-induced resistance and priming of plant defenses. J Chem Ecol 38(6):651–664. https://doi.org/10.1007/s10886-012-0134-6

Juntahum S, Kuyper TW, Boonlue S (2022) Distribution of arbuscular mycorrhizal fungi in sugarcane rhizosphere from various agricultural management practices in Northeast Thailand. Curr Res Environ Appl Mycol 12:44–55. https://doi.org/10.5943/cream/12/1/4

Kadam SB, Pable AA, Barvkar VT (2020) Mycorrhiza-induced resistance (MIR): a defense developed through the synergistic engagement of phytohormones, metabolites and rhizosphere. Funct Plant Biol 47:880–890. https://doi.org/10.1071/FP20035

Kalamulla R, Sandaruwan D, Karunarathna SC, Stephenson S, Tibpromma S, Elgorban A, Al-Rejaie S, Yapa N, Suwannarach N (2022) Assessment of community dynamics of arbuscular mycorrhizal fungi in the Rice (*Oryza sativa* L.) Rhizosphere and potential application as biofertilizer. Sustainability 14(24):16537

Kaldorf M, Kuhn A, Schröder W, Hildebrandt U, Bothe H (1999) Selective element deposits in maize colonized by a heavy metal tolerance conferring arbuscular mycorrhizal fungus. J Plant Physiol 154:718–728

Kameoka H, Maeda T, Okuma N, Kawaguchi M (2019) Structure-specific regulation of nutrient transport and metabolism in arbuscular mycorrhizal fungi. Plant Cell Physiol 60:2272. https://doi.org/10.1093/pcp/pcz122

Karasawa T, Takebe M (2012) Temporal or spatial arrangements of cover crops to promote arbuscular mycorrhizal colonization and P uptake of upland crops grown after non-mycorrhizal crops. Plant Soil 353:355–366. https://doi.org/10.1007/s11104-011-1036-z

Karima B, Samia T (2020) Native arbuscular mycorrhizal fungi and Agro-industries in Arid Lands: productions, applications strategies and challenges. In: Mycorrhizal fungi—utilization in agriculture and industry. Intech Open, London

Khalloufi M, Martínez-Andújar C, Lachaâl M, Karray-Bouraoui N, Pérez-Alfocea F, Albacete A (2017) The interaction between foliar GA3 application and arbuscular mycorrhizal fungi inoculation improves growth in salinized tomato *Solanum lycopersicum* L. plants by modifying the hormonal balance. J Plant Physiol 214:134–144. https://doi.org/10.1016/j.jplph.2017.04.012

Leigh J, Hodge A, Fitter AH (2008) Arbuscular mycorrhizal fungi can transfer substantial amounts of nitrogen to the host plant from organic material. New Phytol 181:199–207

Ma Y, Látr A, Rocha I, Freitas H, Vosatka M, Oliveira R (2019) Delivery of inoculum of *Rhizophagus irregularis* via seed coating in combination with *Pseudomonas libanensis* for cowpea production. Agronomy 9(1):33. https://doi.org/10.3390/agronomy9010033

Ma K, Wang Y, Jin X, Zhao Y, Yan H, Zhang H, Zhou X, Lu G, Deng Y (2022) Application of organic fertilizer changes the rhizosphere microbial communities of a gramineous grass on Qinghai–Tibet Plateau. Microorganisms 10:1148. https://doi.org/10.3390/microorganisms10061148

Marulanda A, Porcel R, Barea JM (2007) Drought tolerance and antioxidant activities in lavender, plants colonized by native drought-tolerant or drought-sensitive Glomus species. Microb Ecol 54:543–552

Mathur S, Jajoo A (2020) Arbuscular mycorrhizal fungi protects maize plants from high temperature stress by regulating photosystem II heterogeneity. Ind Crop Prod 143:111934

Mathur S, Sharma MP, Jajoo A (2016) Improved photosynthetic efficacy of maize *Zea mays* plants with arbuscular mycorrhizal fungi (AMF) under high temperature stress. J Photochem Photobiol B 180:149–154. https://doi.org/10.1016/j.jphotobiol.2018.02.002

Meena RS, Vijayakumar V, Yadav GS, Mitran T (2018) Response and interaction of *Bradyrhizobium japonicum* and arbuscular mycorrhizal fungi in the soybean rhizosphere. Plant Growth Regul 84:207–223

Nazari M, Riebeling S, Banfield CC, Akale A, Crosta M, Mason-Jones K, Dippold MA, Ahmed MA (2020) Mucilage polysaccharide composition and exudation in maize from contrasting climatic regions. Front Plant Sci 11:587610. https://doi.org/10.3389/fpls.2020.587610

Neumann E, George E (2010) Nutrient uptake: the arbuscular mycorrhiza fungal symbiosis as a plant nutrient acquisition strategy. In: Arbuscular mycorrhizas: physiology and function, pp 137–167. https://doi.org/10.1007/978-90-481-9489-6_7

Newsham KK, Fitter AH, Watkinson AR (1995) Arbuscular mycorrhiza protect an annual grass from root pathogenic fungi in the field. J Ecol 83:991–1000

Nurrobifahmi FAN, Citraresmini A et al (2021) Delivery of arbuscular mycorrhiza fungus spores via seed coating with biodegradable binders for enhancement of the spores viability and their beneficial properties in maize. Akta Agrosia 24(1):1–8

Oehl F, Sieverding E, Ineichen K, Mader P, Boller T, Wiemken A (2003) Impact of land use intensity on the species diversity of arbuscular mycorrhizal fungi in agroecosystems of Central Europe. Appl Environ Microbiol 69(5):2816–2824. https://doi.org/10.1128/AEM.69.5.2816-2824.2003

Oliveira RS, Rocha I, Ma Y, Vosatka M, Freitas H (2016) Seed coating with arbuscular mycorrhizal fungi as an ecotechnological approach for sustainable agricultural production of common wheat (*Triticum aestivum* L.). J Toxicol Env Heal A 79(7):329–337

Ortas I, Rafique M (2017) The mechanisms of nutrient uptake by arbuscular mycorrhizae. In: Varma A et al (eds) Mycorrhiza—nutrient uptake, biocontrol, ecorestoration. Springer, Cham. https://doi.org/10.1007/978-3-319-68867-1_1

Pal A, Pandey S (2016) Role of arbuscular mycorrhizal fungi on plant growth and reclamation of barren soil with wheat (*Triticum aestivum* L.) crop. Int J Soil Sci 12:25–31. https://doi.org/10.3923/ijss.2017.25.31

Porcel R, Redondo-Gómez S, Mateos-Naranjo E, Aroca R, Garcia R, Ruiz-Lozano JM (2015) Arbuscular mycorrhizal symbiosis ameliorates the optimum quantum yield of photosystem II and reduces non-photochemical quenching in rice plants subjected to salt stress. J Plant Physiol 185:75–83. https://doi.org/10.1016/j.jplph.2015.08.008

Qi S, Wang J, Wan L, Dai Z, Matos DMS, Du D, Egan S, Bonser SP, Thomas T, Moles AT (2022) Arbuscular mycorrhizal fungi contribute to phosphorous uptake and allocation strategies of Solidago canadensis in a phosphorous-deficient environment. Front Plant Sci 13:831654. https://doi.org/10.3389/fpls.2022.831654

Rani B (2016) Effect of arbuscular mycorrhiza fungi on biochemical parameters in wheat *Triticum aestivum* L. under drought conditions. Doctoral dissertation, CCSHAU, Hisar

Riaz M, Kamran M, Fang Y, Wang Q, Cao H, Yang G, Deng L, Wang Y, Zhou Y, Anastopoulos I, Wang X (2021) Arbuscular mycorrhizal fungi-induced mitigation of heavy metal phytotoxicity in metal contaminated soils: a critical review. J Hazard Mater 402:123919. https://doi.org/10.1016/j.jhazmat.2020.123919

Rillig MC, Mummey DL (2006) Mycorrhizas and soil structure. New Phytol 171:41–53

Ruiz-Lozano JM, Aroca R (2010) Modulation of aquaporin genes by the arbuscular mycorrhizal symbiosis in relation to osmotic stress tolerance: aquaporin in AM plants under osmotic stress. Symbioses and stress. Joint Vent Biol:357–374

Sahodaran NK, Ray JG (2018) Heavy metal contamination in "chemicalised" green revolution banana fields in southern India. Environ Sci Pollut Res 25:26874–26886. https://doi.org/10.1007/s11356-018-2729-0

Sahu PK, Brahmaprakash GP (2016) Formulation of biofertilizer - approaches and advances. In: Singh DP, Singh HB, Prabha R (eds) Microbial inoculants in sustainable agricultural productivity. Springer, New Delhi, pp 179–198

Schaefer DA, Gui H, Mortimer P, Xu J (2021) Arbuscular mycorrhiza and sustainable agriculture. Circ Agric Syst 1(1):1–7

Sene G, Thiao M, Sy O, Mbaye MS, Sylla SN (2021) Seed coating with mycorrhizal fungal spores and LEIFSONIA bacteria: a tool for microbiological fertilization and a seed protection strategy from insect damage. Proc Natl Acad Sci India Sect B Biol Sci 91:909–918

Sharma S, Kumar Sharma A, Prasad R, Varma A (2017) Arbuscular mycorrhiza: a tool for enhancing crop production. In: Varma A et al (eds) Mycorrhiza—nutrient uptake. Biocontrol, Ecorestoration. Springer, Cham. https://doi.org/10.1007/978-3-319-68867-1_12

Shtark O, Borisov A, Zhukov V, Provorov N, Tikhonovich I (2010) Intimate associations of beneficial soil microbes with host plants. In: Dixon GR, Tilston EL (eds) Soil microbiology and sustainable crop production. Springer, Dordrecht. https://doi.org/10.1007/978-90-481-9479-7_5

Silveira R, Mello TRB, Sartori MRS, Alves GSC, Fonseca FCA, Vizzotto CS, Kruger RH, Bustamante MMC (2021) Seasonal and long-term effects of nutrient additions and liming on the nifH gene in cerrado soils under native vegetation. iScience 24:102349. https://doi.org/10.1016/j.isci.2021.102349

Singh SR, Singh U, Chaubey AK, Shiv R (2010) Mycorrhizal fungi for sustainable agriculture-a review. Agric Rev 31(2):79–92

Smith SE, Read DJ (2010) Mycorrhizal symbiosis, 3rd edn. Academic, Cambridge, MA

Sohali M, Pirzada T, Opperman CH, Khan SA (2022) Recent advances in seed coating technologies: transitioning toward sustainable agriculture. Green Chem 16(24):6052–6085

Sparks EE, Benfey PN (2017) The contribution of root systems to plant nutrient acquisition. In: Plant macronutrient use efficiency. Academic, Cambridge, MA, pp 83–92. https://doi.org/10.1016/b978-0-12-811308-0.00005-3

Srisom K, Tittabutr P, Teaumroong N, Lapwong Y, Phatthanakun R, Sirivisoot S, Kuntanawat P (2020) New method for arbuscular mycorrhizal fungus spore separation using a microfluidic device based on manual temporary flow diversion. Mycorrhiza 30:789–796

St-Arnaud M, Hamel C, Vimard B, Caron M, Fortin JA (1996) Enhanced hyphal growth and spore production of the arbuscular mycorrhizal fungus *Glomus intraradices* in an in vitro system in the absence of host roots. Mycol Res 100(3):328–332. https://doi.org/10.1016/S0953-7562(96)80164-X

Toljander JF, Lindahl BD, Paul LR, Elfstrand M, Finlay RD (2007) Influence of arbuscular mycorrhizal exudates on soil bacterial growth and community structure. FEMS Microbiol Ecol 61:295–304

Torrecillas E, Torres P, Alguacil MM, Querejeta JI, Roldan A (2013) Influence of habitat and climate variables on arbuscular mycorrhizal fungus community distribution, as revealed by a case study of facultative plant epiphytism under semiarid conditions. Appl Environ Microbiol 79:7203–7209. https://doi.org/10.1128/AEM.02466-13

Verbruggen E, Toby Kiers E (2010) Evolutionary ecology of mycorrhizal functional diversity in agricultural systems. Evol Appl 3(5-6):547–560

Wattenburger CJ, Gutknecht J, Zhang Q, Brutnell TP (2020) The rhizosphere and cropping system, but not arbuscular mycorrhizae, affect ammonia oxidizing archaea and bacteria abundances in two agricultural soils. Appl Soil Ecol 151:103540

Willis A, Rodrigues BF, Harris PJC (2013) The ecology of arbuscular mycorrhizal fungi. Crit Rev Plant Sci 32(1):1–20. https://doi.org/10.1080/07352689.2012.683375

Yadav SK, Babu S, Yadav MK, Singh K, Yadav GS, Pal S (2013) A review of organic farming for sustainable agriculture in Northern India. Int J Agron 2013:718145. https://doi.org/10.1155/2013/718145

Yadav RS, Singh V, Pal S, Meena SK (2018) Seed bio-priming of baby corn emerged as a viable strategy for reducing mineral fertilizer use and increasing productivity. Sci Hortic 241:93–99

Zhang F, Jia-Dong HE, Qiu-Dan NI, Wu QS (2018) Enhancement of drought tolerance in trifoliate orange by mycorrhiza: changes in root sucrose and proline metabolisms. Not Bot Horti Agrobot Cluj-Napoca 46:270. https://doi.org/10.15835/nbha46110983

Zhang Z, Mallik A, Zhang J, Huang Y (2019) Effects of arbuscular mycorrhizal fungi on inoculated seedling growth and rhizosphere soil aggregates. Soil Tillage Res 194:104340. https://doi.org/10.1016/j.still.2019.104340

Zhao R, Guo W, Bi N, Guo J (2015) Arbuscular mycorrhizal fungi affect the growth, nutrient uptake and water status of maize (*Zea mays*, L.) grown in two types of coal mine spoils under drought stress. Appl Soil Ecol 88:41–49. https://doi.org/10.1016/j.apsoil.2014.11.016

Zhu C, Ling N, Guo J, Wang M, Guo S, Shen Q (2016) The impact of fertilization regimes on arbuscular mycorrhizal fungal (AMF) community composition was correlated with organic matter composition in maize rhizosphere soil. Front Microbiol 7:1840. https://doi.org/10.3389/fmicb.2016.01840

Correction to: Arbuscular Mycorrhizal Fungi and Higher Plants

Golam Jalal Ahammed and Roghieh Hajiboland

Correction to:
G. J. Ahammed, R. Hajiboland (eds.), *Arbuscular Mycorrhizal Fungi and Higher Plants*, https://doi.org/10.1007/978-981-99-8220-2

The online version of this book was inadvertently published as Open Access. The book has been reposted as non-Open Access with this erratum. Chapter 13 alone remains as Open Access.

The updated version of this book can be found at
https://doi.org/10.1007/978-981-99-8220-2

G. J. Ahammed, R. Hajiboland (eds.), *Arbuscular Mycorrhizal Fungi and Higher Plants*, https://doi.org/10.1007/978-981-99-8220-2_15

MIX
Papier aus verantwortungsvollen Quellen
Paper from responsible sources
FSC® C105338

If you have any concerns about our products,
you can contact us on
ProductSafety@springernature.com

In case Publisher is established outside the EU,
the EU authorized representative is:
Springer Nature Customer Service Center GmbH
Europaplatz 3, 69115 Heidelberg, Germany

Printed by Libri Plureos GmbH
in Hamburg, Germany